Universitext

Universitext

Universitext is a series of textbooks that presents material from a wide variety of mathematical disciplines at master's level and beyond. The books, often well class-tested by their author, may have an informal, personal even experimental approach to their subject matter. Some of the most successful and established books in the series have evolved through several editions, always following the evolution of teaching curricula, to very polished texts.

Thus as research topics trickle down into graduate-level teaching, first textbooks written for new, cutting-edge courses may make their way into Universitext.

More information about this series at http://www.springer.com/series/223

Peter K. Friz • Martin Hairer

A Course on Rough Paths

With an Introduction to Regularity Structures

Second Edition

 Springer

Peter K. Friz
Institut für Mathematik
Technische Universität Berlin
Berlin, Germany

Weierstraß-Institut für Angewandte
Analysis und Stochastik
Berlin, Germany

Martin Hairer
Department of Mathematics
Imperial College London
London, UK

ISSN 0172-5939   ISSN 2191-6675   (electronic)
Universitext
ISBN 978-3-030-41555-6   ISBN 978-3-030-41556-3   (eBook)
https://doi.org/10.1007/978-3-030-41556-3

Mathematics Subject Classification (2020): 60Lxx, 60Hxx, 34F05, 35R60, 93E03

This Springer imprint is published by the registered company Springer Nature Switzerland AG
The registered company address is: Gewerbestrasse 11, 6330 Cham, Switzerland

To Waltraud and Rudolf Friz

and

To Xue-Mei

Preface to the Second Edition

It has been a joy seeing the subject of "rough analysis" flourish over the last few years. As far as this book is concerned, this comes at the price of an increasingly long list of (important) omissions. A systematic presentation of higher-level *geometric* and then *branched* (possibly càdlàg) rough paths remains beyond the scope of this book, despite being an excellent preparation for the algebraic thinking later required for regularity structures. (The references [LCL07, FV10b, CF19] and [Gub10, HK15, FZ18, BCFP19] partially make up for this.) Also absent remains a systematic mathematical study of *signatures*. This topic, together with recent applications to data science and machine learning, may well fill a book in its own right; until then the reader may consult Lyons' ICM article [Lyo14] and the survey [CK16].

The theory of *regularity structures*, a major extension of rough path theory, has, since the appearance of the first edition of this book, grown into an essentially complete solution theory for general singular, subcritical semilinear (and quasilinear) stochastic partial differential equations. Despite this progress, our running example of a singular SPDE remains the KPZ equation, originally solved with rough paths [Hai13], later also with the Gubinelli–Imkeller–Perkowksi theory of *paracontrolled distributions* [GIP15, GP15, GP17], another topic that deserves a book in its own right.

As far as the content of this second edition is concerned, we added many new examples and updated notations throughout in order to bring it closer to current practice in the literature. Our short incursion into low regularity (a.k.a. higher order) rough paths in Section 2.4 has been expanded, the recently obtained *stochastic sewing lemma* is presented in Section 4.6. Section 9.4 shows how the Laplace method allows one to elegantly obtain precise asymptotics in the large deviation principle, while Section 12.1 contains a detailed discussion of rough transport equations. We also expanded and updated large parts of Chapters 13-15 dealing with regularity structures. In particular, we give a more modern and self-contained proof of the reconstruction theorem (not relying on wavelet bases anymore), as well as a thorough discussion of an application of regularity structures to a "rough" stochastic volatility model in Section 14.5, and a detailed description of the KPZ structure and renormalisation groups in Sections 15.3 and 15.5.

We also take the opportunity here to thank, in addition to those friends and colleagues already named in the first edition, Yvain Bruned, Ajay Chandra, Ilya Chevyrev, Rosa Preiß, and Lorenzo Zambotti for many interesting discussions over the last few years. Of the many people who communicated to us lists of typos and minor issues we especially thank Christian Litterer. We also thank Carlo Bellingeri, Oleg Butkovsky, Andris Gerasimovics, Mate Gerencsér, Tom Klose, Khoa Lê, Mario Maurelli and Nikolas Tapia for feedback on various aspects of the new content. The first author also thanks ETH Zürich (FIM) for its hospitality during the finalisation of this second edition.

Last but not least, we would like to acknowledge financial support: PKF is supported by the European Research Council under the European Union's Horizon 2020 research and innovation programme through Consolidator Grant 683164 (GPSART), by DFG research unit FOR2402, and the Einstein Foundation Berlin through an Einstein professorship. MH was supported by the European Research Council under the European Union's Seventh Framework Programme through Consolidator Grant 615897 (CRITICAL), by the Leverhulme trust through a leadership award, and by the Royal Society through a research professorship.

Berlin and London,   *Peter K. Friz*
March 2020   *Martin Hairer*

Preface to the First Edition

Since its original development in the mid-nineties by Terry Lyons, culminating in the landmark paper [Lyo98], the theory of rough paths has grown into a mature and widely applicable mathematical theory, and there are by now several monographs dedicated to the subject, notably Lyons–Qian [LQ02], Lyons et al [LCL07] and Friz–Victoir [FV10b]. So why do we believe that there is room for yet another book on this matter? Our reasons for writing this book are twofold.

First, the theory of rough paths has gathered the reputation of being difficult to access for "mainstream" probabilists because it relies on some non-trivial algebraic and / or geometric machinery. It is true that if one wishes to apply it to signals of arbitrary roughness, the general theory relies on several objects (in particular on the Hopf-algebraic properties of the free tensor algebra and the free nilpotent group embedded in it) that are unfamiliar to most probabilists. However, in our opinion, some of the most interesting applications of the theory arise in the context of stochastic differential equations, where the driving signal is Brownian motion. In this case, the theory simplifies dramatically and essentially no non-trivial algebraic or geometric objects are required at all. This simplification is certainly not novel. Indeed, early notes by Lyons, and then of Davie and Gubinelli, all took place in this simpler setting (which allows to incorporate Brownian motion and Lévy's area). However, it does appear to us that all these ideas can nowadays be put together in unprecedented simplicity, and we made a conscious choice to restrict ourselves to this simpler case throughout most of this book.

The second and main *raison d'être* of this book is that the scope of the theory has expanded dramatically over the past few years and that, in this process, the point of view has slightly shifted from the one exposed in the aforementioned monographs. While Lyons' theory was built on the integration of 1-forms, Gubinelli gave a natural extension to the integration of so-called "controlled rough paths". As a benefit, differential equations driven by rough paths can now be solved by fixed point arguments in *linear* Banach spaces which contain a sufficiently accurate (second order) local description of the solution.

This shift in perspective has first enabled the use of rough paths to provide solution theories for a number of classically ill-posed stochastic partial differential equations with *one-dimensional* spatial variables, including equations of Burgers type and the KPZ equation. More recently, the perspective which emphasises linear spaces containing sufficiently accurate local descriptions modelled on some (rough) input, spurred the development of the theory of "regularity structures" which allows to give consistent interpretations for a number of ill-posed equations, also in higher dimensions. It can be viewed as an extension of the theory of controlled rough paths, although its formulation is somewhat different. In the last chapters of this book, we give a short and rather informal (i.e. very few proofs) introduction to that theory, which in particular also sheds new light on some of the definitions of the theory of rough paths.

This book does not have the ambition to provide an exhaustive description of the theory of rough paths, but rather to complement the existing literature on the subject. As a consequence, there are a number of aspects that we chose not to touch, or to do so only barely. One omission is the study of rough paths of arbitrarily low regularity: we do provide hints at the general theory at the end of several chapters, but these are self-contained and can be skipped without impacting the understanding of the rest of the book. Another serious omission concerns the systematic study of *signatures*, that is the collection of all iterated integrals over a fixed interval associated to a sufficiently regular path, providing an intriguing nonlinear characterisation.

We have used several parts of this book for lectures and mini-courses. In particular, over the last years, the material on rough paths was given repeatedly by the first author at TU Berlin (Chapters 1-12, in the form of a 4h/week, full semester lecture for an audience of beginning graduate students in stochastics) and in some mini-courses (Vienna, Columbia, Rennes, Toulouse; e.g. Chapters 1-5 with a selection of further topics). The material of Chapters 13-15 originates in a number of minicourses by the second author (Bonn, ETHZ, Toulouse, Columbia, XVII Brazilian School of Probability, 44th St. Flour School of Probability, etc). The "KPZ and rough paths" summer school in Rennes (2013) was a particularly good opportunity to try out much of the material here in joint mini-course form – we are very grateful to the organisers for their efforts. Chapters 13-15 are, arguably, a little harder to present in a classroom. Jointly with Paul Gassiat, the first author gave this material as full lecture at TU Berlin (with examples classes run by Joscha Diehl, and more background material on Schwartz distributions, Hölder spaces and wavelet theory than what is found in this book); we also started to use consistently colours on our handouts. We felt the resulting improvement in readability was significant enough to try it out also in the present book and take the opportunity to thank Jörg Sixt from Springer for making this possible, aside from his professional assistance concerning all other aspects of this book project. We are very grateful for all the feedback we received from participants at all theses courses. Furthermore, we would like to thank Bruce Driver, Paul Gassiat, Massimilliano Gubinelli, Terry Lyons, Etienne Pardoux, Jeremy Quastel and Hendrik Weber for many interesting discussions on how to present this material. In addition, Khalil Chouk, Joscha Diehl and Sebastian Riedel kindly offered to partially proofread the final manuscript.

At last, we would like to acknowledge financial support: PKF was supported by the European Research Council under the European Union's Seventh Framework

Programme (FP7/2007-2013) / ERC grant agreement nr. 258237 and DFG, SPP 1324. MH was supported by the Leverhulme trust through a leadership award and by the Royal Society through a Wolfson research award.

Berlin and Coventry,   *Peter K. Friz*
June 2014   *Martin Hairer*

Contents

1 Introduction 1
  1.1   What is it all about? 1
  1.2   Analogies with other branches of mathematics 6
  1.3   Regularity structures 8
  1.4   Frequently used notations 10
  1.5   Rough path theory works in infinite dimensions 13

2 The space of rough paths 15
  2.1   Basic definitions 15
  2.2   The space of geometric rough paths 20
  2.3   Rough paths as Lie group valued paths 21
  2.4   Geometric rough paths of low regularity 25
  2.5   Exercises 28
  2.6   Comments 37

3 Brownian motion as a rough path 39
  3.1   Kolmogorov criterion for rough paths 39
  3.2   Itô Brownian motion 43
  3.3   Stratonovich Brownian motion 44
  3.4   Brownian motion in a magnetic field 46
  3.5   Cubature on Wiener Space 50
  3.6   Scaling limits of random walks 52
  3.7   Exercises 54
  3.8   Comments 58

4 Integration against rough paths 61
  4.1   Introduction 61
  4.2   Integration of one-forms 62
  4.3   Integration of controlled rough paths 69
  4.4   Stability I: rough integration 74
  4.5   Controlled rough paths of lower regularity 76

4.6   Stochastic sewing 77
4.7   Exercises 79
4.8   Comments 88

5   Stochastic integration and Itô's formula 89
5.1   Itô integration 89
5.2   Stratonovich integration 91
5.3   Itô's formula and Föllmer 92
5.4   Backward integration 97
5.5   Exercises 102
5.6   Comments 105

6   Doob–Meyer type decomposition for rough paths 107
6.1   Motivation from stochastic analysis 107
6.2   Uniqueness of the Gubinelli derivative and Doob–Meyer 109
6.3   Brownian motion is truly rough 111
6.4   A deterministic Norris' lemma 112
6.5   Brownian motion is Hölder rough 114
6.6   Exercises 117
6.7   Comments 117

7   Operations on controlled rough paths 119
7.1   Relation between rough paths and controlled rough paths 119
7.2   Lifting of regular paths. 120
7.3   Composition with regular functions. 121
7.4   Stability II: Regular functions of controlled rough paths 122
7.5   Itô's formula revisited 125
7.6   Controlled rough paths of low regularity 127
7.7   Exercises 128
7.8   Comments 129

8   Solutions to rough differential equations 131
8.1   Introduction 131
8.2   Review of the Young case: a priori estimates 132
8.3   Review of the Young case: Picard iteration 133
8.4   Rough differential equations: a priori estimates 134
8.5   Rough differential equations 137
8.6   Stability III: Continuity of the Itô–Lyons map 141
8.7   Davie's definition and numerical schemes 143
8.8   Lyons' original definition 144
8.9   Linear rough differential equations 145
8.10   Stability IV: Flows 148
8.11   Exercises 148
8.12   Comments 150

9   Stochastic differential equations 153
9.1   Itô and Stratonovich equations................................. 153
9.2   The Wong–Zakai theorem 154
9.3   Support theorem and large deviations 155
9.4   Laplace method 156
9.5   Exercises 160
9.6   Comments 161

10   Gaussian rough paths 165
10.1 A simple criterion for Hölder regularity 165
10.2 Stochastic integration and variation regularity of the covariance 167
10.3 Fractional Brownian motion and beyond 175
10.4 Exercises 178
10.5 Comments 183

11   Cameron–Martin regularity and applications 185
11.1 Complementary Young regularity 185
11.2 Concentration of measure................................... 190
      11.2.1 Borell's inequality 190
      11.2.2 Fernique theorem for Gaussian rough paths 191
      11.2.3 Integrability of rough integrals and related topics.......... 192
11.3 Malliavin calculus for rough differential equations 196
      11.3.1 Bouleau–Hirsch criterion and Hörmander's theorem 196
      11.3.2 Calculus of variations for ODEs and RDEs.............. 197
      11.3.3 Hörmander's theorem for Gaussian RDEs................ 200
11.4 Exercises 202
11.5 Comments 204

12   Stochastic partial differential equations 207
12.1 First order rough partial differential equations 207
      12.1.1 Rough transport equation............................. 207
      12.1.2 Continuity equation and analytically weak formulation 211
12.2 Second order rough partial differential equations 214
      12.2.1 Linear theory: Feynman–Kac 214
      12.2.2 Mild solutions to semilinear RPDEs 219
      12.2.3 Fully nonlinear equations with semilinear rough noise 223
      12.2.4 Rough viscosity solutions 228
12.3 Stochastic heat equation as a rough path...................... 230
      12.3.1 The linear stochastic heat equation 232
12.4 Exercises 236
12.5 Comments 239

13 Introduction to regularity structures 243
　13.1 Introduction 243
　13.2 Definition of a regularity structure and first examples 244
　　　13.2.1 The polynomial structure 246
　　　13.2.2 The rough path structure 247
　13.3 Definition of a model and first examples 249
　　　13.3.1 The polynomial model 252
　　　13.3.2 The rough path model 254
　13.4 Proof of the reconstruction theorem 256
　13.5 Exercises 260
　13.6 Comments 262

14 Operations on modelled distributions 263
　14.1 Differentiation 263
　14.2 Products and composition by regular functions 264
　14.3 Classical Schauder estimates 267
　14.4 Multilevel Schauder estimates and admissible models 272
　14.5 Rough volatility and robust Itô integration revisited 276
　14.6 Exercises 285
　14.7 Comments 288

15 Application to the KPZ equation 289
　15.1 Formulation of the main result 289
　15.2 Construction of the associated regularity structure 293
　15.3 The structure group and positive renormalisation 297
　15.4 Reconstruction for canonical lifts 301
　15.5 Renormalisation of the KPZ equation 302
　　　15.5.1 The renormalisation group 303
　　　15.5.2 The renormalised equations 309
　　　15.5.3 Convergence of the renormalised models 310
　15.6 The KPZ equation and rough paths 315
　15.7 Exercises 316
　15.8 Comments 320

References 323

Index 343

Chapter 1
Introduction

We give a short overview of the scopes of both the theory of rough paths and the theory of regularity structures. The main ideas are introduced and we point out some analogies with other branches of mathematics.

1.1 What is it all about?

Differential equations are omnipresent in modern pure and applied mathematics; many "pure" disciplines in fact originate in attempts to analyse differential equations from various application areas. Classical ordinary differential equations (ODEs) are of the form $\dot{Y}_t = f(Y_t, t)$; an important subclass is given by *controlled ODEs* of the form

$$\dot{Y}_t = f_0(Y_t) + f(Y_t)\dot{X}_t , \qquad (1.1)$$

where X models the input (taking values in $\mathbf{R}^d$, say), and Y is the output (in $\mathbf{R}^e$, say) of some system modelled by nonlinear functions f_0 and f, and by the initial state Y_0. The need for a non-smooth theory arises naturally when the system is subject to white noise, which can be understood as the scaling limit as $h \to 0$ of the discrete evolution equation

$$Y_{i+1} = Y_i + h f_0(Y_i) + \sqrt{h} f(Y_i) \xi_{i+1} , \qquad (1.2)$$

where the (ξ_i) are i.i.d. standard Gaussian random variables. Based on martingale theory, Itô's *stochastic differential equations* (SDEs) have provided a rigorous and extremely useful mathematical framework for all this. And yet, stability is lost in the passage to continuous time: while it is trivial to solve (1.2) for a fixed realisation of $\xi_i(\omega)$, after all $(\xi_1, \ldots \xi_T; Y_0) \mapsto Y_i$ is surely a continuous map, the continuity of the solution as a function of the driving noise is lost in the limit.

Taking $\dot{X} = \xi$ to be white noise in time (which amounts to say that X is a Brownian motion, say B), the solution map $S \colon B \mapsto Y$ to (1.1), known as *Itô map*, is a measurable map which in general lacks continuity, whatever norm one uses to

© Springer Nature Switzerland AG 2020
P. K. Friz, M. Hairer, *A Course on Rough Paths*, Universitext,
https://doi.org/10.1007/978-3-030-41556-3_1

equip the space of realisations of B. [1] Actually, one can show the following negative result (see [Lyo91, LCL07] as well as Exercise 5.7 below):

Proposition 1.1. *There exists no separable Banach space $\mathcal{B} \subset \mathcal{C}([0,1])$ with the following properties:*

1. *Sample paths of Brownian motions lie in $\mathcal{B}$ almost surely.*
2. *The map $(f,g) \mapsto \int_0^\cdot f(t)\dot{g}(t)\,dt$ defined on smooth functions extends to a continuous map from $\mathcal{B} \times \mathcal{B}$ into the space of continuous functions on $[0,1]$.*

Since, for any two distinct indices i and j, the map

$$B \mapsto \int_0^\cdot B^i(t)\,\dot{B}^j(t)\,dt\,, \tag{1.3}$$

is itself the solution of one of the simplest possible differential equations driven by B (take $Y \in \mathbf{R}^2$ solving $\dot{Y}^1 = \dot{B}^i$ and $\dot{Y}^2 = Y^1\,\dot{B}^j$), this shows that it takes very little for S to lack continuity. In this sense, solving SDEs is an analytically ill-posed task! On the other hand, there are well-known probabilistic well-posedness results for SDEs of the form [2]

$$dY_t = f_0(Y_t)dt + f(Y_t) \circ dB_t\,, \tag{1.4}$$

(see e.g. [INY78, Thm 4.1]), which imply for instance

Theorem 1.2. *Let $\xi_\varepsilon = \delta_\varepsilon * \xi$ denote the regularisation of white noise in time with a compactly supported smooth mollifier δ_ε. Denote by Y^ε the solutions to (1.1) driven by $\dot{X} = \xi_\varepsilon$. Then Y^ε converges in probability (uniformly on compact sets). The limiting process does not depend on the choice of mollifier δ_ε, and in fact is the Stratonovich solution to (1.4).*

There are many variations on such "Wong–Zakai" results, another popular choice being $\xi_\varepsilon = \dot{B}^{(\varepsilon)}$ where $B^{(\varepsilon)}$ is a piecewise linear approximation (of mesh size $\sim \varepsilon$) to Brownian motion. However, as consequence of the aforementioned lack of continuity of the Itô-map, there are also reasonable approximations to white noise for which the above convergence fails. (We shall see an explicit example in Section 3.4.)

Perhaps rather surprisingly, it turns out that well-posedness is restored via the iterated integrals (1.3) which are in fact the only data that is missing to turn S into a continuous map. The role of (1.3) was already appreciated in [INY78, Thm 4.1] and related works in the seventies, but statements at the time were probabilistic in nature, such as Theorem 1.2 above. *Rough path analysis* introduced by Terry Lyons in the seminal article [Lyo98] and by now exposed in several monographs [LQ02, LCL07, FV10b], provides the following remarkable insight: Itô's solution map can be factorised into a measurable "universal" map Ψ and a *continuous* solution map $\hat{S}$ as

---

[1] This lack of regularity is the raison d'être for *Malliavin calculus,* a Sobolev type theory of $\mathcal{C}([0,T])$ equipped with *Wiener measure*, the law of Brownian motion.

[2] For the purpose of this introduction, all coefficients are assumed to be sufficiently nice.

$$B(\omega) \overset{\Psi}{\mapsto} (B, \mathbb{B})(\omega) \overset{\hat{S}}{\mapsto} Y(\omega). \tag{1.5}$$

The map Ψ is universal in the sense that it depends neither on the initial condition, nor on the vector fields driving the stochastic differential equation, but merely consists of *enhancing* Brownian motion with iterated integrals of the form

$$\mathbb{B}^{i,j}(s,t) = \int_s^t \left(B^i(r) - B^i(s) \right) dB^j(r) . \tag{1.6}$$

At this stage, the choice of stochastic integration in (1.6) (e.g. Itô or Stratonovich) *does* matter and probabilistic techniques are required for the construction of Ψ. Indeed, the map Ψ is only measurable and usually requires the use of some sort of stochastic integration theory (or some equivalent construction, see for example Section 10 below for a general construction in a Gaussian, non-semimartingale context).

The solution map $\hat{S}$ on the other hand, the solution map to a *rough differential equation (RDE)*, also known as *Itô–Lyons map* and discussed in Section 8.1, is purely deterministic and only makes use of analytical constructions. More precisely, it allows input signals to be arbitrary *rough paths* which, as discussed in Chapter 2, are objects (thought of as *enhanced paths*) of the form $(X, \mathbb{X})$, defined via certain algebraic properties (which mimic the interplay between a path and its iterated integrals) and certain analytical, Hölder-type regularity conditions. In Chapter 3 these conditions will be seen to hold true a.s. for $(B, \mathbb{B})$; a typical realisation is thus called *Brownian rough path*.

The Itô–Lyons map turns out, cf. Section 8.6, to be "nice" in the sense that it is a continuous map of both its initial condition and the driving noise $(X, \mathbb{X})$, provided that the dependency on the latter is measured in a suitable "rough path" metric. In other words, rough path analysis allows for a pathwise solution theory for SDEs, i.e. for a fixed realisation of the Brownian rough path. The solution map $\hat{S}$ is however a much richer object than the original Itô map, since its construction is completely independent of the choice of stochastic integral and even of the knowledge that the driving path is Brownian. For example, if we denote by Ψ^I (resp. Ψ^S) the maps $B \mapsto (B, \mathbb{B})$ obtained by Itô (resp. Stratonovich) integration, then we have the almost sure identities

$$S^I = \hat{S} \circ \Psi^I , \qquad S^S = \hat{S} \circ \Psi^S ,$$

where S^I (resp. S^S) denotes the solution to (1.4) interpreted in the Itô (resp. Stratonovich) sense. Returning to Theorem 1.2, we see that the convergence there is really a deterministic consequence of the probabilistic question whether or not $\Psi^S(B^\varepsilon) \to \Psi^S(B)$ in probability and rough path topology, with $\dot{B}^\varepsilon = \xi^\varepsilon$. This can be shown to hold in the case of mollifier, piecewise linear, and many other approximations.

So how is this Itô–Lyons map $\hat{S}$ built? In order to solve (1.1), we need to be able to make sense of the expression

$$\int_0^t f(Y_s)\, dX_s \, , \tag{1.7}$$

where Y is itself the as yet unknown solution. Here is where the usual pathwise approach breaks down: as we have seen in Proposition 1.1 it is in general impossible, even in the simplest cases, to find a Banach space of functions containing Brownian sample paths and in which (1.7) makes sense. Actually, if we measure regularity in terms of Hölder exponents, then (1.7) makes sense as a limit of Riemann sums for X and Y that are arbitrary α-Hölder continuous functions if and only if $\alpha > \frac{1}{2}$. The key word here is *arbitrary*: in our case the function Y is anything but arbitrary! Actually, since the function Y solves (1.1), one would expect the small-scale fluctuations of Y to look exactly like a scaled version of the small-scale fluctuations of X in the sense that one would expect that

$$Y_{s,t} = f(Y_s)X_{s,t} + R_{s,t}$$

where, for any path F with values in a linear space, we set $F_{s,t} = F_t - F_s$, and where $R_{s,t}$ is some remainder that one would expect to be "of higher order" in the sense that $|R_{s,t}| \lesssim |t - s|^{\beta}$ for some $\beta > \alpha$. (We will see later that $\beta = 2\alpha$ is a natural choice.)

Suppose now that X is a "rough path", which is to say that it has been "enhanced" with a two-parameter function $\mathbb{X}$ which should be interpreted as postulating the values for

$$\mathbb{X}^{i,j}(s,t) = \int_s^t X^i_{s,r}\, dX^j_r \, . \tag{1.8}$$

Note here that this identity should be read in the reverse order from what one may be used to: it is the right-hand side that is *defined* by the left-hand side and not the other way around! The idea here is that if X is too rough, then we do not a priori know how to define the integral of X against itself, so we simply postulate its values. Of course, $\mathbb{X}$ cannot just be anything, but should satisfy a number of natural algebraic identities and analytical bounds, which will be discussed in detail in Chapter 2.

Anyway, assuming that we are provided with the data $(X, \mathbb{X})$, then we know how to give meaning to the integral of components of X against other components of X: this is precisely what $\mathbb{X}$ encodes. Intuitively, this suggests that if we similarly encode the fact that Y "looks like X at small scales", then one should be able to extend the definition of (1.7) to a large enough class of integrands to include solutions to (1.1), even when $\alpha < \frac{1}{2}$. One of the achievements of rough path theory is to make this intuition precise. Indeed, in the framework of rough integration sketched here and made precise in Chapter 4, the barrier $\alpha = \frac{1}{2}$ can be lowered to $\alpha = \frac{1}{3}$. In principle, this can be lowered further by further enhancing X with iterated integrals of higher order, but we decided to focus on the first non-trivial case for the sake of simplicity and because it already covers the most important case when X is given by a Brownian motion, or a stochastic process with properties similar to those of Brownian motion. We do however indicate briefly in Sections 2.4, 4.5 and 7.6 how the theory can be modified to cover the case $\alpha \leq \frac{1}{3}$, at least in the "geometric" case when X is a limit of smooth paths.

The simplest way for Y to "look like X" is when $Y = G(X)$ for some sufficiently regular function G. Despite what one might guess, it turns out that this particular class of functions Y is already sufficiently rich so that knowing how to define integrals of the form $\int_0^t G(X_s) \, dX_s$ for (non-gradient) functions G allows to give a meaning to equations of the type (1.1), which is the approach originally developed in [Lyo98]. A few years later, Gubinelli realised in [Gub04] that, in order to be able to give a meaning to $\int_0^t Y_s \, dX_s$ given the data $(X, \mathbb{X})$, it is sufficient that Y admits a "derivative" Y' such that

$$Y_{s,t} = Y'_s X_{s,t} + R_{s,t} \,,$$

with a remainder satisfying $R_{s,t} = \mathrm{O}(|t - s|^{2\alpha})$. This extension of the original theory turns out to be quite convenient, especially when applying it to problems other than the resolution of evolution equations of the type (1.1).

An intriguing question is to what extent rough path theory, essentially a theory of controlled ordinary differential equations, can be extended to partial differential equations. In the case of finite-dimensional noise, and very loosely stated, one has for instance a statement of the following type. (See [CF09, CFO11, FO14, GT10, Tei11, DGT12] as well as Section 12.2 below.)

Theorem 1.3. *Classes of SPDEs of the form $du = F[u] \, dt + H[u] \circ dB$, with second and first order differential operators F and H, respectively, and driven by finite-dimensional noise, with the Zakai equation from filtering and stochastic Hamilton–Jacobi–Bellman (HJB) equations as examples, can be solved pathwise, i.e. for a fixed realisation of the Brownian rough path. As in the SDE case, the SPDE solution map factorises as $S^S = \hat{S} \circ \Psi^S$ where $\hat{S}$, the solution map to a rough partial differential equation (RPDE) is continuous in the rough path topology.*

As a consequence, if $\xi_\varepsilon = \delta_\varepsilon * \xi$ denotes the regularisation of white noise in time with a compactly supported smooth mollifier δ_ε that is scaled by ε, and if u^ε denotes the random PDE solutions driven by $\xi_\varepsilon dt$ (instead of $\circ dB$) then u^ε converges in probability. The limiting process does not depend on the choice of mollifier δ_ε, and is viewed as Stratonovich SPDE solution. The same conclusion holds whenever $\Psi^S(B^\varepsilon) \to \Psi^S(B)$ in probability and rough path topology.

The case of SPDEs driven by infinite-dimensional noise poses entirely different problems. Already the stochastic heat equation in space dimension one has not enough spatial regularity for additional nonlinearities of the type $g(u)\partial_x u$ (which arises in applications from path sampling [Hai11b, HW13]) or $(\partial_x u)^2$ (the Kardar–Parisi–Zhang equation) to be well-defined. In space dimension one, "spatial" rough paths indexed by x, rather than t, have proved useful here and the quest to handle dimension larger than one led to the general theory of regularity structures, see Section 1.3 below.

Rather than trying to survey all applications to date of rough paths to stochastics, let us say that the past few years have seen an explosion of results made possible by the use of rough paths theory. New stimulus to the field was given by its use in rather diverse mathematical fields, including for example quantum field theory [GL09], nonlinear PDEs [Gub12], Malliavin calculus [CFV09], non-Markovian Hörmander and ergodic theory, [CF10, HP13, CHLT15] and the multiscale analysis of chaotic behaviour in fast-slow systems [KM16, KM17, CFK+19b].

In view of these developments, we believe that it is an opportune time to try to summarise some of the main results of the theory in a way that is as elementary as possible, yet sufficiently precise to provide a technical working knowledge of the theory. We therefore include elementary but essentially complete proofs of several of the main results, including the continuity and definition of the Itô–Lyons map, the lifting of a class of Gaussian processes to the space of rough paths, etc. In contrast to the available textbook literature [LQ02, LCL07, FV10b], we emphasise Gubinelli's view on rough integration [Gub04, Gub10] which allows to linearise many considerations and to simplify the exposition. That said, the resulting theory of rough differential equations is (immediately) seen to be equivalent to Davie's definition [Dav08] and, generally, we have tried to give a good idea what other perspectives one can take on what amounts to essentially the same objects.

1.2 Analogies with other branches of mathematics

As we have just seen, the main idea of the theory of rough paths is to "enhance" a path X with some additional data $\mathbb{X}$, namely the integral of X against itself, in order to restore continuity of the Itô map. The general idea of building a larger object containing additional information in order to restore the continuity of some nonlinear transformation is of course very old and there are several other theories that have a similar "flavour" to the theory of rough paths, one of them being the theory of *Young measures* (see for example the notes [Bal00]) where the value of a function is replaced by a probability measure, thus allowing to describe limits of highly oscillatory functions.

Nevertheless, when first confronted with some of the notions just outlined, the first reaction of the reader might be that simply postulating the values for the right-hand side of (1.8) makes no sense. Indeed, if X is smooth, then we "know" that there is only one "reasonable" choice for the integral $\mathbb{X}$ of X against itself, and this is the Riemann integral. How could this be replaced by something else and how can one expect to still get a consistent theory with a natural interpretation? These questions will of course be fully answered in these notes.

For the moment, let us draw an analogy with a very well established branch of geometric measure theory, namely the theory of *varifolds* [Alm66, LY02].

Varifolds arise as natural extensions of submanifolds in the context of certain variational problems. We are not going into details here, but loosely speaking a k-dimensional varifold in $\mathbf{R}^n$ is a (Radon) measure $\mathbf{v}$ on $\mathbf{R}^n \times \mathcal{G}(k, n)$, where $\mathcal{G}(k, n)$ denotes the space of all k-dimensional subspaces of $\mathbf{R}^n$. Here, one should interpret $\mathcal{G}(k, n)$ as the space of all possible tangent spaces at any given point for a k-dimensional submanifold of $\mathbf{R}^n$. The projection of $\mathbf{v}$ onto $\mathbf{R}^n$ should then be interpreted as a generalisation of the natural "surface measure" of a submanifold, while the conditional (probability) measure on $\mathcal{G}(k,n)$ induced at almost every point by disintegration should be interpreted as selecting a (possibly random) tangent space at each point. Why is this a reasonable extension of the notion of submanifold? Consider the following sequence M_ε of one-dimensional submanifolds of $\mathbf{R}^2$:

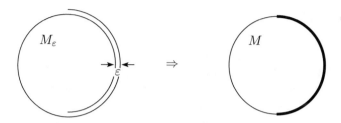

It is intuitively clear that, as $\varepsilon \to 0$, this converges to a circle, but the right half has twice as much "weight" as the left half so that, if we were to describe the limit M simply as a manifold, we would have lost some information about the convergence of the surface measures in the process. More dramatically, there are situations where one has a sequence of smooth manifolds such that the limit is again a smooth manifold, but with a limiting "tangent space" which has nothing to do with the actual tangent space of the limit! Indeed, consider the sequence of one-dimensional submanifolds of $\mathbf{R}^2$ given by

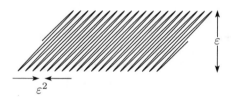

This time, the limit is a piece of straight line, which is in principle a perfectly nice smooth submanifold, but the limiting tangent space is deterministic and makes a 45° angle with the canonical tangent space associated to the limit.

The situation here is philosophically very similar to that of the theory of rough paths: a subset $M \subset \mathbf{R}^n$ may be sufficiently "rough" so that there is no way of canonically associating to it either a k-dimensional Riemannian volume element, or a k-dimensional tangent space, so we simply *postulate* them. The two examples given above show that even in situations where M *is* a nice smooth manifold, it still makes sense to associate to it a volume element and / or tangent space that are different from the ones that one would construct canonically. A similar situation arises in the theory of rough paths. Indeed, it may so happen that X is actually given by a smooth function. Even so, this does *not* automatically mean that the right-hand side of (1.8) is given by the usual Riemann integral of X against itself. An explicit example illustrating this fact is given in Exercise 2.10 below. Similarly to the examples of "non-canonical" varifolds given above, "non-canonical" rough paths can also be constructed as limits of ordinary smooth paths (with the second-order term $\mathbb{X}$ defined by (1.8) where the integral *is* the usual Riemann integral), provided that one takes limits in a suitably weak topology.

1.3 Regularity structures

Recently, a new theory of "regularity structures" was introduced [Hai14b], unifying various flavours of the theory of rough paths (including Gubinelli's controlled rough paths [Gub04], as well as his branched rough paths [Gub10]), as well as the usual Taylor expansions. While it has its conceptual roots in the theory of rough paths, the main advantage of this new theory is that it is no longer tied to the one-dimensionality of the time parameter, which makes it also suitable for the description of solutions to stochastic *partial* differential equations, rather than just stochastic ordinary differential equations.

The main achievement of the theory of regularity structures is that it allows to give a (pathwise!) meaning to ill-posed stochastic PDEs that arise naturally when trying to describe the macroscopic behaviour of models from statistical mechanics near criticality. One example of such an equation is the KPZ equation arising as a natural model for one-dimensional interface motion [KPZ86, BG97, Hai13]:

$$\partial_t h = \partial_x^2 h + (\partial_x h)^2 - C + \xi \,. \tag{1.9}$$

The problem with this equation is that, if anything, one has $(\partial_x h)^2 = +\infty$ (a consequence of the roughness of $(1 + 1)$-dimensional space-time white noise) and one would have to compensate with $C = +\infty$. It has instead become customary to *define* the solution of the KPZ equation as the logarithm of the (multiplicative) stochastic heat equation $\partial_t u = \partial_x^2 u + u\xi$, essentially ignoring the (infinite) Itô-correction term.[3] The so-constructed solutions are called *Hopf–Cole solutions* and, to cite J. Quastel [Qua11],

> The evidence for the Hopf–Cole solutions is now overwhelming. Whatever the physicists mean by KPZ, it is them.

It should emphasised that previous to [Hai13], to be discussed in Chapter 15, no direct mathematical meaning had been given to the actual KPZ equation.

Another example is the dynamical Φ_3^4 model arising for example in the stochastic quantisation of Euclidean quantum field theory [PW81, JLM85, AR91, DPD03, Hai14b], as well as a universal model for phase coexistence near the critical point [GLP99]:

$$\partial_t \Phi = \Delta \Phi + C\Phi - \Phi^3 + \xi \,. \tag{1.10}$$

---

[3] This requires one of course to know that solutions to $\partial_t u = \partial_x^2 u + u\xi$ stay strictly positive with probability one, provided $u_0 > 0$ a.s., but this turns out to be the case.

Here, ξ denotes $(3+1)$-dimensional space-time white noise. In contrast to the KPZ equation where the Hopf–Cole solution is a Hölder continuous random field, here Φ is at best a random Schwartz distribution, making the term Φ^3 ill-defined. Again, one formally needs to set $C = \infty$ to create suitable cancellations and so, again, the stochastic partial differential equation (1.10) has no "naïve" mathematical meaning.

Loosely speaking, the type of well-posedness results that can be proven with the help of the theory of regularity structures can be formulated as follows.

Theorem 1.4. *Consider KPZ and Φ_3^4 on a bounded square spatial domain with periodic boundary conditions. Let $\xi_\varepsilon = \delta_\varepsilon * \xi$ denote the regularisation of space-time white noise with a compactly supported smooth mollifier δ_ε that is scaled by ε in the spatial direction(s) and by ε^2 in the time direction. Denote by h_ε and Φ_ε the solutions to*

$$\partial_t h_\varepsilon = \partial_x^2 h_\varepsilon + (\partial_x h_\varepsilon)^2 - C_\varepsilon + \xi_\varepsilon \,,$$
$$\partial_t \Phi_\varepsilon = \Delta \Phi_\varepsilon + \tilde{C}_\varepsilon \Phi_\varepsilon - \Phi_\varepsilon^3 + \xi_\varepsilon \,.$$

Then, there exist choices of constants C_ε and $\tilde{C}_\varepsilon$ diverging as $\varepsilon \to 0$, as well as processes h and Φ such that $h_\varepsilon \to h$ and $\Phi_\varepsilon \to \Phi$ in probability. Furthermore, while the constants C_ε and $\tilde{C}_\varepsilon$ do depend crucially on the choice of mollifiers δ_ε, the limits h and Φ do not depend on them.

In the case of the KPZ equation, the topology in which one obtains convergence is that of convergence in probability in a suitable space of space-time Hölder continuous functions. Let us also emphasise that in this case the resulting *renormalised* solutions coincide indeed with the Hopf–Cole solutions.

In the case of the dynamical Φ_3^4 model, convergence takes place instead in some space of space-time distributions. One caveat that also has to be dealt with in the latter case is that the limiting process Φ may in principle explode in finite time for some instances of the driving noise. (Although this is of course not expected.)

Chapters 13 and 14 of this book gives a short and mostly self-contained introduction to the theory of regularity structures and the last chapter shows how it can be used to provide a robust solution theory for the KPZ equation. The material in these chapters differs significantly in presentation from the remainder of the book. Indeed, since a detailed and rigorous exposition of this material would require an entire book by itself (see the rather lengthy articles [Hai13] and [Hai14b]), we made a conscious decision to keep the exposition mostly at an intuitive level. We therefore omit virtually all proofs (with the notable exception of the proof of the reconstruction theorem, Theorem 13.12, which is the fundamental result on which the theory builds) and instead merely give short glimpses of the main ideas involved.

1.4 Frequently used notations

Basics: Natural numbers, including zero, are denoted by $\mathbf{N}$, integers by $\mathbf{Z}$, real and complex numbers are denoted by $\mathbf{R}$ and $\mathbf{C}$, respectively. Strictly positive reals are denoted by $\mathbf{R}_+$. For x real, $\lfloor x \rfloor$ (resp. $\lceil x \rceil$) is the largest (resp. smallest) integer n such that $n \leq x$ (resp. $n \geq x$). We also write $\{x\} \in (0, 1]$ for the non-zero fractional part so that $x - \{x\} \in \mathbf{Z}$. A d-dimensional multi-index is an element $k \in \mathbf{N}^d$, and given $x \in \mathbf{R}^d$, we write x^k as a shorthand for $x_1^{k_1} \cdots x_d^{k_d}$ and $k!$ as a shorthand for $k_1! \cdots k_d!$.

Tensors: We shall deal with paths with values in, as well as maps between, Banach spaces, typically denoted by V, W, equipped with their respective norms, always written as $|\cdot|$. Continuous linear maps from V to W form a Banach space, denoted by $\mathcal{L}(V, W)$. It will be important to consider tensor products of Banach spaces. If V, W are finite-dimensional, say $V \cong \mathbf{R}^m$ and $W \cong \mathbf{R}^n$, the tensor product $V \otimes W$ can be identified with the matrix space $\mathbf{R}^{m \times n}$. Indeed, if $(e_i : 1 \leq i \leq m)$ [resp. $(f_j : 1 \leq j \leq n)$] is a basis of V [resp. W], then $(e_i \otimes f_j : 1 \leq i \leq m, 1 \leq j \leq n)$ is a basis of $V \otimes W$. If V and W are Hilbert spaces and (e_i) and (f_j) are *orthonormal* bases it is natural to define a Euclidean structure on $V \otimes W$ by declaring the $(e_i \otimes f_j)$ to be orthonormal. This induces a norm on $V \otimes W$, also denoted by $|\cdot|$, which is *compatible* in the sense that $|v \otimes w| \leq |v| \cdot |w|$ for all $v \in V$, $w \in W$. This tensor norm is furthermore *symmetric*, namely $|u \otimes v| = |v \otimes u|$, equivalently expressed as invariance under transposition $x \mapsto x^T$.

We also introduce the symmetric and antisymmetric parts of $x \in V \otimes V$:

$$\mathrm{Sym}(x) = \tfrac{1}{2}(x + x^T), \qquad \mathrm{Anti}(x) = \tfrac{1}{2}(x - x^T) .$$

A useful feature of tensor product spaces is their ability to linearise bilinear maps,[4]

$$\mathcal{L}^{(2)}(V \times \bar{V}, W) \cong \mathcal{L}\big(V, \mathcal{L}(\bar{V}, W)\big) \cong \mathcal{L}(V \otimes \bar{V}, W) . \tag{1.11}$$

We briefly discuss the extension to infinite dimensions. Given Banach spaces $V, \bar{V}$ one completes the algebraic tensor product $V \otimes_a \bar{V}$ under a compatible tensor norm to obtain a Banach space $V \otimes \bar{V}$. By [Rya02, Thm 2.9], the second[5] identification in (1.11) requires one to work with the *projective tensor norm*

$$|x|_{\mathrm{proj}} \overset{\mathrm{def}}{=} \inf \Big\{ \sum_i |v_i| |\bar{v}_i| : x = \sum_i v^i \otimes \bar{v}^i \Big\} ,$$

where all sums are finite and $|\cdot|$ stands for either norm in V or $\bar{V}$. This norm is obviously compatible and symmetric. Symmetric and antisymmetric part of $x \in V \otimes V$ are defined as before (note that the transposition map $V \otimes W \to W \otimes V$ given by $v \otimes w \mapsto w \otimes v$ defined on the algebraic tensor product uniquely extends

---

[4] This will arise naturally, with $\bar{V} = V$, when pairing the second Fréchet derivatives (of some $F : V \to W$) with second iterated integrals with values in $V \otimes V$.

[5] The first identification holds for general Banach spaces.

to its completion for any symmetric compatible tensor norm). Without going into further detail, we note that the projective tensor norm is the largest compatible tensor norm (by the triangle inequality) satisfying $|u \otimes v| = |u| \cdot |v|$, and thus produces the smallest Banach tensor product space.

Differentiable maps: Given (possibly infinite-dimensional) Banach spaces V, W we write $C^n = C^n(V; W)$, $n \in \mathbf{N}$, for the space of continuous maps from V to W which are n times continuously differentiable in Fréchet sense. A Banach space $C_b^n \subset C^n$ is given by those $F \in C^n$ with

$$\|F\|_{C_b^n} \overset{\text{def}}{=} \|F\|_\infty + \|DF\|_\infty + \ldots + \|D^n F\|_\infty < \infty, \qquad (1.12)$$

where we recall $DF(v) \in \mathcal{L}(V, W)$, $D^2 F \in \mathcal{L}(V, \mathcal{L}(V, W)) \cong \mathcal{L}^{(2)}(V \times V, W)$, the space of continuous bilinear maps from $V \times V$ to W. For $\gamma \in (0, 1)$, we say that F is locally γ-*Hölder continuous*, in symbols $F \in C^\gamma$, if for every $x \in V$ there exists a neighbourhood $N = N(x)$ and constant $C = C(x)$, such that for all $y, z \in N$, $|F(z) - F(y)| \leq C|z - y|^\gamma$. (In finite dimensions, one equivalently demands this estimate to hold on bounded sets.) The case $\gamma = 1$ is meaningful (*"locally Lipschitz"*) but not denoted by C^1 for the sake of notational consistency.[6] More generally, we say $F \in C^\gamma$, for non-integer $\gamma = n + \{\gamma\}$, with fractional part $\{\gamma\} \in (0, 1)$, when $F \in C^n$ and $D^n F \in C^{\{\gamma\}}$. A Banach space $C_b^\gamma \subset C^\gamma$ is introduced via

$$\|F\|_{C_b^\gamma} \overset{\text{def}}{=} \|F\|_{C_b^n} + \sup_{y \neq z} \frac{|D^n F(z) - D^n F|}{|z - y|^{\{\gamma\}}} < \infty. \qquad (1.13)$$

The spaces C^γ and C_b^γ satisfy the obvious inclusions and continuous embeddings, respectively. **Warning.** The Lip^γ-spaces frequently seen in the rough path literature are precisely our C_b^γ-spaces for $\gamma \notin \mathbf{N}$ (at least when V is finite-dimensional), whereas $F \in \mathrm{Lip}^{n+1}$ means $F \in C^n$ with globally Lipschitz $D^n F$; some authors also interpret C_b^γ-spaces, for integer γ, in this way.

Path spaces: We say that $X : [0, T] \to V$ is continuously (Fréchet) differentiable if $X_\cdot = X_0 + \int_0^\cdot \dot{X}_t dt$, for some continuous path $\dot{X} : [0, T] \to V$, the derivative of X. We say that X is smooth, $X \in C^\infty = C^\infty([0, T], V)$, if X and all its derivatives are continuously differentiable. The Banach space $C^\alpha = C^\alpha([0, T], V)$, for $\alpha \in (0, 1)$, consists of α-Hölder paths, with finite α-Hölder seminorm,

$$\|X\|_\alpha \overset{\text{def}}{=} \sup_{s,t \in [0,T]} \frac{|X_{s,t}|}{|t - s|^\alpha} < \infty,$$

where we define the *path increment* $X_{s,t} \overset{\text{def}}{=} X_t - X_s$ (and also use the convention $0/0 \overset{\text{def}}{=} 0$); we also write δX for the map $(s, t) \mapsto X_t - X_s$. This seminorm fails to separate constants, the norm on C^α is then given by $\|X\|_{C^\alpha} \overset{\text{def}}{=} |X_0| + \|X\|_\alpha$. We write $C^{0,\alpha} \subset C^\alpha$ for the closure of smooth paths. As above, C^1 is potentially ambiguous,

---

[6] One checks that every $F \in C^1$ is locally Lipschitz (though not necessarily C_b^1 on bounded sets).

and we adopt (in the context of paths) the Lipschitz (1-Hölder) interpretation; this is convenient e.g. to include piecewise smooth approximations of less regular paths. We will sometimes say that $X \in \mathcal{C}^{\alpha^-}$ as a shorthand for the statement that $X \in \mathcal{C}^\beta$ for every $\beta < \alpha$. This abuse of notation will also be used for other scales of "Hölder-type" spaces depending on a regularity index α. Similarly, one introduces the Banach space $\mathcal{C}^{p\text{-var}}([0,T],V)$, for $p \in [1,\infty)$, of continuous paths with finite p-variation seminorm,

$$\|X\|_{p\text{-var};[0,T]} \overset{\text{def}}{=} \left(\sup_{\mathcal{P}} \sum_{[s,t] \in \mathcal{P}} |X_{s,t}|^p \right)^{\frac{1}{p}} < \infty .$$

Here one works with partitions or dissections of $[0,T]$; since every *dissection* $\mathcal{D} = \{0 = t_0 < t_1 < \ldots < t_n = T\} \subset [0,T]$ can be thought of as a *partition* of $[0,T]$ into (essentially) disjoint intervals, $\mathcal{P} = \{[t_{i-1},t_i] : i = 1, \ldots n\}$, and vice-versa, we shall use whatever is (notationally) more convenient.

We further recall that $\lim_{|\mathcal{P}| \to 0}$, typically defined via *nets*, means convergence along any sequence $(\mathcal{P}_k)$ with mesh $|\mathcal{P}_k| \to 0$, with identical limit along each such sequence. Here, the mesh $|\mathcal{P}|$ of a partition $\mathcal{P}$ is the length of its largest element, i.e. $|\mathcal{P}| = \sup_{k \in \{1,\ldots,n\}} |t_k - t_{k-1}|$ if $\mathcal{P}$ is as above.

Two parameter spaces: Every V-valued path X gives rise to its increment function $\delta X : (s,t) \mapsto X_{s,t} = X_t - X_s$. More generally, consider $(s,t) \mapsto \Xi_{s,t}$, with some sort of "on-diagonal" β-Hölder regularity, formalised by the Banach space $\mathcal{C}_2^\beta$ of maps $\Xi : [0,T] \to V$ with finite norm,

$$\|\Xi\|_\beta \overset{\text{def}}{=} \sup_{s,t \in [0,T]} \frac{|\Xi_{s,t}|}{|t-s|^\beta} < \infty .$$

(Note $X \in \mathcal{C}^\beta$ if and only if $\delta X \in \mathcal{C}_2^\beta$.)

Rough path spaces: The symbols $\mathscr{C}^\alpha$, $\mathscr{D}_X^\alpha$ etc. refer to spaces of rough paths and controlled rough paths, respectively. In the given order, $\mathscr{L}(\mathcal{C}^\infty) \subset \mathscr{C}_g^{0,\alpha} \subset \mathscr{C}_g^\alpha$ denote the spaces of canonically lifted smooth paths, geometric and weakly geometric rough paths; $\mathscr{C}^\infty$ is the space of smooth rough paths. Every level-2- rough path $\mathbf{X} \in \mathscr{C}^\alpha$ admits a *bracket* $[\mathbf{X}]$ that quantifies deviations from the classical chain rule.

Hölder spaces and distributions: Local and global regularity of maps $f : \mathbf{R}^d \to \mathbf{R}$ can be measured in the above-mentioned Hölder space scale $\mathcal{C}^\gamma$ and $\mathcal{C}_b^\gamma$, for $\gamma \in \mathbf{R}_+$. We write $\mathcal{D}(\mathbf{R}^d)$ or $\mathcal{D}$ for $\mathcal{C}_c^\infty$, the space of smooth, compactly supported functions. Upon equipping $\mathcal{D}$ with a suitable topology, the topological dual $\mathcal{D}' = \mathcal{D}'(\mathbf{R}^d)$ is the space of *generalised functions* or *distributions*. The Hölder scale extends to negative γ, and then agrees (for non-integer γ) with the *Zygmund spaces* $\mathcal{Z}^\gamma$, precise definitions are left to Chapter 14.

Stochastic analysis: We expect the reader is familiar with (d-dimensional standard) Brownian motion $B = B(t,\omega)$ and basics of Itô calculus for (continuous) semimartingales as exposed e.g. in [RY99]. In particular, for (continuous) semimartingales X, Y, the angle-bracket $\langle X, Y \rangle$ denotes the usual quadratic covariation process, so that $\langle X, Y \rangle_T = \lim \sum_{[s,t] \in \mathcal{P}} X_{s,t} Y_{s,t}$ along any sequence $(\mathcal{P}_n)$ of deterministic partitions of $[0,T]$ with mesh $|\mathcal{P}_n| \to 0$. We also write $\langle X \rangle = \langle X, X \rangle$. The square-bracket $[X,Y]$ is understood in Föllmer sense (and tacitly depends on a fixed sequence partitions); here too we write $[X] = [X,X]$.

Miscellaneous: We will use the notation $A = \mathrm{O}(x)$ if there exists a constant C such that the bound $|A| \leq Cx$ holds for every $x \leq 1$ (or every $x \geq 1$, depending on the context). Similarly, we write $A = \mathrm{o}(x)$ if the constant C can be made arbitrarily small as $x \to 0$ (or as $x \to \infty$, depending on the context). We will also occasionally write C for a generic constant that only depends on the data of the problem under consideration and which can change value from one line to the other without further notice. We further write $x \lesssim y$ for two positive quantities to express an estimate $x \leq Cy$; and $x \asymp y$ if in addition $y \lesssim x$. Dependence on a parameter δ may be indicated by writing $\asymp_\delta$. We often consider quantities $A = A_{s,t}$ and $B = B_{s,t}$ with $s,t \in [0,T]$, for fixed $T > 0$, and then write $A \overset{\gamma}{=} B$ for $|A_{s,t} - B_{s,t}| \lesssim |t-s|^\gamma$, with (hidden) constant uniform in $s,t \in [0,T]$.

We write $\mathrm{int}(A), \mathrm{cl}(A)$ for the interior and closure of a subset A in (some topological space) X.

Exercises: The difficulty of exercises is indicated with the same convention as in [RY99]: one star $*$ denotes difficult ones and two stars $**$ denote very difficult ones. The symbol $\natural$ denotes exercises that are important for the comprehension of subsequent material.

1.5 Rough path theory works in infinite dimensions

Unless explicitly otherwise stated, all rough path results in this book are valid in infinite dimensions. This is rather obvious in the case of Young integration, say with $\mathcal{L}(V,W)$-valued integrand and V-valued integrator, for general Banach spaces V and W. In the case of rough integration, Section 4.2, of a $\mathcal{L}(V,W)$-valued one-form F, against a $V \oplus V^{\otimes 2}$-valued rough path, the pairing of DF, with values in $\mathcal{L}(V, \mathcal{L}(V,W))$, with $V^{\otimes 2}$ is crucial and requires (1.11). As was explained there, this is guaranteed by equipping $V \otimes V$ with the projective norm which will be our *standing assumption* for the rest of this text, unless otherwise stated.

Alternatively, Lyons [Lyo98], [LQ02, pp. 28, 110] or [LCL07, pp.75] adjusts the notion of C^γ-regularity required for F in a way that basically forces DF to take values in $\mathcal{L}(V \otimes V, W)$, with the consequence that the regularity condition on F then depends on the chosen tensor product norm. This modification entails no changes in subsequent arguments. Of course, there is no difference whatsoever when $\dim V < \infty$.

The same remarks apply to solving rough differential equations. The Young case of Section 8.3 is not affected by tensor norms, whereas the typical second order approximation for RDEs, as e.g. seen in 8.13 later on, immediately points to the need for a well-defined pairing of the form (1.11). This is ensured by having $V \otimes V$

equipped with the projective norm. Alternatively, and as before, it is possible to replace the projective norm by weaker compatible norms, but this then forces one to think more carefully about the necessary modifications on the precise assumptions on the space of vector fields when solving RDEs. This can be important when the existence of the $V^{\otimes 2}$-valued rough path in the projective tensor product space is problematic, as happens in the case of Banach valued Brownian motion (e.g. [LLQ02]] and Exercise 3.5, used e.g. in [IK06, IK07]). See also [Lyo98, Def. 1.2.4] or [KM17, Proof of Thm 3.3], where dim $W < \infty$ is noted to be helpful, and [LCL07, pp.19–20] and [LQ02, pp. 28, 111] for more information.

Chapter 2
The space of rough paths

We define the space of (Hölder continuous) rough paths, as well as the subspace of "geometric" rough paths which preserve the usual rules of calculus. The latter can be interpreted in a natural way as paths with values in a certain nilpotent Lie group. At the end of the chapter, we give a short discussion showing how these definitions should be generalised to treat paths of arbitrarily low regularity.

2.1 Basic definitions

In this section, we give a practical definition of the space of Hölder continuous rough paths. Our choice of Hölder spaces is chiefly motivated by our hope that most readers will already be familiar with the classical Hölder spaces from real analysis. We could in the sequel have replaced "α-Hölder continuous" by "finite p-variation" for $p = 1/\alpha$ in many statements. This choice would also have been quite natural, due to the fact that one of our primary goals will be to give meaning to integrals of the form $\int f(X)\, dX$ or solutions to controlled differential equations of the form $dY = f(Y)\, dX$ for rough paths X. The value of such an integral / solution does not depend on the parametrisation of X, which dovetails nicely with the fact that the p-variation of a function is also independent of its parametrisation. This motivated its choice in the original development of the theory. In some other applications however (like the solution theory to rough stochastic *partial* differential equations developed in [Hai11b, HW13, Hai13] and more generally the theory of regularity structures [Hai14b] exposed in the last chapters), parametrisation-independence is lost and the choice of Hölder norms is more natural.

A rough path on an interval $[0, T]$ with values in a Banach space V then consists of a continuous function $X\colon [0, T] \to V$, as well as a continuous "second order process" $\mathbb{X}\colon [0, T]^2 \to V \otimes V$, subject to certain algebraic and analytical conditions. Regarding the former, the behaviour of iterated integrals, such as (2.2) below, suggests to impose the algebraic relation ("*Chen's relation*"),

© Springer Nature Switzerland AG 2020
P. K. Friz, M. Hairer, *A Course on Rough Paths*, Universitext,
https://doi.org/10.1007/978-3-030-41556-3_2

$$\mathbb{X}_{s,t} - \mathbb{X}_{s,u} - \mathbb{X}_{u,t} = X_{s,u} \otimes X_{u,t} \,, \tag{2.1}$$

which we assume to hold for every triplet of times (s, u, t). Since $X_{t,t} = 0$, it immediately follows (take $s = u = t$) that we also have $\mathbb{X}_{t,t} = 0$ for every t. As already mentioned in the introduction, one should think of $\mathbb{X}$ as *postulating* the value of the quantity

$$\int_s^t X_{s,r} \otimes dX_r \overset{\text{def}}{=} \mathbb{X}_{s,t} \,, \tag{2.2}$$

where we take the right-hand side as a *definition* for the left-hand side. (And not the other way around!) We insist (cf. Exercise 2.4 below) that as a consequence of (2.1), knowledge of the path $t \mapsto (X_{0,t}, \mathbb{X}_{0,t})$ already determines the entire second order process $\mathbb{X}$. In this sense, the pair $(X, \mathbb{X})$ is indeed a path, and not some two-parameter object, although it is often more convenient to consider it as one. If X is a smooth function and we read (2.2) from right to left, then it is straightforward to verify (see Exercise 2.1 below) that the relation (2.1) does indeed hold. Furthermore, one can convince oneself that if $f \mapsto \int f \, dX$ denotes any form of "integration" which is linear in f, has the property that $\int_s^t dX_r = X_{s,t}$, and is such that $\int_s^t f(r) \, dX_r + \int_t^u f(r) \, dX_r = \int_s^u f(r) \, dX_r$ for any admissible integrand f, and if we use such a notion of "integral" to define $\mathbb{X}$ via (2.2), then (2.1) does automatically hold. This makes it a very natural postulate in our setting.

Note that the algebraic relations (2.1) are by themselves not sufficient to determine $\mathbb{X}$ as a function of X. Indeed, for any $V \otimes V$-valued function F, the substitution $\mathbb{X}_{s,t} \mapsto \mathbb{X}_{s,t} + F_t - F_s$ leaves the left-hand side of (2.1) invariant. We will see later on how one should interpret such a substitution. It remains to discuss what are the natural analytical conditions one should impose for $\mathbb{X}$. We are going to assume that the path X itself is α-Hölder continuous, so that $|X_{s,t}| \lesssim |t-s|^\alpha$. The archetype of an α-Hölder continuous function is one which is self-similar with index α, so that $X_{\lambda s, \lambda t} \sim \lambda^\alpha X_{s,t}$.

(We intentionally do not give any mathematical definition of self-similarity here, just think of $\sim$ as having the vague meaning of "looks like".) Given (2.2), it is then very natural to expect $\mathbb{X}$ to also be self-similar, but with $\mathbb{X}_{\lambda s, \lambda t} \sim \lambda^{2\alpha} \mathbb{X}_{s,t}$. This discussion motivates the following definition of our basic spaces of rough paths.

Definition 2.1. For $\alpha \in (\frac{1}{3}, \frac{1}{2}]$, define the space of α-Hölder rough paths (over V), in symbols $\mathscr{C}^\alpha([0,T], V)$, as those pairs $(X, \mathbb{X}) =: \mathbf{X}$ such that

$$\|X\|_\alpha \overset{\text{def}}{=} \sup_{s \neq t \in [0,T]} \frac{|X_{s,t}|}{|t-s|^\alpha} < \infty \,, \qquad \|\mathbb{X}\|_{2\alpha} \overset{\text{def}}{=} \sup_{s \neq t \in [0,T]} \frac{|\mathbb{X}_{s,t}|}{|t-s|^{2\alpha}} < \infty \,, \tag{2.3}$$

and such that the algebraic constraint (2.1) is satisfied.

The obvious example is the *canonical rough path lift* of a smooth path X, of the form $(X, \int X \otimes dX)$, and we write $\mathscr{L}(\mathcal{C}^\infty)$ for the class of rough paths obtained in this way.[1] We have the strict inclusion $\mathcal{L}(C^\infty) \subset \mathscr{C}^\infty$, the class of *smooth rough paths*,[2] by which we mean a genuine rough path with the additional property that the V-valued (resp. $V \otimes V$-valued) maps X. and $\mathbb{X}_{s,\cdot}$ are smooth, for every basepoint s. For instance, $\mathbf{X} \equiv (0,0)$ is the trivial canonical rough path associated to the scalar zero path, as opposed to the smooth "pure second level" rough path (over $\mathbf{R}$) given by $(s,t) \mapsto (0, t-s)$; see also Exercise 2.10 for a natural example with $\dim V > 1$.

Remark 2.2. Any *scalar* path $X \in C^\alpha$ can be lifted to a rough path (over $\mathbf{R}$), simply by setting $\mathbb{X}_{s,t} := (X_{s,t})^2/2$. However, for a vector-valued path $X \in C^\alpha$, with values in some Banach space V, it is far from obvious that one can find suitable "second order increments" $\mathbb{X}$ such that X lifts to a rough path $(X, \mathbb{X}) \in \mathscr{C}^\alpha$. The *Lyons–Victoir extension theorem* (Exercise 2.14) asserts that this can always be done, even in a *continuous* fashion, provided that $1/\alpha \notin \mathbf{N}$ which means $\alpha \in (\frac{1}{3}, \frac{1}{2})$ in our present discussion. (A counterexample for $\alpha = \frac{1}{2}$ is hinted on in Exercise 2.13). The reader may wonder how this continuity property dovetails with Proposition 1.1. The point is that if we define $X \mapsto \mathbf{X}$ by an application of the Lyons–Victoir extension theorem, this map restricted to smooth paths does in general *not* coincide with the Riemann–Stieltjes integral of X against itself.

Remark 2.3. In typical applications to stochastic processes with α-Hölder continuous sample paths, $\alpha \in (\frac{1}{3}, \frac{1}{2})$, such as Brownian motion, rough path lift(s) are constructed via probability, and one does not rely on the extension theorem. In many cases, one has a "canonical" (a.k.a. Stratonovich, Wong-Zakai) lift of a process given as limit (in probability and rough path topology) of canonically lifted sample path mollification of the process. Examples where such a construction works include a large class of Gaussian processes, in particular Brownian motion, and more generally fractional Brownian motion for every Hurst parameter $H > \frac{1}{4}$, cf. Section 10. However, this may not be the only meaningful construction: already in Section 3, we will discuss three natural, but different, ways to lift Brownian motion to a rough path. For a detailed discussion of Markov (with uniformly elliptic generator in divergence form) and semimartingale rough paths we refer to [FV10b].

If one ignores the nonlinear constraint (2.1), the quantities defined in (2.3) suggest to think of $(X, \mathbb{X})$ as an element of the Banach space $C^\alpha \oplus C_2^{2\alpha}$ with (semi-)norm $\|X\|_\alpha + \|\mathbb{X}\|_{2\alpha}$ (which vanishes when X is constant and $\mathbb{X} \equiv 0$). However, taking into account (2.1) we see that $\mathscr{C}^\alpha$ is not a linear space, although it is a closed subset of the aforementioned Banach space; see Exercise 2.7. We will need (some sort of) a norm and metric on $\mathscr{C}^\alpha$. The induced "natural" norm on $\mathscr{C}^\alpha$ given by $\|X\|_\alpha + \|\mathbb{X}\|_{2\alpha}$ fails to respect the structure of (2.1) which is homogeneous with respect to a natural *dilation* on $\mathscr{C}^\alpha$, given by $\delta_\lambda : (X, \mathbb{X}) \mapsto (\lambda X, \lambda^2 \mathbb{X})$. This suggests to introduce the α-Hölder *homogeneous rough path norm*

---

[1] We note immediately that "smooth" can be replaced by "sufficiently smooth", such as C^1 and even C^α, with $\alpha > 1/2$, in view of Young integration, Section 4.1.

[2] We deviate here from the early rough path literature, including [LQ02], where smooth rough paths meant canonical rough paths. Instead, we are aligned with the terminology of regularity structures, where (canonical, smooth) *models* generalise the corresponding notions of rough paths.

$$\|\mathbf{X}\|_\alpha \overset{\text{def}}{=} \|X\|_\alpha + \sqrt{\|\mathbb{X}\|_{2\alpha}} \, , \tag{2.4}$$

which, although not a norm in the usual sense of normed linear spaces, is a very adequate concept for the rough path $\mathbf{X} = (X, \mathbb{X})$. On the other hand, (2.3) leads to a natural notion of rough path metric (and then rough path topology).

Definition 2.4. Given rough paths $\mathbf{X}, \mathbf{Y} \in \mathscr{C}^\alpha([0, T], V)$, we define the (inhomogeneous) α-Hölder rough path metric [3]

$$\varrho_\alpha(\mathbf{X}, \mathbf{Y}) := \sup_{s \neq t \in [0,T]} \frac{|X_{s,t} - Y_{s,t}|}{|t - s|^\alpha} + \sup_{s \neq t \in [0,T]} \frac{|\mathbb{X}_{s,t} - \mathbb{Y}_{s,t}|}{|t - s|^{2\alpha}} .$$

The perhaps easiest way to show convergence with respect to this rough path metric is based on interpolation: in essence, it is enough to establish pointwise convergence together with uniform "rough path" bounds of the form (2.3); see Exercise 2.9. Let us also note that $\mathscr{C}^\alpha([0, T], V)$ endowed with this distance is a complete metric space; the reader is asked to work out the details in Exercise 2.7.

We conclude this part with two important remarks. First, we can ask ourselves up to which point the relations (2.1) are already sufficient to determine $\mathbb{X}$. Assume that we can associate to a given function X two different second order processes $\mathbb{X}$ and $\bar{\mathbb{X}}$, and set $G_{s,t} = \mathbb{X}_{s,t} - \bar{\mathbb{X}}_{s,t}$. It then follows immediately from (2.1) that

$$G_{s,t} = G_{u,t} + G_{s,u} \, ,$$

so that in particular $G_{s,t} = G_{0,t} - G_{0,s}$. Since, conversely, we already noted that setting $\bar{\mathbb{X}}_{s,t} = \mathbb{X}_{s,t} + F_t - F_s$ for an arbitrary continuous function F does not change the left-hand side of (2.1), we conclude that $\mathbb{X}$ is in general determined only up to the increments of some function $F \in \mathcal{C}^{2\alpha}(V \otimes V)$. The choice of F *does* usually matter and there is in general no obvious canonical choice.

The second remark is that this construction can possibly be useful only if $\alpha \leq \frac{1}{2}$. Indeed, if $\alpha > \frac{1}{2}$, then a canonical choice of $\mathbb{X}$ is given by reading (2.2) from right to left and interpreting the left-hand side as a simple Young integral [You36]. Furthermore, it is clear in this case that $\mathbb{X}$ must be unique, since any additional increment should be 2α-Hölder continuous by (2.3), which is of course only possible if $\alpha \leq \frac{1}{2}$. Let us stress once more however that this is *not* to say that $\mathbb{X}$ is uniquely determined by X if the latter is smooth, when it is interpreted as an element of $\mathscr{C}^\alpha$ for some $\alpha \leq \frac{1}{2}$. Indeed, if $\alpha \leq \frac{1}{2}$, F is any 2α-Hölder continuous function with values in $V \otimes V$ and $\mathbb{X}_{s,t} = F_t - F_s$, then the path $(0, \mathbb{X})$ is a perfectly "legal" element of $\mathscr{C}^\alpha$, even though one cannot get any smoother than the function 0. The impact of perturbing $\mathbb{X}$ by some $F \in \mathcal{C}^{2\alpha}$ in the context of integration is considered

---

[3] As was already emphasised, $\mathscr{C}^\alpha$ is not a linear space but is naturally embedded in the Banach space $\mathcal{C}^\alpha \oplus \mathcal{C}_2^{2\alpha}$ (cf. Exercise 2.7), the (inhomogeneous) rough path metric is then essentially the induced metric. While this may not appear intrinsic (the situation is somewhat similar to using the (restricted) Euclidean metric on $\mathbf{R}^3$ on the 2-sphere), the ultimate justification is that the Itô map will turn out to be locally Lipschitz continuous in this metric.

in Example 4.14 below. In Chapter 5, we shall use this for a pathwise understanding of how exactly Itô and Stratonovich integrals differ.

Remark 2.5. There are some simple variations on the definition of a rough path, and it can be very helpful to switch from one view-point to the other. (The analytic conditions are not affected by this.)

a) From the "full increment" view point one has $(X, \mathbb{X}) : [0, T]^2 \to V \oplus V^{\otimes 2}$, $(s, t) \mapsto (X_{s,t}, \mathbb{X}_{s,t})$ subject to the "full" Chen relation

$$X_{s,t} = X_{s,u} + X_{u,t}, \quad \mathbb{X}_{s,t} = \mathbb{X}_{s,u} + \mathbb{X}_{u,t} + X_{s,u} \otimes X_{u,t}. \tag{2.5}$$

Every path $X : [0, T] \to V$ induces (vector) increments $X_{s,t} \equiv (\delta X)_{s,t} = X_t - X_s$ for which the first equality is a triviality. Conversely, increments determine a path modulo constants. In particular, $X_t = X_0 + X_{0,t}$ and this definition is equivalent to what we had in Definition 2.1), *if* restricted to paths with $X_0 = 0$ (or, less rigidly, by identifying paths $X, \bar{X}$ for which $\bar{X} - X$ is constant). In many situations, notably differential equations driven by $(X, \mathbb{X})$, this difference does not matter. (This increment view point is also closest to "models" (Π, Γ) in the theory of regularity structures, Section 13.3, where s is regarded as base-point and one is given a collection of functions $(X_{s,\cdot}, \mathbb{X}_{s,\cdot})$. The Chen relation (2.5) then has the interpretation of shifting the base-point.)

b) The "full path" view point starts with $\mathbf{X} : [0, T] \to \{1\} \times V \oplus V^{\otimes 2} \equiv T_1^{(2)}(V)$, a Lie group under the (truncated) tensor product, the details of which are left to Section 2.3 below. Every such path has *group increments* defined by

$$\mathbf{X}_s^{-1} \otimes \mathbf{X}_t =: \mathbf{X}_{s,t} =: (X_{s,t}, \mathbb{X}_{s,t}).$$

Chen's relation (2.5) is nothing but the trivial identity $\mathbf{X}_{s,u} \otimes \mathbf{X}_{u,t} = \mathbf{X}_{s,t}$ so that any such group-valued path $\mathbf{X}$ induces an increment map $(X, \mathbb{X})$, of the form discussed in a). Conversely, such increments determine $\mathbf{X}$ modulo constants as seen from $\mathbf{X}_t = \mathbf{X}_0 \otimes \mathbf{X}_{0,t}$. If we restrict to $\mathbf{X}_0 = \mathbf{1} = (1, 0, 0)$, or identify paths $\mathbf{X}, \tilde{\mathbf{X}}$ for which $\tilde{\mathbf{X}} \otimes \mathbf{X}^{-1}$ is constant, then there is no difference. (Such a "base-point free" object corresponds to "fat" Π in the theory of regularity structures and induces a model (Π, Γ) in great generality.)

c) Our Definition 2.1 is a compromise in the sense that we want to start from a familiar object, namely a path $X : [0, T] \to V$, together with minimal second level increment information to define (in Section 4.2) the prototypical rough integral $\int F(X) d(X, \mathbb{X})$. From the "increment" view point, we have thus supplied more than necessary (namely X_0), whereas from the "full path" view point, we have supplied $\mathbf{X}$, with $\mathbf{X}_0 = (1, X_0, *)$ specified on the first level only. (Of course, this affects in no way the second level increments $\mathbb{X}_{s,t}$.)

2.2 The space of geometric rough paths

While (2.1) does capture the most basic (additivity) property that one expects any decent theory of integration to respect, it does *not* imply any form of integration by parts / chain rule. Now, if one looks for a first order calculus setting, such as is valid in the context of smooth paths or the Stratonovich stochastic calculus, then for any pair e_i^*, e_j^* of elements in V^*, writing $X_t^i = e_i^*(X_t)$ and $\mathbb{X}_{s,t}^{ij} = (e_i^* \otimes e_j^*)(\mathbb{X}_{s,t})$, one would expect to have the identity

$$
\mathbb{X}_{s,t}^{ij} + \mathbb{X}_{s,t}^{ji} \; \text{"="} \; \int_s^t X_{s,r}^i \, dX_r^j + \int_s^t X_{s,r}^j \, dX_r^i
$$

$$
= \int_s^t d(X^i X^j)_r - X_s^i \, X_{s,t}^j - X_s^j \, X_{s,t}^i
$$

$$
= (X^i X^j)_{s,t} - X_s^i \, X_{s,t}^j - X_s^j \, X_{s,t}^i = X_{s,t}^i \, X_{s,t}^j \, ,
$$

so that the symmetric part of $\mathbb{X}$ is determined by X. In other words, for all times s, t we have the "first order calculus" condition

$$
\text{Sym}(\mathbb{X}_{s,t}) = \frac{1}{2} X_{s,t} \otimes X_{s,t} \, . \tag{2.6}
$$

However, if we take X to be an n-dimensional Brownian path and define $\mathbb{X}$ by Itô integration, then (2.1) still holds, but (2.6) certainly does not.

There are two natural ways to define a set of "geometric" rough paths for which (2.6) holds. On the one hand, we can define the space of *weakly geometric (α-Hölder) rough paths*.

$$
\mathscr{C}_g^\alpha([0,T], V) \subset \mathscr{C}^\alpha([0,T], V) \, ,
$$

by stipulating that $(X, \mathbb{X}) \in \mathscr{C}_g^\alpha$ if and only if $(X, \mathbb{X}) \in \mathscr{C}^\alpha$ and (2.6) holds as equality in $V \otimes V$, for every $s, t \in [0, T]$. Note that $\mathscr{C}_g^\alpha$ is a closed subset of $\mathscr{C}^\alpha$.

On the other hand, we have already seen that every smooth path can be lifted canonically to an element in $\mathscr{L}(C^\infty) \subset \mathscr{C}^\alpha$ by reading the definition (2.2) from right to left. This choice of $\mathbb{X}$ then obviously satisfies (2.6) and we can define the space of *geometric (α-Hölder) rough paths*,

$$
\mathscr{C}_g^{0,\alpha}([0,T], V) \subset \mathscr{C}^\alpha([0,T], V) \, ,
$$

as the closure of $\mathscr{L}(C^\infty)$ in $\mathscr{C}^\alpha$. We leave it as exercise to the reader to see that C^∞ here may be replaced by C^1 paths without changing the resulting space of geometric rough paths.

One has the obvious inclusion $\mathscr{C}_g^{0,\alpha} \subset \mathscr{C}_g^\alpha$, which turns out to be strict. In fact, $\mathscr{C}_g^{0,\alpha}$ is separable (provided V is separable), whereas $\mathscr{C}_g^\alpha$ is not, cf. Exercise 2.8 below. The situation is similar to the classical situation of the set of α-Hölder continuous functions being strictly larger than the closure of smooth functions under the α-Hölder norm. (Or the set of bounded measurable functions being strictly larger than $\mathcal{C}$, the closure of smooth functions under the supremum norm.) In practice, at least when $\dim V < \infty$, the distinction between weakly and "genuinely" geometric rough paths rarely matters for the following reason: similar to classical Hölder spaces, one has the converse inclusion $\mathscr{C}_g^\beta \subset \mathscr{C}_g^{0,\alpha}$ whenever $\beta > \alpha$, see Proposition 2.8 below and also Exercise 2.12. For this reason, we will often casually speak of "geometric rough paths", even when we mean weakly geometric rough paths. (There is no confusion in precise statements when we write $\mathscr{C}_g^{0,\alpha}$ or $\mathscr{C}_g^\alpha$.) Let us finally mention that non-geometric rough paths can always be embedded in a space of geometric rough paths at the expense of adding new components; in the present (level-2) setting this can be accomplished in terms of a *rough path bracket*, see Exercise 2.11 and also Section 5.3.

2.3 Rough paths as Lie group valued paths

We now present a very fruitful view of rough paths, taken over a Banach space V. Consider $X : [0,T] \to V$, $\mathbb{X} : [0,T]^2 \to V^{\otimes 2}$ subject to (2.1) and define, with $X_{s,t} = X_t - X_s$ as usual,

$$\mathbf{X}_{s,t} := (1, X_{s,t}, \mathbb{X}_{s,t}) \in \mathbf{R} \oplus V \oplus V^{\otimes 2} \overset{\text{def}}{=} T^{(2)}(V). \tag{2.7}$$

The space $T^{(2)}(V)$ is itself a Banach space, with the norm of an element (a,b,c) given by $|a| + |b| + |c|$, where in abusive notation $|\cdot|$ standards for any of the norms in $\mathbf{R}$, V and $V \otimes V$, the norm on the latter assumed compatible and symmetric, cf. Section 1.4 . More interestingly for our purposes, this space is a Banach algebra, non-commutative when $\dim V > 1$ and with unit element $(1,0,0)$, when endowed with the product

$$(a,b,c) \otimes (a',b',c') \overset{\text{def}}{=} (aa', ab' + a'b, ac' + a'c + b \otimes b') .$$

We call $T^{(2)}(V)$ the *step-2 truncated tensor algebra over* V. This multiplicative structure is very well adapted to our needs since Chen's relation (2.1), combined with the obvious identity $X_{s,t} = X_{s,u} + X_{u,t}$, takes the elegant form

$$\mathbf{X}_{s,t} = \mathbf{X}_{s,u} \otimes \mathbf{X}_{u,t} . \tag{2.8}$$

Set $T_a^{(2)}(V) \overset{\text{def}}{=} \{(a,b,c) \colon b \in V,\ c \in V \otimes V\}$. As suggested by (2.7), the affine subspace $T_1^{(2)}(V)$ will play a special role for us. We remark that each of its elements has an inverse given by

$$(1,b,c) \otimes (1,-b,-c+b \otimes b) = (1,-b,-c+b \otimes b) \otimes (1,b,c) = (1,0,0) , \tag{2.9}$$

so that $T_1^{(2)}(V)$ is a *Lie group*.[4] It follows that $\mathbf{X}_{s,t} = \mathbf{X}_{0,s}^{-1} \otimes \mathbf{X}_{0,t}$ are the natural increments of the group valued path $t \mapsto \mathbf{X}_{0,t} =: \mathbf{X}_t$.

---

[4] The Lie group $T_1^{(2)}(V)$ is finite-dimensional if and only if $\dim V < \infty$.

Identifying $1, b, c$ with elements $(1, 0, 0), (0, b, 0), (0, 0, c) \in T^{(2)}(V)$, we may write $(1, b, c) = 1 + b + c$. The resulting calculus is familiar from formal power series in non-commuting indeterminates. For instance, the usual power series $(1 + x)^{-1} = 1 - x + x^2 - \ldots$ leads to, omitting tensors of order 3 and higher,

$$\begin{aligned}
(1 + b + c)^{-1} &= 1 - (b + c) + (b + c) \otimes (b + c) \\
&= 1 - b - c + b \otimes b,
\end{aligned}$$

allowing us to recover (2.9). We also introduce the *dilation* operator δ_λ on $T^{(2)}(V)$, with $\lambda \in \mathbf{R}$, which acts by multiplication with λ^n on the nth tensor level $V^{\otimes n}$, namely

$$\delta_\lambda : (a, b, c) \mapsto \left(a, \lambda b, \lambda^2 c\right).$$

Having identified $T_1^{(2)}(V)$ as the natural state space of (step-2) rough paths, we now equip it with a *homogeneous, symmetric* and *subadditive* norm. For $\mathbf{x} = (1, b, c)$,

$$\|\mathbf{x}\| \stackrel{\text{def}}{=} \tfrac{1}{2}\left(N(\mathbf{x}) + N(\mathbf{x}^{-1})\right) \quad \text{with } N(\mathbf{x}) = \max\{|b|, \sqrt{2|c|}\}, \tag{2.10}$$

noting $\|\delta_\lambda \mathbf{x}\| = |\lambda| \|\mathbf{x}\|$, homogeneity with respect to dilation, and $\|\mathbf{x} \otimes \mathbf{x}'\| \leq \|\mathbf{x}\| + \|\mathbf{x}'\|$, a consequence of subadditivity for $N(\cdot)$ which requires a short argument left to the reader. It is clear that

$$(\mathbf{x}, \mathbf{x}') \mapsto \|\mathbf{x}^{-1} \otimes \mathbf{x}'\| \stackrel{\text{def}}{=} d(\mathbf{x}, \mathbf{x}')$$

defines a bona fide (left-invariant) metric on the group $T_1^{(2)}(V)$. Important for us, the graded Hölder regularity (2.3) of $\mathbf{X} = (X, \mathbb{X})$, part of the definition of a rough path, can now be condensed to demand the "metric" Hölder seminorm

$$\sup_{s \neq t \in [0,T]} \frac{d(\mathbf{X}_s, \mathbf{X}_t)}{|t - s|^\alpha} \asymp \|X\|_\alpha + \sqrt{\|\mathbb{X}\|_{2\alpha}} = \|\mathbf{X}\|_{\alpha;[0,T]} \tag{2.11}$$

to be finite. To summarise, we arrived at the following appealing characterisation of (Banach space valued) rough paths.

Proposition 2.6. *(Hölder continuity is with respect to the left-invariant metric d.)*

a) *Assume $(X, \mathbb{X}) \in \mathscr{C}^\alpha([0, T], V)$. Then the path $t \mapsto \mathbf{X}_t = (1, X_{0,t}, \mathbb{X}_{0,t})$, with values in $T_1^{(2)}(V)$ is α-Hölder continuous.*

b) *Conversely, if $[0, T] \ni t \mapsto \mathbf{X}_t$ is a $T_1^{(2)}(V)$-valued and α-Hölder continuous path, then $(X, \mathbb{X}) \in \mathscr{C}^\alpha([0, T], V)$ with $(1, X_{s,t}, \mathbb{X}_{s,t}) := \mathbf{X}_s^{-1} \otimes \mathbf{X}_t$.*

The usual power series and / or basic Lie group theory suggest to define

$$\log(1 + b + c) \stackrel{\text{def}}{=} b + c - \frac{1}{2} b \otimes b, \tag{2.12}$$

$$\exp(b + c) \stackrel{\text{def}}{=} 1 + b + c + \frac{1}{2} b \otimes b, \tag{2.13}$$

which allow us to identify $T_0^{(2)}(V) \cong V \oplus V^{\otimes 2}$ with $T_1^{(2)}(V) = \exp(V \oplus V^{\otimes 2})$. The following *Lie bracket* makes $T_0^{(2)}(V)$ a *Lie algebra*. For $b, b' \in V, c, c' \in V^{\otimes 2}$,

$$[b + c, b' + c'] \stackrel{\text{def}}{=} b \otimes b' - b' \otimes b ,$$

and $T_0^{(2)}(V)$ is *step-2 nilpotent* in the sense that all iterated brackets of length 2 vanish. Define $\mathfrak{g}^{(2)}(V) \subset T_0^{(2)}(V)$ as the closed Lie subalgebra generated by $V \subset T_0^{(2)}(V)$, explicitly given by

$$\mathfrak{g}^{(2)}(V) = V \oplus [V, V] \text{ with } [V, V] \stackrel{\text{def}}{=} \mathrm{cl}(\mathrm{span}\{[v, w] : v, w \in V\}) ,$$

called the *free step-2 nilpotent Lie algebra over V*. Note that in finite dimensions, say $V = \mathbf{R}^d$, the closing procedure is unnecessary and $[V, V]$ is nothing but the space of antisymmetric $d \times d$ matrices, with linear basis $([e_i, e_j] : 1 \le i < j \le d)$, where $(e_i : 1 \le i \le d)$ denotes the standard basis of $\mathbf{R}^d$. Thanks to step-2 nilpotency, one checks by hand the *Baker–Campbell–Hausdorff formula*

$$\exp(b + c) \otimes \exp(b' + c') = \exp(b + b' + c + c' + \tfrac{1}{2}[b, b']) .$$

The image of $\mathfrak{g}^{(2)}$ under the exponential map then defines a closed Lie subgroup,

$$G^{(2)}(V) \stackrel{\text{def}}{=} \exp\big(\mathfrak{g}^{(2)}(V)\big) \subset T_1^{(2)}(V) ,$$

called the *free step-2 nilpotent group over V*. These considerations provide us with an elegant characterisation of weakly geometric rough paths. (The proof is immediate from the previous proposition and rewriting (2.6) as $\mathbb{X}_{s,t} - \tfrac{1}{2} X_{s,t} \otimes X_{s,t} \in [V, V]$.)

Proposition 2.7 (Weakly geometric case).

a) Assume $(X, \mathbb{X}) \in \mathscr{C}_g^\alpha([0, T], V)$. Then the path $t \mapsto \mathbf{X}_t = (1, X_{0,t}, \mathbb{X}_{0,t})$, with values in $G^{(2)}(V)$ is α-Hölder continuous (with respect to the metric d.)

b) Conversely, if $[0, T] \ni t \mapsto \mathbf{X}_t$ is a $G^{(2)}(V)$-valued and α-Hölder continuous path, then $(X, \mathbb{X}) \in \mathscr{C}_g^\alpha([0, T], V)$ with $(1, X_{s,t}, \mathbb{X}_{s,t}) := \mathbf{X}_s^{-1} \otimes \mathbf{X}_t$.

It is clear from the discussion in Section 2.2 that any sufficiently smooth path, say $\gamma \in \mathcal{C}^1([0, 1], V)$, produces an element in $G^{(2)}(V)$ by iterated integration, namely

$$S^{(2)}(\gamma) = \left(1, \int_0^1 d\gamma(t), \int_0^1 \int_0^t d\gamma(s) \otimes d\gamma(t)\right) \in G^{(2)}(V) .$$

The map $S^{(2)}$, which maps (sufficiently regular) paths on a fixed interval, here $[0, 1]$, into the above collection of tensors is know as *step-2 signature map*. We note in passing that Chen's relation here has the pretty interpretation that the signature map is a morphism from the space of paths, equipped with concatenation product, to the tensor algebra. The inclusion $S^{(2)}(\mathcal{C}^1) \subset G^{(2)}$ becomes an equality in finite dimensions,

$$\{S^{(2)}(\gamma) : \gamma \in \mathcal{C}^1([0, 1], \mathbf{R}^d)\} = G^{(2)}(\mathbf{R}^d) . \tag{2.14}$$

To see this, fix $b + c \in \mathfrak{g}^{(2)}(\mathbf{R}^d)$ and try to find finitely many, say n, affine linear paths γ_i, with each signature determined by the direction $\gamma_i(1) - \gamma_i(0) = v_i \in \mathbf{R}^d$, so that

$$\exp(v_1) \otimes \ldots \otimes \exp(v_n) = \exp(b + c) \,.$$

Properly applied, the Baker–Campbell–Hausdorff formula allows to "break up" the exponential $\exp(\sum_i b^i e_i + \sum_{j,k} c^{jk}[e_j, e_k])$. In conjunction with the identity $e^{[v,w]} = e^{-w} \otimes e^{-v} \otimes e^w \otimes e^v$ it is easy to find a possible choice of $v_1, \ldots, v_n$. By concatenation of the γ_i's one has constructed a path γ with prescribed signature $S^{(2)}(\gamma) = \exp(b + c)$. This path is clearly in $\mathcal{C}^1$, the space of Lipschitz paths.[5] This gives a very natural way to introduce another (homogeneous, symmetric, subadditive) norm on $G^{(2)}(\mathbf{R}^d)$, namely

$$\|\mathbf{x}\|_{\mathrm{C}} \stackrel{\text{def}}{=} \inf\left\{ \int_0^1 |\dot{\gamma}(t)|\, dt \; : \; \gamma \in \mathcal{C}^1([0,1], \mathbf{R}^d)\,, \quad S^{(2)}(\gamma) = \mathbf{x} \right\}, \qquad (2.15)$$

known as *Carnot–Carathéodory norm*. (In infinite dimensions, there is no guarantee for the set on the right-hand side to be non-empty.) When equipped with its Euclidean structure, $\mathbf{R}^d$ defines a "horizontal" subspace $\mathbf{R}^d \times \{0\} \subset \mathfrak{g}^{(2)}(\mathbf{R}^d)$, seen as tangent space to $G^{(2)}(\mathbf{R}^d)$ at $(1, 0, 0)$ which in turn induces a left-invariant sub-Riemannian structure on $G^{(2)}(\mathbf{R}^d)$. The associated left-invariant Carnot–Carathéodory metric d_{C} can then be seen as the minimal length of "horizontal" paths connecting two points. Any minimising sequence in (2.15), parametrised by constant speed, is equicontinuous so that by Arzela–Ascoli such minimisers, also called *sub-Riemannian geodesics*, exist and must be in $\mathcal{C}^1$. Such geodesics are a key tool in the approach of Friz–Victoir [FV10b]. The explicit computation of such geodesics (and Carnot–Carathéodory norms) is a difficult problem, with explicit formulae available for $d = 2$, noting that, as Lie groups, $G^{(2)}(\mathbf{R}^2) \cong \mathbf{H}^3$, the 3-dimensional Heisenberg group, see e.g. [Mon02]. Fortunately, a compactness argument, as detailed for example in [FV10b, Sec 7.5], shows that all continuous homogeneous norms are equivalent. Upon checking continuity of the Carnot–Carathéodory norm, one gets, for $\mathbf{x} = \exp(b + c) \in G^{(2)}(\mathbf{R}^d)$,

$$\|\mathbf{x}\|_{\mathrm{C}} \asymp_d |b| + |c|^{1/2} \asymp \max\{|b|, |c|^{1/2}\}\,, \qquad (2.16)$$

which, despite its dependence on the dimension d, is sufficient for many practical purposes. As a useful application, we now state an approximation result for weakly geometric roughs over $\mathbf{R}^d$. With the preparations made, the interested reader will have no trouble to provide a full proof for

Proposition 2.8 (Geodesic approximation). *For every* $(X, \mathbb{X}) \in \mathscr{C}_g^\beta([0, T], \mathbf{R}^d)$, *there exists a sequence of smooth paths* $X^n : [0, T] \to \mathbf{R}^d$ *such that*

---

[5] In fact, by smoothly slowing down speed to zero whenever switching directions, the path γ can also be parametrized to be smooth. In particular, in (2.14) and (2.15) below we could have replaced $\mathcal{C}^1$ by $\mathcal{C}^\infty$.

$$\left(X^n, \mathbb{X}^n\right) := \left(X^n, \int_0^\cdot X_{0,t}^n \otimes dX_t^n\right) \to (X, \mathbb{X}) \ \textit{uniformly on } [0, T]$$

with uniform rough path bound $\sup_{n \geq 1} \|X^n, \mathbb{X}^n\|_\beta \lesssim \|X, \mathbb{X}\|_\beta$. *By interpolation, convergence holds in* $\mathscr{C}^\alpha$, *for any* $\alpha < \beta$.

Remark 2.9. By definition, every geometric rough path $\mathbf{X} \in \mathscr{C}_g^{0,\beta}$ is the limit of canonical rough path lifts $(X^n, \mathbb{X}^n) = \mathbf{X}^n$; trivially then, $\|\mathbf{X}^n\|_\beta \to \|\mathbf{X}\|_\beta$. This is not true for a generic weakly geometric rough path $\mathbf{X} \in \mathscr{C}_g^\beta$. However, the above proposition supplies approximations $(\mathbf{X}^n)$, which converge uniformly with uniform rough paths bounds. In such a case, $\|\mathbf{X}\|_\beta \leq \liminf_{n \geq 1} \|\mathbf{X}^n\|_\beta$ and this can be strict. This lower-semicontinuous behaviour of the rough path norm is reminiscent of norms on Hilbert spaces under weak convergence and led to the terminology of "weakly" geometric rough paths.

2.4 Geometric rough paths of low regularity

The interpretation given above gives a strong hint on how to construct geometric rough paths with α-Hölder regularity for $\alpha \leq \frac{1}{3}$: setting $N = \lfloor 1/\alpha \rfloor$, one defines the step-$N$ truncated tensor algebra over a Banach space V

$$T^{(N)}(V) \overset{\text{def}}{=} \bigoplus_{n=0}^N (V)^{\otimes n} \ ,$$

with the natural convention that $(V)^{\otimes 0} = \mathbf{R}$. The product in $T^{(N)}(V)$ is simply the tensor product $\otimes$, but we truncate it in a natural way by postulating that $a \otimes b = 0$ for $a \in (V)^{\otimes k}, b \in (V)^{\otimes \ell}$ with $k + \ell > N$. A homogeneous, symmetric and subadditive norm which generalises (2.10) to the step-N case is given by

$$\|\mathbf{x}\| \overset{\text{def}}{=} \tfrac{1}{2}\left(N(\mathbf{x}) + N(\mathbf{x}^{-1})\right) \quad \text{with } N(\mathbf{x}) = \max_{n=1,\ldots,N} (n!|\mathbf{x}^n|)^{1/n} \ , \tag{2.17}$$

where every $\mathbf{x} = (1, \mathbf{x}^1, \ldots, \mathbf{x}^N) \in T_1^{(N)}(V)$, element with scalar component 1, is invertible, and where $|\cdot|$ denotes any of the tensor norms on $(V)^{\otimes n}$, assumed compatible and symmetric (permutation invariant).[6].

Proposition 2.6 suggests the naïve definition of an α-*Hölder rough path* over V as a path $\mathbf{X}$, on $[0, T]$ say, with values in the group $T_1^{(N)}(V)$ which is α-Hölder continuous with respect to $d(\mathbf{x}, \mathbf{x}') = \|\mathbf{x}^{-1} \otimes \mathbf{x}'\|$. Modulo knowledge of $\mathbf{X}_0$ this is equivalent to a multiplicative map $(s, t) \mapsto \mathbf{X}_{s,t} \in T_1^{(N)}(V)$, *multiplicative in the sense that Chen's relation holds*,

$$\mathbf{X}_{s,t} = \mathbf{X}_{s,u} \otimes \mathbf{X}_{u,t} \ , \tag{2.18}$$

---

[6] The definitions from Section 1.4 for $N = 2$ extend easily to $N > 2$, see also [LCL07, Def 1.25]

for every triplet of times (s, u, t), and with graded Hölder regularity,

$$|\mathbf{X}_{s,t}^n| \lesssim |t - s|^{k\alpha}, \qquad n = 1, \ldots, N,$$

uniformly over $s, t \in [0, T]$. The interpretation of rough paths discussed at length in the step-2 setting is unchanged and $\mathbf{X}_{s,t}^n \in V^{\otimes n}$ should be thought of as a substitute for the (possibly ill-defined) n-fold integral $\int dX_{u_1} \otimes \cdots \otimes dX_{u_n}$ over the n-simplex $\{s < u_1 < \ldots < u_n < t\}$. Such a notion of naïve higher oder rough path is sometimes sufficient, e.g. for solving linear rough differential equations, see also Exercise 4.18, but does not contain the necessary information to deal with non-linearities, already seen in the simple example of the form $\int_s^t (X_r - X_s)^{\otimes 2} \otimes dX_r$.

Higher order (weakly) geometric rough paths resolve this problem by imposing a chain rule. In the above example, $(\delta X)^{\otimes 2}/2 = \mathrm{Sym}(\mathbf{X}^2)$, formerly written as $\mathrm{Sym}(\mathbb{X})$, and the situation is reduced to (a linear combination of) 3-fold iterated integrals. To proceed in a systematic fashion, we first introduce the correct state space as the *free step-N nilpotent Lie group over V*

$$G^{(N)}(V) \overset{\text{def}}{=} \exp(\mathfrak{g}^{(N)}(V)) \subset T_1^{(N)}(V)$$

where the exponential map is defined via its power series and $\mathfrak{g}^{(N)}(V) \subset T_0^{(N)}(V)$ is the (closed) Lie algebra generated by all elements of the form $(0, c, 0, \ldots, 0)$ with $c \in V$ via the natural Lie bracket $[a, b] = a \otimes b - b \otimes a$. The neutral element $\mathbf{1} \in G^{(N)}(V)$ is given by $\mathbf{1} = (1, 0, \ldots, 0)$. Given any $\alpha \in (0, 1]$ and $N = \lfloor 1/\alpha \rfloor$ as the number of "levels", Proposition 2.7 now suggests the definition of a *weakly geometric α-Hölder rough path* over V as a path $\mathbf{X}$, on $[0, T]$ say, with values in the group $G^{(N)}(V)$ which is α-Hölder continuous with respect to $d(\mathbf{x}, \mathbf{x}') = \|\mathbf{x}^{-1} \otimes \mathbf{x}'\|$. Modulo knowledge of $\mathbf{X}_0$ this is equivalent to a multiplicative map $(s, t) \mapsto \mathbf{X}_{s,t} \in G^{(N)}(V)$ with graded Hölder regularity, uniformly over $s, t \in [0, T]$,

$$|\mathbf{X}_{s,t}^n| \lesssim |t - s|^{n\alpha}, \qquad n = 1, \ldots, N.$$

Here, again multiplicative means validity of Chen's relation as spelled out in (2.18) above.

We now assume, for notationally convenience, $V = \mathbf{R}^d$, which allows us to think of components of some fixed rough path increment $\mathbf{X}_{s,t} \in T_1^{(N)}(\mathbf{R}^d)$ as being indexed by words w of length at most N with letters in the alphabet $\{1, \ldots, d\}$. Similarly to before, given a word $w = w_1 \cdots w_n$, the corresponding component $\mathbf{X}^w$, which we also write as $\langle \mathbf{X}, w \rangle$, is then interpreted as the n-fold integral

$$\langle \mathbf{X}_{s,t}, w \rangle = \int_s^t \int_s^{s_n} \cdots \int_s^{s_1} dX_{s_1}^{w_1} \cdots dX_{s_n}^{w_n}, \tag{2.19}$$

and $\|\mathbf{X}_{s,t}\| \lesssim |t - s|^\alpha$ is equivalent to, for all words with length $|w| \le \lfloor 1/\alpha \rfloor$,

$$|\langle \mathbf{X}_{s,t}, w \rangle| \lesssim |t - s|^{\alpha|w|}. \tag{2.20}$$

In order to describe the constraints imposed on these iterated integrals by the chain rule, we define the *shuffle product* ⊔⊔ between two words as the formal sum over all possible ways of interleaving them. For example, one has

$$a \amalg x = ax + xa , \quad ab \amalg xy = abxy + axby + xaby + axyb + xayb + xyab ,$$

with the empty word acting as the neutral element. With this notation at hand, it was already remarked by Ree [Ree58] (see also [Che71]) that the chain rule implies the identity

$$\langle \mathbf{X}_{s,t}, v \rangle \langle \mathbf{X}_{s,t}, w \rangle = \langle \mathbf{X}_{s,t}, v \amalg w \rangle . \tag{2.21}$$

(The reader is asked to show this in Exercise 2.2.) It is a remarkable fact that the algebraic properties of the tensor and shuffle algebras combine in such a way that the set of elements $\mathbf{X} \in T^{(N)}$ satisfying (2.21) is not only stable under the product $\otimes$, but forms a group, which in turn was shown in [Ree58] to be nothing but the group $G^{(N)}(\mathbf{R}^d)$. In the language of Hopf algebras, this group is exactly the *character group* for the (truncated) shuffle Hopf algebra.

In general, one may decide to forego the chain rule (after all, it doesn't hold in the context of Itô integration, as is manifest in Itô's formula) in which case there is no reason to impose (2.21). In this case, considering a rough path as an enhancement of a path X by iterated integrals of the type (2.19) no longer provides sufficient additional data. Indeed, in order to solve differential equations driven by X, one would like to give meaning to expressions like for example

$$\int_s^t \left(\int_s^r dX_u^i \right) \left(\int_s^r dX_v^j \right) dX_r^k =: \langle \mathbf{X}_{s,t}, {}^i \mathsf{Y}_k^j \rangle . \tag{2.22}$$

We already remarked earlier, that in the (weakly) geometric case, the assumed chain rule (now in the form of (2.21)) allows to reduce such expressions to linear combinations of iterated integrals. In general, one should define a rough path as the enhancement of a path X with additional functions that are interpreted as the various formal expressions that can be formed by the two operations "multiplication" and "integration against X". The resulting algebraic construction is more involved and gives rise to the concept of *branched rough path* X due to Gubinelli [Gub10]. The terminology comes from the fact that the natural way of indexing the components of such an object is no longer given by words, but by labelled trees, as suggested in (2.22) above with labels $i, j, k \in \{1, \dots, d\}$. As detailed in [Gub10], see also [HK15, BCFP19], branched rough paths take values in the character group of the *Connes–Kreimer Hopf algebra* of trees [CK00], also known as the *Butcher group* [But72]. A concise description of the branched rough path regularity via an explicit homogeneous subadditive norms on this Lie group, similar to (2.17), can be found in [TZ18], cf. also [HS90].

2.5 Exercises

♯ **Exercise 2.1** *Let X be a smooth V-valued path.*

a) *Show that $\mathbb{X}_{s,t} := \int_s^t X_{s,r} \otimes \dot{X}_r \, dr$ satisfy Chen's relation (2.1).*
b) *Consider the collection of all iterated integrals over $[s, t]$,*

$$\mathbf{X}_{s,t} := \left(1, X_{s,t}, \mathbb{X}_{s,t}, \int_{\Delta_{s,t}^{(3)}} dX_{u_1} \otimes dX_{u_2} \otimes dX_{u_3}, \dots \right) \in T((V)) \, , \quad (2.23)$$

where $\Delta_{s,t}^{(3)} = \{u : s < u_1 < u_2 < u_3 < t\}$ and $T((V)) \overset{def}{=} \prod_{k=0}^{\infty} V^{\otimes k}$ is the space of tensor series over V, equipped with the obvious algebra structure (cf. Section 2.4). Show that the following general form of Chen's relation holds:

$$\mathbf{X}_{s,t} = \mathbf{X}_{s,u} \otimes \mathbf{X}_{u,t} \, .$$

The element $\mathbf{X}_{s,t} \in T((V))$ is known as the signature of X on the interval $[s, t]$.
c) *Show that the indefinite signature $\mathbf{S} := \mathbf{X}_{0,\cdot}$ solves the linear differential equation*

$$d\mathbf{S} = \mathbf{S} \otimes dX \, , \qquad \mathbf{S}_0 = \mathbf{1} \, .$$

We will see later (Exercises 4.6 and 8.9) that the signature can be defined for every rough path.

Hint: *For point (b), it suffices to consider the projection of $\mathbf{X}_{s,t}$ to $V^{\otimes n}$, for an arbitrary integer n, given by the n-fold integral of $dX_{u_1} \otimes \dots \otimes dX_{u_n}$ over the simplex $\{s < u_1 < \dots < u_n < t\}$.*

♯ **Exercise 2.2 (Shuffle)** *Let $V = \mathbf{R}^d$. As discussed in (2.19), the collection $\mathbf{X}_{s,t}$ of all iterated integrals over a fixed interval $[s, t]$ can also be viewed as*

$$\left\{ \mathbf{X}_{s,t}^w = \langle \mathbf{X}_{s,t}, w \rangle : w \text{ word on } \mathcal{A} \right\} \, ,$$

with alphabet $\mathcal{A} = \{1, \dots, d\}$, where we recall that a word on $\mathcal{A}$ is a finite sequence of elements of $\mathcal{A}$, including the empty sequence $\emptyset$, called the empty word. By convention, $\mathbf{X}_{s,t}^{\emptyset} = 1$. Write uv for the concatenation of two words u and v, and accordingly ui for attaching a letter $i \in \mathcal{A}$ to the right of u. The linear span of such words (which can be identified with polynomials in d non-commuting indeterminates) carries an important commutative product known as the shuffle product. It is defined recursively by requiring $\emptyset$ to be the neutral element, ie. $u \sqcup\!\sqcup \emptyset = \emptyset \sqcup\!\sqcup u = u$, and then

$$ui \sqcup\!\sqcup vj = (u \sqcup\!\sqcup vj)i + (ui \sqcup\!\sqcup v)j \, .$$

Let $\mathbf{X}_{s,t}$ be the signature of a smooth path X, as given in (2.23). Show that, for all words u, v,

$$\langle \mathbf{X}_{s,t}, u \sqcup\!\sqcup v \rangle = \langle \mathbf{X}_{s,t}, u \rangle \langle \mathbf{X}_{s,t}, v \rangle \, . \quad (2.24)$$

The case of single letter words $w = i, v = j$ gives $i \sqcup\!\sqcup j = ij + ji$ and expresses precisely the product rule from calculus, which leads us to the level-2 geometricity condition (2.6).

Hint: *Proceed by induction in joint length: express $\langle \mathbf{X}_{s,t}, ui \rangle \langle \mathbf{X}_{s,t}, vj \rangle$ by the product rule as an integral over $[s, t]$ and use the hypothesis for words of joint length $|u| + |v| + 1 < |ui| + |vj|$.*

∗ **Exercise 2.3** *Call a tensor series $\mathbf{x} \in T((\mathbf{R}^d))$ group-like, in symbols $\mathbf{x} \in G((\mathbf{R}^d))$, if for all words u, v,*

$$\langle \mathbf{x}, u \sqcup\!\sqcup v \rangle = \langle \mathbf{x}, u \rangle \langle \mathbf{x}, v \rangle . \tag{2.25}$$

An element in $T((\mathbf{R}^d))$ is called a Lie series if, for all $N \in \mathbf{N}$, its projection to $T^{(N)} = T^{(N)}(\mathbf{R}^d)$ is a Lie polynomial, i.e. an element of $\mathfrak{g}^{(N)}$, which was defined in Section 2.4 as the Lie algebra generated by $\mathbf{R}^d \subset T_0^{(N)}$. Given $\mathbf{x} \in T((\mathbf{R}^d))$, show that $\mathbf{x}$ is group-like, i.e. $\mathbf{x} \in G((\mathbf{R}^d))$, if and only if $\log \mathbf{x}$ is a Lie series.

♯ **Exercise 2.4**

a) *It is common to define the $(V \otimes V)$-valued map $\mathbb{X}$ on $\Delta_{0,T} := \{(s, t) : 0 \leq s \leq t \leq T\}$ rather than $[0, T]^2$. There is no difference however: if $\mathbb{X}_{s,t}$ is only defined for $s \leq t$, show that the relation (2.1) implies*

$$\mathbb{X}_{t,s} = -\mathbb{X}_{s,t} + X_{s,t} \otimes X_{s,t} .$$

b) *In fact, show that knowledge of the path $t \mapsto (X_{0,t}, \mathbb{X}_{0,t})$ already determines the entire second order process $\mathbb{X}$. In this sense $(X, \mathbb{X})$ is indeed a path, and not some two-parameter object, cf. Remark 2.5.*

c) *Specialise to the case of geometric rough path and show the identity $\mathbb{X}_{t,s} = \mathbb{X}_{t,s}^T$ where $(\ldots)^T$ denotes the transpose. (When $\dim V = 1$, so that X is scalar valued, this is a trivial consequence of $\mathbb{X}_{s,t} = X_{s,t}^2/2$.)*

Exercise 2.5 *Consider $s \equiv \tau_0 < \tau_1 < \ldots < \tau_N \equiv t$. Show that (2.1) implies*

$$\mathbb{X}_{s,t} = \sum_{0 \leq i < N} \mathbb{X}_{\tau_i, \tau_{i+1}} + \sum_{0 \leq j < i < N} X_{\tau_j, \tau_{j+1}} \otimes X_{\tau_i, \tau_{i+1}}$$

$$= \sum_{i=0}^{N-1} \left(\mathbb{X}_{\tau_i, \tau_{i+1}} + X_{s, \tau_i} \otimes X_{\tau_i, \tau_{i+1}} \right). \tag{2.26}$$

This identity effectively compares $\mathbb{X}_{s,t}$ with a left-point Riemann-Stieltjes approximation $\sum_{i=0}^{N-1} X_{s, \tau_i} \otimes X_{\tau_i, \tau_{i+1}}$ of the "motivating" integral expression in (2.2).

Exercise 2.6 *Following Section 2.3 and Exercise 2.4, view $\mathbf{X} \in \mathscr{C}^\alpha([0, T], V)$ as a one-parameter path and define the (time T) time reversal of $\mathbf{X}$ in the "naïve" way as*

$$\overleftarrow{\mathbf{X}}_t = \mathbf{X}_{T-t}, \qquad 0 \leq t \leq T .$$

Verify that $\overleftarrow{\mathbf{X}}$ is again a rough path, i.e. $\overleftarrow{\mathbf{X}} \in \mathscr{C}^\alpha$. Show furthermore that $\overleftarrow{\mathbf{X}}$ is geometric if and only if $\mathbf{X}$ is geometric.

♯ **Exercise 2.7** *Let V be a Banach space.*

a) *Let $\alpha \in (0,1]$. Show that the linear space of all continuous maps $\mathbb{X} : [0,T]^2 \to V \otimes V$ s.t. $\|\mathbb{X}\| := \sup |\mathbb{X}_{s,t}|/|t-s|^{2\alpha} < \infty$ is a Banach space, denoted by $C_2^{2\alpha}$. Deduce that $C^\alpha \oplus C_2^{2\alpha}$ is also Banach, with seminorm $\|\cdot, \cdot\|_{\alpha, 2\alpha} = \|\cdot\|_\alpha + \|\cdot\|_{2\alpha}$. (A genuine norm is given by $(X, \mathbb{X}) \mapsto |X_0| + \|X, \mathbb{X}\|_{\alpha, 2\alpha}$.)*

b) *Show that the rough path spaces $\mathscr{C}_g^\alpha$ and $\mathscr{C}^\alpha$ are complete metric spaces. In fact, both are closed subspaces, defined through (nonlinear) algebraic relations, of the infinite-dimensional Banach space $C^\alpha \oplus C_2^{2\alpha}$.*

c) *Show that the rough path spaces $\mathscr{C}_g^\alpha$ and $\mathscr{C}^\alpha$ over $V = \mathbf{R}$ (and a fortiori every $V \neq 0$) are* **not** *separable. (You may use the well-known fact that the Hölder spaces $C^\alpha([0,T], \mathbf{R})$ are non-separable.)*

Exercise 2.8 (Separable rough path spaces) *Let V be a separable Banach space and $\alpha \in (\frac{1}{3}, \frac{1}{2}]$.*

a) *Show separability of the space of geometric (α-Hölder) rough paths*

$$\mathscr{C}_g^{0,\alpha}([0,T], V) \stackrel{def}{=} \mathrm{cl}(\mathscr{L}(\mathcal{C}^\infty)) \subset \mathscr{C}^\alpha([0,T], V) .$$

Together with Exercise 2.7, b), this shows that $\mathscr{C}_g^{0,\alpha}$ is Polish.

b) *Show that the closure of smooth rough paths,*

$$\mathscr{C}^{0,\alpha}([0,T], V) \stackrel{def}{=} \mathrm{cl}(\mathscr{C}^\infty) \subset \mathscr{C}^\alpha([0,T], V) ,$$

is also separable (and hence Polish).

Solution. (a) Let Q be a countable, dense subset of V and consider the space Λ_n of paths which are piecewise linear between level-n dyadic rationals $\mathbb{D}^n := \{kT/2^n : 0 \leq k \leq 2^n\}$, and, at level-$n$ dyadic points, take values in Q. Clearly $\Lambda = \cup \Lambda_n$ is countable for each Λ_n is in one-to-one correspondence with the $(2^n + 1)$-fold Cartesian product of Q. It is easy to see that each smooth X is the limit in $\mathcal{C}^1$ of some sequence $(X^n) \subset \Lambda$. Indeed, one can take X^n to be the piecewise linear dyadic approximation, modified such that $X^n|_{\mathbb{D}^n}$ takes values in Q and such that $|(X^n - X)|_{\mathbb{D}^n}| < 1/n$. By continuity of the map $X \in \mathcal{C}^1 \mapsto (X, \int X \otimes dX) \in \mathscr{C}^\alpha$ in the respective topologies (we could even take $\alpha = 1$), we have more than enough to assert that every lifted smooth path, $(X, \int X \otimes dX)$, is the limit in $\mathscr{C}^\alpha$ of lifted paths in Λ. It is then easy to see that every limit point of lifted smooth paths is also the limit of lifted paths in Λ.

♯ **Exercise 2.9 (Interpolation)** *Assume that $\mathbf{X}^n \in \mathscr{C}^\beta$, for $1/3 < \alpha < \beta$, with uniform bounds*

$$\sup_n \|X^n\|_\beta < \infty \qquad and \qquad \sup_n \|\mathbb{X}^n\|_{2\beta} < \infty$$

and uniform convergence $X_{s,t}^n \to X_{s,t}$ and $\mathbb{X}_{s,t}^n \to \mathbb{X}_{s,t}$, i.e. uniformly over $s,t \in [0,T]$. Show that this implies $\mathbf{X} \in \mathscr{C}^\beta$ and $\mathbf{X}^n \to \mathbf{X}$ in $\mathscr{C}^\alpha$. Show furthermore that the assumption of uniform convergence can be weakened to pointwise convergence:

$$\forall t \in [0,T]: \quad X_{0,t}^n \to X_{0,t} \quad \text{and} \quad \mathbb{X}_{0,t}^n \to \mathbb{X}_{0,t} \ .$$

Solution. Using the uniform bounds and *pointwise* convergence, there exists C such that uniformly in s,t

$$|X_{s,t}| = \lim_n |X_{s,t}^n| \le C|t-s|^\beta \ , \qquad |\mathbb{X}_{s,t}| = \lim_n |\mathbb{X}_{s,t}^n| \le C|t-s|^{2\beta} \ .$$

It readily follows that $\mathbf{X} = (X,\mathbb{X}) \in \mathscr{C}^\beta$. In combination with the assumed uniform convergence, there exists $\varepsilon_n \to 0$, such that, uniformly in s,t,

$$|X_{s,t} - X_{s,t}^n| \le \varepsilon_n \ , \qquad |X_{s,t} - X_{s,t}^n| \le 2C|t-s|^\beta \ ,$$
$$|\mathbb{X}_{s,t}^n - \mathbb{X}_{s,t}| \le \varepsilon_n \ , \qquad |\mathbb{X}_{s,t}^n - \mathbb{X}_{s,t}| \le 2C|t-s|^{2\beta} \ .$$

By geometric interpolation ($a \wedge b \le a^{1-\theta} b^\theta$ when $a, b > 0$ and $0 < \theta < 1$) with $\theta = \alpha/\beta$ we have

$$|X_{s,t} - X_{s,t}^n| \lesssim \varepsilon_n^{1-\alpha/\beta} |t-s|^\alpha \ , \qquad |\mathbb{X}_{s,t}^n - \mathbb{X}_{s,t}| \lesssim \varepsilon_n^{1-\alpha/\beta} |t-s|^{2\alpha} \ ,$$

and the desired convergence in $\mathscr{C}^\alpha$ follows.

It remains to weaken the assumption to pointwise convergence. By Chen's relation, pointwise convergence of $\mathbf{X}_{0,t}^n$ for all t actually implies pointwise convergence of $\mathbf{X}_{s,t}^n$ for all s,t. We claim that, thanks to the uniform Hölder bounds, this implies uniform convergence. Indeed, given $\varepsilon > 0$, pick a (finite) dissection D of $[0,T]$ with small enough mesh so that $C|D|^\beta < \varepsilon/8$. Given $s,t \in [0,T]$ write $\hat{s}, \hat{t}$ for the nearest points in D and note that

$$|X_{s,t} - X_{s,t}^n| \le |X_{\hat{s},\hat{t}} - X_{\hat{s},\hat{t}}^n| + |X_{s,\hat{s}}| + |X_{s,\hat{s}}^n| + |X_{t,\hat{t}}| + |X_{t,\hat{t}}^n|$$
$$\le |X_{\hat{s},\hat{t}} - X_{\hat{s},\hat{t}}^n| + \varepsilon/2 \ .$$

By picking n large enough, $|X_{\hat{s},\hat{t}} - X_{\hat{s},\hat{t}}^n|$ can also be bounded by $\varepsilon/2$, uniformly over the (finitely many!) points in D, so that $X^n \to X$ uniformly. Although the second level is handled similarly, the non-additivity of $(s,t) \mapsto \mathbb{X}_{s,t}$ requires some extra care, (2.1). For simplicity of notation only, we assume $s < \hat{s} < t = \hat{t}$ so that

$$|\mathbb{X}_{s,t} - \mathbb{X}_{s,t}^n| \le |\mathbb{X}_{s,\hat{s}} - \mathbb{X}_{\hat{s},t}^n| + |\mathbb{X}_{\hat{s},t}| + |X_{s,\hat{s}} \otimes X_{\hat{s},t} - X_{s,\hat{s}}^n \otimes X_{\hat{s},t}^n| \ .$$

It remains to write the last summand as $|X_{s,\hat{s}} \otimes (X_{\hat{s},t} - X_{\hat{s},t}^n) - (X_{s,\hat{s}}^n - X_{s,\hat{s}}) \otimes X_{\hat{s},t}^n|$ and to repeat the same reasoning as in the first level.

♯ **Exercise 2.10 (Pure area rough path)** *Identify* $\mathbf{R}^2$ *with the complex numbers and consider*

$$[0,1] \ni t \mapsto n^{-1} \exp\left(2\pi i n^2 t\right) \equiv X^n.$$

a) Set $\mathbb{X}_{s,t}^n := \int_s^t X_{s,r}^n \otimes dX_r^n$. *Show that, for fixed* $s < t$,

$$X^n_{s,t} \to 0, \qquad \mathbb{X}^n_{s,t} \to \pi(t-s)\begin{pmatrix} 0 & 1 \\ -1 & 0 \end{pmatrix}. \qquad (2.27)$$

b) *Establish the uniform bounds* $\sup_n \|X^n\|_{1/2} < \infty$ *and* $\sup_n \|\mathbb{X}^n\|_1 < \infty$.

c) *Conclude that* $(X^n, \mathbb{X}^n)$ *converges in* $\mathscr{C}^\alpha$, *any* $\alpha < 1/2$.

Solution.   a) Obviously, $X^n_{s,t} = O(1/n) \to 0$ uniformly in s, t. Then

$$\mathbb{X}^n_{s,t} = \frac{1}{2} X^n_{s,t} \otimes X^n_{s,t} + A^n_{s,t} = O(1/n^2) + A^n_{s,t}$$

where $A^n_{s,t} \in \mathfrak{so}(2)$ is the antisymmetric part of $\mathbb{X}^n_{s,t}$. To avoid cumbersome notation, we identify

$$\begin{pmatrix} 0 & a \\ -a & 0 \end{pmatrix} \in \mathfrak{so}(2) \leftrightarrow a \in \mathbf{R}.$$

$A^n_{s,t}$ then represents the signed area between the curve $(X^n_r : s \le r \le t)$ and the straight chord from X^n_t to X^n_s. (This is a simple consequence of Stokes theorem: the exterior derivative of the 1-form $\frac{1}{2}(x\,dy - y\,dx)$ which vanishes along straight chords, is the volume form $dx \wedge dy$.) With $s < t$, $(X^n_r : s \le r \le t)$ makes $\lfloor n^2(t-s) \rfloor$ full spins around the origin, at radius $1/n$. Each full spin contributes area $\pi(1/n)^2$, while the final incomplete spin contributes some area less than $\pi(1/n)^2$. The total signed area, with multiplicity, is thus

$$A^n_{s,t} = \left(n^2(t-s) + O(1)\right)\frac{\pi}{n^2} = \pi(t-s) + \frac{C_{s,t}}{n^2},$$

where $|C_{s,t}| \le \pi$ uniformly in s, t. It follows that

$$\mathbb{X}^n_{s,t} = \pi(t-s)\begin{pmatrix} 0 & 1 \\ -1 & 0 \end{pmatrix} + O(1/n^2) \qquad (2.28)$$

and the claimed uniform convergence follows.

b) The following two estimates for path increments of $n^{-1}\exp\left(2\pi i n^2 t\right) \equiv X^n_t$ hold true:

$$\left|X^n_{s,t}\right| \le \left|\dot{X}^n\right|_\infty |t-s| \le n|t-s|, \qquad \left|X^n_{s,t}\right| \le 2\left|X^n\right|_\infty = 2/n.$$

Since $a \wedge b \le \sqrt{ab}$, it immediately follows that

$$\left|X^n_{s,t}\right| \le \sqrt{2|t-s|},$$

uniformly in n, s, t. In other words, $\sup_n \|X^n\|_{1/2} < \infty$. The argument for the uniform bounds on $\mathbb{X}_{s,t}$ is similar. On the one hand, we have the bound (2.28). On the other hand, we also have

$$|\mathbb{X}^n_{s,t}| = \left| \int \int_{s<u<v<t} \dot{X}^n_u \otimes \dot{X}^n_v \, du \, dv \right| \le |\dot{X}^n|^2_\infty \frac{|t-s|^2}{2} \le \frac{n^2}{2}|t-s|^2 \, .$$

The required uniform bound on $\|\mathbb{X}\|_1$ follows by using (2.28) for $n^2|t-s| > 1$ and the above bound for $n^2|t-s| \le 1$.

c) The interpolation argument is left to the reader.

Exercise 2.11 (Second order translation and bracket) *Fix* $\alpha \in (\frac{1}{3}, \frac{1}{2}]$ *and* $\mathbf{X} = (X, \mathbb{X}) \in \mathscr{C}^\alpha([0,T], V)$. *Define the (second order) translation of* $\mathbf{X}$ *in direction* $\mathbb{H} \in \mathcal{C}^{2\alpha}([0,T], V \otimes V)$ *by*

$$T_{\mathbb{H}}(\mathbf{X}) \overset{def}{=} (X, \mathbb{X} + \delta\mathbb{H}) \, ,$$

where $(\delta\mathbb{H})$ *denotes the map* $(s,t) \mapsto \mathbb{H}_t - \mathbb{H}_s$.

a) *Show that* $T_{\mathbb{H}}(\mathbf{X}) \in \mathscr{C}^\alpha$. *In fact, show that the (linear) space* $\mathcal{C}^{2\alpha}$ *acts freely on the (nonlinear) rough path space* $\mathscr{C}^\alpha$ *in the sense that, for all* $\mathbb{G}, \mathbb{H} \in \mathcal{C}^{2\alpha}$, *we have*

$$T_{\mathbb{G}}(T_{\mathbb{H}}(\mathbf{X})) = (T_{\mathbb{G}} \circ T_{\mathbb{H}})(\mathbf{X}) = T_{\mathbb{G}+\mathbb{H}}(\mathbf{X}) \, .$$

Fix $\mathbf{X} \in \mathscr{C}^\alpha$. *Is* $\mathbb{H} \mapsto T_{\mathbb{H}}(\mathbf{X})$ *is injective?*
b) *When does* $T_{\mathbb{H}}$ *preserve the space* $\mathscr{C}^\alpha_g([0,T], V)$?
c) *Show that any* $\mathbf{X} = (X, \mathbb{X}) \in \mathscr{C}^\alpha([0,T], V)$ *can be written, in a unique way, as* $T_{\mathbb{H}}(\mathbf{X}_g)$, *where* $\mathbf{X}_g \in \mathscr{C}^\alpha_g([0,T], V)$ *for some* $\mathbb{H} \in \mathcal{C}^{2\alpha}([0,T], \operatorname{Sym}(V \otimes V))$, *so that we have the bijection*

$$\mathscr{C}^\alpha([0,T], V) \leftrightarrow \mathscr{C}^\alpha_g([0,T], V) \times \mathcal{C}^{2\alpha}([0,T], \operatorname{Sym}(V \otimes V)).$$

Show that $2\delta\mathbb{H} = (\delta X)^{\otimes 2} - 2\operatorname{Sym}(\mathbb{X}) =: [\mathbf{X}]$, *called* bracket *of the rough path* $\mathbf{X}$, *further studied in Section 5.3.*

Exercise 2.12 (Vanishing Hölder oscillation) a) *Let* $X \in \mathcal{C}^\alpha([0,T], V)$ *with Hölder exponent* $\alpha \in (0,1]$. *Define the space of Hölder path with "vanishing Hölder oscillation",*

$$\mathcal{C}^{\mathrm{van},\alpha} \overset{def}{=} \left\{ X \in \mathcal{C}^\alpha : \sup_{s,t:|t-s|<\varepsilon} \frac{|X_{s,t}|}{|t-s|^\alpha} \to 0, \text{ as } \varepsilon \to 0 \right\}.$$

Show that for $\alpha \in (0,1)$ *we have* $\mathcal{C}^{\mathrm{van},\alpha} = \mathcal{C}^{0,\alpha}$, *the closure of smooth paths in* $\mathcal{C}^\alpha$. *(For* $\alpha = 1$ *this fails, why?) Show by explicit example that the inclusion* $\mathcal{C}^{0,\alpha} \subset \mathcal{C}^\alpha$ *is strict. (Hint: consider the function* $t \mapsto t^\alpha$.)
b) *Let* $\mathbf{X} = (X, \mathbb{X}) \in \mathscr{C}^\alpha_g([0,T], V)$ *with* $\alpha \in (\frac{1}{3}, \frac{1}{2}]$. *Define the space of Hölder rough paths with "vanishing Hölder oscillation",*

$$\mathscr{C}^{\mathrm{van},\alpha}_g \overset{def}{=} \left\{ \mathbf{X} \in \mathscr{C}^\alpha_g : \sup_{|t-s|<\varepsilon} \frac{|X_{s,t}|}{|t-s|^\alpha} + \sup_{|t-s|<\varepsilon} \frac{|\mathbb{X}_{s,t}|}{|t-s|^{2\alpha}} \to 0 \text{ as } \varepsilon \to 0 \right\}.$$

    i) *Show the inclusions $\mathscr{C}_g^{0,\alpha} \subset \mathscr{C}_g^{\text{van},\alpha}$ and also $\mathscr{C}_g^{\beta} \subset \mathscr{C}_g^{\text{van},\alpha}$, whenever $\alpha < \beta$. Show that the inclusion $\mathscr{C}_g^{\text{van},\alpha} \subset \mathscr{C}_g^{\alpha}$ is strict.*

    ii) *Assume $\dim V < \infty$ from here on. Show $\mathscr{C}_g^{0,\alpha} = \mathscr{C}_g^{\text{van},\alpha}$ (Hint: use the "geodesic" approximations from Proposition 2.8.)*

    iii) *From ii) we have $\mathscr{C}_g^{\beta} \subset \mathscr{C}_g^{0,\alpha} \subset \mathscr{C}_g^{\alpha}$, whenever $\frac{1}{3} < \alpha < \beta \le \frac{1}{2}$. Show that one has the compact embedding (Hint: Arzela–Ascoli)*

$$\mathscr{C}_g^{\beta} \hookrightarrow \mathscr{C}_g^{0,\alpha}.$$

c) *Discuss similar statements for non-geometric rough path spaces. In particular, discuss the validity of*

$$\mathscr{C}^{0,\alpha} \overset{\text{def}}{=} \text{cl}(\mathscr{C}^{\infty}) = \mathscr{C}^{\text{van},\alpha},$$

and also, cf. Exercise 2.11, c),

$$\mathscr{C}^{0,\alpha} \leftrightarrow \mathscr{C}_g^{0,\alpha} \times C^{0,2\alpha};$$

for $\alpha = 1/2$ this fails, why?

Remark: *This is essentially taken from [FV06a], for a recent extension to*

∗ **Exercise 2.13** *Show that for every geometric $1/2$-Hölder rough path, $\mathbf{X} \in \mathscr{C}_g^{0,1/2}$, $\mathbb{X}$ is necessarily the iterated Riemann–Stieltjes integral of the underlying path $X \in C^{0,1/2}$. Show also that there exists $X \in C^{0,1/2}$ (with values in $\mathbf{R}^2$) such that the iterated Riemann–Stieltjes integrals do not exist. This further shows that the Lyons–Victoir extension (Exercise 2.14, part d) can fail for α-Hölder rough paths when $1/\alpha \in \mathbf{N}$.*

Solution. We use $\mathscr{C}_g^{0,\alpha} \subset \mathscr{C}_g^{\text{van},\alpha}$ Exercise 2.12, for $\alpha = 1/2$. Consider a dissection $\{s = \tau_0 < \tau_1 < \ldots < \tau_N = t\}$ with mesh $\le \varepsilon$. It follows from Chen's relation (2.1), in the form (2.26),

$$\left| \mathbb{X}_{s,t} - \sum_{0 \le i < N} X_{s,\tau_i} \otimes X_{\tau_i, \tau_{i+1}} \right| = \left| \sum_{0 \le i < N} \mathbb{X}_{\tau_i, \tau_{i+1}} \right|$$

$$\le C(\varepsilon) \sum_{0 \le i < N} |\tau_{i+1} - \tau_i|^{2\alpha} = T C(\varepsilon).$$

It follows that $\mathbb{X}_{s,t}$ is the limit of the above Riemann–Stieltjes sum.

    Regarding the second question, a counterexample is found in [FV10b, Ex.9.14 (iii)].

♯∗ **Exercise 2.14 (Lyons–Victoir extension [LV07])** *Let $\alpha \in (0, 1/2)$ and consider $X \in C^{\alpha}([0,T], L(V, W)), Y \in C^{\alpha}([0,T], V)$ and $\mathbb{Z} \in C_2^{2\alpha}([0,T], W)$. We omit $[0,T]$ and the precise target space in what follows. We here say that Chen's relation holds if, for every triple of times (s, t, u),*

$$\mathbb{Z}_{s,u} = \mathbb{Z}_{s,t} + \mathbb{Z}_{t,u} + Y_{s,t} X_{t,u}.$$

(This is the algebraic relation satisfied by $(s,t) \mapsto \int_s^t Y_{s,r} dX_r$ whenever $X \in C^1$.)

a) *Show that here exists a bilinear continuous map $\Phi : C^\alpha \times C^\alpha \to C_2^{2\alpha}$,*

$$(Y, X) \mapsto \mathbb{Z} := \Phi(Y, X)$$

   such that Chen's relation holds.

b) *Show that the restriction of Φ to Hölder paths with exponent $\beta \in (1/2, 1)$ cannot possibly be a continuous as map $C^\beta \times C^\beta \to C_2^{2\beta}$. (Hint: the Chen relation would force $\Phi(Y, X)$ to coincide with the Young integral $\int Y dX$. In particular, $\Phi_{0,.}$ would have to coincide with $\int_0^\cdot Y(t) \dot{X}(t) dt$ in case of smooth path. Proposition 1.1 then allows to conclude.)*

c) *Show however that Φ can be constructed such that its restriction to a map $C^\beta \times C^\beta \to C^\beta$, where the image is now regarded as path $t \mapsto \Phi(Y, X)_{0,t}$, is a bilinear continuous map.*

d) *Let $\alpha \in (1/3, 1/2)$. Show that every path $X \in C^\alpha([0, T], V)$ admits a (if so desired: geometric) rough path lift $(X, \mathbb{X}) \in \mathscr{C}^\alpha([0, T], V)$.*

e) *Conclude that the nonlinear rough path space $\mathscr{C}^\alpha([0, T], V)$ is in (non-canonical) one-one correspondence with the linear space $C^\alpha([0, T], V) \oplus C^{2\alpha}([0, T], V \otimes V)$. (For a generalisation of this to rough paths of low regularity see [TZ18].)*

Solution. We show a) and c) together; d) is really a variation / consequence of a) and we leave b) and e) to the reader. Without loss of generality, $T = 1$. Write $\mathbb{Z}_{(s,t]} \equiv \mathbb{Z}_{s,t}$ and similarly for the path increments of Y, X. We want to construct $\mathbb{Z}$ such that

$$\mathbb{Z}_I = \mathbb{Z}_L + \mathbb{Z}_R + Y_L \otimes X_R$$

whenever $I = (s, t]$ is the union of two adjacent "left and right" intervals L and R, and such that

$$|\mathbb{Z}_I| \lesssim |I|^{2\alpha} \qquad\qquad (\star)$$

where $|I| = |t - s|$. By a continuity and chaining argument (see the proof of Theorem 3.1 below), it is enough to do so for dyadic times, i.e. $s, t \in \bigcup_{n \geq 0} \mathbf{D}_n$ where $\mathbf{D}_0 = \{(0, 1]\}$, $\mathbf{D}_1 = \{(0, 1/2], (1/2, 1]\}$ and so on. We start with the (ad-hoc!) choice $\mathbb{Z}_{0,1} \equiv \mathbb{Z}_{(0,1]} = 0$ and note its (trivial) bilinearity in (Y, X). Assume now $\mathbb{Z}_I$ for $I \in \mathbf{D}_{n-1}$ has been constructed. Write I as the union of two nth level dyadic intervals, $I = L \cup R$. Make the (ad-hoc) imposition $\mathbb{Z}_L = \mathbb{Z}_R$ which leads to

$$\mathbb{Z}_L = \mathbb{Z}_R = \frac{1}{2}(\mathbb{Z}_I - Y_L \otimes X_R).$$

(Note that bilinear dependence in Y, X is preserved.) On the analytic side, we have

$$|\mathbb{Z}_L| = |\mathbb{Z}_R| = \frac{1}{2}|\mathbb{Z}_I - Y_L \otimes X_R| \leq \frac{1}{2}|\mathbb{Z}_I| + \frac{1}{2}|Y_L| \cdot |X_R|$$

and, setting $a_n := \sup_{J \in \mathbf{D}_n} |\mathbb{Z}_J|/|J|^{2\alpha} = 2^{2n\alpha} \sup_{J \in \mathbf{D}_n} |\mathbb{Z}_J|$, it follows that

$$a_n \leqslant 2^{-(1-2\alpha)} a_{n-1} + \frac{1}{2} \|Y\|_\alpha \|X\|_\alpha,$$

so that the sequence (a_n) is bounded since $1 - 2\alpha > 0$. In fact, one easily obtains the bound

$$\sup_{n \geqslant 0} |a_n| \lesssim \|Y\|_\alpha \|X\|_\alpha,$$

with proportionality constant only depending on $\alpha < 1/2$. This implies the estimate $(\star)$ and also settles continuity of $\Phi = \Phi(Y, X)$. It remains to show that $t \mapsto \mathbb{Z}_{0,t} \in \mathcal{C}^\beta$ whenever $Y, X \in \mathcal{C}^\beta$ and $\beta \in (1/2, 1)$. But this is an immediate consequence of the bound

$$|\mathbb{Z}_{0,t} - \mathbb{Z}_{0,s}| \leqslant |\mathbb{Z}_{s,t}| + |X_{0,s}| \cdot |X_{s,t}|,$$

noting that, thanks to the first part of the theorem, $|\mathbb{Z}_{s,t}| \lesssim |t - s|^{2\alpha}$ for all $2\alpha < 1$.

Exercise 2.15 (Translation of rough paths) *Fix* $\alpha \in (\frac{1}{3}, \frac{1}{2}]$ *and* $\mathbf{X} = (X, \mathbb{X}) \in \mathcal{C}^\alpha([0, T], \mathbf{R}^d)$. *For sufficiently smooth* $h : [0, T] \to \mathbf{R}^d$, *the translation of* $\mathbf{X}$ *in direction* h *is given by*

$$T_h(\mathbf{X}) \stackrel{def}{=} \left(X^h, \mathbb{X}^h \right),$$

where $X^h := X + h$ *and*

$$\mathbb{X}^h_{s,t} := \mathbb{X}_{s,t} + \int_s^t h_{s,r} \otimes dX_r + \int_s^t X_{s,r} \otimes dh_r + \int_s^t h_{s,r} \otimes dh_r . \qquad (2.29)$$

a) *Assume* $h \in \mathcal{C}^1$. *(In particular, the last three integrals above are well-defined Riemann–Stieltjes integrals.) Show that for fixed* h, *the translation operator* $T_h : \mathbf{X} \mapsto T_h(\mathbf{X})$ *is a continuous map from* $\mathcal{C}^\alpha$ *into itself.*

b) *By convention,* $h \in \mathcal{C}^1$ *means Lipschitz or equivalently* $h \in W^{1,\infty}$, *where* $W^{1,q}$ *denotes the space of absolutely continuous paths* h *with derivative* $\dot{h} \in L^q$. *Weaken the assumption on* h *by only requiring* $\dot{h} \in L^q$, *for suitable* $q = q(\alpha)$. *Show that* $q = 2$ *("Cameron–Martin paths of Brownian motion") works for all* $\alpha \leq 1/2$. *(As a matter of fact, the integrals appearing in (2.29) make sense for every* $q \geq 1$, *but the resulting translated "rough path" falls out of the class of Hölder rough paths. One can resolve this issue by switching to* $(1/\alpha)$-*variation rough paths.)*

c) *Call any* $\mathbf{h} = (h, \mathbb{H}) : [0, T] \to \mathbf{R}^d \oplus (\mathbf{R}^d)^{\otimes 2} = T_0^{(2)}$, *with* $h \in W^{1,2}$ *and* $\mathbb{H} \in \mathcal{C}^{2\alpha}$ *an admissible perturbation. With some notational overloading,* T *is also used for the second order translation introduced in Exercise 2.11, show that*

$$T_{\mathbf{h}} := T_h \circ T_{\mathbb{H}} = T_{\mathbb{H}} \circ T_h$$

is a well-defined action on $\mathcal{C}^\alpha$, *in the sense of* $T_{\mathbf{g}} \circ T_{\mathbf{h}} = T_{\mathbf{g}+\mathbf{h}}$. *Show that for any fixed* $(a, b) \in T_0^{(2)}$, *the constant speed perturbation* $t \mapsto (at, bt)$ *is admissible, which then yields an action of* $T_0^{(2)}$ *with its additive structure on* $\mathcal{C}^\alpha$. *Show that these statements remain true for* $\mathcal{C}^\alpha_g$ *provided admissible perturbations take values in the Lie algebra* $\mathfrak{g}^{(2)} = \mathbf{R}^d \oplus \mathfrak{so}(d)$ *as introduced in Section 2.3.*

Remark: *Some far-reaching extensions of this are found in [BCFP19]. Constant speed perturbations respect stationarity of the noise (stationary increments of the process) and thus serve as elementary examples of (algebraic) renormalisation of models in regularity structures. The (abelian) groups $(\mathfrak{g}^{(2)}, +)$ and $(T_0^{(2)}, +)$ together with their action $\mathbf{h} \mapsto T_{\mathbf{h}}$, are examples of a renormalisation group in the sense of Section 15.5.1.*

2.6 Comments

Many early works in stochastic analysis starting from Itô (and then in no particular order Kunita, Yamato, Sugita, Azencott, Ben Arous [BA89], etc) and in control theory (Magnus, Brocket, Sussmann, Fliess [FNC82], etc) have recognised the importance of iterated integrals of the driving noise / signal; many references are given [Lyo98] and the books [LQ02, LCL07, FV10b].

The notion of rough path is due to Lyons and was introduced in [Lyo98] in p-variation sense, $p \in [1, \infty)$, and over Banach spaces. Earlier notes [Lyo94, Lyo95] already dealt with α-Hölder rough paths for $\alpha \in \left(\frac{1}{3}, \frac{1}{2}\right]$.

The analytical aspects of rough paths are related to Young's seminal work [You36], revisited in Chapter 4. On the algebraic side, Chen's relation is rooted in [Che54, Che57] and encodes abstractly basic additivity properties of iterated integrals. A key observation of Chen [Che57, Che58] was that log signatures are Lie series, the description via shuffles (cf. Section 2.4) is due to Ree [Ree58] (see also [Che71]). It follows from the works of Chow and Rashevskii [Cho39, Ras38], also [Che57, Che58], that this map is, upon truncation, onto: for every element in $\mathbf{x} \in G^{(N)}(\mathbf{R}^d) := \exp(\mathfrak{g}^{(N)}(\mathbf{R}^d))$ there exists a smooth path $\gamma : [0, 1] \to \mathbf{R}^d$ with prescribed signature $\mathbf{x} = S^{(N)}(\gamma)$. The shortest such path can be viewed as sub-Riemannian geodesic, concatenation of such geodesics is then a natural way to approximate weakly geometric rough paths (cf. Proposition 2.8) and underlies the geometric approach of Friz–Victoir [FV05, FV10b], surveyed from a sub Riemannian perspective in [FG16a]. The polynomial nature of (truncated) shuffle relations and log Lie conditions recently led Améndola, Friz and Sturmfels [AFS19] to the study of *signature varieties* in computational algebraic geometry.

Up to equivalence under a generalised notion of reparameterisation of paths known as *treelike equivalence*, the "full" signature map $\gamma \mapsto S(\gamma) \in G((V)) \subset T((V))$ was shown to be injective by Chen [Che58] in case of piecewise smooths paths, Hambly–Lyons [HL10] in case of rectifiable paths, and Boedihardjo et al. [BGLY16] in case of weakly geometric rough paths of arbitrarily low regularity, see also Boedihardjo, Ni and Qian [BNQ14]. The inversion problem "signature $\mapsto$ path" is studied by Lyons–Xu [LX17, LX18] and [AFS19]. All this is part of the mathematical justification of the *signature method* in machine learning, see e.g. Lyons' ICM article [Lyo14] and the survey [CK16].

For some constructions of level-2 geometric rough paths motivated from harmonic analysis see Hara–Lyons [HL07] and Lyons–Yang [LY13], see also the comments

Section 3.8 for some martingale constructions related to harmonic analysis. Lyons–Qian, in their monograph [LQ02] work with geometric rough paths (over a Banach space V), per definition limits of canonically lifted smooth paths. The strict inclusion "geometric $\subset$ weakly geometeric" was somewhat blurred in the earlier rough paths literature. For $\dim V < \infty$, matters were clarified in [FV06a]. For a discussion of weakly geometric rough paths over Banach spaces in their own right, see e.g. in [CDLL16], see also the supplementary appendix [BGLY15] of [BGLY16]. The discussion in Section 2.4, the "shuffle" view on weakly geometric rough paths and then Gubinelli's branched rough paths [Gub10], also extends from $V = \mathbf{R}^d$ to infinite dimension but setting up basis-independent notations is somewhat more involved. See for example [CW16, CCHS20] for some recent results in this direction.

"Naïve" higher order non-geometric rough paths with values in $T_1^{(N)}(V)$ are called in [Lyo98] *multiplicative functionals* (with α-Hölder or p-variation regularity, $\lfloor p \rfloor = N$), insisting on their inability to handle nonlinearities when $N \geq 3$. The notion of branched rough path, for any $\alpha \in (0, 1]$, further studied in [HK15, FZ18, BCFP19, BC19, TZ18] provides the required extra information when $N \geq 3$; for $N = \lfloor 1/\alpha \rfloor = 2$ there is no difference. It is possible to embed spaces of non-geometric rough paths of low regularity into suitable spaces of geometric rough paths, see [LV06] or Exercise 2.11 part c) when $N = 2$. The case of very low regularities, with N large, is much more involved and studied by Hairer–Kelly [HK15] and later Boedihardjo–Chevyrev [BC19].

Rough paths with jumps, in p-variation scale, are studied in [Wil01, FS17, FZ18, CF19], previously introduced *discrete rough paths* [Kel16] are also accomodated e.g. by the *càdlàg rough path* setting of [FZ18]. See also the comment Sections 4.8, 5.6 and 9.6. Rough paths in a geometric ambient space have been studied by Cass, Driver, Litterer and Lyons in [CLL12, CDL15], see also Bailleul [Bai19] for rough paths on Banach manifolds.

Chapter 3
Brownian motion as a rough path

In this chapter, we consider the most important example of a rough path, which is the one associated to Brownian motion. We discuss the difference, at the level of rough paths, between Itô and Stratonovich Brownian motion. We also provide a natural example of approximation to Brownian motion which converges to neither of them.

3.1 Kolmogorov criterion for rough paths

Consider random $X(\omega) : [0, T] \to V$ and $\mathbb{X}(\omega) : [0, T]^2 \to V \otimes V$, subject to (2.1). Equivalently, following Exercise 2.4, we can think of

$$\mathbf{X}(\omega) \equiv (X, \mathbb{X})(\omega) : [0, T] \to V \oplus (V \otimes V)$$

as a (random) path. The basic example, of course, is that of d-dimensional standard Brownian motion B enhanced with

$$\mathbb{B}_{s,t} \stackrel{\text{def}}{=} \int_s^t B_{s,r} \otimes dB_r \in \mathbf{R}^d \otimes \mathbf{R}^d \cong \mathbf{R}^{d \times d} . \tag{3.1}$$

The integration here is understood either in Itô or Stratonovich sense (in the latter case, we would write $\circ dB$); sometimes we indicate this by writing $\mathbb{B}^{\text{Itô}}$ resp. $\mathbb{B}^{\text{Strat}}$. It should be noted that the antisymmetric part of $\mathbb{B}$, also known as *Lévy's stochastic area*, with values in $\mathfrak{so}(d)$, is not affected by the choice of stochastic integration. Condition (2.1) is seen to be valid with either choice, while condition (2.6) only holds in the Stratonovich case. We now address the question of α- resp. 2α-Hölder regularity of X resp. $\mathbb{X}$ by a suitable extension of the classical Kolmogorov criterion; the application to Brownian motion is then carried out in detail in the following subsection.

Recalling that $B \in \mathcal{C}^\alpha([0, T], \mathbf{R}^d)$, a.s. for any $\alpha < 1/2$, we now address the question of 2α-Hölder regularity for $\mathbb{B}$.

© Springer Nature Switzerland AG 2020
P. K. Friz, M. Hairer, *A Course on Rough Paths*, Universitext,
https://doi.org/10.1007/978-3-030-41556-3_3

Using Brownian scaling and exponential integrability of $\mathbb{B}_{0,1}$, which is an immediate consequence of the integrability properties of the second Wiener chaos, the following result applies with $\beta = 1/2$ and all $q < \infty$. It gives the desired 2α-Hölder regularity for $\mathbb{B}$, a.s. for any $\alpha < 1/2$. As a consequence, $(B, \mathbb{B}) \in \mathscr{C}^\alpha$ almost surely, where we may take any $\alpha \in \left(\frac{1}{3}, \frac{1}{2}\right)$ and $\mathbf{B} \equiv (B, \mathbb{B})$ is known as *Brownian rough path* or *enhanced Brownian motion*. In the Stratonovich case, thanks to (2.6), we obtain a *geometric* rough path, i.e. $(B, \mathbb{B}^{\text{Strat}}) \in \mathscr{C}_g^\alpha$.

Theorem 3.1 (Kolmogorov criterion for rough paths). *Let $q \geq 2, \beta > 1/q$. Assume, for all s, t in $[0, T]$*

$$|X_{s,t}|_{L^q} \leq C|t - s|^\beta, \qquad |\mathbb{X}_{s,t}|_{L^{q/2}} \leq C|t - s|^{2\beta}, \tag{3.2}$$

for some constant $C < \infty$. Then, for all $\alpha \in [0, \beta - 1/q)$, there exists a modification of $(X, \mathbb{X})$ (also denoted by $(X, \mathbb{X})$) and random variables $K_\alpha \in L^q, \mathbb{K}_\alpha \in L^{q/2}$ such that, for all s, t in $[0, T]$

$$|X_{s,t}| \leq K_\alpha(\omega)|t - s|^\alpha, \qquad |\mathbb{X}_{s,t}| \leq \mathbb{K}_\alpha(\omega)|t - s|^{2\alpha}. \tag{3.3}$$

In particular, if $\beta - \frac{1}{q} > \frac{1}{3}$ then, for every $\alpha \in \left(\frac{1}{3}, \beta - \frac{1}{q}\right)$, we have homogeneous rough path norm $\|\mathbf{X}\|_\alpha \in L^q$ and hence $\mathbf{X} = (X, \mathbb{X}) \in \mathscr{C}^\alpha$ almost surely.

Proof. The proof is almost the same as the classical proof of Kolmogorov's continuity criterion, as exposed for example in [RY99]. Without loss of generality take $T = 1$ and let D_n denote the set of integer multiples of 2^{-n} in $[0, 1)$. As in the usual criterion, it suffices to consider $s, t \in \bigcup_n D_n$, with the values at the remaining times filled in using continuity. (This is why in general one ends up with a modification.) Note that the number of elements in D_n is given by $\#D_n = 1/|D_n| = 2^n$. Set

$$K_n = \sup_{t \in D_n} \left|X_{t, t+2^{-n}}\right|, \qquad \mathbb{K}_n = \sup_{t \in D_n} \left|\mathbb{X}_{t, t+2^{-n}}\right|.$$

It follows from (3.2) that

$$\mathbf{E}\left(K_n^q\right) \leq \mathbf{E} \sum_{t \in D_n} \left|X_{t, t+2^{-n}}\right|^q \leq \frac{1}{|D_n|} C^q |D_n|^{\beta q} = C^q |D_n|^{\beta q - 1},$$

$$\mathbf{E}\left(\mathbb{K}_n^{q/2}\right) \leq \mathbf{E} \sum_{t \in D_n} \left|\mathbb{X}_{t, t+2^{-n}}\right|^{q/2} \leq \frac{1}{|D_n|} C^{q/2} |D_n|^{2\beta q/2} = C^{q/2} |D_n|^{\beta q - 1}.$$

Fix $s < t$ in $\bigcup_n D_n$ and choose $m : |D_{m+1}| < t - s \leq |D_m|$. The interval $[s, t]$ can be expressed as the finite disjoint union of intervals of the form $[u, v] \in D_n$ with $n \geq m + 1$ and where no three intervals have the same length. In other words, we have a partition of $[s, t]$ of the form

$$s = \tau_0 < \tau_1 < \ldots < \tau_N = t,$$

where $(\tau_i, \tau_{i+1}) \in D_n$ for some $n \geq m+1$, and for each fixed $n \geq m+1$ there are at most two such intervals taken from D_n. In this context, such a type of multiscale decomposition is sometimes called a "chaining argument". It follows that

$$|X_{s,t}| \leq \max_{0 \leq i < N} |X_{s,\tau_{i+1}}| \leq \sum_{i=0}^{N-1} |X_{\tau_i,\tau_{i+1}}| \leq 2 \sum_{n \geq m+1} K_n \,,$$

and similarly,

$$|\mathbb{X}_{s,t}| = \left| \sum_{i=0}^{N-1} \mathbb{X}_{\tau_i,\tau_{i+1}} + X_{s,\tau_i} \otimes X_{\tau_i,\tau_{i+1}} \right| \leq \sum_{i=0}^{N-1} \left(|\mathbb{X}_{\tau_i,\tau_{i+1}}| + |X_{s,\tau_i}||X_{\tau_i,\tau_{i+1}}| \right)$$

$$\leq \sum_{i=0}^{N-1} |\mathbb{X}_{\tau_i,\tau_{i+1}}| + \max_{0 \leq i < N} |X_{s,\tau_{i+1}}| \sum_{j=0}^{N-1} |X_{\tau_j,\tau_{j+1}}|$$

$$\leq 2 \sum_{n \geq m+1} \mathbb{K}_n + \left(2 \sum_{n \geq m+1} K_n \right)^2 .$$

We thus obtain

$$\frac{|X_{s,t}|}{|t-s|^\alpha} \leq \sum_{n \geq m+1} \frac{1}{|D_{m+1}|^\alpha} 2K_n \leq \sum_{n \geq m+1} \frac{2K_n}{|D_n|^\alpha} \leq K_\alpha \,,$$

where $K_\alpha := 2 \sum_{n \geq 0} K_n / |D_n|^\alpha$ is in L^q. Indeed, since $\alpha < \beta - 1/q$ by assumption and $|D_n|$ to any positive power is summable, we have

$$\|K_\alpha\|_{L^q} \leq \sum_{n \geq 0} \frac{2}{|D_n|^\alpha} |\mathbf{E}(K_n^q)|^{1/q} \leq \sum_{n \geq 0} \frac{2C}{|D_n|^\alpha} |D_n|^{\beta - 1/q} < \infty \,.$$

Similarly,

$$\frac{|\mathbb{X}_{s,t}|}{|t-s|^{2\alpha}} \leq \sum_{n \geq m+1} \frac{1}{|D_{m+1}|^{2\alpha}} 2\mathbb{K}_n + \left(\sum_{n \geq m+1} \frac{1}{|D_{m+1}|^\alpha} 2K_n \right)^2 \leq \mathbb{K}_\alpha + K_\alpha^2 \,,$$

where $\mathbb{K}_\alpha := 2 \sum_{n \geq 0} \mathbb{K}_n / |D_n|^{2\alpha}$ is in $L^{q/2}$. Indeed,

$$\|\mathbb{K}_\alpha\|_{L^{q/2}} \leq \sum_{n \geq 0} \frac{2}{|D_n|^{2\alpha}} \left| \mathbf{E}\left(\mathbb{K}_n^{q/2} \right) \right|^{2/q} \leq \sum_{n \geq 0} \frac{2C}{|D_n|^{2\alpha}} |D_n|^{2\beta - 2/q} < \infty \,,$$

thus concluding the proof. $\square$

The reader will notice that the classical Kolmogorov criterion (KC) is contained in the above proof and theorem by simply ignoring all considerations related to the second-order process $\mathbb{X}$. Let us also note in this context that the classical KC works for processes $(\mathbf{X}_t : 0 \leq t \leq 1)$ with values in an arbitrary (separable) metric space (it suffices to replace $|X_{s,t}|$ by $d(\mathbf{X}_s, \mathbf{X}_t)$ in the argument). This observation actually gives an alternative and immediate proof of Theorem 3.1. All we have to do is to remember from Proposition 2.6 that rough paths can always be viewed as bona fide paths with values in a metric space, namely $T_1^{(2)}$, equipped with the homogeneous left-invariance metric $d(\mathbf{X}_s, \mathbf{X}_t) \asymp |X_{s,t}| + |\mathbb{X}_{s,t}|^{1/2}$. The moment assumption (3.2) is then equivalent to $|d(\mathbf{X}_s, \mathbf{X}_t)|_{L^q} \leq C|t - s|^\beta$ and we can conclude with the "metric" form of KC. From Section 2.4, a version of this KC for "level-N" low regularity rough paths is then also immediate. The reason we still like the pedestrian step-2 proof is that it is easily tweaked, e.g. to the case of the $\mathbf{R}^2$-valued process (B^H, B) the pair of a fractional and standard Brownian motion, independent say, with Itô second level $\mathbb{B}^H := \int B^H dB$, in the rough regime $H \in (0, 1/2]$. In this case β should be replaced by the vector $(\beta_1, \beta_2) = (H, 1/2)$ of regularities, and the conclusion can be stated with α resp. 2α replaced by the vector $(\alpha_1, \alpha_2) = (H^-, 1/2^-)$ resp. $(H + 1/2)^-$.

Remark 3.2 (Warning). It is not possible to obtain (3.3) by applying the classical KC to the $(V \otimes V)$-valued process $(\mathbb{X}_{0,t} : 0 \leq t \leq T)$. Doing so only gives $|\mathbb{X}_{s,t}| = O(|t - s|^\alpha)$ a.s. since one misses a crucial cancellation inherent in (cf. (2.1))

$$\mathbb{X}_{s,t} = \mathbb{X}_{0,t} - \mathbb{X}_{0,s} - X_{0,s} \otimes X_{s,t}.$$

That said, it is possible [Fri05] (but tedious) to use a 2-parameter version of the KC to see that $(s, t) \mapsto \mathbb{X}_{s,t}/|t - s|^{2\alpha}$ admits a continuous modification, which implies that $\|\mathbb{X}\|_{2\alpha}$ is finite almost surely.

Here is a similar result for rough path distances, say between $\mathbf{X}$ and $\tilde{\mathbf{X}}$. Note that, due to the nonlinear structure of rough path spaces, one cannot simply apply Theorem 3.1 to the "difference" of two rough paths. Indeed, if we consider $\tilde{\mathbf{X}} - \mathbf{X}$, where addition is taken in the ambient Banach space $\mathcal{C}_\alpha \oplus \mathcal{C}_2^{2\alpha}$, then Chen's relation is in general not satisfied.

Theorem 3.3 (Kolmogorov criterion for rough path distance). *Let α, β, q be as above in Kolmogorov's criterion (KC), Theorem 3.1. Assume that both $\tilde{\mathbf{X}} = (\tilde{X}, \tilde{\mathbb{X}})$ and $\mathbf{X} = (X, \mathbb{X})$ satisfy the moment condition in the statement of KC with some constant C. Set*

$$\Delta X := \tilde{X} - X, \qquad \Delta\mathbb{X} := \tilde{\mathbb{X}} - \mathbb{X},$$

and assume that for some $\varepsilon > 0$ and all $s, t \in [0, T]$

$$|\Delta X_{s,t}|_{L^q} \leq C\varepsilon|t - s|^\beta, \qquad |\Delta\mathbb{X}_{s,t}|_{L^{q/2}} \leq C\varepsilon|t - s|^{2\beta}.$$

Then there exists M, depending increasingly on C, so that

$$\left|\|\Delta X\|_\alpha\right|_{L^q} \leq M\varepsilon, \qquad \left|\|\Delta\mathbb{X}\|_{2\alpha}\right|_{L^{q/2}} \leq M\varepsilon.$$

In particular, if $\beta - \frac{1}{q} > \frac{1}{3}$ then, for every $\alpha \in \left(\frac{1}{3}, \beta - \frac{1}{q}\right)$ we have $\|\tilde{\mathbf{X}}\|_\alpha, \|\mathbf{X}\|_\alpha \in L^q$ and

$$\left|\varrho_\alpha(\tilde{\mathbf{X}}, \mathbf{X})\right|_{L^{q/2}} \leq M\varepsilon.$$

Proof. The proof is a straightforward modification of the proof of Theorem 3.1 and is left as an exercise to the reader. □

Often one has a sequence of (random) rough paths $\{\mathbf{X}^n \equiv (X^n, \mathbb{X}^n) : 1 \leq n \leq \infty\}$, such that the moment conditions in the statement of Kolmogorov's criterion hold with a constant C, uniformly over $1 \leq n \leq \infty$, and such that $\varepsilon = \varepsilon_n \to 0$. Theorem 3.3 now quantifies the convergence $\mathbf{X}^n \to \mathbf{X}^\infty$, with rates given by

$$|\varrho_\alpha(\mathbf{X}^n, \mathbf{X}^\infty)|_{L^{q/2}} \lesssim \varepsilon_n .$$

Of course, when ε_n decays sufficiently fast, a Borel–Cantelli argument also gives almost sure convergence with suitable rates.

3.2 Itô Brownian motion

Consider a d-dimensional standard Brownian motion B enhanced with its iterated integrals

$$\mathbb{B}_{s,t} \stackrel{\text{def}}{=} \int_s^t B_{s,r} \otimes dB_r \in \mathbf{R}^d \otimes \mathbf{R}^d \cong \mathbf{R}^{d\times d} , \tag{3.4}$$

where the stochastic integration is understood in the sense of Itô. Sometimes we indicate this by writing $\mathbb{B}^{\text{Itô}}$. We shall assume straight away that B_t and $\mathbb{B}_{s,t}$ are continuous in t and s, t respectively, with probability one. For instance, if one takes as granted that, almost surely, Brownian motion and indefinite Itô integrals against Brownian motion (such as $\mathbb{B}_{0,\cdot}$) are continuous, then it suffices to (re)define the second order increments as $\mathbb{B}_{s,t} = \mathbb{B}_{0,t} - \mathbb{B}_{0,s} - B_s \otimes B_{s,t}$. Of course, by additivity of the Itô integral, this coincides a.s. with the earlier definition. En passant, (2.1) it then immediately satisfied, for all times, on a common set of probability one.

Proposition 3.4. *For any* $\alpha \in \left(\frac{1}{3}, \frac{1}{2}\right)$*, with probability one,*

$$\mathbf{B}^{\text{Itô}} = (B, \mathbb{B}^{\text{Itô}}) \in \mathscr{C}^\alpha([0, T], \mathbf{R}^d) .$$

In fact, the homogeneous rough path norm $\|\mathbf{B}^{\text{Itô}}\|_\alpha$ *has Gaussian tails.*

Proof. Using Brownian scaling and finite moments of $\mathbb{B}_{0,1}$, which are immediate from integrability properties of the (homogeneous) second Wiener–Itô chaos, the KC for rough paths applies with $\beta = 1/2$ and all $q < \infty$. (As an exercise, the reader may want to show finite moments of $\mathbb{B}_{0,1}$ without chaos arguments; an elementary way to do so is via conditioning, Itô isometry, and reflection principle.) The integrability $\|\mathbf{B}^{\text{Itô}}\|_\alpha \in L^q$, any $q < \infty$, is clear from KC. The Gaussian integrability (and hence tails) can be obtained by carefully tracking the moment growth in Theorem 3.1 applied to $\mathbf{B}^{\text{Itô}}$; alternatively see Theorem 11.9 below for an elegant Gaussian argument). □

Observe that Brownian motion enhanced with its iterated Itô integrals (2nd order calculus!) yields a (random) rough path but not a *geometric* rough path which is, by definition, an object with hardwired first order behaviour. Indeed, Itô formula yields the identity

$$d(B^i B^j) = B^i dB^j + B^j dB^i + \langle B^i, B^j \rangle dt \,, \qquad i, j = 1, \dots, d \,,$$

so that, writing Id for the identity matrix in d dimensions, we have for $s < t$,

$$\text{Sym}\left(\mathbb{B}^{\text{Itô}}_{s,t}\right) = \frac{1}{2} B_{s,t} \otimes B_{s,t} - \frac{1}{2} \text{Id}(t-s) \neq \frac{1}{2} B_{s,t} \otimes B_{s,t} \,,$$

in contradiction with (2.6).

Let us finally mention that Brownian motion with values in infinite-dimensional spaces can also be lifted to rough paths, see the exercise section.

3.3 Stratonovich Brownian motion

In the previous section we defined $\mathbb{B}^{\text{Itô}}$ by *Itô integration* of d-dimensional Brownian motion B against itself. Now, for (scalar) continuous semimartingales, M, N say, the *Stratonovich integral* is defined as

$$\int_0^t M \circ dN := \int_0^t M dN + \frac{1}{2} \langle M, N \rangle_t$$

and has the advantage of a first order calculus. For instance, one has the *first order product rule*

$$d(MN) = M \circ dN + N \circ dM \,.$$

One can then define $\mathbb{B}^{\text{Strat}}$ by (component-wise) *Stratonovich integration* of Brownian motion against itself. Using basic results on quadratic variation of Brownian motion, namely $d\langle B^i, B^j \rangle_t = \delta^{i,j} dt$ where $\delta^{i,j} = 1$ if $i = j$, zero else, we see that

$$\mathbb{B}^{\text{Strat}}_{s,t} = \mathbb{B}^{\text{Itô}}_{s,t} + \frac{1}{2} \text{Id}(t-s) \,. \tag{3.5}$$

Note that the difference between $\mathbb{B}^{\text{Strat}}$ and $\mathbb{B}^{\text{Itô}}$ is symmetric, so that the antisymmetric parts of the two processes (Lévy's stochastic area) are identical.

Proposition 3.5. *For any $\alpha \in (1/2, 1/3)$, with probability one,*

$$\mathbf{B}^{\text{Strat}} = (B, \mathbb{B}^{\text{Strat}}) \in \mathscr{C}_g^\alpha([0, T], \mathbf{R}^d) \,,$$

and here again the homogeneous rough path norm $\|\mathbf{B}^{\text{Strat}}\|_\alpha$ has Gaussian tails.

Proof. Using (3.5), rough path regularity of $\mathbf{B}^{\text{Strat}}$ is immediately reduced to the already established Itô case. (Alternatively, one can use again the Kolmogorov criterion for rough paths; the only – insignificant – difference is that now $\mathbb{B}_{0,1}^{\text{Strat}}$ takes values in the inhomogeneous second chaos, due to the deterministic part $\text{Id}/2$.) At last, $\mathbf{B}(\omega)$ is *geometric* since

$$\text{Sym}\left(\mathbb{B}_{s,t}^{\text{Strat}}\right) = \frac{1}{2} B_{s,t} \otimes B_{s,t},$$

an immediate consequence of the first order product rule. Finally, integrability of $\mathbb{B}^{\text{Strat}}$ is clear from the already seen integrability of $\mathbb{B}^{\text{Itô}}$, proving the final claim. $\square$

A typical realisation $\mathbf{B}(\omega)$ is called *Brownian rough path*, as a process $\mathbf{B} = \mathbf{B}^{\text{Strat}}$ is a.k.a. (Stratonovich) *enhanced Brownian motion*. It is a *deterministic* feature of every weakly geometric rough path $(X, \mathbb{X})$ that it can be approximated – in the precise sense of Proposition 2.8 – by smooth paths in the rough path topology. Such approximations require knowledge not only of the underlying path X, but of the entire *rough path*, including the second-order information $\mathbb{X}$.

In contrast, one has the *probabilistic* statement that piecewise linear, mollifier and many other "obvious" approximations still converge in rough path sense. More specifically, in the present context of d-dimensional standard Brownian motion, we now give an elegant proof of this based on (discrete-time!) martingale arguments.

Proposition 3.6. *Consider dyadic piecewise linear approximations $(B^{(n)})$ to B on $[0, T]$. That is, $B_t^{(n)} = B_t$ whenever $t = iT/2^n$ for some integer i, and linearly interpololated on intervals $[iT/2^n, (i+1)T/2^n]$. Then, with probability one,*

$$\left(B^{(n)}, \int_0^{\cdot} B^{(n)} \otimes dB^{(n)}\right) \to \left(B, \mathbb{B}^{\text{Strat}}\right) \quad in \quad \mathscr{C}_g^\alpha .$$

(The integral on the left-hand side is understood as classical Riemann–Stieltjes integral.)

Remark 3.7. With Theorem 3.3, one can see rough path convergence (in probability, and actually L^q, any $q < \infty$) of piecewise linear approximation along any sequence of dissections with mesh tending to zero. Moreover, this approach will give the rate θ, any $\theta < 1/2 - \alpha$.

Proof. It is easy to check that B gives $B^{(n)}$ via conditioning on B at dyadic times,

$$B^{(n)} = \mathbf{E}(B \mid \sigma\{B_{kT2^{-n}} : 0 \le k \le 2^n\}).$$

By independence of the components B^i, B^j for $i \ne j$, the same holds for $\mathbb{B}^{\text{Strat}}$ *off-diagonal*; the on-diagonal terms require no further attention since $\mathbb{B}_{s,t}^{\text{Strat};i,i} = \frac{1}{2}(B_{s,t}^i)^2$. Almost sure pointwise convergence then readily follows from martingale convergence. Furthermore, Theorem 3.1 implies

$$\left|B_{s,t}^i\right| \le K_\alpha(\omega)|t - s|^\alpha , \qquad \left|\mathbb{B}_{s,t}^{\text{Strat};i,j}\right| \le \mathbb{K}_\alpha(\omega)|t - s|^{2\alpha} ,$$

and upon conditioning with respect to $\sigma\{B_{kT2^{-n}} : 0 \le k \le 2^n\}$, the same bounds hold for $B^{(n);i}$ and for $\int_0^{\cdot} B^{(n);i} dB^{(n);j}$. In fact, $K_\alpha, \mathbb{K}_\alpha$ have (more than enough) integrability to apply Doob's maximal inequality. This leads, with probability one, to the bound

$$\sup_{n} \left\| B^{(n)}, \int_0^{\cdot} B^{(n)} \otimes dB^{(n)} \right\|_{2\alpha} < \infty \,.$$

Together with a.s. pointwise convergence, a (deterministic) interpolation argument shows a.s. convergence with respect to the α-Hölder rough path metric ϱ_α. $\quad\square$

The reader should be warned that there are perfectly smooth and uniform approximations to Brownian motion, which do not converge to Stratonovich enhanced Brownian motion, but instead to some different geometric (random) rough path, such as

$$\mathbf{\bar{B}} = \left(B, \bar{\mathbb{B}} \right) \,, \quad \text{where} \quad \bar{\mathbb{B}}_{s,t} = \mathbb{B}_{s,t}^{\mathrm{Strat}} + (t-s)A \,, \qquad A \in \mathfrak{so}(d) \,.$$

Note that the difference between $\bar{\mathbb{B}}$ and $\mathbb{B}^{\mathrm{Strat}}$ is now *antisymmetric*, i.e. $\mathbf{\bar{B}}$ has a stochastic area that is different from Lévy's area. To construct such approximations, it suffices to include oscillations (at small scales) such as to create the desired effect in the area, while they do not affect the limiting path, see Exercise 2.10. (In the context of Brownian motion and SDEs driven by Brownian motion such approximations were studied by McShean, Ikeda–Watanabe and others, see [McS72, IW89].) Although such "twisted" approximations do not seem to be the most obvious way to approximate Brownian motion, they also arise naturally in some perfectly reasonable situations.

3.4 Brownian motion in a magnetic field

Newton's second law for a particle in $\mathbf{R}^3$ with mass m, and position $x = x(t)$, (for simplicity: constant) frictions $\alpha_1, \alpha_2, \alpha_3 > 0$ in orthonormal directions, subject to a (3-dimensional) white noise in time, i.e. the distributional derivative of a 3-dimensional Brownian motion B, reads

$$m\ddot{x} = -M\dot{x} + \dot{B}, \qquad (3.6)$$

assuming M symmetric with spectrum $\alpha_1, \alpha_2, \alpha_3$. The process $x(t)$ describes what is known as *physical Brownian motion*. It is well known that in small mass regime, $m \ll 1$, of obvious physical relevance when dealing with particles, a good approximation is given by (mathematical) *Brownian motion* (with non-standard covariance). To see this formally, it suffices to take $m = 0$ in (3.6) in which case $x = M^{-1}B$.

Let us now assume that our particle (with position x and momentum $m\dot{x}$) carries a non-zero electric charge and moves in a magnetic field which we assume to be constant. Recall that such a particle experiences a sideways force ("Lorentz force") that is proportional to the strength of the magnetic field, the component of the velocity that is perpendicular to the magnetic field and the charge of the particle. In terms of our assumptions, this simply means that a non-zero antisymmetric component is added to M. We shall hence drop the assumption of symmetry, and instead consider for M a general square matrix with

$$\text{Real}\{\sigma(M)\} \subset (0, \infty).$$

Note that these second order dynamics can be rewritten as evolution equation for the *momentum* $p(t) = m\dot{x}(t)$,

$$\dot{p} = -M\dot{x} + \dot{B} = -\frac{1}{m}Mp + \dot{B}.$$

As we shall see $X = X^m$, indexed by "mass" m, converges in a quite non-trivial way to Brownian motion on the level of rough paths. In fact, the correct limit in rough path sense is $\bar{\mathbf{B}} = (B, \bar{\mathbb{B}})$, where

$$\bar{\mathbb{B}}_{s,t} = \mathbb{B}_{s,t}^{\text{Strat}} + (t - s)A, \tag{3.7}$$

in terms of an *antisymmetric* matrix A; written explicitly as $A = \frac{1}{2}(M\Sigma - \Sigma M^*) \in \mathfrak{so}(d)$, where

$$\Sigma = \int_0^\infty e^{-Ms}e^{-M^*s}ds.$$

When M is normal, i.e. $M^*M = MM^*$, it is an exercise in linear algebra to show that this expression simplifies to

$$A = \frac{1}{2}\text{Anti}(M)\text{Sym}(M)^{-1},$$

where $\text{Anti}(M)$ denotes the antisymmetric part of a matrix and $\text{Sym}(M)$ its symmetric part. We can now state the result in full detail.

Theorem 3.8. *Let $M \in \mathbf{R}^{d \times d}$ be a square matrix in dimension d such that all its eigenvalues have strictly positive real part. Let B be a d-dimensional standard Brownian motion, $m > 0$, and consider the stochastic differential equations*

$$dX = \frac{1}{m}P\,dt, \qquad dP = -\frac{1}{m}MP\,dt + dB.$$

with zero initial position X and momentum P. Then, for any $q \geq 1$ and $\alpha \in (1/3, 1/2)$, as mass $m \to 0$,

$$\left(MX, \int MX \otimes d(MX)\right) \to \bar{\mathbf{B}} \quad \text{in } \mathscr{C}^\alpha \text{ and } L^q.$$

Proof. Step 1. (Pointwise convergence in L^q.) In order to exploit Brownian scaling, it is convenient to set $m = \varepsilon^2$ and then Y^ε as rescaled momentum,

$$Y_t^\varepsilon = P_t/\varepsilon.$$

We shall also write $X^\varepsilon = X$, to emphasise dependence on ε. We then have

$$dY_t^\varepsilon = -\varepsilon^{-2} M Y_t^\varepsilon \, dt + \varepsilon^{-1} dB_t , \qquad dX_t^\varepsilon = \varepsilon^{-1} Y_t^\varepsilon \, dt .$$

By assumption, there exists $\lambda > 0$ such that the real part of every eigenvalue of M is (strictly) bigger than λ. For later reference, we note that this implies the estimate $|\exp(-\tau M)| = O(\exp(-\lambda \tau))$ as $\tau \to \infty$. For fixed ε, define the Brownian motion $\tilde{B} = \varepsilon^{-1} B_{\varepsilon^2}$. so that $\varepsilon^{-1} dB_t = d\tilde{B}_{\varepsilon^{-2} t}$, and consider the SDEs

$$d\tilde{Y}_t = -M \tilde{Y}_t \, dt + d\tilde{B}_t , \qquad d\tilde{X}_t = \tilde{Y}_t \, dt .$$

Note that the law of the solutions does not depend on ε. Furthermore, when solved with identical initial data, we have pathwise equality

$$\left(Y_t^\varepsilon, \varepsilon^{-1} X_t^\varepsilon \right) = \left(\tilde{Y}_{\varepsilon^{-2} t}, \tilde{X}_{\varepsilon^{-2} t} \right) . \tag{3.8}$$

Thanks to our assumption on M, $\tilde{Y}$ is ergodic; the stationary solution has (zero mean, Gaussian) law $\nu = \mathcal{N}(0, \Sigma)$ for some covariance matrix Σ. To compute it, write down the stationary solution

$$\tilde{Y}_t^{\text{stat}} = \int_{-\infty}^{t} e^{-M(t-s)} dB_s .$$

For each t (and in particular for $t = 0$), the law of $\tilde{Y}_t^{\text{stat}}$ is precisely ν. We then see that

$$\Sigma = \mathbf{E}\left(\tilde{Y}_0^{\text{stat}} \otimes \tilde{Y}_0^{\text{stat}} \right) = \int_{-\infty}^{0} e^{-M(-s)} e^{-M^*(-s)} ds = \int_{0}^{\infty} e^{-Ms} e^{-M^* s} ds.$$

Since $\sup_{0 \le t < \infty} \mathbf{E}|\tilde{Y}_t^2| < \infty$, it is clear that $\varepsilon \tilde{Y}_{\varepsilon^{-2} t} = \varepsilon Y_t^\varepsilon \to 0$ in L^2 uniformly in t (and hence in L^q for any $q < \infty$). Noting that $M X_t^\varepsilon = B_t - \varepsilon Y_{0,t}^\varepsilon$, the first part of the proposition is now obvious. Moreover, by the ergodic theorem[1],

$$\int_0^t f(Y_t^\varepsilon) \, dt \to t \int f(y) \nu(dy) , \qquad \text{in } L^q \text{ for any } q < \infty, \tag{3.9}$$

for all reasonable test functions f; we shall only use it for quadratics. Using $dX^\varepsilon = \varepsilon^{-1} Y^\varepsilon dt$ we can then write

$$\int_0^t M X_s^\varepsilon \otimes d(M X^\varepsilon)_s = \int_0^t M X_s^\varepsilon \otimes dB_s - \varepsilon \int_0^t M X_s^\varepsilon \otimes dY_s^\varepsilon$$

$$= \int_0^t M X_s^\varepsilon \otimes dB_s - M X_t^\varepsilon \otimes (\varepsilon Y_t^\varepsilon) + \varepsilon \int_0^t d(M X^\varepsilon)_s \otimes Y_s^\varepsilon$$

$$= \int_0^t M X_s^\varepsilon \otimes dB_s - M X_t^\varepsilon \otimes (\varepsilon Y_t^\varepsilon) + \int_0^t M Y_s^\varepsilon \otimes Y_s^\varepsilon ds$$

---

[1] In its standard form, see e.g. Stroock [Str11] or Kallenberg [Kal02], test functions are assumed to be bounded. In our setting an easy truncation argument yields the extension to quadratics.

$$\rightarrow \int_0^t B_s \otimes dB_s - 0 + t \int (My \otimes y)\, \nu(dy)$$

$$= \int_0^t B_s \otimes dB_s + tM\Sigma = \mathbb{B}_{0,t} + t\left(M\Sigma - \frac{1}{2}\mathrm{Id}\right),$$

where the convergence is in L^q for any $q \geq 2$. By considering the symmetric part of the above equation,

$$\frac{1}{2}(MX_t^\varepsilon) \otimes (MX_t^\varepsilon) \rightarrow \frac{1}{2}B_t \otimes B_t + \mathrm{Sym}\left(M\Sigma - \frac{1}{2}\mathrm{Id}\right),$$

we see that $M\Sigma - \frac{1}{2}I$ is antisymmetric, and hence also equals $\frac{1}{2}(M\Sigma - \Sigma M^*)$. This settles pointwise convergence, in the sense that

$$S(MX^\varepsilon)_t := \left(MX_t^\varepsilon, \int_0^t MX_s^\varepsilon \otimes d(MX^\varepsilon)_s\right) \rightarrow (B_t, \bar{\mathbb{B}}_{0,t}).$$

Step 2. (Uniform rough path bounds in L^q.) We claim that, for any $q < \infty$,

$$\sup_{\varepsilon \in (0,1]} \mathbf{E}[\|MX^\varepsilon\|_\alpha^q] < \infty, \qquad \sup_{\varepsilon \in (0,1]} \mathbf{E}\left[\left\|\int MX^\varepsilon \otimes d(MX^\varepsilon)\right\|_{2\alpha}^q\right] < \infty,$$

which, in view of Theorem 3.1, is an immediate consequence of the bounds

$$\sup_{\varepsilon \in (0,1]} \mathbf{E}\left[|X_{s,t}^\varepsilon|^q\right] \lesssim |t-s|^{\frac{q}{2}}, \qquad \sup_{\varepsilon \in (0,1]} \mathbf{E}\left[\left|\int_s^t X_{s,\cdot}^\varepsilon \otimes dX^\varepsilon\right|^q\right] \lesssim |t-s|^q.$$

Since X is Gaussian, it follows from integrability properties of the first two Wiener–Itô chaoses that it is enough to show these bounds for $q = 2$. Furthermore, we note that the desired estimates are a consequence of the bounds

$$\mathbf{E}\left[|\tilde{X}_{s,t}|^2\right] \lesssim |t-s|, \tag{3.10}$$

$$\mathbf{E}\left[\left|\int_s^t \tilde{X}_{s,u} \otimes d\tilde{X}_u\right|^2\right] \lesssim |t-s|^2, \tag{3.11}$$

where the implied proportionality constants are uniform over $t, s \in (0, \infty)$. Indeed, this follows directly from writing

$$\mathbf{E}\left[|X_{s,t}^\varepsilon|^2\right] = \mathbf{E}\left[|\varepsilon \tilde{X}_{\varepsilon^{-2}s,\varepsilon^{-2}t}|^2\right] \lesssim \varepsilon^2 |\varepsilon^{-2}t - \varepsilon^{-2}s| = |t-s|,$$

(note the uniformity in ε), and similarly for the second moment of the iterated integral.

In order to check (3.10), it is enough to note that $M\tilde{X}_{s,t} = \tilde{B}_{s,t} - \tilde{Y}_{s,t}$, combined with the estimate

$$\mathbf{E}\left[|\tilde{Y}_{s,t}|^2\right] = \mathbf{E}\left[\left|(e^{-M(t-s)} - I)\tilde{Y}_s\right|^2\right] + \int_s^t \mathrm{Tr}(e^{-Mu}e^{-M^*u})\,du \lesssim |t-s| ,$$

where we used the fact that $\mathrm{Real}\{\sigma(M)\} \subset (0, \infty)$ to get a uniform bound. In order to control (3.11), we consider one of the components and write

$$\mathbf{E}\left[\left|\int_s^t \tilde{X}_{s,u}^i\,d\tilde{X}_u^j\right|^2\right] = \mathbf{E}\left[\left|\int_s^t \int_s^u \tilde{Y}_r^i \tilde{Y}_u^j\,dr\,du\right|^2\right]$$

$$= \int_{[s,t]^4} \mathbf{E}\left[\tilde{Y}_r^i \tilde{Y}_u^j \tilde{Y}_q^i \tilde{Y}_v^j\right] \mathbf{1}_{\{r\leq u;q\leq v\}}\,dr\,du\,dq\,dv$$

$$\leq \int_{[s,t]^4} \left(\left|\mathbf{E}\left[\tilde{Y}_r^i \tilde{Y}_u^j\right]\right| \left|\mathbf{E}\left[\tilde{Y}_q^i \tilde{Y}_v^j\right]\right| + \left|\mathbf{E}\left[\tilde{Y}_r^i \tilde{Y}_q^i\right]\right| \left|\mathbf{E}\left[\tilde{Y}_u^j \tilde{Y}_v^j\right]\right|\right.$$

$$\left. + \left|\mathbf{E}\left[\tilde{Y}_r^i \tilde{Y}_v^j\right]\right| \left|\mathbf{E}\left[\tilde{Y}_u^j \tilde{Y}_q^i\right]\right|\right)\,dr\,du\,dq\,dv$$

$$\lesssim \left(\int_{[s,t]^2} \left|\mathbf{E}\left[\tilde{Y}_r \otimes \tilde{Y}_u\right]\right|\,dr\,du\right)^2$$

$$\lesssim \left(\int_{[s,t]^2} \left|\mathbf{E}\left[\tilde{Y}_r \otimes \tilde{Y}_u\right]\right| \mathbf{1}_{\{r\leq u\}}\,dr\,du\right)^2 ,$$

where we have used the fact that $\tilde{Y}$ is Gaussian (which yields Wick's formula for the expectation of products) in order to get the bound on the third line. But for $r \leq u$, $\mathbf{E}\left[\tilde{Y}_u | \tilde{Y}_r\right] = e^{-M(u-r)}\tilde{Y}_r$, so that

$$\int_{[s,t]^2} \left|\mathbf{E}\left[\tilde{Y}_r \otimes \tilde{Y}_u\right]\right| \mathbf{1}_{\{r\leq u\}}\,dr\,du = \int_{[s,t]^2} \left|\mathbf{E}\left[\tilde{Y}_r \otimes e^{-M(u-r)}\tilde{Y}_r\right]\right| \mathbf{1}_{\{r\leq u\}}\,dr\,du$$

$$\lesssim \int_s^t \left(\int_r^t e^{-\lambda(u-r)}\,du\right) \mathbf{E}\left[|\tilde{Y}_r|^2\right]\,dr \lesssim |t-s| .$$

It now suffices to recall that $|\exp(-\tau M)| = \mathrm{O}(\exp(-\lambda\tau))$ to conclude the proof of (3.11).

Step 3. (Rough path convergence in L^q.) The remainder of the proof is an easy application of interpolation, along the lines of Exercise 2.9. □

3.5 Cubature on Wiener Space

Quadrature rules replace Lebesgue measure λ on $[0, 1]$ by a finite, convex linear combination of point masses, say $\mu = \sum a_i \delta_{x_i}$, where weights (a_i) and points (x_i) are chosen such that all monomials (and hence all polynomials) up to degree N are correctly evaluated. In other words, one first computes the moments of λ, namely

$$\int_0^1 x^n d\lambda(x) = \frac{1}{n+1} ,$$

for all $n \geq 0$. One then looks for a measure μ such that $\int_0^1 x^n d\mu(x) = 1/(n+1)$ for all $n \in \{0, 1, \ldots, N\}$. The same can be done on Wiener space: the monomial x^n is then replaced by the n-fold iterated integrals (in the sense of Stratonovich), integration is on $C([0, T], \mathbf{R}^d)$ against standard d-dimensional Wiener measure. In order to find such cubature formulae, the mandatory first step, on which we focus here, is the computation of the expectations of the n-fold iterated integrals[2]

$$\mathbf{E}\left(\int_{0<t_1<\ldots<t_n<T} \circ dB \otimes \ldots \otimes \circ dB\right) .$$

Let us combine all of these integrals into one single object, also known as (Stratonovich) *signature of Brownian motion*, by writing

$$S(B)_{0,T} = 1 + \sum_{n \geq 1} \int_{0<t_1<\ldots<t_n<T} \circ dB \otimes \ldots \otimes \circ dB .$$

The signature $S(B)_{0,T}$ naturally takes values in the algebra of infinite formal tensor series $T((\mathbf{R}^d))$, effectively the closure of the space of tensor polynomials given by $\bigoplus_{n \geq 0} (\mathbf{R}^d)^{\otimes n}$. It turns out that in the case of Brownian motion, the expected signature can be expressed in a particularly concise and elegant form.

Theorem 3.9 (Fawcett). *Consider $S(B)_{0,T}$ as above as a $T((\mathbf{R}^d))$-valued random variable. Then*

$$\mathbf{E}S(B)_{0,T} = \exp\left(\frac{T}{2} \sum_{i=1}^d e_i \otimes e_i\right).$$

Proof. (Shekhar) Set $\varphi_t := \mathbf{E}S(B)_{0,t}$. (It is not hard to see, by Wiener–Itô chaos integrability or otherwise, that all involved iterated integrals are integrable so that φ is well-defined.) By Chen's formula (in its general form, see Exercise 2.1) and the independence of Brownian increments, one has the identity

$$\varphi_{t+s} = \varphi_t \otimes \varphi_s .$$

Since $\varphi_t \otimes \varphi_s = \varphi_s \otimes \varphi_t$, we have $[\varphi_s, \varphi_t] = 0$, so that

$$\log \varphi_{t+s} = \log \varphi_t + \log \varphi_s .$$

For integers m, n we have $\log \varphi_m = n \log \varphi_{m/n}$ and $\log \varphi_m = m \log \varphi_1$. It follows that

$$\log \varphi_t = t \log \varphi_1 ,$$

---

[2] We remark that all n-fold iterated Stratonovich integrals can be obtained from the "level-2" rough path $(B(\omega), \mathbb{B}^{\text{Strat}}(\omega)) \in \mathscr{C}_g^\alpha$ by a continuous map. In fact, this so-called *Lyons lift*, allows to view any geometric rough path as a "level-n" rough path for arbitrary $n \geq 2$.

first for $t = \frac{m}{n} \in \mathbf{Q}$, then for any real t by continuity. On the other hand, for $t > 0$, Brownian scaling implies that $\varphi_t = \delta_{\sqrt{t}} \varphi_1$ where δ_λ is the dilation operator, which acts by multiplication with λ^n on the nth tensor level, $(\mathbf{R}^d)^{\otimes n}$. Since δ_λ commutes with $\otimes$ (and thus also with log, defined as power series),

$$\log \varphi_t = \delta_{\sqrt{t}} \log \varphi_1$$

and it follows that one necessarily has

$$\log \varphi_1 \in \left(\mathbf{R}^d\right)^{\otimes 2} .$$

It remains to identify $\log \varphi_1$ with $\frac{1}{2} \sum_{i=1}^d e_i \otimes e_i$. To this end it suffices to compute the expected signature up to level two, which yields

$$\mathbf{E}S^{(2)}(B) = \mathbf{E}\left(1 + B_{0,1} + \int_0^1 B \otimes \circ dB\right) = 1 + \frac{1}{2} \sum_{i=1}^d e_i \otimes e_i .$$

Recall that in this expression, "1" is identified with $(1, 0, 0)$ in the truncated tensor algebra, and similarly for the other summands, and addition also takes place in $T^{(2)}(\mathbf{R}^d)$. Taking the logarithm (in the tensor algebra truncated beyond level 2; in this case $\log (1 + a + b) = a + \left(b - \frac{1}{2} a \otimes a\right)$ if a is a 1-tensor, b a 2-tensor) then immediately gives the desired identification. $\square$

The (constructive) existence of *cubature formulae*, a finite family of piecewise smooth paths with associated probabilities, such as to mimic the behaviour of the expected signature up to a given level is not a trivial problem (although much has been achieved to date), the reader can explore a simple case in Exercise 3.11 below.

3.6 Scaling limits of random walks

Consider a family of continuous processes $\mathbf{X}^n = (X^n, \mathbb{X}^n)$, with values in $V \oplus (V \otimes V)$ where dim $V < \infty$. Assume $\mathbf{X}_0^n = (0, 0)$ for all n. We leave the proof of the following result as exercise.

Theorem 3.10 (Kolmogorov tightness criterion for rough paths). *Let $q \geq 2, \beta > 1/q$. Assume, for all s, t in $[0, T]$*

$$\mathbf{E}_n |X_{s,t}^n|^q \leq C|t - s|^{\beta q} , \qquad \mathbf{E}_n |\mathbb{X}_{s,t}^n|^{q/2} \leq C|t - s|^{\beta q} , \qquad (3.12)$$

for some constant $C < \infty$. Assume $\beta - \frac{1}{q} > \frac{1}{3}$. Then for every $\alpha \in \left(\frac{1}{3}, \beta - \frac{1}{q}\right)$, the $\mathbf{X}^n$'s are tight in $\mathscr{C}^{0,\alpha}$.

In typical applications, the X^n are only defined for discrete times, such as $s = j/n, t = k/n$ for integers j, k. The non-trivial work then consists, for a suitable choice of $\mathbb{X}^n$, in checking the following discrete tightness estimates,

$$\mathbf{E}_n \left| X^n_{\frac{j}{n}, \frac{k}{n}} \right|^q \le C \left| \frac{j-k}{n} \right|^{\beta q} , \qquad \mathbf{E}_n \left| \mathbb{X}^n_{\frac{j}{n}, \frac{k}{n}} \right|^{q/2} \le C \left| \frac{j-k}{n} \right|^{\beta q} . \qquad (3.13)$$

The analogous continuous tightness estimates are typically obtained by suitable extension of $\mathbf{X}^n$ to continuous times (e.g. piecewise geodesic).

Proposition 3.11. *Consider a d-dimensional random walk $(X_j : j \in \mathbf{N})$, with i.i.d. increments of zero mean, finite moments of any order $q < \infty$, and unit covariance matrix. Extend the rescaled random walk*

$$X^n_{\frac{j}{n}} := \frac{1}{\sqrt{n}} X_j ,$$

defined on discrete times only, by piecewise linear interpolation to all times and construct $\mathbf{X}^n = (X^n, \mathbb{X}^n)$ by iterated (Riemann–Stieltjes) integration. Then the tightness estimates in Theorem 3.10 hold with $\beta = 1/2$ and all $q < \infty$.

Proof. The iterated integrals of a linear (or affine) path with increment $v \in \mathbf{R}^d$ takes the simple form $\exp(v)$ in terms of the tensor exponential introduced in (2.13). Chen's relation then implies

$$\mathbf{X}^n_{\frac{j}{n}, \frac{k}{n}} = \exp \left(X^n_{\frac{j}{n}, \frac{j+1}{n}} \right) \otimes \ldots \otimes \exp \left(X^n_{\frac{k-1}{n}, \frac{k}{n}} \right) . \qquad (3.14)$$

The simple calculus on the level-2 tensor algebra $T^{(2)}(\mathbf{R}^d)$ leads to an explicit expression for $\mathbb{X}^n_{\frac{j}{n}, \frac{k}{n}}$, to which one can apply the (discrete) Burkholder–Davis–Gundy inequality in order to get the discrete tightness estimates (3.13). The extension to all times is straightforward. Details are left to the reader (see e.g. [BF13]). An alternative argument, not restricted to level 2, is found in Breuillard et al. [BFH09]. $\square$

Note that $\mathbf{X}^n$, as constructed above, is a (random) geometric rough path. Recall that such rough paths can be viewed as genuine paths with values in the Lie group $G^{(2)}(\mathbf{R}^d) \subset T^{(2)}(\mathbf{R}^d)$. On the other hand, from (3.14), we see that $\mathbf{X}^n$ restricted to discrete times $\{ \frac{j}{n} : j \in \mathbf{N} \}$ is a Lie group valued random walk, rescaled with the aid of the dilation operator. By using central limit theorems available on such Lie groups, one can see that $\mathbf{X}^n$ at unit time converges weakly to Brownian motion, enhanced with its iterated integrals in the Stratonovich sense. Under the additional assumption that $\mathbf{E}(X \otimes X) = \text{Id}$, the identity matrix, this Brownian motion is in fact a standard Brownian motion. This is enough to characterise the finite-dimensional distributions of any weak limit point and one has the following "Donsker" type result.

Theorem 3.12. *In the rescaled random walk setting of Proposition 3.11, and under the additional assumption that $\mathbf{E}(X \otimes X) = \text{Id}$, we have the weak convergence*

$$\mathbf{X}^n \implies \mathbf{B}^{\text{Strat}}$$

in the rough path space $\mathscr{C}^\alpha([0,T], \mathbf{R}^d)$, any $\alpha < 1/2$.

Recall that, by definition, weak convergence is stable under pushforward by continuous maps. The interest in this result is therefore clearly given by the fact that stochastic integrals and the Itô map can be viewed as continuous maps on rough path spaces, as will be discussed in later chapters.

3.7 Exercises

Exercise 3.1 *Complete the proof of Theorem 3.3.*

Exercise 3.2 *Bypass the use of Wiener–Itô chaos integrability in Proposition 3.4 by showing directly that the matrix-valued random variable $\mathbb{B}_{0,1}^{\text{Itô}}$ has moments of all orders.* **Hint:** *This is trivial for the on-diagonal entries, for the off-diagonal entries one can argue via conditioning, Itô isometry, and reflection principle.*

♯ **Exercise 3.3** *Show that d-dimensional Brownian motion B enhanced with Lévy's stochastic area is a degenerate diffusion process and find its generator.*

Exercise 3.4 (Q-Wiener process as rough path) *Given a separable Hilbert space H with orthonormal basis (e_k), $(\lambda_k) \in l^1$, $\lambda_k > 0$ for all k, and a countable sequence (β^k) of independent standard Brownian motions, the limit*

$$X_t := \sum_{k=1}^{\infty} \lambda_k^{1/2} \beta_t^k e_k$$

exists a.s. and in L^2, uniformly on compacts. This defines a Q-Wiener process in the sense of [DPZ92], where $Q = \sum_k \lambda_k \langle e_k, \cdot \rangle e_k$ is symmetric, non-negative and trace-class; conversely, any such operator Q on H can be written in this form and thus gives rise to a Q-Wiener process. Show that

$$\mathbb{X}_{s,t} := \sum_{j,k=1}^{\infty} \lambda_j^{1/2} \lambda_k^{1/2} \int_s^t \beta_s^j \, d\beta_s^k \, e_j \otimes e_k$$

exists a.s. and in L^2, uniformly on compacts and so defines $\mathbb{X}$ with values in $H \otimes_{\text{HS}} H$, the closure of the algebraic tensor product $H \otimes_a H$ under the Hilbert–Schmidt norm. Consider both the case of Itô and Stratonovich integration and verify that with either choice, $(X, \mathbb{X}) \in \mathscr{C}^\alpha$ a.s. for any $\alpha < 1/2$.

∗ **Exercise 3.5 (Banach-valued Brownian motion as rough path [LLQ02])** *Given a separable Banach space V equipped with a centred Gaussian measure μ, a standard construction (cf. [Led96]) gives rise to a so-called abstract Wiener space (V, H, μ), with $H \subset V$ the Cameron–Martin space of μ. (Examples to have in mind are $V = H = \mathbf{R}^d$ with $\mu = N(0, I)$, or the usual Wiener space $V = \mathcal{C}([0, 1])$ equipped with Wiener measure, H is then the space of absolutely continuous paths starting at zero with L^2-derivative.) There then exists a V-valued Brownian motion $(B_t : t \in [0, T])$ such that*

- $B_0 = 0$,
- B has independent increments,
- $\langle B_{s,t}, v^* \rangle \sim N\big(0, (t-s)|v^*|_H^2\big)$ whenever $0 \le s < t \le T$ and $v^* \in V^* \hookrightarrow$ $H^* \cong H$.

We assume that $V \otimes V$ is equipped with an exact tensor norm (with respect to μ) in the sense that there exists $\gamma \in [1/2, 1)$ and a constant $C > 0$ such that for any sequence $\{G_k \otimes \tilde{G}_k : k \ge 1\}$ of independent V-valued Gaussian random variables with identical distribution μ,

$$\mathbf{E}\left(\left|\sum_{k=1}^{N} G_k \otimes \tilde{G}_k\right|_{V \otimes V}^2\right) \le C N^{2\gamma} = \mathrm{o}(N).$$

a) *Verify that exactness holds with $\gamma = 1/2$ whenever $\dim V < \infty$. (More generally, exactness with $\gamma = 1/2$ always holds true if one works with the injective tensor product space, $V \otimes_{\mathrm{inj}} V$, the injective norm being the smallest possible. For the largest possible norm, the projective norm, the $\mathrm{o}(N)$-estimate remains true but can be as slow as one wishes. Exactness may then fail, see for example [LLQ02]. Exactness of the usual Wiener space, with uniform or Hölder norm, is also known to be true.)*

b) *Fix $\alpha < 1/2$. Show that dyadic piecewise linear approximations B^n, enhanced with $\mathbb{B}^n = \int B^n \otimes dB^n$, converge in α-Hölder rough path metric to a limit $\mathbf{B}$ in $\mathscr{C}^\alpha([0,T], V)$. More precisely, use the previous exercise to show that the sequence $\mathbf{B}^n = (B^n, \mathbb{B}^n)$ is Cauchy in the sense that*

$$|\varrho_a(\mathbf{B}^n, \mathbf{B}^m)|_{L^q} \to 0 \quad \text{with} \quad n, m \to \infty.$$

Conclude that $\mathbf{B}^n$ converges in $\mathscr{C}^\alpha$ and L^q to some limit $\mathbf{B} \in \mathscr{C}^\alpha([0,T], V)$ a.s.

c) *Show that $\mathbf{B}$ is the L^q-limit in α-Hölder rough path metric for all piecewise linear approximations, say B^{D_n}, as long as mesh $|D_n| \to 0$ with $n \to \infty$. Show that the convergence is almost sure if $|D_n| \sim 2^{-n}$ and also $|D_n| \sim 1/n$.*

Solution. We only sketch the main step in the proof of b). Without loss of generality, we set $T = 1$. The crux of the matter is to show that $\mathbb{B}^n_{0,1}$ converges in $V \otimes V$. The rest follows from scaling and equivalence of moments in the first two Wiener chaoses. Set $t_k^n = k/2^n$. Then

$$\left|\mathbb{B}^{n+1}_{0,1} - \mathbb{B}^n_{0,1}\right|_{L^2}^2 \sim \mathbf{E}\left|\sum_{k=1}^{2^n} B_{t^{n+1}_{2k-2}, t^{n+1}_{2k-1}} \otimes B_{t^{n+1}_{2k-1}, t^{n+1}_{2k}}\right|_{V \otimes V}^2$$

$$\sim \frac{1}{2^{2n+2}} \mathbf{E}\left|\sum_{k=1}^{2^n} 2^{\frac{n+1}{2}} B_{t^{n+1}_{2k-2}, t^{n+1}_{2k-1}} \otimes 2^{\frac{n+1}{2}} B_{t^{n+1}_{2k-1}, t^{n+1}_{2k}}\right|_{V \otimes V}^2$$

$$\sim 2^{-2n-2} \mathbf{E}\left|\sum_{k=1}^{2^n} G_k \otimes \tilde{G}_k\right|_{V \otimes V}^2 \lesssim 2^{-2n-2} 2^{2\gamma n}$$

$$\sim 2^{-2n(1-\gamma)} \, ,$$

where the penultimate bound was obtained by exactness. By definition of exactness $1 - \gamma > 0$ and so $\mathbb{B}_{0,1}^n$ is Cauchy in the L^2-space of $V \otimes V$-valued random variables.

Exercise 3.6 *In the context of Theorem 3.8, show that for M normal the Lévy area correction takes the form*

$$A = \frac{1}{2} \operatorname{Anti}(M) \operatorname{Sym}(M)^{-1} \, .$$

Conclude that the correction vanishes if and only if M is symmetric. Is this also true without the assumption that M is normal?

Exercise 3.7 *In the context of Theorem 3.8, show that "physical Brownian motion with mass m" converges as $m \to 0$, in ϱ_α and L^q, $\alpha \in (1/2, 1/3)$ and $q < \infty$, with rate*

$$O\left(\frac{1}{m^\theta}\right), \quad any \ \theta < 1/2 - \alpha.$$

Hint: *Use Theorem 3.3 to show rough path convergence. (The computations are a little longer, but of similar type, with the additional feature that the use of the ergodic theorem can be avoided.)*

Exercise 3.8 *Consider physical Brownian motion in dimension $d = 2$, with*

$$M = I - \alpha \begin{pmatrix} 0 & -1 \\ 1 & 0 \end{pmatrix}, \quad \alpha \in \mathbf{R}.$$

Show that the area correction of X^m, in the (small mass) limit $m \to 0$, is given by

$$\frac{\alpha}{2(1+\alpha^2)} \begin{pmatrix} 0 & -1 \\ 1 & 0 \end{pmatrix}.$$

(This correction is computed by multiscale / homogenisation techniques in [PS08]).

Exercise 3.9 *Consider $X_t = bt + \sigma B_t$ where $b \in \mathbf{R}^d$, $a = \sigma\sigma^* \in (\mathbf{R}^d)^{\otimes 2}$. In other words, X is a Lévy process with triplet $(a, b, 0)$. Show that the expected signature of X over $[0, T]$ is given by*

$$\mathbf{E}S(X)_{0,T} = \exp\left(T\left(b + \frac{1}{2}a\right)\right).$$

Here, the exponential should be interpreted as the exponential in the tensor algebra, i.e.

$$\exp(u) = 1 + u + \frac{1}{2!}u \otimes u + \frac{1}{3!}u \otimes u \otimes u + \dots$$

Exercise 3.10 (Expected signature for Lévy processes [FS17]) *Consider a compound Poisson process Y with intensity λ and jumps distributed like $J = J(\omega) \sim \nu$.*

in other words, Y is Lévy with triplet $(0, 0, K)$ where the Lévy measure is given by $K = \lambda\nu$. A sample path of Y gives rise to piecewise linear, continuous path; simply by connecting J_1, $J_1 + J_2$ etc. Show that, under a suitable integrability condition for J,

$$\mathbf{E}S(Y)_{0,T} = \exp T\lambda\mathbf{E}(e^J - 1).$$

Can you handle the case of a general Lévy process?

Exercise 3.11 (Level-3 cubature formula) *Define a measure μ on $C([0,1], \mathbf{R}^d)$ by assigning equal weight 2^{-d} to each of the paths*

$$t \mapsto t \begin{pmatrix} \pm 1 \\ \pm 1 \\ \cdots \\ \pm 1 \end{pmatrix} \in \mathbf{R}^d.$$

Call the resulting process $(X_t(\omega) : t \in [0,1])$ and compute the expected signature up to level 3, that is

$$\mathbf{E}\left(1, X_{0,1}, \int_{0<t_1<t_2<1} dX_{t_1} \otimes dX_{t_2}, \int_{0<t_1<t_2<t_3<1} dX_{t_1} \otimes dX_{t_2} \otimes dX_{t_3}\right).$$

Compare with expected signature of Brownian motion, the tensor exponential $\exp(\frac{1}{2}I)$, projected to the first 3 levels.

Solution. One can write $X_t(\omega) = t \sum_i Z_i(\omega)e_i$ with i.i.d. random variables Z_i taking values $+1, -1$ with equal probability. Clearly,

$$\mathbf{E}\int_{0<t_1<1} dX_{t_1} = \mathbf{E}X_{t_1} = 0.$$

Then,

$$\int_{0<t_1<t_2<1} dX_{t_1} \otimes dX_{t_2} = \frac{1}{2}\sum_{i,j} Z_i Z_j e_i \otimes e_j = \frac{1}{2}\mathrm{Id} + \text{(zero mean)}$$

and so the expected value at level 2 matches $\pi_2\left(\exp(\frac{1}{2}I)\right) = \frac{1}{2}\mathrm{Id}$. A similar expansion on level 3 shows that every summand either contains, for some i, a factor $\mathbf{E}Z_{t_1}^i = 0$ or $\mathbf{E}\left(Z_{t_1}^i\right)^3 = 0$. In other words, the expected signature at level 3 is zero, in agreement with $\pi_3\left(\exp(\frac{1}{2}\mathrm{Id})\right) = 0$. We conclude that the expected signatures, of μ on the one hand and Wiener measure on the other hand, agree up to level 3.

Exercise 3.12 *Prove the Kolmogorov tightness criterion, Theorem 3.10.*

3.8 Comments

The modification of Kolmogorov's criterion for rough paths (Theorem 3.1) is a minor variation on a rather well-known theme. Rough path regularity of Brownian motion was first established in the thesis of Sipiläinen, [Sip93].

For extensions to infinite-dimensional Wiener processes (and also convergence of piecewise linear approximations in rough path sense) see Ledoux, Lyons and Qian [LLQ02] and Dereich [Der10]; much of the interest here is to go beyond the Hilbert space setting. The resulting stochastic integration theory against Banach space valued Brownian motion, which in essence cannot be done by classical methods, has proven crucial in some recent applications (cf. the works of Kawabi–Inahama [IK06], Dereich [Der10]).

Early proofs of Brownian rough path regularity were typically established by convergence of dyadic piecewise linear approximations to $(B, \mathbb{B}^{\text{Strat}})$ in (p-variation) rough path metric; see e.g. Lyons–Qian [LQ02]. Many other "obvious" (but as we have seen: not all reasonable) approximations are seen to yield the same Brownian rough path limit. The discussion of Brownian motion in a magnetic field follows closely Friz, Gassiat and Lyons [FGL15]. Semimartingales [CL05, FV08a, LP18, CF19] and large classes of Markovian processes [Lej06, FV08c] lift in a natural way to random rough paths. For Gaussian rough paths see Chapter 10. Infinite dimensional rough path constructions from free probability include [CDM01, Vic04].

Friz–Victoir [FV08a] extend Lépingle's classical p-variation Burkholder–Davis–Gundy (BDG) inequality [Lep76] for martingales to continuous martingale rough paths (a.k.a. enhanced martingales). This was further extended to càdlàg martingale rough paths by Chevyrev–Friz [CF19] and a precise "off-diagonal" variation estimate for $\int M dN$, two martingales, was given by Kovač and Zorin–Kranich [KZK19], extending a variational estimate of Do, Musalu and Thiele [DMT12], with motivation from harmonic analysis.

Lyons–Zeitouni [LZ99] use rough paths to bound Stratonovich iterated stochastic integrals under conditioning, with application to Onsager-Machlup functionals. The componentwise expectation of (Stratonovich) iterated integrals, expected signature of Brownian motion, was first computed in the thesis of Fawcett [Faw04]; different proofs were then given by Lyons–Victoir, Baudoin and Friz–Shekhar, [LV04, Bau04, FS17]. Fawcett's formula is central to the Kusuoka–Lyons–Victoir cubature method [Kus01, LV04]. More generally, expected signatures capture important aspects of the law of a stochastic process, see Chevyrev–Lyons [CL16]. The computation of expected signatures of large classes of stochastic processes including fractional Brownian motion, Schramm–Loewner trace, stopped Brownian motion and Lévy processes has been pursued by a number of people including Baudoin [Bau04], Werness [Wer12], Lyons–Ni [LN15], Friz–Shekhar [FS17]. The Donsker type theorem, Theorem 3.12, in uniform topology, is a consequence of Stroock–Varadhan [SV73]; the rough path case is due to Breuillard, Friz and Huesmann [BFH09]. Applications to cubature are discussed in [BF13]. Several authors have studied functional CLTs in rough paths topology in more complicated settings, including [LS17, LS18, LO18], see also [IKN18]. The case of random walks in random environments is a consequence of a Kipnis–Varadhan view on additive functionals as rough paths [DOP19]. Convergence to Brownian rough paths, with area anomaly, is also generic in the context of homogenisation, Section 9.6 contains precise references. Chevyrev [Che18] considers random walks and Lévy processes on homogeneous groups from a rough path point of view.

Chapter 4
Integration against rough paths

The aim of this chapter is to give a meaning to the expression $\int Y_t \, dX_t$ for a suitable class of integrands Y, integrated against a rough path X. We first discuss the case originally studied by Lyons where $Y = F(X)$. We then introduce the notion of a controlled rough path and show that this forms a natural class of integrands.

4.1 Introduction

We consider the problem of giving a meaning to the expression $\int Y_t \, dX_t$, for $X \in \mathscr{C}^\alpha([0, T], V)$ and Y some continuous function with values in $\mathcal{L}(V, W)$, the space of bounded linear operators from V into some other Banach space W. Of course, such an integral cannot be defined for arbitrary continuous functions Y, especially if we want the map $(X, Y) \mapsto \int Y \, dX$ to be continuous in the relevant topologies. We therefore also want to identify a "good" class of integrands Y for the rough path X.

A natural approach would be to try to define the integral as a limit of Riemann–Stieltjes sums, that is

$$\int_0^1 Y_t \, dX_t = \lim_{|\mathcal{P}| \to 0} \sum_{[s,t] \in \mathcal{P}} Y_s \, X_{s,t} \,, \tag{4.1}$$

where $\mathcal{P}$ denotes a partition of $[0, 1]$ (interpreted as a finite collection of essentially disjoint intervals such that $\bigcup \mathcal{P} = [0, 1)$) and $|\mathcal{P}|$ denotes the length of the largest element of $\mathcal{P}$. Such a definition – the *Young integral* – was studied in detail in the seminal paper by Young [You36], where it was shown that such a sum converges if $X \in \mathcal{C}^\alpha$ and $Y \in \mathcal{C}^\beta$, provided $\alpha + \beta > 1$, and that the resulting bilinear map is continuous. This result is sharp in the sense that one can construct sequences of smooth functions Y^n and X^n such that $Y^n \to 0$ and $X^n \to 0$ in $\mathcal{C}^{1/2}([0, 1], \mathbf{R})$, but such that $\int Y^n \, dX^n \to \infty$.

As a consequence of Young's inequality [You36], one has the bound

© Springer Nature Switzerland AG 2020
P. K. Friz, M. Hairer, *A Course on Rough Paths*, Universitext,
https://doi.org/10.1007/978-3-030-41556-3_4

$$\left| \int_0^1 (Y_r - Y_0) \, dX_r \right| \le C \|Y\|_{\beta;[0,1]} \|X\|_{\alpha;[0,1]} \, , \tag{4.2}$$

with C depending on $\alpha + \beta > 1$. Given paths X, Y defined on $[s, t]$ rather than $[0, 1]$ it is an easy consequence of the scaling properties of Hölder seminorms, that

$$\left| \int_s^t Y_r dX_r - Y_s X_{s,t} \right| \le C \|Y\|_\beta \|X\|_\alpha |t - s|^{\alpha+\beta} \, . \tag{4.3}$$

In particular, when $\alpha = \beta > 1/2$, the right-hand side is proportional to $|t - s|^{2\alpha} = o(|t - s|)$ which is to be compared with the estimate (4.22) below.

The main insight of the theory of rough paths is that this seemingly unsurmountable barrier of $\alpha + \beta > 1$ (which reduces to $\alpha > 1/2$ in the case $\alpha = \beta$ which is our main interest[1]) can be broken by adding additional structure to the problem. Indeed, for a rough path X, we *postulate* the values $\mathbb{X}_{s,t}$ of the integral of X against itself, see (2.2). It is then intuitively clear that one should be able to define $\int Y \, dX$ in a consistent way, provided that Y "looks like X", at least on very small scales (in the precise sense of (4.18) below). The easiest way for a function Y to "look like X" is to have $Y_t = F(X_t)$ for some sufficiently smooth $F\colon V \to \mathcal{L}(V, W)$, called a *one-form*.

4.2 Integration of one-forms

We aim to integrate $Y = F(X)$ against $\mathbf{X} = (X, \mathbb{X}) \in \mathscr{C}^\alpha$. When $F\colon V \to \mathcal{L}(V, W)$ is in C^1, or better, a Taylor approximation gives

$$F(X_r) \approx F(X_s) + DF(X_s)X_{s,r}, \tag{4.4}$$

for r in some (small) interval $[s, t]$, say. Recall (see sections 1.4 and 1.5 concerning the infinite-dimensional case) that[2]

$$\mathcal{L}(V, \mathcal{L}(V, W)) \cong \mathcal{L}(V \otimes V, W) \, ,$$

so that $DF(X_s)$ may be regarded as element in $\mathcal{L}(V \otimes V, W)$. Since the Young integral defined in (4.1), when applied to $Y = F(X)$, is effectively based on the approximation $F(X_r) \approx F(X_s)$, for $r \in [s, t]$, it is natural to hope, with a motivating look at (2.2), that the *compensated Riemann–Stieltjes sum* appearing at the right-hand

---

[1] ...but see Exercise 4.7.

[2] In coordinates, when $\dim V, \dim W < \infty$, $G = DF(X_s)$ takes the form of a $(1, 2)$-tensor $(G^k_{i,j})$ and the identification amounts to

$$v \mapsto \left(\tilde{v} \mapsto \left(\sum_{i,j} G^k_{i,j} v^i \tilde{v}^j \right)_k \right) \quad \text{versus} \quad M \mapsto \left(\sum_{i,j} G^k_{i,j} M^{i,j} \right)_k .$$

side of

$$\int_0^1 F(X_s)\,d\mathbf{X}_s \approx \sum_{[s,t]\in\mathcal{P}} \left(F(X_s)X_{s,t} + DF(X_s)\mathbb{X}_{s,t} \right), \qquad (4.5)$$

provides a good enough approximation (say, is Cauchy as $|\mathcal{P}| \to 0$) even when X ceases to have α-Hölder regularity for $\alpha > 1/2$ (as required by Young theory), but assuming instead $\mathbf{X} = (X,\mathbb{X}) \in \mathscr{C}^\alpha$, $\alpha \in \left(\frac{1}{3}, \frac{1}{2}\right]$. Why should this be good enough? The intuition is as follows: given $\alpha \in \left(\frac{1}{3}, \frac{1}{2}\right]$ neither $|X_{s,t}| \sim |t-s|^\alpha$ nor $|\mathbb{X}_{s,t}| \sim |t-s|^{2\alpha}$ in the above sum will be negligible as $|\mathcal{P}| \to 0$. Continuing in the same fashion, one expects (in fact one can show it) that the third iterated integral $\mathbb{X}_{s,t}^{(3)}$ is of order $\mathbb{X}_{s,t}^{(3)} \sim |t-s|^{3\alpha} = \mathrm{o}(|t-s|)$, so that adding a third term of the form $D^2 F(X_s)\mathbb{X}_{s,t}^{(3)}$ in the sum of (4.5), at the very least, will not affect any limit, should it exist. In the following, we will see that this limit,[3]

$$\int_0^1 F(X_s)\,d\mathbf{X}_s = \lim_{|\mathcal{P}|\to 0} \sum_{[s,t]\in\mathcal{P}} \left(F(X_s)X_{s,t} + DF(X_s)\mathbb{X}_{s,t} \right), \qquad (4.6)$$

does exist and call it *rough integral*.[4] In fact, in this section we shall construct the (indefinite) rough integral $Z = \int_0^\cdot F(X)d\mathbf{X}$ as element in $\mathcal{C}^\alpha$, i.e. as *path*, similar to the construction of stochastic integrals *as processes* rather than random variables. Even this may not be sufficient in applications – one often wants to have an extended meaning of the rough integral, such as $(Z, \mathbf{Z}) \in \mathscr{C}^\alpha$, point of view emphasised in [Lyo98, LQ02, LCL07], or something similar (such as "Z controlled by X" in the sense of Definition 4.6 below, to be discussed in the next section).

Lemma 4.1. *Let* $F\colon V \to \mathcal{L}(V,W)$ *be a* C_b^2 *function and let* $(X,\mathbb{X}) \in \mathscr{C}^\alpha$ *for some* $\alpha > \frac{1}{3}$. *Set* $Y_s := F(X_s)$, $Y_s' := DF(X_s)$ *and* $R_{s,t}^Y := Y_{s,t} - Y_s' X_{s,t}$. *Then*

$$Y, Y' \in \mathcal{C}^\alpha \quad and \quad R^Y \in \mathcal{C}_2^{2\alpha}. \qquad (4.7)$$

(In the terminology of the forthcoming Definition 4.6: "Y is controlled by X with Gubinelli derivative Y'; in symbols $(Y,Y') \in \mathscr{D}_X^{2\alpha}$*".) More precisely, we have the estimates*

$$\|Y\|_\alpha \le \|DF\|_\infty \|X\|_\alpha,$$
$$\|Y'\|_\alpha \le \|D^2 F\|_\infty \|X\|_\alpha,$$
$$\|R^Y\|_{2\alpha} \le \frac{1}{2}\|D^2 F\|_\infty \|X\|_\alpha^2.$$

---

[3] Recall that $\lim_{|\mathcal{P}|\to 0}$ means convergence along any sequence $(\mathcal{P}_n)$ with mesh $|\mathcal{P}_n| \to 0$, with identical limit along each such sequence. In particular, it is not enough to establish convergence along a particular sequence $(\mathcal{P}_n)$, although a particular sequence may be used to identify the limit.

[4] Of course, we can and will consider intervals other than $[0, 1]$. Without further notice, $\mathcal{P}$ always denotes a partition of the interval under consideration.

Proof. $\mathcal{C}_b^2$ regularity of F implies that F and DF are both Lipschitz continuous with Lipschitz constants $\|DF\|_\infty$ and $\|D^2F\|_\infty$ respectively. The α-Hölder bounds on Y and Y' are then immediate. For the remainder term, consider the function

$$[0,1] \ni \xi \mapsto F(X_s + \xi X_{s,t}) \,.$$

A Taylor expansion, with intermediate value remainder, yields $\xi \in (0,1)$ such that

$$R_{s,t}^Y = F(X_t) - F(X_s) - DF(X_s)X_{s,t} = \frac{1}{2}D^2F(X_s + \xi X_{s,t})(X_{s,t}, X_{s,t}) \,.$$

The claimed 2α-Hölder estimate, in the sense that $|R_{s,t}^Y| \lesssim |t-s|^{2\alpha}$, then follows at once. $\square$

Before we prove that the rough integral (4.6) exists, we discuss some sort of abstract Riemann integration. In what follows, at first reading, one may have in mind the construction of a Riemann–Stieltjes (or Young) integral $Z_t := \int_0^t Y_r dX_r$. From Young's inequality (4.3), one has (with $Z_{s,t} = Z_t - Z_s$ as usual)

$$Z_{s,t} = Y_s X_{s,t} + \mathrm{o}(|t-s|)$$

and $\Xi_{s,t} := Y_s X_{s,t}$ is a sufficiently good local approximation in the sense that it fully determines the integral Z via the limiting procedure given in (4.1)). In this sense $Z = \mathcal{I}\Xi$ is the well-defined image of Ξ under some *abstract integration map* $\mathcal{I}$. Note that $Z_{s,t} = Z_{s,u} + Z_{u,t}$, i.e. increments are additive (or "multiplicative" if one regards $+$ as group operation[5]) whereas a similar property fails for Ξ. In the language of [Lyo98], such a Ξ corresponds to a "almost multiplicative functional" and it is a key result in the theory that there is a unique associated "multiplicative functional" (here: $Z = \mathcal{I}\Xi$). Following [FdLP06] we call "sewing" the step from a (good enough) local approximation Ξ to some (abstract) integral $\mathcal{I}\Xi$; the concrete estimate which quantifies how well $\mathcal{I}\Xi$ is approximated by Ξ will be called "sewing lemma". It plays an analogous role to Davie's lemma (cf. Section 8.7) in the context of (rough) differential equations.

We now formalise what we mean by Ξ being a good enough local approximation. For this, we introduce the space $\mathcal{C}_2^{\alpha,\beta}([0,T],W)$ of functions Ξ from the 2-simplex $\{(s,t): 0 \le s \le t \le T\}$ into W such that $\Xi_{t,t} = 0$ and such that

$$\|\Xi\|_{\alpha,\beta} \overset{\text{def}}{=} \|\Xi\|_\alpha + \|\delta\Xi\|_\beta < \infty \,, \tag{4.8}$$

where $\|\Xi\|_\alpha = \sup_{s<t} \frac{|\Xi_{s,t}|}{|t-s|^\alpha}$ as usual, and also

$$\delta\Xi_{s,u,t} \overset{\text{def}}{=} \Xi_{s,t} - \Xi_{s,u} - \Xi_{u,t}, \quad \|\delta\Xi\|_\beta \overset{\text{def}}{=} \sup_{s<u<t} \frac{|\delta\Xi_{s,u,t}|}{|t-s|^\beta} \,.$$

---

[5] This terminology becomes natural if one considers Z together with its iterated integrals as group-valued path, increments of which satisfy Chen's "multiplicative" relation, see (2.8).

Provided that $\beta > 1$, it turns out that such functions are "almost" of the form $\Xi_{s,t} = F_t - F_s$, for some α-Hölder continuous function F (they would be if and only if $\delta\Xi = 0$). Indeed, it is possible to construct in a canonical way a function $\hat{\Xi}$ with $\delta\hat{\Xi} = 0$ and such that $\hat{\Xi}_{s,t} \approx \Xi_{s,t}$ for $|t - s| \ll 1$:

Lemma 4.2 (Sewing lemma). *Let α and β be such that $0 < \alpha \leq 1 < \beta$. Then, there exists a unique continuous linear map $\mathcal{I} \colon C_2^{\alpha,\beta}([0,T], W) \to C^\alpha([0,T], W)$ such that $(\mathcal{I}\Xi)_0 = 0$ and*

$$\left| (\mathcal{I}\Xi)_{s,t} - \Xi_{s,t} \right| \leq C|t - s|^\beta . \tag{4.9}$$

where C only depends on β and $\|\delta\Xi\|_\beta$. (The α-Hölder norm of $\mathcal{I}\Xi$ also depends on $\|\Xi\|_\alpha$ and hence on $\|\Xi\|_{\alpha,\beta}$.)

Proof. As linear map, continuity of $\mathcal{I}$ will be an immediate consequence of its boundedness. We shall construct the path $\mathcal{I}\Xi =: I$, with $I_0 = 0$, via its increments $I_{s,t} = I_t - I_s$. Additivity of these increments ($\delta I = 0$) is an important aspect of the proof. Uniqueness of $\mathcal{I}$ is immediate: assuming two paths I and $\bar{I}$ both satisfy (4.9), it follows that $I - \bar{I}$ satisfies $(I - \bar{I})_0 = 0$ and $|(I - \bar{I})_{s,t}| = |(I - \bar{I})_t - (I - \bar{I})_s| \lesssim |t - s|^\beta$. Since $\beta > 1$ by assumption, we conclude that $I - \bar{I}$ vanishes identically. In fact, (4.9) shows that I is necessarily given as Riemann-type limit: writing $\mathcal{P}$ for a partition of $[s, t]$ and $|\mathcal{P}|$ for its mesh size, we have

$$\left| I_{s,t} - \sum_{[u,v]\in\mathcal{P}} \Xi_{u,v} \right| = \left| \sum_{[u,v]\in\mathcal{P}} \left(I_{u,v} - \Xi_{u,v} \right) \right| = \mathrm{O}(|\mathcal{P}|^{\beta-1})$$

which is nothing but a quantitative form of

$$(\mathcal{I}\Xi)_{s,t} = \lim_{|\mathcal{P}|\to 0} \sum_{[u,v]\in\mathcal{P}} \Xi_{u,v} . \tag{4.10}$$

Because of its importance we give *two* independent but related arguments. The *first argument* is based on successive (dyadic) refinement to construct $I_{s,t}$ with the desired bound (4.9), followed by an argument for additivity. Fix $[s, t] \subset [0, T]$ and let $\mathcal{P}_n$ be the level-n dyadic partion of $[s, t]$, which contains 2^n intervals, each of length $2^{-n}|t - s|$, starting with the trivial partition $\mathcal{P}_0 = \{[s, t]\}$. Define $I_{s,t}^0 = \Xi_{s,t}$ and then the nth level approximation by

$$I_{s,t}^{n+1} \stackrel{\text{def}}{=} \sum_{[u,v]\in\mathcal{P}_{n+1}} \Xi_{u,v} = I_{s,t}^n - \sum_{[u,v]\in\mathcal{P}_n} \delta\Xi_{u,m,v} ,$$

where it is a straightforward exercise to check that the second equality holds. It then follows immediately from the definition of $\|\delta\Xi\|_\beta$ that

$$\left| I_{s,t}^{n+1} - I_{s,t}^n \right| \leq 2^{n(1-\beta)}|t - s|^\beta \|\delta\Xi\|_\beta .$$

Since $\beta > 1$, these terms are summable whence we conclude that the sequence $(I_{s,t}^n : n \in \mathbf{N})$ is Cauchy. Its limit $I_{s,t}$ is such that, summing up the bound above,

$$\left| I_{s,t} - \Xi_{s,t} \right| \le \sum_{n \ge 0} \left| I_{s,t}^{n+1} - I_{s,t}^n \right| \le C \|\delta\Xi\|_\beta |t - s|^\beta , \qquad (4.11)$$

for some universal constant C depending only on β, which is precisely the required bound (4.9). Unfortunately, additivity of I is no consequence of this argument so we have to be a little smarter (but see Remark 4.3). Taking $T = 1$ without loss of generality (and for notational simplicity only), we restrict the previous construction to *elementary* dyadic intervals of the form $[s,t] = 2^{-k}[\ell, \ell+1]$ for some $k \ge 0$ and $\ell \in \{0, \dots, 2^k - 1\}$. The advantage is that now mid-point additivity holds in the sense that

$$I_{s,t} = I_{s,u} + I_{u,t} , \qquad u = \frac{s+t}{2} , \qquad (4.12)$$

as a simple consequence of taking limits in the identity $I_{s,t}^{n+1} = I_{s,u}^n + I_{u,t}^n$. The natural additive extension of I to non-elementary dyadic intervals $2^{-k}[\ell, m]$ is then given by *postulating* that

$$I_{2^{-k}\ell, 2^{-k}m} = \sum_{j=\ell}^{m-1} I_{2^{-k}j, 2^{-k}(j+1)} , \qquad (4.13)$$

which is indeed well-defined (note that $2^{-k}[\ell, m] = 2^{-k-1}[2\ell, 2m]$ for example so (4.13) can be written in several ways) by (4.12). This defines $I_{s,t}$ for all dyadic numbers s, t and the construction guarantees additivity. We leave the fact that $I_{s,t}$ satisfies (4.9) for all dyadic s, t (and therefore for all $s, t \in [0,1]$ by continuous extension) as Exercise 4.3.

The *second* argument, which is essentially due to Young, yields immediately the convergence (4.10), as $|\mathcal{P}| \to 0$, i.e. the same limit is obtained along any sequence $\mathcal{P}_n$ with mesh tending to zero. This has the important consequence that additivity of increment ($\delta I = 0$) is a consequence of (4.10) and requires no additional argument. (Another advantage of Young's construction is that it also works under variation - rather than Hölder type assumption and thus in application allows to deal with jumps.) Consider a partition $\mathcal{P}$ of $[s,t]$ and let $r \ge 1$ be the number of intervals in $\mathcal{P}$. When $r \ge 2$ there exists $u \in [s,t]$ such that $[u_-, u], [u, u_+] \in \mathcal{P}$ and

$$|u_+ - u_-| \le \frac{2}{r-1} |t - s|.$$

Indeed, assuming otherwise gives the contradiction $2|t-s| \ge \sum_{u \in \mathcal{P}^\circ} |u_+ - u_-| > 2|t-s|$. Hence, $|\int_{\mathcal{P}\setminus\{u\}} \Xi - \int_{\mathcal{P}} \Xi| = |\delta\Xi_{u_-,u,u_+}| \le \|\delta\Xi\|_\beta \left(2|t-s|/(r-1)\right)^\beta$ and by iterating this procedure until the partition is reduced to $\mathcal{P} = \{[s,t]\}$, we arrive at the *maximal inequality*,

$$\sup_{\mathcal{P}\subset[s,t]} \left| \Xi_{s,t} - \int_{\mathcal{P}} \Xi \right| \leq 2^{\beta} \|\delta\Xi\|_{\beta} \zeta(\beta) |t-s|^{\beta} ,$$

where ζ denotes the classical ζ function. It then remains to show that

$$\sup_{|\mathcal{P}|\vee|\mathcal{P}'|<\varepsilon} \left| \int_{\mathcal{P}} \Xi - \int_{\mathcal{P}'} \Xi \right| \to 0 \quad \text{as } \varepsilon \downarrow 0, \tag{4.14}$$

which implies existence of $\mathcal{I}\Xi$ as the limit $\lim_{|\mathcal{P}|\to 0} \int_{\mathcal{P}} \Xi$. To this end, at the price of adding / subtracting $\mathcal{P} \cup \mathcal{P}'$, we can assume without loss of generality that $\mathcal{P}'$ refines $\mathcal{P}$. In particular, then $|\mathcal{P}| \vee |\mathcal{P}'| = |\mathcal{P}|$ and

$$\int_{\mathcal{P}} \Xi - \int_{\mathcal{P}'} \Xi = \sum_{[u,v]\in\mathcal{P}} \left(\Xi_{u,v} - \int_{\mathcal{P}'\cap[u,v]} \Xi \right).$$

But then, for any $\mathcal{P}$ with $|\mathcal{P}| \leq \varepsilon$ we can use the maximal inequality to see that

$$\left| \int_{\mathcal{P}} \Xi - \int_{\mathcal{P}'} \Xi \right| \leq 2^{\beta} \zeta(\beta) \|\delta\Xi\|_{\beta} \sum_{[u,v]\in\mathcal{P}} |v-u|^{\beta} = O\big(|\mathcal{P}|^{\beta-1}\big) = O(\varepsilon^{\beta-1}).$$

This concludes the Young argument (with no hidden tedium left to the reader).   □

Remark 4.3. The first argument ultimately suffered from the tedium of checking the additivity property $\delta\mathcal{I}\Xi = 0$. In some situations this extra step can be avoided, notably in the case where all one wants are *uniform* rough path estimates for classical Riemann–Stieltjes integrals. More precisely, consider the case that $X : [0,T] \to V$ is smooth, $\mathbb{X} = \int X \otimes dX$, and one is only interested in an error estimate for second order approximations of Riemann–Stieltjes integrals, of the form

$$\left| \int_s^t F(X_r) \, dX_r - F(X_s) X_{s,t} - DF(X_s) \mathbb{X}_{s,t} \right| \leq O(|t-s|^{3\alpha}),$$

uniform over all (smooth) paths X with $\|X\|_{\alpha} + \|\mathbb{X}\|_{2\alpha}$ bounded. In the context of the above proof, this estimate is contained in the first step, applied with (cf. the proof of Theorem 4.4)

$$\Xi_{s,t} = F(X_s) X_{s,t} + DF(X_s) \mathbb{X}_{s,t} .$$

But here we know already from classical Riemann integration theory that $(\mathcal{I}\Xi)_{s,t}$, constructed as limit of dyadic partitions of $[s,t]$, is precisely the Riemann–Stieltjes integral $\int_s^t F(X_r) \, dX_r$ and therefore additive. (The contribution of $DF(X)\mathbb{X}$ effectively constitutes a higher-order approximation and surely does not affect the limit, as can be seen from the estimate $|\mathbb{X}_{u,v}| \lesssim |v-u|^2$, thanks to smoothness of X.)

We now apply the sewing lemma to the construction of (4.6).

Theorem 4.4 (Lyons). *Let* $\mathbf{X} = (X, \mathbb{X}) \in \mathscr{C}^{\alpha}([0,T], V)$ *for some* $T > 0$ *and* $\alpha > \frac{1}{3}$, *and let* $F \colon V \to L(V, W)$ *be a* C_b^2 *function. Then, the rough integral defined in (4.6) exists and one has the bound*

$$\left| \int_s^t F(X_r)\, d\mathbf{X}_r - F(X_s) X_{s,t} - DF(X_s) \mathbb{X}_{s,t} \right|$$

$$\lesssim \|F\|_{\mathcal{C}_b^2} \left(\|X\|_\alpha^3 + \|X\|_\alpha \|\mathbb{X}\|_{2\alpha} \right) |t-s|^{3\alpha}, \quad (4.15)$$

where the proportionality constant depends only on α. Furthermore, the indefinite rough integral is α-Hölder continuous on $[0,T]$ and we have the following quantitative estimate,

$$\left\| \int_0^{\cdot} F(X)\, d\mathbf{X} \right\|_\alpha \leq C \|F\|_{\mathcal{C}_b^2} \left(\|\!|\mathbf{X}|\!\|_\alpha \vee \|\!|\mathbf{X}|\!\|_\alpha^{1/\alpha} \right), \quad (4.16)$$

where the constant C only depends on T and α and can be chosen uniformly in $T \leq 1$. Furthermore, $\|\!|\mathbf{X}|\!\|_\alpha = \|X\|_\alpha + \sqrt{\|\mathbb{X}\|_{2\alpha}}$ denotes again the homogeneous α-Hölder rough path norm.

Remark 4.5. We will see in Section 4.4 that the map $(X, \mathbb{X}) \in \mathscr{C}^\alpha \mapsto \int_0^{\cdot} F(X)\, d\mathbf{X} \in \mathscr{C}^\alpha$ is continuous in α-Hölder rough path metric.

Proof. Let us stress the fact that the argument given here only relies on the properties of the integrand $Y = F(X)$ collected in Lemma 4.1 above. In particular, the generalisation to "extended" integrands (Y, Y'), which replace $(F(X), DF(X))$, subject to (4.7), will be immediate. (We shall develop this "Gubinelli" point of view further in Section 4.3 below.)

The result follows as a consequence of Lemma 4.2. With the notation that we just introduced, the classical Young integral [You36] can be defined as the usual limit of Riemann sums by

$$\int_s^t Y_r\, dX_r = \left(\mathcal{I}\Xi \right)_{s,t}, \qquad \Xi_{s,t} = Y_s X_{s,t}.$$

Unfortunately, this definition satisfies the identity

$$\delta \Xi_{s,u,t} = -Y_{s,u} X_{u,t},$$

so that, except in trivial cases, the required bound (4.8) is satisfied only if Y and X are Hölder continuous with Hölder exponents adding up to $\beta > 1$. In order to be able to cover the situation $\alpha < \frac{1}{2}$, it follows that we need to consider a better approximation to the Riemann sums, as discussed above. To this end, we use the notation from Lemma 4.1, namely

$$Y_s := F(X_s), \quad Y_s' := DF(X_s) \quad \text{and} \quad R_{s,t}^Y := Y_{s,t} - Y_s' X_{s,t},$$

and then set $\Xi_{s,t} = Y_s X_{s,t} + Y_s' \mathbb{X}_{s,t}$. Note that, for any $u \in (s,t)$, we have the identity

$$\delta \Xi_{s,u,t} = -R_{s,u}^Y X_{u,t} - Y_{s,u}' \mathbb{X}_{u,t}.$$

Thanks to the α-Hölder regularity of X, Y' and the 2α-regularity of $R, \mathbb{X}$, the triangle inequality shows that (4.8) holds true with the given $\alpha > 1/3$ and $\beta := 3\alpha > 1$. The fact that the integral is well-defined, and the bound

$$\left| \int_s^t Y\,d\mathbf{X} - Y_s X_{s,t} - Y_s' \mathbb{X}_{s,t} \right| \lesssim \left(\|X\|_\alpha \|R^Y\|_{2\alpha} + \|\mathbb{X}\|_{2\alpha} \|Y'\|_\alpha \right) |t - s|^{3\alpha}$$

$$(4.17)$$

then follow immediately from (4.11). Upon substituting the estimate obtained in Lemma 4.1, we obtain (4.15).

We now turn to the proof of (4.16). Writing $Z = \int F(X)d\mathbf{X}$ and using the triangle inequality in (4.15) gives

$$
\begin{aligned}
|Z_{s,t}| &\le \|F\|_\infty |X_{s,t}| + \|DF\|_\infty |\mathbb{X}_{s,t}| \\
&\quad + C\|F\|_{\mathcal{C}_b^2} \left(\|X\|_\alpha^3 + \|X\|_\alpha \|\mathbb{X}\|_{2\alpha} \right) |t - s|^{3\alpha} \\
&\le C\|F\|_{\mathcal{C}_b^2} \left[\mathrm{A}_1 |t - s|^\alpha + \mathrm{A}_2 |t - s|^{2\alpha} + \mathrm{A}_3 |t - s|^{3\alpha} \right] ,
\end{aligned}
$$

with $\mathrm{A}_i \le \|\mathbf{X}\|_\alpha$, for $1 \le i \le 3$. Allowing C to change, this already implies

$$\|Z\|_\alpha \le C\|F\|_{\mathcal{C}_b^2} \left(\|\mathbf{X}\|_\alpha \vee \|\mathbf{X}\|_\alpha^3 \right) ,$$

which is the claimed estimate (4.16) in the limit $\alpha \downarrow 1/3$. However, one can do better by realising that the above estimate is best for $|t - s|$ small, whereas for $t - s$ large it is better to split up $|Z_{s,t}|$ into the sum of small increments. To make this more precise, set $\varrho := \|\mathbf{X}\|_\alpha$ and write (hide factor $C = C(\alpha, T)$ in $\lesssim$ below)

$$
\begin{aligned}
|Z_{s,t}| &\lesssim \varrho|t - s|^\alpha + \varrho^2 |t - s|^{2\alpha} + \varrho^3 |t - s|^{3\alpha} \\
&\le 3\varrho|t - s|^\alpha \text{ for } \varrho^{1/\alpha}|t - s| \le 1.
\end{aligned}
$$

Increments of Z over $[s, t]$ with length greater than $h := \varrho^{-1/\alpha}$ are handled by cutting them into pieces of length h. More precisely (cf. Exercise 4.5) we have $\|Z\|_{\alpha;h} \le 3\varrho$ which entails

$$\|Z\|_\alpha \le 3\varrho\left(1 \vee 2h^{-(1-\alpha)} \right) \le 6\left(\varrho \vee \varrho^{1/\alpha} \right).$$

At last, we note that $C = C(\alpha, T)$ can be chosen uniformly in $T \le 1$. $\square$

4.3 Integration of controlled rough paths

Motivated by Lemma 4.1 and the observation that rough integration essentially relies on the properties (4.7) we introduce the notion of a *controlled* path Y, relative to some "reference" path X, due to Gubinelli [Gub04]. For the sake of the following definition we assume that Y takes values in some Banach space, say $\bar{W}$. When it comes to the definition of a rough integral we typically take $\bar{W} = \mathcal{L}(V, W)$; although other choices can be useful (see e.g. Remark 4.12). In the context of rough differential equations, with solutions in $\bar{W} = W$, we actually need to integrate $f(Y)$, which will be seen to be controlled by X for sufficiently smooth coefficients $f : W \to \mathcal{L}(V, W)$.

Definition 4.6. Given a path $X \in \mathcal{C}^\alpha([0, T], V)$, we say that $Y \in \mathcal{C}^\alpha([0, T], \bar{W})$ is *controlled* by X if there exists $Y' \in \mathcal{C}^\alpha([0, T], \mathcal{L}(V, \bar{W}))$ so that the remainder term R^Y given implicitly through the relation

$$Y_{s,t} = Y_s' X_{s,t} + R_{s,t}^Y , \tag{4.18}$$

satisfies $\|R^Y\|_{2\alpha} < \infty$. This defines the space of *controlled rough paths*,

$$(Y, Y') \in \mathscr{D}_X^{2\alpha}([0, T], \bar{W}).$$

Although Y' is not, in general, uniquely determined from Y (cf. Remark 4.7 and Section 6 below) we call any such Y' the *Gubinelli derivative* of Y (with respect to X).

Here, $R_{s,t}^Y$ takes values in $\bar{W}$, and the norm $\| \cdot \|_{2\alpha}$ for a function with two arguments is given by (2.3) as before. We endow the space $\mathscr{D}_X^{2\alpha}$ with the seminorm

$$\|Y, Y'\|_{X,2\alpha} \overset{\text{def}}{=} \|Y'\|_\alpha + \|R^Y\|_{2\alpha} . \tag{4.19}$$

As in the case of classical Hölder spaces, $\mathscr{D}_X^{2\alpha}$ is a Banach space under the norm $(Y, Y') \mapsto |Y_0| + |Y_0'| + \|Y, Y'\|_{X,2\alpha}$. This quantity also controls the α-Hölder regularity of Y since, uniformly over X bounded in α-Hölder seminorm,

$$\|Y\|_\alpha \le \|Y'\|_\infty \|X\|_\alpha + T^\alpha \|R^Y\|_{2\alpha} \le |Y_0'| \|X\|_\alpha + T^\alpha \{\|Y'\|_\alpha \|X\|_\alpha + \|R^Y\|_{2\alpha}\}$$
$$\le (1 + \|X\|_\alpha)\,(|Y_0'| + T^\alpha \|Y, Y'\|_{X,2\alpha}) \lesssim |Y_0'| + T^\alpha \|Y, Y'\|_{X,2\alpha} . \tag{4.20}$$

Remark 4.7. Since we only assume that $\|Y\|_\alpha < \infty$, but then impose that $\|R^Y\|_{2\alpha} < \infty$, it is in general the case that a genuine cancellation takes place in (4.18). The question arises to what extent Y determines Y'. Somewhat contrary to the classical situation, where a smooth function has a unique derivative, too much regularity of the underlying rough path $\mathbf{X}$ leads to less information about Y'. For instance, if Y is smooth, or in fact in $\mathcal{C}^{2\alpha}$, and the underlying rough path $\mathbf{X}$ happens to have a path component X that is also $\mathcal{C}^{2\alpha}$, then we may take $Y' = 0$, but as a matter of fact any continuous path Y' would satisfy (4.18) with $\|R\|_{2\alpha} < \infty$. On the other hand, if X is far from smooth, i.e. genuinely rough on all (small) scales, uniformly in all directions, then Y' is uniquely determined by Y, cf. Section 6 below.

Remark 4.8. It is important to note that while the space of rough paths $\mathscr{C}^\alpha$ is not even a vector space, the space $\mathscr{D}_X^{2\alpha}$ is a perfectly normal Banach space for any given $\mathbf{X} = (X, \mathbb{X}) \in \mathscr{C}^\alpha$. The twist of course is that the space in question depends in a crucial way on the choice of $\mathbf{X}$. The set of all pairs $(\mathbf{X}; (Y, Y'))$ gives rise to the total space

$$\mathscr{C}^\alpha \ltimes \mathscr{D}^{2\alpha} \overset{\text{def}}{=} \bigsqcup_{\mathbf{X} \in \mathscr{C}^\alpha} \{\mathbf{X}\} \times \mathscr{D}_X^{2\alpha},$$

with base space $\mathscr{C}^\alpha$ and "fibres" $\mathscr{D}_X^{2\alpha}$. We will see in Exercise 4.9 that $\mathscr{C}^\alpha \times \mathscr{D}^{2\alpha}$ is actually a "trivial" infinite-dimensional fibre bundle in the sense that it is homeomorphic to $\mathscr{C}^\alpha \times (C^{2\alpha} \oplus C^\alpha)$, albeit not in a canonical way. (At least when $\alpha \neq \frac{1}{2}$.) At the intuitive level, this clashes with the results of Chapter 6 which suggest that, the rougher the underlying path X, the "smaller" is $\mathscr{D}_X^{2\alpha}$.

Remark 4.9. While the notion of "controlled rough path" has many appealing features, it does not come with a natural approximation theory. To wit, consider $(X, \mathbb{X}) \in \mathscr{C}_g^\alpha([0, T], \mathbf{R}^d)$ as limit of smooth paths $X_n : [0, T] \to \mathbf{R}^d$ in the sense of Proposition 2.8. Then it is natural to approximate $Y = F(X)$ by $Y_n = F(X_n)$, which is again smooth (to the extent that F permits). There is no obvious analogue of this for controlled rough paths. However, there is a non-canonical approximation result, based on the Lyons–Victoir extension, which the reader is invited to explore in Exercise 4.8.

We are now ready to extend Young's integral to that of a path controlled by X against $\mathbf{X} = (X, \mathbb{X})$. Recall from Lemma 4.1 that $Y = F(X)$, with $Y' = DF(X)$, is somewhat the prototype of a controlled rough path. The definition of the rough integral $\int F(X) d\mathbf{X}$ in terms of compensated Riemann sums, cf. (4.6), then immediately suggests to define the integral of Y against $\mathbf{X}$ by[6]

$$\int_0^1 Y \, d\mathbf{X} \overset{\text{def}}{=} \lim_{|\mathcal{P}| \to 0} \sum_{[s,t] \in \mathcal{P}} \left(Y_s \, X_{s,t} + Y_s' \, \mathbb{X}_{s,t} \right), \tag{4.21}$$

where we took $\bar{W} = \mathcal{L}(V, W)$ and used the canonical injection $\mathcal{L}(V, \mathcal{L}(V, W)) \hookrightarrow \mathcal{L}(V \otimes V, W)$ in writing $Y_s' \mathbb{X}_{s,t}$. With these notations, the resulting integral takes values in W.

With these notations at hand, it is now straightforward to prove the following result, which is a slight reformulation of [Gub04, Prop.1]:

Theorem 4.10 (Gubinelli). *Let $T > 0$, let $\mathbf{X} = (X, \mathbb{X}) \in \mathscr{C}^\alpha([0, T], V)$ for some $\alpha \in \left(\frac{1}{3}, \frac{1}{2} \right]$, and let $(Y, Y') \in \mathscr{D}_X^{2\alpha}([0, T], \mathcal{L}(V, W))$. Then there exists a constant C depending only on α such that*

a) The integral defined in (4.21) exists and, for every pair s, t, one has the bound

$$\left| \int_s^t Y_r \, d\mathbf{X}_r - Y_s X_{s,t} - Y_s' \, \mathbb{X}_{s,t} \right| \leq C \left(\|X\|_\alpha \|R^Y\|_{2\alpha} + \|\mathbb{X}\|_{2\alpha} \|Y'\|_\alpha \right) |t - s|^{3\alpha}. \tag{4.22}$$

b) The map from $\mathscr{D}_X^{2\alpha}([0, T], \mathcal{L}(V, W))$ to $\mathscr{D}_X^{2\alpha}([0, T], W)$ given by

$$(Y, Y') \mapsto (Z, Z') := \left(\int_0^\cdot Y_t \, d\mathbf{X}_t, Y \right), \tag{4.23}$$

---

[6] Note the abuse of notation: we hide dependence on Y' which in general affects the limit but is usually clear from the context.

is a continuous linear map between Banach spaces and one has the bound [7]

$$\|Z, Z'\|_{X,2\alpha} \leq \|Y\|_\alpha + \|Y'\|_\infty \|\mathbb{X}\|_{2\alpha} + CT^\alpha \left(\|X\|_\alpha \|R^Y\|_{2\alpha} + \|\mathbb{X}\|_{2\alpha} \|Y'\|_\alpha \right) .$$

Proof. Part a) is an immediate consequence of Lemma 4.2, as already pointed out in the proof of Theorem 4.4. The estimate (4.22) was pointed out explicitly in (4.17).

It remains to show the bound on $\|Z, Z'\|_{X,2\alpha}$. Splitting up the left-hand side of (4.22) after the first term, using the triangle inequality, gives immediately an α Hölder estimate on $\int_s^t Y_r dX_r = Z_{s,t}$, so that $Z \in \mathcal{C}^\alpha$. ($Z' = Y \in \mathcal{C}^\alpha$ is trivial, by the very nature of Y since it is controlled by X.) Similarly, splitting up the left-hand side of (4.22) after the second term, gives a 2α-Hölder type estimate on $\int_s^t Y_r dX_r - Y_s X_{s,t} = Z_{s,t} - Z'_s X_{s,t} =: R^Z_{s,t}$, i.e. on the remainder term in the sense of (4.18). The explicit estimate for $\|Z, Z'\|_{X,2\alpha} = \|Y\|_\alpha + \|R^Z\|_{2\alpha}$ is then obvious. $\square$

Remark 4.11. One actually obtains better information than just $(Z, Z') \in \mathscr{D}^{2\alpha}_X$, namely one has control up to order 3α in the sense that

$$\left| Z_{s,t} - Y_s X_{s,t} - Y'_s \mathbb{X}_{s,t} \right| \lesssim |t - s|^{3\alpha} ,$$

see (4.34). Similar consideration will lead to the more general concept of *modelled distribution* in the theory of regularity structures, see in particular Definition 13.10.

Remark 4.12. As in the above theorem, assume that $(X, \mathbb{X}) \in \mathscr{C}^\alpha([0,T], V)$ and consider Y and Z two paths controlled by X. More precisely, we assume $(Y, Y') \in \mathscr{D}^{2\alpha}_X([0,T], \mathcal{L}(\bar{V}, W))$ and $(Z, Z') \in \mathscr{D}^{2\alpha}_X([0,T], \bar{V})$, where of course $V, \bar{V}, W$ are all Banach spaces. Then, in terms of the abstract integration map $\mathcal{I}$ (cf. the sewing lemma) we may define the integral of Y against Z, with values in W, as follows,

$$\int_s^t Y_u \, dZ_u \stackrel{\text{def}}{=} (\mathcal{I}\Xi)_{s,t} , \quad \Xi_{u,v} = Y_u Z_{u,v} + Y'_u Z'_u \mathbb{X}_{u,v} . \qquad (4.24)$$

Here, we use the fact that $Z'_u \in \mathcal{L}(V, \bar{V})$ can be canonically identified with an operator in $\mathcal{L}(V \otimes V, V \otimes \bar{V})$ by acting only on the second factor, and $Y'_u \in \mathcal{L}(V, \mathcal{L}(\bar{V}, W))$ is identified as before with an operator in $\mathcal{L}(V \otimes \bar{V}, W)$. The reader may be helped to see this spelled out in coordinates, assuming finite dimensions: using indices i, j in $W, \bar{V}$ respectively, and then k, l in V:

$$(\Xi_{u,v})^i = (Y_u)^i_j (Z_{u,v})^j + (Y'_u)^i_{k,j} (Z'_u)^j_l (\mathbb{X}_{u,v})^{k,l}.$$

A short computation, similar to the one that justified the application of the sewing lemma for the construction of the rough integral introduced in (4.21), gives

$$-\delta\Xi_{s,u,t} = R^Y_{s,u} Z_{u,t} + Y'_s X_{s,u} R^Z_{s,u} + Y'_s X_{s,u} Z'_{s,u} X_{u,t} + (Y'Z')_{s,u} \mathbb{X}_{u,t} .$$

---

[7] As in (4.20), this implies $\|Z, Z'\|_{X,2\alpha} \lesssim |Y'_0| + T^\alpha \|Y, Y'\|_{X,2\alpha}$, uniformly over bounded **X**.

It immediately follows that $\|\delta\Xi\|_{3\alpha} < \infty$ so that, since $3\alpha > 1$, the right-hand side of (4.24) is well defined. The sewing lemma furthermore yields the following generalisation of (4.22), with Ξ as given in (4.24),

$$\left| \int_s^t Y\,dZ - \Xi_{s,t} \right| \lesssim (\|R^Y\|_{2a}\|Z\|_\alpha + (*) + \|Y'Z'\|_\alpha\|\mathbb{X}\|_{2\alpha})|t - s|^{3\alpha} , \quad (4.25)$$

and additional term

$$(*) = \|Y'\|_\infty\|X\|_\alpha(\|R^Z\|_{2\alpha} + \|Z'\|_\alpha\|X\|_\alpha) .$$

Note that $(*)$ duly vanishes when $Z = X$ and Z' is the identity operator, since then $R^Z \equiv 0$ and Z', constant in time, has vanishing α-Hölder seminorm. In that case, we recover precisely the previously obtained estimate for the rough integral introduced in (4.21). Furthermore, in the smooth case, one can check that we again recover the usual Riemann / Young integral.

Remark 4.13. If, in the notation of the proof of Theorem 4.4, Ξ and $\tilde{\Xi}$ are such that $\Xi - \tilde{\Xi} \in C_2^\beta$ for some $\beta > 1$, i.e.

$$|\Xi_{s,t} - \tilde{\Xi}_{s,t}| = O(|t - s|^\beta) ,$$

then $\mathcal{I}\Xi = \mathcal{I}\tilde{\Xi}$. Indeed, it is immediate that

$$\sum_{[u,v]\in\mathcal{P}} |\Xi_{u,v} - \tilde{\Xi}_{u,v}| = O(|\mathcal{P}|^{\beta-1}) ,$$

which converges to 0 as $|\mathcal{P}| \to 0$. (This remains true if $O(|t - s|^\beta)$ with $\beta > 1$ is replaced by $o(|t - s|)$.)

This also shows that, if X and Y are smooth functions and $\mathbb{X}$ is defined by (2.2), the integral that we just defined does coincide with the usual Riemann–Stieltjes integral. However, if we change $\mathbb{X}$, then the resulting integral *does* change, as will be seen in the next example.

Example 4.14. Let f be a 2α-Hölder continuous function and let $\mathbf{X} = (X, \mathbb{X})$ and $\bar{\mathbf{X}} = (\bar{X}, \bar{\mathbb{X}})$ be two rough paths such that

$$\bar{X}_t = X_t , \qquad \bar{\mathbb{X}}_{s,t} = \mathbb{X}_{s,t} + f(t) - f(s) .$$

Let furthermore $(Y, Y') \in \mathscr{D}_X^{2\alpha}$ as above. Then also $(\bar{Y}, \bar{Y}') := (Y, Y') \in \mathscr{D}_{\bar{X}}^{2\alpha}$. However, it follows immediately from (4.21) that

$$\int_s^t \bar{Y}_r\,d\bar{\mathbf{X}}_r = \int_s^t Y_r\,d\mathbf{X}_r + \int_s^t Y'_r\,df(r) . \quad (4.26)$$

Here, the second term on the right-hand side is a simple Young integral, which is well-defined since $\alpha + 2\alpha > 1$ by assumption.

Remark 4.15. As we will see in Section 5.2 below, (4.26) can be interpreted as a generalisation of the usual expression relating Itô integrals to Stratonovich integrals.

Remark 4.16. The bound (4.22) does behave in a very natural way under dilations. Indeed, the integral is invariant under the transformation

$$(Y, Y', X, \mathbb{X}) \mapsto (\lambda^{-1} Y, \lambda^{-2} Y', \lambda X, \lambda^2 \mathbb{X}) . \tag{4.27}$$

The same is true for the right-hand side of (4.22), since under this dilation, we also have $R^Y \mapsto \lambda^{-1} R^Y$.

4.4 Stability I: rough integration

Consider $\mathbf{X} = (X, \mathbb{X})$, $\tilde{\mathbf{X}} = (\tilde{X}, \tilde{\mathbb{X}}) \in \mathscr{C}^\alpha$ with $(Y, Y') \in \mathscr{D}_X^{2\alpha}$, $(\tilde{Y}, \tilde{Y}') \in \mathscr{D}_{\tilde{X}}^{2\alpha}$. As earlier, we consider a fixed time horizon $[0, T]$. Although (Y, Y') and $(\tilde{Y}, \tilde{Y}')$ live, in general, in different Banach spaces, the "distance"

$$\|Y, Y'; \tilde{Y}, \tilde{Y}'\|_{X, \tilde{X}, 2\alpha} \overset{\text{def}}{=} \|Y' - \tilde{Y}'\|_\alpha + \|R^Y - R^{\tilde{Y}}\|_{2\alpha} \tag{4.28}$$

will be useful. Even when $X = \tilde{X}$, it is not a proper metric for it fails to separate (Y, Y') and $(Y + cX + \bar{c}, Y' + c)$ for any two constants c and $\bar{c}$. When $X \neq \tilde{X}$, the assertion "zero distance implies $(Y, Y') = (\tilde{Y}, \tilde{Y}')$" does not even make sense. (The two objects live in completely different spaces!) That said, for every fixed $(X, \mathbb{X}) \in \mathscr{C}^\alpha$, one has (with $R^Y_{s,t} = Y_{s,t} - Y'_s X_{s,t}$ as usual), a canonical map

$$\iota_X : (Y, Y') \in \mathcal{C}_X^\alpha \mapsto (Y', R^Y) \in \mathcal{C}^\alpha \oplus \mathcal{C}_2^{2\alpha} .$$

Given $Y_0 = \xi$, this map is injective since one can reconstruct Y by $Y_t = \xi + Y'_0 X_{0,t} + R^Y_{0,t}$. From this point of view, one simply has

$$\|\bullet ; *\|_{X, \tilde{X}, 2\alpha} = \|\iota_X(\bullet) - \iota_{\tilde{X}}(*)\|_{\alpha, 2\alpha} ,$$

and one is back in a normal Banach setting, where $\|\bullet, \bullet\|_{\alpha, 2\alpha} = \|\bullet\|_\alpha + \|\bullet\|_{2\alpha}$ is a natural seminorm on $\mathcal{C}^\alpha \oplus \mathcal{C}_2^{2\alpha}$; cf. Exercise 2.7. Elementary estimates of the form

$$|ab - \tilde{a}\tilde{b}| \leq |a| \, |b - \tilde{b}| + |a - \tilde{a}| \, |\tilde{b}| \tag{4.29}$$

then lead to, with a constant $C = C_R$,

$$
\begin{aligned}
|Y_{s,t} - \tilde{Y}_{s,t}| &= \left| (Y'_{0,s} - Y'_0) X_{s,t} + (\tilde{Y}'_{0,s} + \tilde{Y}'_0) \tilde{X}_{s,t} + R^Y_{s,t} - R^{\tilde{Y}}_{s,t} \right| \\
&\leq C |t - s|^\alpha \left(|Y'_0 - \tilde{Y}'_0| + \|X - \tilde{X}\|_\alpha + \|Y'_{0,\cdot} - \tilde{Y}'_{0,\cdot}\|_\infty + \|R^Y - R^{\tilde{Y}}\|_\alpha \right) \\
&\leq C |t - s|^\alpha \left(|Y'_0 - \tilde{Y}'_0| + \|X - \tilde{X}\|_\alpha + T^\alpha \left(\|Y' - \tilde{Y}'\|_\alpha + \|R^Y - R^{\tilde{Y}}\|_{2\alpha} \right) \right) ,
\end{aligned}
$$

provided $|Y_0'|, \|Y'\|_\infty, \|X\|_\alpha$, and also with tilde, are bounded by R. It follows that

$$\|Y - \tilde{Y}\|_\alpha \leq C\Big(\|X - \tilde{X}\|_\alpha + |Y_0' - \tilde{Y}_0'| + T^\alpha \|Y,Y'; \tilde{Y}, \tilde{Y}'\|_{X,\tilde{X},2\alpha}\Big) . \quad (4.30)$$

An estimate of the proper α-Hölder norm of $Y - \tilde{Y}$ (rather than its seminorm) is obtained by adding $|Y_0 - \tilde{Y}_0|$ to both sides.

Theorem 4.17 (Stability of rough integration). *For $\alpha \in \big(\frac{1}{3}, \frac{1}{2}\big]$ as before, consider* $\mathbf{X} = (X, \mathbb{X}), \tilde{\mathbf{X}} = (\tilde{X}, \tilde{\mathbb{X}}) \in \mathscr{C}^\alpha$, $(Y, Y') \in \mathscr{D}_X^{2\alpha}, (\tilde{Y}, \tilde{Y}') \in \mathscr{D}_{\tilde{X}}^{2\alpha}$ *in a bounded set, in the sense*

$$|Y_0'| + \|Y, Y'\|_{X,2\alpha} \leq M, \quad \varrho_\alpha(0, \mathbf{X}) \equiv \|X\|_\alpha + \|\mathbb{X}\|_{2\alpha} \leq M,$$

with identical bounds for $(\tilde{X}, \tilde{\mathbb{X}})$, $(\tilde{Y}, \tilde{Y}')$, for some $M < \infty$. Define

$$(Z, Z') := \left(\int_0^\cdot Y \, d\mathbf{X}, Y\right) \in \mathscr{D}_X^{2\alpha} ,$$

and similarly for $(\tilde{Z}, \tilde{Z}')$. Then, the following local Lipschitz estimates holds true,

$$\|Z, Z'; \tilde{Z}, \tilde{Z}'\|_{X,\tilde{X},2\alpha} \leq C\Big(\varrho_\alpha(\mathbf{X}, \tilde{\mathbf{X}}) + |Y_0' - \tilde{Y}_0'| + T^\alpha \|Y, Y'; \tilde{Y}, \tilde{Y}'\|_{X,\tilde{X},2\alpha}\Big),$$
$$(4.31)$$

and also

$$\|Z - \tilde{Z}\|_\alpha \leq C\Big(\varrho_\alpha(\mathbf{X}, \tilde{\mathbf{X}}) + |Y_0 - \tilde{Y}_0| + |Y_0' - \tilde{Y}_0'| + T^\alpha \|Y, Y'; \tilde{Y}, \tilde{Y}'\|_{X,\tilde{X},2\alpha}\Big),$$
$$(4.32)$$

where $C = C_M = C(M, \alpha)$ is a suitable constant.

Proof. (The reader is advised to review the proofs of Theorems 4.4, 4.10.) We first note that (4.30) applied to $Z, \tilde{Z}$ (note: $Z_0' - \tilde{Z}_0' = Y_0 - \tilde{Y}$) shows that (4.32) is an immediate consequence of the first estimate (4.31). Thus, we only need to discuss the first estimate. By definition of $d_{X,\tilde{X},2\alpha}$, we need to estimate

$$\|Z' - \tilde{Z}'\|_\alpha + \|R^Z - R^{\tilde{Z}}\|_{2\alpha} = \|Y - \tilde{Y}\|_\alpha + \|R^Z - R^{\tilde{Z}}\|_{2\alpha}.$$

Thanks to (4.30), the first summand is clearly bounded by the right-hand side of (4.31). For the second summand we recall

$$R_{s,t}^Z = Z_{s,t} - Z_s' X_{s,t} = \int_s^t Y \, d\mathbf{X} - Y_s X_{s,t} = (\mathcal{I}\Xi)_{s,t} - \Xi_{s,t} + Y_s' \mathbb{X}_{s,t}$$

where $\Xi_{s,t} = Y_s X_{s,t} + Y_s' \mathbb{X}_{s,t}$ and similar for $R^{\tilde{Z}}$. Setting $\Delta = \Xi - \tilde{\Xi}$, we use (4.11) with $\beta = 3\alpha$ and Ξ replaced by Δ, so that

$$|R_{s,t}^Z - R_{s,t}^{\tilde{Z}}| = |(\mathcal{I}\Delta)_{s,t} - \Delta_{s,t}| + |Y_s' \mathbb{X}_{s,t} - \tilde{Y}_s' \tilde{\mathbb{X}}_{s,t}|$$

$$\leq C\|\delta\Delta\|_{3\alpha}|t-s|^{3\alpha} + \left|Y_s'\mathbb{X}_{s,t} - \tilde{Y}_s'\tilde{\mathbb{X}}_{s,t}\right|,$$

where $\delta\Delta_{s,u,t} = R_{s,u}^{\tilde{Y}}\tilde{X}_{u,t} - R_{s,u}^Y X_{u,t} + \tilde{Y}_{s,u}'\tilde{\mathbb{X}}_{u,t} - Y_{s,u}'\mathbb{X}_{u,t}$. We then conclude with some elementary estimates of the type (4.29), just like in the proof of Theorem 4.10. □

4.5 Controlled rough paths of lower regularity

Recall that we showed in Section 2.3 how an α-Hölder rough path $\mathbf{X}$ could be defined as a path with values in the free step-N nilpotent Lie group $G^{(N)}(\mathbf{R}^d) \subset T^{(N)}(\mathbf{R}^d)$, with $N = \lfloor 1/\alpha \rfloor$. It does not seem obvious at all a priori how one would define a controlled rough path in this context. One way of interpreting Definition 4.6 is as a kind of local "Taylor expansion" up to order 2α. It seems natural in the light of the previous subsections that if $\alpha \leq \frac{1}{3}$, a controlled rough path should have a kind of "Taylor expansion" up to order $N\alpha$.

As a consequence, if we expand $\mathbf{X}_{s,t} \overset{\text{def}}{=} \mathbf{X}_s^{-1} \otimes \mathbf{X}_t$ as

$$\mathbf{X}_{s,t} = \sum_{|w| \leq N} \mathbf{X}_{s,t}^w\, e_w\,,$$

where $|w|$ denotes the length of the word w, one would expect that a controlled rough path should have an expansion of the form

$$\delta Y_{s,t} = \sum_{|w| \leq N-1} Y_s^w\, \mathbf{X}_{s,t}^w + R_{s,t}^Y\,, \tag{4.33}$$

with $|R_{s,t}^Y| \lesssim |t-s|^{N\alpha}$. Here, given a word $w = w_1 \cdots w_k$ with letters in $\{1, \ldots, d\}$, we write $e_w = e_1 \otimes \ldots \otimes e_k$ for the corresponding basis vector of $T^{(N)}(\mathbf{R}^d)$. As in Section 2.4, we then identify the words themselves as the dual basis of $T^{(N)}(\mathbf{R}^d)^*$. Note that $e_\emptyset = 1 \in \mathbf{R} \simeq (\mathbf{R}^d)^{\otimes 0} \subset T^{(N)}(\mathbf{R}^d)$.

Recall that in Definition 4.6 we also needed a regularity condition on the "derivative process" Y'. The equivalent statement in the present context is that the Y_s^w should themselves be described by a local "Taylor expansion", but this time only up to order $(N - |w|)\alpha$. A neat way of packaging this into a compact statement is to view a controlled rough path as a $T^{(N-1)}(\mathbf{R}^d)^*$-valued function. Definition 4.6 then generalises as follows.[8]

Definition 4.18. Let $\alpha \in (0,1)$, let $N = \lfloor 1/\alpha \rfloor$, and let $\mathbf{X}$ be a geometric α-Hölder rough path as defined in Section 2.4. A *controlled* rough path is a $T^{(N-1)}(\mathbf{R}^d)^*$-valued function $\mathbf{Y}$ such that, for every word w with $|w| \leq N - 1$, one has the bound

$$\left|\langle e_w, \mathbf{Y}_t \rangle - \langle \mathbf{X}_{s,t} \otimes e_w, \mathbf{Y}_s \rangle\right| \leq C|t-s|^{(N-|w|)\alpha}\,. \tag{4.34}$$

---

[8] This is for Y with values in $\mathbf{R}$, but the extension to vector-valued Y is straightforward.

We call Y a lift of $Y_t := \langle e_\emptyset, \mathbf{Y}_t \rangle$ and write $\mathscr{D}_{\mathbf{X}}^{N\alpha}$ for the space of such controlled rough paths.

It is convenient to write $\mathbf{Y}_t^w$ instead of $\langle e_w, \mathbf{Y}_t \rangle$. Given such a controlled rough path Y, it is then natural to define its integral against any component X^i by

$$Z_t = \int_0^t Y_s \, dX_s^i \overset{\text{def}}{=} \lim_{|\mathcal{P}| \to 0} \sum_{[r,s] \in \mathcal{P}} \sum_{|w| \le N-1} \mathbf{Y}_r^w \langle \mathbf{X}_{r,s}, wi \rangle , \tag{4.35}$$

where wi denotes the concatenation of w with the letter i. It turns out [Gub10, HK15] that Z can be lifted as controlled rough path Z in the sense of Definition 4.18. It suffices to set $\mathbf{Z}_t^\emptyset = \langle e_\emptyset, \mathbf{Z}_t \rangle \overset{\text{def}}{=} Z_t$,

$$\langle e_w \otimes e_i, \mathbf{Z}_t \rangle \overset{\text{def}}{=} \mathbf{Y}_t^w ,$$

and $\mathbf{Z}_t^w = 0$ for all non-empty words w that do not terminate with the letter i.

4.6 Stochastic sewing

We saw in Theorem 4.10 that suitably controlled integrands, such as $F(B), F \in \mathcal{C}_b^2$ can be integrated against a Brownian rough path $\mathbf{B} = (B, \mathbb{B})$, as constructed in Chapter 3. In this case (see the proof of Theorem 4.4) one applies the sewing lemma with $\tilde{\Xi}(s,t) = F(B_s)B_{s,t} + DF(B_s)\mathbb{B}_{s,t}$, crucially using that $\delta\tilde{\Xi}$ is of order $3\alpha = 1 + \varepsilon > 1$, in the sense that $|\delta\tilde{\Xi}_{sut}| \lesssim |t - s|^{1+\varepsilon}$ uniformly over $s < u < t$ in $[0, T]$. We leave it to Chapter 5 to reconcile this construction *a posteriori* with classical stochastic integration. In the present section we show that stochastic and rough analysis can also be combined *a priori*; the resulting *stochastic sewing lemma* obtained by K. Lê in [Lê18] has proved very useful in a number of recent applications.

The setting is similar as in the sewing lemma, but the to-be-sewed two-parameter function Ξ is now a sufficiently integrable random field. As running example, consider the Itô left point approximation $\Xi_{s,t} = F(B_s)B_{s,t}$. With this choice of Ξ (i.e. without the term $DF(B_s)\mathbb{B}_{s,t}$), classical sewing fails since $\delta\Xi_{s,u,t} = -F(B)_{s,u}B_{u,t}$ is at best of order $2\alpha < 1$. Note however that the martingale property of Brownian motion makes this problem disappear upon inserting a conditional expectation. Indeed, writing $\mathbf{E}_s$ for the conditional expectation with respect to $\mathcal{F}_s$ for some fixed filtration $\mathcal{F} = (\mathcal{F}_t)_{t \le T}$ such that B is $\mathcal{F}$-adapted we have, always with $s < u < t$,

$$\mathbf{E}_s \delta\Xi_{sut} = \mathbf{E}_s \mathbf{E}_u \delta\Xi_{sut} = -\mathbf{E}_s \left(F(B)_{s,u} \mathbf{E}_u B_{u,t} \right) = 0 .$$

This is of course very similar to the reason why classical Itô integration works: even though $\Xi_{s,t}$ is of size about $|t - s|^{1/2}$ so that there is no reason a priori to believe that Riemann sums converge, they do so thanks to the stochastic cancellations encoded in the fact that $\mathbf{E}_s \Xi_{s,t} = 0$. The idea now is to obtain a version of the sewing lemma which combines the "best of both worlds": its assumptions should be strictly weaker than those of Lemma 4.2 and it should exploit improvements from situations in which the conditional expectation of an expression is much smaller than the expression itself.

Throughout this section, we assume that we are working with L^2 random variables on a filtered probability space $(\Omega, (\mathcal{F}_t)_{0 \leq t \leq T}, \mathbf{P})$ and we write L_s^2 for the space of $\mathcal{F}_s$-measurable square integrable random variables. We also write as usual $\|X\|_{L^2} \overset{\text{def}}{=} (\mathbf{E}X^2)^{1/2}$. In fact, using the Burkholder–Davis–Gundy inequality, it is not difficult to extend the following results to an L^q setting with $2 \leq q < \infty$.

Proposition 4.19 (Stochastic Sewing Lemma). *Let $(s,t) \mapsto \Xi_{s,t} \in L_t^2$ for $0 \leq s \leq t \leq T$ be continuous (viewed as a map with values in L^2) with $\Xi_{t,t} = 0$ for all t. Suppose that there are constants $\Gamma_1, \Gamma_2 \geq 0$ and $\varepsilon_1, \varepsilon_2 > 0$ such that for all $0 \leq s \leq u \leq t \leq T$,*

$$\|\delta\Xi_{sut}\|_{L^2} \leq \Gamma_1 |t - s|^{\frac{1}{2} + \varepsilon_1}. \tag{4.36}$$

and

$$\|\mathbf{E}_s \delta\Xi_{sut}\|_{L^2} \leq \Gamma_2 |t - s|^{1 + \varepsilon_2}, \tag{4.37}$$

Then there exists a unique continuous (again as a map $[0, T] \to L^2$) process $t \mapsto X_t \in L_t^2$ with $X_0 = 0$ and a suitable constant C such that, for all $0 \leq s \leq t \leq T$,

$$\|X_t - X_s - \Xi_{s,t}\|_{L^2} \leq C\Gamma_1 |t - s|^{\frac{1}{2} + \varepsilon_1} + C\Gamma_2 |t - s|^{1 + \varepsilon_2} \tag{4.38}$$

and

$$\|\mathbf{E}_s(X_t - X_s - \Xi_{s,t})\|_{L^2} \leq C\Gamma_2 |t - s|^{1 + \varepsilon_2}. \tag{4.39}$$

Proof. (Uniqueness) Assuming there are two adapted processes $X, \bar{X}$ with the stated properties (4.38) and (4.39), we show that $\Delta_t := X_t - \bar{X}_t = 0$ almost surely for every t. Let n be a positive integer and set $t_i = ti/n$. The abusive notation $X_i := X_{t_i, t_{i+1}}$ and similarly for Δ and Ξ is convenient. Note that L^2 estimates for $\Delta_i = (X_i - \Xi_i) - (\bar{X}_i - \Xi_i)$, as well as $\mathbf{E}_{t_i}\Delta_i$ are immediate from (4.38) and (4.39). We have

$$\Delta_t = \sum_{i=0}^{n-1}(\Delta_i - \mathbf{E}_{t_i}\Delta_i) + \sum_{i=0}^{n-1}\mathbf{E}_{t_i}\Delta_i =: \Delta_t^{(1)} + \Delta_t^{(2)},$$

which is nothing but Doob's decomposition of the partial sum process $\sum_i \Delta_i$ into martingale and predictable component. Using the orthogonality of martingale increments, L^2-contraction property of the conditional expectation, and (4.38), we have

$$\|\Delta_t^{(1)}\|_{L^2} = \left(\sum_{i=0}^{n-1} \|(\Delta_i - \mathbf{E}_{t_i}\Delta_i)\|_{L_2}^2 \right)^{\frac{1}{2}} \leq 2\left(\sum_{i=0}^{n-1} \|\Delta_i\|_{L_2}^2 \right)^{\frac{1}{2}}$$

$$\lesssim n^{1/2} \cdot \left(\frac{1}{n}\right)^{1/2 + \varepsilon_1}.$$

Since n is arbitrary, it follows that $\Delta_t^{(1)} = 0$ a.s. The same conclusion for $\Delta_t^{(2)}$ is immediate from the triangle inequality and (4.39), since

$$\|\Delta_t^{(2)}\|_{L^2} \le \sum_i \|\mathbf{E}_{t_i} \Delta_i\|_{L^2} \lesssim n \cdot \left(\frac{1}{n}\right)^{1+\varepsilon_2} .$$

(Existence) The proof follows the "dyadic refinement" proof of the sewing lemma given earlier. Fix $0 \le s < t \le T$ and consider dyadic refinements (t_i^k) of $[s,t]$, so that the kth level approximation is given by

$$I_{s,t}^k = \sum_{i=0}^{2^k-1} \Xi_{t_i^k, t_{i+1}^k} \in L_t^2 .$$

With midpoint $u_i^k \in [t_i^k, t_{i+1}^k]$ and, for fixed k, $\delta\Xi_i := \delta\Xi_{t_i^k, u_i^k, t_{i+1}^k}$, we again work with the Doob decomposition

$$I_{s,t}^{k+1} - I_{s,t}^k = \sum_{i=0}^{2^k-1} \delta\Xi_i = I_{s,t}^{k;(1)} + I_{s,t}^{k;(2)} . \tag{4.40}$$

Arguing as in the uniqueness part, the first (resp. second) term is estimated (in L^2) with (4.36) (resp. (4.37)) and one arrives at

$$\|I_{s,t}^{k+1} - I_{s,t}^k\|_{L^2} \lesssim |t-s|^{\frac{1}{2}+\varepsilon_1} 2^{-k\varepsilon_1} + |t-s|^{1+\varepsilon_2} 2^{-k\varepsilon_2} .$$

which implies existence of $I_{s,t} := \lim_{k\to\infty} I_{s,t}^k$ in L_t^2, uniformly in $0 \le s \le t \le T$, with a local estimate of the form (4.38) with $X_t - X_s$ replaced by $I_{s,t}$. (By assumption Ξ, and hence all I^k, are L^2-continuous, and so is the uniform limit I.) Moreover, since $\mathbf{E}_s I_{s,t}^{k;(1)} = 0$, for all k, a better estimate, of the form (4.39), is obtained for $\mathbf{E}_s I_{s,t} = \lim_{k\to\infty} \mathbf{E}_s I_{s,t}^k$. At last, as in the "dyadic" proof of the deterministic sewing lemma, one needs to argue that I is additive, a non-trivial exercise left to the reader, and hence the increment of a unique L^2-path I started from $I_0 = 0$ which is nothing but the desired square-integrable process $X = X(t, \omega)$. $\quad\square$

4.7 Exercises

Exercise 4.1 *a) In the setting of Young integration, deduce (4.3) from (4.2).*
b) Show that there is a constant C depending only on $T > 0$ and $\alpha + \beta > 1$ such that

$$\left\|\int_0^{\cdot} Y\,dX\right\|_{\alpha;[0,T]} \le C\Big(|Y_0| + \|Y\|_{\beta;[0,T]}\Big)\|X\|_{\alpha;[0,T]}. \tag{4.41}$$

In fact, show that C can be chosen uniformly over $T \in (0, 1]$.

Solution.   a) Given X on $[s, t]$, define $\tilde{X} : [0, 1] \ni u \mapsto X(s + u(t - s))$ and verify $\|\tilde{X}\|_{\alpha;[0,1]} = |t - s|^{\beta} \|X\|_{\beta;[s,t]}$. Proceeding similarly for Y, applying (4.2) to $\tilde{X}, \tilde{Y}$ then gives (4.3).

  b)   Write Z for the indefinite integral. From (4.3), for every $0 \le s < t \le T$,

$$|Z_{s,t}| \le |Y_s||X_{s,t}| + C\|Y\|_{\beta;[s,t]}\|X\|_{\alpha;[s,t]}|t - s|^{\alpha+\beta}$$

$$\le \left(|Y_0| + \|Y\|_{\beta;[0,T]}T^{\beta}\right)|X_{s,t}| + C\|Y\|_{\beta;[0,T]}\|X\|_{\alpha;[0,T]}T^{\beta}|t - s|^{\alpha}$$

$$\le \left[|Y_0| + \|Y\|_{\beta;[0,T]}T^{\beta}(1 + C)\right]\|X\|_{\alpha;[0,T]}|t - s|^{\alpha}.$$

$$\le (1 \vee T)^{\beta}(1 + C)\left[|Y_0| + \|Y\|_{\beta;[0,T]}\right]\|X\|_{\alpha;[0,T]}|t - s|^{\alpha},$$

and this entails the claimed estimates.

Exercise 4.2 *Let* $\mathbf{X} = (X, \mathbb{X}) \in \mathscr{C}^{\alpha}([0,T], V)$, $\alpha \in \left(\frac{1}{3}, \frac{1}{2}\right]$, *and assume that* $F : V \to \mathcal{L}(V, W)$ *is of gradient form, i.e.* $F = DG$ *where* $G : V \to W$ *is sufficiently smooth, say* $\mathcal{C}_b^3$. *Show that the relation*

$$\int_s^t F(X)d\mathbf{X} = G(X_t) - G(X_s),$$

holds true whenever $\mathbf{X}$ *is a geometric rough path. (Hence, from a rough path perspective, integration of gradient 1-forms against geometric rough paths is trivial for the outcome does not depend on* $\mathbb{X}$.) *What about non-geometric rough paths?*

Exercise 4.3 *Complete the first "dyadic" proof of the sewing Lemma 4.2.*

Solution. To show that (4.9) is valid for all intervals $[s, t] \subset [0, 1]$ it suffices to consider $s < t$ dyadic by continuity. As in the proof of the Kolmogorov criterion, Theorem 3.1, we consider a (finite) partition $P = (\tau_i)$ of $[s, t]$, which "efficiently" exhausts $[s, t]$ with dyadic intervals of length $\sim 2^{-n}, n \ge m$, in the sense that no three intervals have the same length. Note that $|P| \equiv \max\{|v - u| : [v, u] \in P\} = 2^{-m} \le |t - s|$ (and in fact $\sim |t - s|$ due to minimal choice of m). Thanks to the additivity of I and (4.9) for dyadic intervals,

$$|I_{s,t} - \Xi_{s,t}| = \left| \sum_{[u,v]\in P} (I_{u,v} - \Xi_{u,v}) - \left(\Xi_{s,t} - \sum_{[u,v]\in P} \Xi_{u,v}\right) \right|$$

$$\lesssim \sum_{[u,v]\in P} |v - u|^{\beta} + \left(\Xi_{s,t} - \sum_{[u,v]\in P} \Xi_{u,v}\right).$$

$$\le |t - s|^{\beta} + \sum_{i=0}^{\infty} \left|\delta\Xi_{s,\tau_{-(i+1)},\tau_{-i}} + \delta\Xi_{\tau_i,\tau_{i+1},t}\right|,$$

where the sum is actually finite. Possibly allowing equality ("$\tau_i = \tau_{i+1}$" for some i), we may assume $|\tau_{i+1} - \tau_i| = |\tau_{-i} - \tau_{-(i+1)}| \lesssim 1/2^{m+i}$, so that

$$|t - \tau_i| = \sum_{j=i}^{\infty} |\tau_{j+1} - \tau_j| \lesssim \sum_{j=i}^{\infty} 1/2^{m+j} \sim 1/2^{m+i} \ ,$$

and similarly, $|\tau_{-i} - s| \lesssim 1/2^{m+i}$. As a consequence, one obtains

$$\sum_{i=0}^{\infty} \left| \delta \Xi_{s,\tau_{-(i+1)},\tau_{-i}} + \delta \Xi_{\tau_i,\tau_{i+1},t} \right| \lesssim 2 \sum_{n \geq m} (1/2^n)^{\beta} \sim 1/2^{m\beta} \sim |t - s|^{\beta} \ ,$$

so that $|I_{s,t} - \Xi_{s,t}| \lesssim |t - s|^{\beta}$, as required.

Exercise 4.4 *Adapt the proof of Theorem 4.4 to obtain Young's estimate (4.3).*

Exercise 4.5 *Fix $\alpha \in (0,1], h > 0$ and $M > 0$. Consider a path $Z : [0,T] \to V$ and show that*

$$\|Z\|_{\alpha;h} \equiv \sup_{\substack{0 \leq s < t \leq T \\ t-s \leq h}} \frac{|Z_{s,t}|}{|t-s|^{\alpha}} \leq M \implies \|Z\|_{\alpha;[0,T]} \leq M \left(1 \vee 2h^{-(1-\alpha)} \right).$$

Solution. By scaling it suffices to consider $M = 1$. Fix $0 \leq s < t \leq T$, we need to show $|Z_{s,t}|/|t-s|^{\alpha}$ is bounded by $1 \vee 2h^{\alpha-1}$. There is nothing to show for $|t-s| \leq h$. We therefore assume $h \leq |t-s|$ and define $t_i = (s+ih) \wedge t$, for $i = 0, 1, \ldots$ noting that $t_N = t$ for $N \geq |t-s|/h$ and also $t_{i+1} - t_i \leq h$ for all i. It then suffices to estimate

$$|Z_{s,t}| \leq \sum_{0 \leq i < |t-s|/h} |Z_{t_i,t_{i+1}}|$$
$$\leq h^{\alpha}(1 + |t-s|/h) = h^{\alpha-1}(h + |t-s|) \leq 2h^{\alpha-1}|t-s|.$$

♯ **Exercise 4.6 (Lyons extension theorem)** *a) Let $X \in \mathcal{C}^1([0,T],V)$, so that the Lipschitz seminorm $\|X\|_1$ is finite, and consider the n-fold iterated (Riemann–Stieltjes) integral with values in $V^{\otimes n}$,*

$$X_{s,t}^{(n)} = \int_{s<t_1<\ldots<t_n<t} dX \otimes \ldots \otimes dX \ .$$

Show that, with $C_n^n = \frac{1}{n!}$, and for all $0 \leq s \leq t \leq T$,

$$|X_{s,t}^{(n)}|^{\frac{1}{n}} \leq C_n \|X\|_1 |t-s| \ .$$

b) Show an analogous result in the Young case i.e. when $X \in \mathcal{C}^{\alpha}([0,T],V), \alpha > \frac{1}{2}$.

c) Fix $\mathbf{X} = (X, \mathbb{X}) \in \mathscr{C}^{\alpha}([0,T],V), \alpha \in (\frac{1}{3}, \frac{1}{2}]$, and define $\mathbf{X}_{s,t}^{(n)} \in V^{\otimes n}$, any $n \geq 1$, by the right-hand side above, via iterated integration of controlled rough paths. Noting $(\mathbf{X}^{(1)}, \mathbf{X}^{(2)}) = (\delta X, \mathbb{X})$, define the $T^{(N)}(V)$-valued extension of $\mathbf{X}$ by

$$\bar{\mathbf{X}} := (1, \mathbf{X}^{(1)}, \mathbf{X}^{(2)}, \mathbf{X}^{(3)}, \ldots, \mathbf{X}^{(N)}) \ ,$$

for any integer $N > \lfloor \frac{1}{\alpha} \rfloor = 2$. Show the validity of Chen's relation, i.e. $\mathbf{X}_{s,t} = \mathbf{X}_{s,u} \otimes \mathbf{X}_{u,t}$, $0 \le s < u < t \le T$, as equation in $T^{(N)}(V)$, and the estimate

$$|\mathbf{X}_{s,t}^{(n)}|^{\frac{1}{n}} \le C_{n,\alpha} \|\mathbf{X}\|_{\alpha} |t - s|^{\alpha} ,$$

for $0 \le s < t \le T$ and $n = 1, \dots, N$. Show that these properties uniquely determine $\bar{\mathbf{X}}$, called (level-N) Lyons lift of $\mathbf{X}$. Show that Lyons' extension map $\mathbf{X} \mapsto \bar{\mathbf{X}}$ is continuous in the appropriate rough path spaces. Is $\bar{\mathbf{X}}$ geometric when $\mathbf{X}$ is?

Hint: *Use induction for the analytic estimate. To get started, note $(\mathbb{X}_{s,\cdot}, X_{s,\cdot}) \in \mathscr{D}_X^{2\alpha}([s,t])$ and, with all norms on $[s,t]$, one has $\|\mathbb{X}_{s,\cdot}, X_{s,\cdot}\|_{2\alpha,X} \equiv \|X_{s,\cdot}\|_{\alpha} + \|R^{\mathbb{X}_{s,\cdot}}\|_{2\alpha} = \|X\|_{\alpha} + \|\mathbb{X}\|_{2\alpha}$. Then*

$$\mathbf{X}_{s,t}^{(3)} = \int_s^t \mathbb{X}_{s,\cdot} \otimes d\mathbf{X} = \int_0^1 \hat{\mathbb{X}}_{0,\tau} \otimes d\hat{\mathbf{X}}_{\tau} = c^3 \int_0^1 \tilde{\mathbb{X}}_{0,\tau} \otimes d\tilde{\mathbf{X}}_{\tau} ,$$

in terms of $\hat{\mathbf{X}} : \tau \mapsto \mathbf{X}(s + \tau(t - s))$, noting $\|\hat{\mathbf{X}}\|_{\alpha;[0,1]} = \|\mathbf{X}\|_{\alpha;[s,t]} |t - s|^{\alpha} =: c$, and then "unit size" $\tilde{\mathbf{X}} = \delta_{1/c} \hat{\mathbf{X}}$, with "$\asymp$ 1-estimate" for the final rough integral.

Remark: *One knows that $C_{n,\alpha}^n$ is of order $1/(n\alpha)! = 1/\Gamma(n\alpha+1)$ as a consequence of the Lyons–Hara–Hino neo-classical inequality [Lyo98, HH10]. For continuity of the extension map, uniform over $n \in \mathbf{N}$, see also [LX13]. For extensions to branched rough paths see [Gub10, Boe18].*

Exercise 4.7 *Show that the assumption on $Y \in \mathscr{D}_X^{2\alpha}$ can be weakend to $Y \in \mathscr{D}_X^{2\alpha'}$, $\alpha' < \alpha$, provided $\alpha + 2\alpha' > 1$, and reformulate Theorem 4.10 accordingly. In particular, show that the estimate (4.22) holds upon replacing the final factor $|t - s|^{3\alpha}$ by $|t - s|^{\alpha+2\alpha'}$, and $\|Y'\|_{\alpha}$ (resp. $\|R^Y\|_{2\alpha}$) by $\|Y'\|_{\alpha'}$ (resp. $\|R^Y\|_{2\alpha'}$).*

** Exercise 4.8 (Approximation of controlled rough paths) *Let $\alpha \in \left(\frac{1}{3}, \frac{1}{2}\right)$. Assume $X \in \mathcal{C}^{\alpha}$ and $(Y, Y') \in \mathscr{D}_X^{2\alpha}$. Consider smooth approximations X_{ε} such that $X_{\varepsilon} \to X$ in $\mathcal{C}^{\alpha}$. Show that there then exist smooth paths*

$$(Y_{\varepsilon}, Y_{\varepsilon}') \in \mathscr{D}_{X_{\varepsilon}}^{2\alpha}$$

such that $(Y_{\varepsilon}, Y_{\varepsilon}') \to (Y, Y')$ uniformly with uniform bounds in $\mathscr{D}_{X_{\varepsilon}}^{2\alpha}$. By interpolation, for any $\alpha' < \alpha$,

$$\|Y_{\varepsilon}' - Y'\|_{\alpha'} + \|R^{Y_{\varepsilon}} - R^Y\|_{2\alpha'} \to 0.$$

(Such an approximation result was first suggested in [GH19, Rem 5.5], for a generalisation to modelled distributions in the theory of regularity structures see [ST18].)

Solution. Let $\Phi : \mathcal{C}^{\alpha} \times \mathcal{C}^{\alpha} \to \mathcal{C}_2^{2\alpha}$ be the map constructed in part a) of Exercise 2.14. Set $\mathbb{Z} := \Phi(Y', X) \in \mathcal{C}_2^{2\alpha}$ and also $\bar{Y} := \mathbb{Z}_{0,\cdot} \in \mathcal{C}^{\alpha}$. From the properties of Φ ("Chen's relation")

$$\bar{Y}_t = \bar{Y}_s + Y'_s X_{s,t} + \mathbb{Z}_{s,t}$$

which shows $(\bar{Y}, Y') \in \mathscr{D}_X^{2\alpha}$. On the other hand, $(Y, Y') \in \mathscr{D}_X^{2\alpha}$ means that $Y_t = Y_s + Y'_s X_{s,t} + R^Y_{s,t}$ with remainder of order 2α. Upon taking the difference we see that $\Gamma := R^Y - \mathbb{Z} \in C_2^{2\alpha}$ can be written as

$$Y_t - Y_s - (\bar{Y}_t - \bar{Y}_s) = \Gamma_{s,t}$$

which identifies Γ as the the the increment of a path; we write $\Gamma \in C^{2\alpha}$ accordingly. Let ψ_ε be an approximation of the identity so that

$$Y'_\varepsilon := Y' * \psi_\varepsilon \in C^\infty$$

converges uniformly, with uniform α-Hölder bounds, to Y'. (By interpolation, this entails convergence in C^{α^-}.) On the other hand, thanks to part c) of Exercise 2.14,

$$\bar{Y}_\varepsilon := \Phi(Y'_\varepsilon, X_\varepsilon)_{0,\cdot} \in C^{1^-},$$

and also, thanks to the first part of that theorem, with $\bar{R}^\varepsilon := \Phi(Y'_\varepsilon, X_\varepsilon)$, uniformly in $C_2^{2\alpha}$,

$$\bar{Y}_\varepsilon(t) = \bar{Y}_\varepsilon(s) + Y'_\varepsilon(s) X_\varepsilon(s,t) + \bar{R}^\varepsilon_{s,t}.$$

By continuity of Φ, it is clear that $\bar{R}^\varepsilon \to \Phi(Y', X) \in C_2^{2\alpha}$, uniformly, with uniform 2α-Hölder bounds. (As before, this entails $C_2^{2\alpha^-}$-convergence.) It remains to deal with the (mostly cosmetic) problem that $\bar{Y}_\varepsilon$ is not smooth. But then $Y_\varepsilon := \bar{Y}_\varepsilon * \psi_\varepsilon \in C^\infty$ converges uniformly with uniform 1^--Hölder bounds and from

$$R^\varepsilon_{s,t} := Y_\varepsilon(s,t) - Y'_\varepsilon(s) X_\varepsilon(s,t) = \bar{R}^\varepsilon_{s,t} + Y_\varepsilon(s,t) - \bar{Y}_\varepsilon(s,t)$$

we see that $R^\varepsilon - \bar{R}^\varepsilon \to 0$ uniformly, also with uniform 1^--Hölder bounds (and hence $R^\varepsilon \to \Phi(Y', X)$ with uniform 2α-Hölder bounds).

* **Exercise 4.9** *For $\alpha \in (\frac{1}{3}, \frac{1}{2})$, consider the space $\mathscr{C}^\alpha \ltimes \mathscr{D}^{2\alpha}$ as in Remark 4.8 endowed with the distance*

$$d(\mathbf{X}, (Y, Y'); \bar{\mathbf{X}}, (\bar{Y}, \bar{Y}')) = \varrho_\alpha(\mathbf{X}, \bar{\mathbf{X}}) + \|Y, Y'; \tilde{Y}, \tilde{Y}'\|_{X, \tilde{X}, 2\alpha},$$

see (4.28). Show that this space is homeomorphic to $\mathscr{C}^\alpha \times (C^{2\alpha} \oplus C^\alpha)$. (Here, $C^{2\alpha}$ denotes the usual space of 2α-Hölder functions in one variable. See also [TZ18, BH19] for generalisations of this statement.)

Solution. As in the solution to the previous exercise, we can use the Lyons–Victoir extension theorem (see Exercise 2.14), to find a continuous map $\mathcal{I}: \mathscr{C}^\alpha \times C^\alpha \to C^\alpha$ with the property that $Z = \mathcal{I}(\mathbf{X}, Y')$ satisfies $(Z, Y') \in \mathscr{D}_X^{2\alpha}$. (One should think of $\mathcal{I}(\mathbf{X}, Y')$ as being a "plausible candidate" for $\int_0^\cdot Y'_s \, dX_s$, which is of course ill-defined since we do not assume that Y' is controlled by X.)

In particular, the map $\tilde{\mathcal{I}}: (\mathbf{X}, \tilde{Y}, Y') \mapsto (\mathbf{X}, \tilde{Y} + \mathcal{I}(\mathbf{X}, Y'), Y')$ is continuous from $\mathscr{C}^\alpha \times (C^{2\alpha} \oplus C^\alpha)$ to $\mathscr{C}^\alpha \ltimes \mathscr{D}^{2\alpha}$. Its inverse map is given by

$$(\mathbf{X}, Y, Y') \mapsto (\mathbf{X}, Y - \mathcal{I}(\mathbf{X}, Y'), Y') \, ,$$

which concludes the proof. Note that this construction is far from being canonical due to the lack of a canonical map $\mathcal{I}$ having these properties.

Exercise 4.10 (Rough Fubini) *Let* $\mathbf{X} = (X, \mathbb{X}) \in \mathscr{C}^{\alpha}([0, T], V), \alpha > \frac{1}{3}$ *and consider a measurable map from some measure space* $(\Omega, \mathcal{F}, \mu)$ *to* $\mathscr{D}_X^{2\alpha}$, *so that*

$$\{\omega \mapsto |Y_0^{\omega}| + \|Y^{\omega}, (Y^{\omega})'\|_{2\alpha, X}\} \in L^1(\Omega, \mathcal{F}, \mu).$$

With a pointwise definition of the μ-integrated controlled rough path on the right-hand side, show that both sides are well-defined and equality holds,

$$\int_{\Omega} \left(\int_0^T (Y^{\omega}, (Y^{\omega})') d\mathbf{X} \right) \mu(d\omega) = \int_0^T \left(\int_{\Omega} (Y^{\omega}, (Y^{\omega})') \mu(d\omega) \right) d\mathbf{X}.$$

Exercise 4.11 (Rough Fubini d'après [GH19])

a) *As a warmup, consider a real-valued càdlàg path X of bounded variation on $[0, T]$, so that integration of càglàd integrands against dX can be understood equivalently in Lebesgue or Riemann–Stieltjes sense. Write $[X]_t$ for the sumsquare of all jumps at times in $(0, t]$. Given two real-valued càglàd paths $Y, \tilde{Y}$, set $Z_{s,t} := Y_s \tilde{Y}_t$ and show*

$$\int_0^T \int_0^t Z_{s,t} dX_s dX_t = \int_0^T \int_s^T Z_{s,t} dX_t dX_s + \int_0^T Z_{t,t} d[X]_t.$$

Hint: *Apply the integration by parts formula for bounded variation paths to the indefinite integrals of Y and $\tilde{Y}$ against X.*

b) *Let now $\mathbf{X} = (X, \mathbb{X}) \in \mathscr{C}^{\alpha}([0, T])$ for some $\alpha > 1/3$, and $(Y, Y'), (\tilde{Y}, \tilde{Y}') \in \mathscr{D}_X^{2\alpha}$. Set $Z_{s,t} := Y_s \otimes Y_t$. Show that*

$$\int_0^T \int_0^t Z_{s,t} d\mathbf{X}_s d\mathbf{X}_t = \int_0^T \int_s^T Z_{s,t} d\mathbf{X}_t d\mathbf{X}_s + \int_0^T Z_{t,t} d[\mathbf{X}]_t \, ,$$

where the final integral is a Young integral against $[\mathbf{X}] \in \mathcal{C}^{2\alpha}$, the bracket introduced in Exercise 2.11.

Hint: *If $\mathbf{X}$ is the canonical lift of some smooth X, then both $[X]$ and $[\mathbf{X}]$ vanish and the equality follows from part a) and consistency of rough with Riemann–Stieltjes integration in case of smooth integrators. Treat the case of $\mathbf{X} \in \mathscr{C}_g^{\alpha}$ with the approximation result of Exercise 4.8 and then $\mathbf{X} \in \mathscr{C}^{\alpha}$ as "second level perturbation", as in Exercise 2.11.*

Exercise 4.12 (Singular rough paths, improper rough integration [BFG20])

a) *(Young case) Consider $0 < \alpha \leq 1$ and $\eta \leq \alpha$ and a path Y defined on $(0, T]$. Show that*

$$\|Y\|_{\alpha,\eta} \overset{def}{=} \sup_{0<s<t\leq T} \frac{|Y_t - Y_s|}{s^{\eta-\alpha}|t-s|^{\alpha}} < \infty$$

if and only if $\|Y\|_{\alpha;[\varepsilon,T]} = O(\varepsilon^{\eta-\alpha})$ *as* $\varepsilon \downarrow 0$, *and write* $Y \in \mathcal{C}^{\alpha,\eta}((0,T])$ *for the resulting class of "singular" Hölder paths. Fix* $X \in \mathcal{C}^{\alpha}([0,T]), \alpha > 1/2$ *and assume* $\eta + \alpha > 0, \eta \neq 0$. *Show that the improper Young integral*

$$Z_t := \int_{0+}^t Y\,dX \overset{def}{=} \lim_{\varepsilon \downarrow 0} \int_{\varepsilon}^t Y\,dX, \quad 0 < t \leq T,$$

exists and defines a singular Hölder path $Z \in \mathcal{C}^{\alpha,\eta \wedge 0 + \alpha}((0,T])$.

Hint: *For a start, apply the Young estimate*

$$\left| \int_s^t Y\,dX \right| \lesssim |Y_s||t-s|^{\alpha} + \|Y\|_{\alpha;[s,T]}|t-s|^{2\alpha}$$

with $s = 2^{-(n+1)}, t = 2^{-n}$ *and show that* $I_n := \int_{2^{-n}}^T Y\,dX$ *is a Cauchy sequence.*

b) *(Rough path case) Let* $\mathbf{X} = (X, \mathbb{X}) \in \mathscr{C}^{\alpha}([0,T])$ *for some* $\alpha > 1/3$, *and let* (Y, Y') *be defined on* $(0,T]$ *so that, for some* $\eta \leq 2\alpha$,

$$\|Y, Y'\|_{X,2\alpha;[\varepsilon,1]} = O(\varepsilon^{\eta-2\alpha}), \quad \varepsilon \downarrow 0 .$$

Show that this estimate is equivalent to finiteness of

$$\|Y, Y'\|_{X,2\alpha,\eta} \overset{def}{=} \sup_{0\leq s<t\leq T} \frac{|Y_t' - Y_s'|}{s^{\eta-2\alpha}|t-s|^{\alpha}} + \sup_{0\leq s<t\leq T} \frac{|Y_t - Y_s - Y_s' X_{s,t}|}{s^{\eta-2\alpha}|t-s|^{2\alpha}}$$

and write $(Y, Y') \in \mathscr{D}_X^{2\alpha,\eta}$ *for the resulting class of singular controlled rough paths. Show that under this condition, provided that* $-\alpha < \eta \leq 2\alpha$ *and* $\eta \neq 0$, *the improper rough integral*

$$Z_t := \int_{0+}^t Y\,d\mathbf{X} \overset{def}{=} \lim_{\varepsilon \downarrow 0} \int_{\varepsilon}^t Y\,d\mathbf{X},$$

exists and defines a singular Hölder path $Z \in \mathcal{C}^{\alpha,\eta \wedge 0 + \alpha}((0,T])$. *In fact, show that* $(Z, Z') := (\int_{0+} Y\,d\mathbf{X}, Y) \in \mathscr{D}_X^{2\alpha,\eta \wedge 0 + \alpha}$. *(Such singular controlled rough paths are examples of singular modelled distributions in the theory of regularity structures, [Hai14b, Ch. 6].)*

Exercise 4.13 *Check that Definition 4.18 is consistent with Definition 4.6 in the case when* $\alpha \in \left(\frac{1}{3}, \frac{1}{2}\right]$. *Check also that if one takes* $w = \emptyset$, *the empty word, then (4.34) reduces to (4.33) with* $|R_{s,t}^Y| \lesssim |t-s|^{N\alpha}$.

Exercise 4.14 (From [Lê18]) *Let* B *be a Brownian motion. Assume* F *is bounded and* ε-*Hölder continuous for some* $\varepsilon > 0$. *Apply the stochastic sewing lemma with*

$\Xi_{s,t} = F(B_s)B_{s,t}$ and identify the resulting process X as the indefinite Itô integral $\int F(B)dB$.

Exercise 4.15 (Hybrid stochastic rough integral) *Let B be a Brownian motion and $\mathbf{X} = (X, \mathbb{X}) \in \mathscr{C}^\alpha([0, T], V)$ a (deterministic) rough path, $\alpha \in \left(\frac{1}{3}, \frac{1}{2}\right]$. Apply the stochastic sewing lemma with*

$$\Xi_{s,t} = F(B_s + X_s)X_{s,t} + DF(B_s + X_s)\mathbb{X}_{s,t}$$

to define the stochastic rough *integral*

$$\int F(B_t + X_t)d\mathbf{X}_t \ .$$

Detail the assumptions on F. Since $\int F(B_t + X_t)dB_t$ is automatically well-defined as Itô integral this settles integration against "$B + \mathbf{X}$".

Exercise 4.16 (Mild sewing, [GT10, GH19]) *Consider a strongly continuous semigroup $(S_t)_{t\geq 0}$ acting on a scale of Hilbert spaces $(H_\alpha : \alpha \in \mathbf{R})$ with $H_\alpha \subset H_\beta$ densely whenever $\alpha \geq \beta$, such that, for all $\alpha \geq \beta$ and $\gamma \in [0, 1]$, one has*

$$\|S_t u\|_{H_\alpha} \lesssim t^{\beta-\alpha}\|u\|_{H_\beta} \ , \qquad \|S_t u - u\|_{H_{\beta-\gamma}} \lesssim t^\gamma \|u\|_{H_\beta} \ , \tag{4.42}$$

uniformly over $t \in (0, 1]$ and $u \in H_\beta$. (This situation is typical when S is an analytic semigroup, for example generated by a self-adjoint operator, cf. Section 12.2.2.) We define $\hat{\mathcal{C}}_2^{\gamma,\mu}([0, T], H_\alpha)$ as functions Ξ from the simplex $\{0 \leq s < t \leq T\}$ into H_α such that

$$\|\Xi\|_\gamma + \|\hat{\delta}\Xi\|_\mu < \infty \ ,$$

where we used the modified second order increment operator

$$\hat{\delta}\Xi_{s,u,t} := \Xi_{s,t} - S_{u,t}\Xi_{s,u} - \Xi_{u,t} \ .$$

a) *Let $0 < \gamma \leq 1 < \mu$. Show that there exists a unique continuous linear map $\mathcal{I} : \hat{\mathcal{C}}_2^{\gamma,\mu}([0, T], H_\alpha) \to \mathcal{C}^\gamma([0, T], H_\alpha)$ such that $(\mathcal{I}\Xi)_0 = 0$ and*

$$\|(\mathcal{I}\Xi)_{s,t} - \Xi_{s,t}\|_{H_\alpha} \lesssim |t - s|^\mu. \tag{4.43}$$

Hint: $(\mathcal{I}\Xi)_{s,t} = \lim_{|\mathcal{P}|\to 0}\sum_{[u,v]\in\mathcal{P}} S_{t-v}\Xi_{u,v}.$

b) *If in addition $\hat{\delta}\Xi_{u,m,v} = S_{v-m}\tilde{\Xi}_{u,m,v}$ for some H_α-valued function $\tilde{\Xi} = \tilde{\Xi}(u, m, v)$, with $0 \leq u < m < v \leq T$, for which there exists $M > 0$ with*

$$\|\tilde{\Xi}_{u,m,v}\|_{H_\alpha} \leq M|v - m|^{\mu-1}|v - u| \ , \tag{4.44}$$

then for every $\beta \in [0, \mu)$ the following inequality holds:

$$\|\mathcal{I}\Xi_{s,t} - \Xi_{s,t}\|_{H_{\alpha+\beta}} \lesssim_{\mu,\beta} M|t - s|^{\mu-\beta}. \tag{4.45}$$

Exercise 4.17 (Rough convolution, [GT10, GH19]) *We continue in the Hilbert/
semigroup setting of Exercise 4.16 and fix $\alpha \in \mathbf{R}$. Consider a rough path $\mathbf{X} =
(X, \mathbb{X}) \in \mathscr{C}^\gamma([0, T], \mathbf{R}^d)$ for some $\gamma \in (1/3, 1/2]$ and take $d = 1$ for notational
simplicity only. In the semigroup setting of the previous exercise, write $Y \in \hat{\mathcal{C}}^\gamma H_\alpha$ if
$Y : [0, T] \to H_\alpha$ with $\|Y\|_{\gamma;\alpha}^\wedge := \sup \frac{\|\hat{\delta}Y_{s,t}\|_{H_\alpha}}{|t-s|^\gamma} < \infty$, where $\hat{\delta}Y_{s,t} = Y_t - S_{t-s}Y_s$.
We say that $(Y, Y') \in \mathscr{D}_{S,X}^{2\gamma}([0, T], H_\alpha)$ and call it a mildly controlled rough path if
$(Y, Y') \in \hat{\mathcal{C}}^\gamma H_\alpha \times \hat{\mathcal{C}}^\gamma H_\alpha$ and*

$$R_{s,t}^Y := \hat{\delta}Y_{s,t} - S_{t-s}Y_s' X_{s,t}\,, \tag{4.46}$$

*belongs to $\mathcal{C}_2^{2\gamma} H_\alpha$. With $\|Y, Y'\|_{X,2\gamma;\alpha}^\wedge := \|Y'\|_{\gamma;\alpha}^\wedge + \|R^Y\|_{2\gamma;\alpha}$ a seminorm is
defined on $\mathscr{D}_{S,X}^{2\gamma}$. We show that mildly controlled rough paths are stable under rough
convolution.*

a) *Apply the modified sewing lemma of Exercise 4.16 to show existence of the rough
convolution integral*

$$\int_s^t S_{t-u} Y_u d\mathbf{X}_u := \lim_{|\mathcal{P}| \to 0} \sum_{[u,v] \in \mathcal{P}} S_{t-u}(Y_u X_{u,v} + Y_u' \mathbb{X}_{u,v}), \tag{4.47}$$

exists as an element of $\hat{\mathcal{C}}^\gamma H_\alpha$ and satisfies for every $0 \le \beta < 3\gamma$:

$$\left\| \int_s^t S_{t-u} Y_u d\mathbf{X}_u - S_{t-s} Y_s X_{s,t} - S_{t-s} Y_s' \mathbb{X}_{s,t} \right\|_{H_{\alpha+\beta}} \tag{4.48}$$
$$\lesssim \left(\|R^Y\|_{2\gamma;\alpha} \|X\|_\gamma + \|Y'\|_{\gamma;\alpha}^\wedge \|\mathbb{X}\|_{2\gamma} \right) |t-s|^{3\gamma-\beta}.$$

b) *Show that the map $(Y, Y') \mapsto (Z, Z') := \left(\int_0^\cdot S_{\cdot-u} Y_u d\mathbf{X}_u, Y \right)$ is continuous
from $\mathscr{D}_{S,X}^{2\gamma}([0, T], H_\alpha)$ to $\mathscr{D}_{S,X}^{2\gamma}([0, T], H_\alpha)$ and one has the bound:*

$$\|Z, Z'\|_{X,2\gamma;\alpha}^\wedge \lesssim \|Y\|_{\gamma;\alpha}^\wedge + (\|Y_0'\|_{H_\alpha} + \|(Y, Y')\|_{X,2\gamma;\alpha}^\wedge)(\|X\|_\gamma + \|\mathbb{X}\|_{2\gamma}). \tag{4.49}$$

c) *Make the (notational) adjustment to handle general $d \in \mathbf{N}$.*

Exercise 4.18 (Integration against step-N rough paths) *Any path $\mathbf{X} : [0, T] \to
T_1^{(N)}(\mathbf{R}^d)$ gives rise to increments $\mathbf{X}_s^{-1} \otimes \mathbf{X}_t =: \mathbf{X}_{s,t}$ so that Chen's relation becomes
a tautology. Assume also $|\langle \mathbf{X}_{s,t}, w \rangle| \lesssim |t-s|^{\alpha|w|}, |w| \le N = \lfloor 1/\alpha \rfloor$. (These
are the naïve higher order rough paths introduced in Section 2.4.) Show that the
rough integral $\int Y \, d\mathbf{X}$ defined as in (4.35) is well-defined and detail its structure.
(Naïve rough paths are ill-suited to integrate $f(\mathbf{Y})$ with regular but non-linear f, in
Section 7.6 this is resolved for geometric rough paths.)*

4.8 Comments

Young integration [You36], which can be seen as level-1 rough integration, was a key inspiration for the analytical aspects of Lyons' rough integration [Lyo94, Lyo98], and has remained the "entrance test" for every subsequent (re)interpretation of rough integration, including [Gub04, FdLP06, Pic08, HN09, GIP15, GIP16, FS17]. From a harmonic analysis perspective, the here presented Young integration *in Hölder scale* implies that the product of smooth functions extends naturally to $\mathcal{C}^\beta \times \mathcal{C}^{-\alpha}$ into $\mathcal{D}'(\mathbf{R})$ if and only if $\beta > \alpha$. Similar statements, replacing one-dimensional space ($[0, T] \subset \mathbf{R}$, "time") by $\mathbf{R}^d$ are well known, cf. e.g. [BCD11, Thm 2.52] and Theorem 13.18 later on in the book. Young (and later rough) integration is naturally formulated in *p-variation scale*, examples with $p < 2$ are plentiful and range from Schramm–Loewner trace [Wer12, FT17], fractional Brownian motion (cf. Section 10.3) with Hurst parameter $H > 1/2$ to Lévy processes and homogenisation problems [CFKM19]. Of course, $p = 2^+$ is the correct scale for semimartingales, also in the càdlàg setting, see Section 3.8. The *sewing lemma*, obtained independently by Feyel–De La Pradelle (in an early version of [FdLP06]) and Gubinelli [Gub04], formalises abstract Riemann–Young integration and is a flexible real analysis lemma, with many variations found in [FDM08, GT10, BL19, Yas18, GH19, GHN19] and also [FS17, FZ18] for a sewing lemma, and subsequent integration theory, with jumps. An application of sewing to level sets in the Heisenberg group is given in [MST18]. The applications of Lê's important *stochastic sewing* lemma [Lê18], Section 4.6, include regularisation by noise [Lê18], the construction of rough Markov diffusions [FHL20] by solving hybrid Itô-rough differential equation in the spirit of Section 12.2.1, and an averaging result for SDEs driven by fractional Brownian motion [HL19].

Integration of one-forms against continuous p-variation geometric rough paths for any $p \in [1, \infty)$ was developed by Lyons [Lyo98]; see also [LQ02, LCL07, FV10b, LY15]. For a careful discussion of the integration of weakly geometric rough paths in infinite dimensions we refer to Cass et al. [CDLL16].

Rough integration against *controlled paths* is due to Gubinelli, see [Gub04] where it is developed in an α-Hölder setting, $\alpha > \frac{1}{3}$. Loosely speaking, it allows to "linearise" many considerations (the space of controlled paths is a Banach space, while a typical space of rough paths is not). This point of view has been generalised to arbitrary α (both in the geometric and the non-geometric setting) in [Gub10], see also [HK15, FZ18]. Rough convolution, Exercise 4.17, follows [GT10, GH19], crucial for "mild" RPDE solution, cf. Section 12.5.

The controlled rough path integration point of view can be pushed even further and, as a matter of fact, the theory of regularity structures developed in [Hai14b] and exposed in Chapter 13 onwards, provides a unified framework in which the Gubinelli derivative and the regular derivatives are but two examples of a more general theory of objects behaving "like Taylor expansions" and allowing to describe the small-scale structure of a function and / or distribution in terms of "known" objects (polynomials in the case of Taylor expansions, the underlying rough path in the case of controlled paths).

Chapter 5
Stochastic integration and Itô's formula

In this chapter, we compare the integration theory developed in the previous chapter to the usual theories of stochastic integration, be it in the Itô or the Stratonovich sense.

5.1 Itô integration

Recall from Section 3 that Brownian motion B can be enhanced to a (random) rough path $\mathbf{B} = (B, \mathbb{B})$. Presently our focus is the case when $\mathbb{B}$ is given by the iterated Itô integral [1]

$$\mathbb{B}_{s,t} = \mathbb{B}_{s,t}^{\text{Itô}} \stackrel{\text{def}}{=} \int_s^t B_{s,u} \otimes dB_u$$

and the so enhanced Brownian motion has almost surely (non-geometric) α-Hölder rough sample paths, for any $\alpha \in \left(\frac{1}{3}, \frac{1}{2}\right)$. That is, $\mathbf{B}(\omega) = (B(\omega), \mathbb{B}(\omega)) \in \mathscr{C}^\alpha$ for every $\omega \in N_1^c$ where, here and in the sequel, $N_i, i = 1, 2, \ldots$ denote suitable null sets. We now show that rough integrals (against $\mathbf{B} = \mathbf{B}^{\text{Itô}}$) and Itô integrals, whenever both are well-defined, coincide.

Proposition 5.1. *Assume* $(Y(\omega), Y'(\omega)) \in \mathscr{D}_{B(\omega)}^{2\alpha}$ *for every* $\omega \in N_2^c$. *Set* $N_3 = N_1 \cup N_2$. *Then the rough integral*

$$\int_0^T Y \, d\mathbf{B} = \lim_{n \to \infty} \sum_{[u,v] \in \mathcal{P}_n} \left(Y_u B_{u,v} + Y_u' \mathbb{B}_{u,v} \right)$$

exists, for each fixed $\omega \in N_3^c$, *along any sequence* $(\mathcal{P}_n)$ *with mesh* $|\mathcal{P}_n| \downarrow 0$. *If* Y, Y' *are adapted then, almost surely,*

$$\int_0^T Y \, d\mathbf{B} = \int_0^T Y \, dB .$$

---

[1] The case when $\mathbb{B}$ is given via iterated Stratonovich integration is left to Section 5.2 below.

© Springer Nature Switzerland AG 2020
P. K. Friz, M. Hairer, *A Course on Rough Paths*, Universitext,
https://doi.org/10.1007/978-3-030-41556-3_5

Proof. Without loss of generality $T = 1$. The existence of the rough integral for $\omega \in N_3^c$ under the stated assumptions is immediate from Theorem 4.10, applied to $Y(\omega)$, controlled by $B(\omega)$, for $\omega \in N_2^c$ fixed. Recall (e.g. [RY99]) that for any continuous, adapted process Y the Itô integral against Brownian motion has the representation

$$\int_0^1 Y \, dB = \lim_{n \to \infty} \sum_{[u,v] \in \mathcal{P}_n} Y_u B_{u,v} \quad \text{(in probability)}$$

along any sequence $(\mathcal{P}_n)$ with mesh $|\mathcal{P}_n| \downarrow 0$. By switching to a subsequence, if necessary, we can assume that the convergence holds almost surely, say on N_4^c. Set $N_5 := N_3 \cup N_4$. We shall complete the proof under the assumption that there exists a (deterministic) constant $M > 0$ such that

$$\sup_{\omega \in N_5^c} |Y'(\omega)|_\infty \leq M \ .$$

(This is the case in the "model" situation $Y = F(X), Y' = DF(X)$ where F was in particular assumed to have bounded derivatives; the general case is obtained by localisation and left to Exercise 5.1.)

The claim is that the rough and Itô integral coincide on N_5^c. With a look at the respective Riemann-sums, convergent away from N_5, basic analysis tells us that

$$\forall \omega \in N_5^c : \ \exists \lim_n \sum_{[u,v] \in \mathcal{P}_n} Y_u' \mathbb{B}_{u,v} \ ,$$

and that this limit equals the difference of rough and Itô integrals (on N_5^c, a set of full measure). Of course, $|\mathcal{P}_n| \downarrow 0$, and to see that the above limit is indeed zero (at least on a set of full measure), it will be enough to show that

$$\left\| \sum_{[u,v] \in \mathcal{P}} Y_u' \mathbb{B}_{u,v} \right\|_{L^2}^2 = \mathrm{O}(|\mathcal{P}|) \ . \tag{5.1}$$

To this end, assume the partition is of the form $\mathcal{P} = \{0 = \tau_0 < \ldots < \tau_N = 1\}$ and define a (discrete-time) martingale started at $S_0 := 0$ with increments $S_{k+1} - S_k = Y_{\tau_k}' \mathbb{B}_{\tau_k, \tau_{k+1}}$. Since $|\mathbb{B}_{\tau_k, \tau_{k+1}}|_{L^2}^2$ is proportional to $|\tau_{k+1} - \tau_k|^2$, as may be seen from Brownian scaling, we then have

$$\left| \sum_{[u,v] \in \mathcal{P}} Y_u' \mathbb{B}_{u,v} \right|_{L^2}^2 = \left| \sum_{k=0}^{N-1} (S_{k+1} - S_k) \right|_{L^2}^2 = \sum_{k=0}^{N-1} |S_{k+1} - S_k|_{L^2}^2$$

$$\leq M^2 \sum_{k=0}^{N-1} |\mathbb{B}_{\tau_k, \tau_{k+1}}|_{L^2}^2 = \mathrm{O}(|\mathcal{P}|) \ ,$$

as desired.   □

5.2 Stratonovich integration

We could equally well have enhanced Brownian motion with

$$\mathbb{B}^{\text{Strat}}_{s,t} := \int_s^t B_{s,u} \otimes \circ dB_u = \mathbb{B}^{\text{Itô}}_{s,t} + \frac{1}{2}(t-s)I .$$

Almost surely, this construction then yields *geometric* α-Hölder rough sample paths, for any $\alpha \in \left(\frac{1}{3}, \frac{1}{2}\right)$. Recall that, by definition, the Stratonovich integral is given by

$$\int_0^T Y \circ dB \stackrel{\text{def}}{=} \int_0^T Y dB + \frac{1}{2}[Y, B]_T$$

whenever the Itô integral is well-defined and the *quadratic covariation* of Y and B exists in the sense that $[Y, B]_T := \lim_{|\mathcal{P}| \to 0} \sum_{[u,v] \in \mathcal{P}} Y_{u,v} B_{u,v}$ exists as limit in probability.

In complete analogy to the Itô case, we now show that rough integration against *Stratonovich* enhanced Brownian motion coincides with usual Stratonovich integration against Brownian motion under some natural assumptions guaranteeing that both notions of integral are well-defined.

Corollary 5.2. *As above, assume* $Y = Y(\omega) \in \mathcal{C}^\alpha_{B(\omega)}$ *for every* $\omega \in N_2^c$. *Set* $N_3 = N_1 \cup N_2$. *Then the rough integral of* Y *against* $\mathbf{B} = \mathbf{B}^{\text{Strat}}$ *exists,*

$$\int_0^T Y d\mathbf{B} = \lim_{n \to \infty} \sum_{[u,v] \in \mathcal{P}_n} \left(Y_u B_{u,v} + Y_u' \mathbb{B}^{\text{Strat}}_{u,v}\right).$$

Moreover, if Y, Y' *are adapted, the quadratic covariation of* Y *and* B *exists and, almost surely,*

$$\int_0^T Y d\mathbf{B} = \int_0^T Y \circ dB.$$

Proof. $\mathbb{B}^{\text{Strat}}_{s,t} = \mathbb{B}^{\text{Itô}}_{s,t} + f_{s,t}$ where $f(t) = \frac{t}{2}\text{Id}$. This entails, as was discussed in Example 4.14,

$$\int_0^1 Y d\mathbf{B}^{\text{Strat}} = \int_0^1 Y d\mathbf{B}^{\text{Itô}} + \int_0^1 Y' df.$$

Thanks to Proposition 5.1, it only remains to identify $2\int_0^1 Y' df = \int_0^1 Y_t' dt$ with $[Y, B]_1$. To see this, write

$$\sum_{[u,v] \in \mathcal{P}} Y_{u,v} B_{u,v} = \sum_{[u,v] \in \mathcal{P}} \left((Y_{u,v}' B_{u,v}) B_{u,v} + R_{u,v} B_{u,v}\right)$$

$$= \left(\sum_{[u,v] \in \mathcal{P}} Y_{u,v}' (B_{u,v} \otimes B_{u,v})\right) + O(|\mathcal{P}|^{3\alpha-1}) ,$$

where we used that $\sum R_{u,v} B_{u,v} = O(|\mathcal{P}|^{3\alpha-1})$ thanks to $R \in \mathcal{C}_2^{2\alpha}$ and $B \in \mathcal{C}^\alpha$. Note that

$$B_{u,v} \otimes B_{u,v} = 2 \operatorname{Sym}\left(\mathbb{B}_{u,v}^{\text{Strat}}\right) = 2 \operatorname{Sym}\left(\mathbb{B}_{u,v}^{\text{Itô}}\right) + (v-u)I.$$

We have seen in the proof of Proposition 5.1 that any limit (in probability, say) of

$$\sum_{[u,v] \in \mathcal{P}} Y'_{u,v} \mathbb{B}_{u,v}^{\text{Itô}}$$

must be zero. In fact, a look at the argument reveals that this remains true with $\mathbb{B}_{u,v}^{\text{Itô}}$ replaced by $\operatorname{Sym}\left(\mathbb{B}_{u,v}^{\text{Itô}}\right)$. It follows that

$$\lim_{|\mathcal{P}| \to 0} \sum_{[u,v] \in \mathcal{P}} Y_{u,v} B_{u,v} = \lim_{|\mathcal{P}| \to 0} \left(\sum_{[u,v] \in \mathcal{P}} Y'_{u,v} (v-u) \right) = \int_0^1 Y'_t \, dt \,,$$

thus concluding the proof.   □

5.3 Itô's formula and Föllmer

Given a smooth path $X : [0,T] \to V$ and a map $F : V \to W$ in $\mathcal{C}_b^1$, where V, W are Banach spaces as usual, the chain rule from classical "first oder" calculus tells us that

$$F(X_t) = F(X_0) + \int_0^t DF(X_s) dX_s, \quad 0 \leq t \leq T.$$

Unsurprisingly, the same change of variables formula holds for *geometric* rough paths $\mathbf{X} = (X, \mathbb{X})$, which are essentially limits of smooth paths, and it is not hard to figure out, in view of Example 4.14, that a "second order" correction, involving $D^2 F$, appears in the non-geometric case. In other words, one can write down Itô formulae for rough paths.

Before doing so, however, an important preliminary discussion is in order. Namely, much of our effort so far was devoted to the understanding of (rough) integration against 1-forms, say $G = G(X)$ and indeed we found

$$\int G(X) d\mathbf{X} \approx \sum_{[s,t] \in \mathcal{P}} \left(G(X_s) X_{s,t} + DG(X_s) \mathbb{X}_{s,t} \right)$$

in the sense that the *compensated* Riemann-Stieltjes sums appearing on the right-hand side converge with mesh $|\mathcal{P}| \to 0$. Let us split $\mathbb{X}$ into symmetric part, $\mathbb{S}_{s,t} := \operatorname{Sym}(\mathbb{X}_{s,t})$, and antisymmetric ("area") part, $\operatorname{Anti}(\mathbb{X}_{s,t}) := \mathbb{A}_{s,t}$. Then

$$DG(X_s) \mathbb{X}_{s,t} = DG(X_s) \mathbb{S}_{s,t} + DG(X_s) \mathbb{A}_{s,t}$$

and the final term disappears in the gradient case, i.e. when $G = DF$. Indeed, the contraction of a symmetric tensor (here: $D^2 F$) with an antisymmetric tensor (here: $\mathbb{A}$) always vanishes. In other words, area matters very much for general integrals of 1-forms but not at all for gradient 1-forms. Note also that, contrary to $\mathbb{A}$, the symmetric part $\mathbb{S}$ is a nice function of the underlying path X. For instance, for Itô enhanced Brownian motion in $\mathbf{R}^d$, one has the identity

$$\mathbb{S}_{s,t}^{i,j} = \int_s^t B_{s,r}^i \, dB_r^j = \frac{1}{2}\left(B_{s,t}^i B_{s,t}^j - \delta^{ij}(t - s) \right), \qquad 1 \leq i, j \leq d \, .$$

These considerations suggest that the following definition encapsulates all the data required for the integration of gradient 1-forms.

Definition 5.3. We call $\mathbf{X} = (X, \mathbb{S})$ a *reduced rough path*, in symbols $\mathbf{X} \in \mathscr{C}_r^\alpha([0,T], V)$, if $X = X_t$ takes values in a Banach space V, $\mathbb{S} = \mathbb{S}_{s,t}$ takes values in $\mathrm{Sym}\,(V \otimes V)$, and the following hold:

i) a "reduced" Chen relation

$$\mathbb{S}_{s,t} - \mathbb{S}_{s,u} - \mathbb{S}_{u,t} = \mathrm{Sym}\,(X_{s,u} \otimes X_{u,t}), \qquad 0 \leq s, t, u \leq T \, ,$$

ii) the usual analytical conditions, $X_{s,t} = \mathrm{O}(|t - s|^\alpha)$, $\mathbb{S}_{s,t} = \mathrm{O}(|t - s|^{2\alpha})$, for some $\alpha > 1/3$.

Clearly, any $\mathbf{X} = (X, \mathbb{X}) \in \mathscr{C}^\alpha([0,T], V)$ induces a reduced rough path by ignoring its area $\mathbb{A} = \mathrm{Anti}\,(\mathbb{X})$. More importantly, and in stark contrast to the general rough path case, a lift of a path $X \in \mathcal{C}^\alpha$ to a reduced rough path can be trivially obtained via its square-increments $\frac{1}{2}X_{s,t} \otimes X_{s,t}$. We have the following result.

Lemma 5.4. *Given* $X \in \mathcal{C}^\alpha$, $\alpha \in (1/3, 1/2]$, *the "geometric" choice* $\bar{\mathbb{S}}_{s,t} = \frac{1}{2}X_{s,t} \otimes X_{s,t}$ *yields a reduced rough path, i.e.* $(X, \bar{\mathbb{S}}) \in \mathscr{C}_r^\alpha$. *Moreover, for any* 2α-*Hölder path* γ *with values in* $\mathrm{Sym}\,(V \otimes V)$, *the perturbation*

$$\mathbb{S}_{s,t} = \bar{\mathbb{S}}_{s,t} + \frac{1}{2}(\gamma_t - \gamma_s) = \frac{1}{2}(X_{s,t} \otimes X_{s,t} + \gamma_{s,t})$$

also yields a reduced rough path $(X, \mathbb{S})$. *Finally, all reduced rough path lifts of* X *are obtained in this fashion.*

Proof. A simple exercise for the reader. □

The previous lemma gives in particular a one-one correspondence between $\mathbb{S}$ and γ. We thus formalise the role of γ.

Definition 5.5 (Bracket of a reduced rough path). Given $\mathbf{X} = (X, \mathbb{S}) \in \mathscr{C}_r^\alpha(V)$, we define the bracket

$$[\mathbf{X}] : [0, T] \to \mathrm{Sym}\,(V \otimes V)$$
$$t \mapsto [\mathbf{X}]_t \overset{\mathrm{def}}{=} X_{0,t} \otimes X_{0,t} - 2\mathbb{S}_{0,t} \, .$$

Note that, as consequence of the previous lemma, $[\mathbf{X}] \in \mathcal{C}^{2\alpha}$. Furthermore, if one defines

$$[\mathbf{X}]_{s,t} \overset{\text{def}}{=} X_{s,t} \otimes X_{s,t} - 2\mathbb{S}_{s,t} ,$$

then one has the identity $[\mathbf{X}]_{s,t} = [\mathbf{X}]_{0,t} - [\mathbf{X}]_{0,s}$ for any two times s, t.

Remark 5.6. We already encountered $[\mathbf{X}]$ as a way to decompose $\mathbf{X} \in \mathcal{C}^\alpha$ into a geometric rough paths plus extra information (Exercise 2.11). Our motivation here is different in that we explore that the fact that $[\mathbf{X}]$ requires **no** knowledge of the area Anti $(\mathbb{X}) := \mathbb{A}$, a central object for rough path theory.

Remark 5.7. While this notion of bracket does not rely on any sort of "quadratic variation", it is consistent with the product (a.k.a. integration by parts) formula from Itô calculus. Indeed, for any semimartingale $X = X(t, \omega)$, with $X_0 = 0$ say, we have

$$\int_0^t X_s^i dX_s^j + \int_0^t X_s^j dX_s^i = X_t^i X_t^j - \langle X^i, X^j \rangle_t ; \tag{5.2}$$

from a rough path perspective, the left-hand side is precisely $\mathbb{X}_{0,t}^{i,j} + \mathbb{X}_{0,t}^{j,i} = 2\mathbb{S}_{0,t}^{i,j}$.

Proposition 5.8 (Itô formula for reduced rough paths). *Let* $F : V \to W$ *be of class* $\mathcal{C}_b^3$ *and let* $\mathbf{X} = (X, \mathbb{S}) \in \mathcal{C}_r^\alpha([0, T], V)$ *with* $\alpha > 1/3$. *Then*

$$F(X_t) = F(X_0) + \int_0^t DF(X_s)d\mathbf{X}_s + \frac{1}{2}\int_0^t D^2F(X_s)d[\mathbf{X}]_s, \quad 0 \le t \le T.$$

Here, writing $\mathcal{P}$ *for partitions of* $[0, t]$, *the first integral is given by*[2]

$$\int_0^t DF(X_s)d\mathbf{X}_s \overset{\text{def}}{=} \lim_{|\mathcal{P}| \to 0} \sum_{[u,v] \in \mathcal{P}} \left(DF(X_u)X_{u,v} + D^2F(X_u)\mathbb{S}_{u,v} \right), \tag{5.3}$$

while the second integral is a well-defined Young integral.

Proof. Consider first the geometric case, $\mathbb{S} = \bar{\mathbb{S}}$, in which case the bracket is zero. The proof is straightforward. Indeed, thanks to α-Hölder regularity of X with $\alpha > 1/3$, we obtain

$$
\begin{aligned}
F(X_T) - F(X_0) &= \sum_{[u,v] \in \mathcal{P}} \left(F(X_v) - F(X_u) \right) \\
&= \sum_{[u,v] \in \mathcal{P}} \left(DF(X_u)X_{u,v} + \frac{1}{2}D^2F(X_u)(X_{u,v}, X_{u,v}) \right. \\
&\qquad\qquad \left. + \mathrm{o}(|v - u|) \right) \\
&= \sum_{[u,v] \in \mathcal{P}} \left(DF(X_u)X_{u,v} + D^2F(X_u), \bar{\mathbb{S}}_{u,v} + \mathrm{o}(|v - u|) \right) .
\end{aligned}
$$

---

[2] Note consistency with the rough integral when $\mathbf{X} \in \mathcal{C}^\alpha$.

We conclude by taking the limit $|\mathcal{P}| \to 0$, also noting that $\sum_{[u,v] \in \mathcal{P}} o(|v - u|) \to 0$. For the non-geometric situation, just substitute

$$\bar{\mathbb{S}}_{u,v} = \mathbb{S}_{u,v} + \frac{1}{2}[\mathbf{X}]_{u,v}.$$

Since $D^2 F$ is Lipschitz, $D^2 F(X_\cdot) \in C^\alpha$ and we can split-up the "bracket" term and note that

$$\sum_{[u,v] \in \mathcal{P}} D^2 F(X_u)[\mathbf{X}]_{u,v} \to \int_0^t D^2 F(X_u) d[\mathbf{X}]_u ,$$

where the convergence to the Young integral follows from $[\mathbf{X}] \in C^{2\alpha}$. The rest is now obvious. □

Example 5.9. Consider the case when $\mathbf{X} = \mathbf{B}$, Itô enhanced Brownian motion. Then $\mathbb{X}$ is given by iterated Itô integrals and, thanks to the Itô product rule (5.2),

$$2\mathbb{S}_{0,t}^{i,j} = \int_0^t \left(B^i dB^j + B^j dB^i \right) = B_t^i B_t^j - \left\langle B^i, B^j \right\rangle_t.$$

The usual Itô formula is then recovered from the fact that

$$[\mathbf{B}]_t^{i,j} = B_{0,t}^i B_{0,t}^j - 2\mathbb{S}_{0,t}^{i,j} = \left\langle B^i, B^j \right\rangle_{0,t} = \delta^{i,j} t .$$

We conclude this section with a short discussion on Föllmer's *calcul d'Itô sans probabilités* [Föl81]. For simplicity of notation, we take $V = \mathbf{R}^d, W = \mathbf{R}^e$ in what follows. With regard to (5.3), let us insist that the compensation is necessary and one cannot, in general, separate the sum into two convergent sums. On the other hand, we can combine the converging sums and write

$$
\begin{aligned}
F(X)_{0,t} &= \lim_{|\mathcal{P}| \to 0} \sum_{[u,v] \in \mathcal{P}} \left(DF(X_u)X_{u,v} + D^2 F(X_u)\mathbb{S}_{u,v} \right. \\
&\qquad\qquad \left. + \frac{1}{2} \sum_{[u,v] \in \mathcal{P}} D^2 F(X_u)[\mathbf{X}]_{u,v} \right) \\
&= \lim_{|\mathcal{P}| \to 0} \sum_{[u,v] \in \mathcal{P}} \left(DF(X_u)X_{u,v} + \frac{1}{2} D^2 F(X_u)(X_{u,v}, X_{u,v}) \right) .
\end{aligned}
\tag{5.4}
$$

We now put forward an assumption that allows to break up the above sum.

Definition 5.10. Let $\pi = (\mathcal{P}_n)_{n \geq 0}$ be a sequence of partitions of $[0, T]$ with mesh $|\mathcal{P}_n| \to 0$. We say that $X : [0, T] \to \mathbf{R}^d$ has finite quadratic variation in the sense of Föllmer along π if, for every $t \in [0, T]$ and $1 \leq i, j \leq d$ the limit

$$\left[X^i, X^j\right]_t^\pi := \lim_{n \to \infty} \sum_{[u,v] \in \mathcal{P}_n} \left(X_{v \wedge t}^i - X_{u \wedge t}^i\right)\left(X_{v \wedge t}^j - X_{u \wedge t}^j\right)$$

exists. Write $[X, X]^\pi$ for the resulting path with values in $\mathrm{Sym}\left(\mathbf{R}^d \otimes \mathbf{R}^d\right)$, i.e. the space of symmetric $d \times d$ matrices.

Lemma 5.11. *Assume $X : [0, T] \to \mathbf{R}^d$ is continuous and has finite quadratic variation in the sense of Föllmer, along $\pi = (\mathcal{P}_n)_{n \geq 0}$. Then the map $t \mapsto [X, X]_t^\pi$ is of bounded variation on $[0, T]$ and, for any continuous $G : [0, T] \to \mathcal{L}^{(2)}\left(\mathbf{R}^d \times \mathbf{R}^d, \mathbf{R}^e\right)$,*

$$\lim_{n \to \infty} \sum_{\substack{[u,v] \in \mathcal{P}_n \\ u < t}} G(u)\left(X_{u,v}, X_{u,v}\right) = \int_0^t G(u)d[X, X]_u^\pi \in \mathbf{R}^e .$$

Proof. For the first statement, it is enough to argue component by component. Set $[X^i]^\pi := [X^i, X^i]^\pi$. By polarisation,

$$[X^i, X^j]_t^\pi = \frac{1}{2}[X^i + X^j]_t^\pi - [X^i]_t^\pi - [X^j]_t^\pi .$$

Since each term on the right-hand side is monotone in t, we see that $t \mapsto [X^i, X^j]_t^\pi$ is indeed of bounded variation.

Regarding the second statement, it is enough to check that, for continuous $g : [0, T] \to \mathbf{R}$ and Y of finite quadratic variation, with continuous bracket $t \mapsto [Y]_t^\pi$,

$$\lim_{n \to \infty} \sum_{\substack{[u,v] \in \mathcal{P}_n \\ u < t}} g(u)Y_{u,v}^2 = \int_0^t g(u)d[Y]_u^\pi. \tag{5.5}$$

Indeed, we can apply this for each component, with $g = G_{i,j}^k$ and

$$Y \in \left\{\left(X^i + X^j\right), X^i, X^j\right\},$$

which then also gives, by polarisation,

$$\sum_{\substack{[u,v] \in \mathcal{P}_n \\ u < t}} G_{i,j}^k(u)X_{u,v}^i X_{u,v}^j \to \int_0^t G_{i,j}^k(u)d[X^i, X^j]_u^\pi.$$

To see that (5.5) holds, write $\sum_{[u,v] \in \mathcal{P}_n, u<t} g(u)Y_{u,v}^2 = \int_{[0,t)} g(u)d\mu_n(u)$ with

$$\mu_n = \sum_{[u,v] \in \mathcal{P}_n, u<t} Y_{u,v}^2 \, \delta_u .$$

Note that μ_n is a finite measure on $[0, t)$ with distribution function

$$F_n(s) := \mu_n([0, s]) = \sum_{\substack{[u,v] \in \mathcal{P}_n \\ u \leq s}} Y_{u,v}^2.$$

As $n \to \infty$, $F_n(s) \to [Y]_s^\pi$ for any $s \le t$ by continuity of Y. Pointwise convergence of the distribution functions implies weak convergence of the measures μ_n to the measure $d[Y]^\pi$ on $[0, t)$, with distribution function the right-continuous modification of $[Y]^\pi$. Since $g|_{[0,t)}$ is continuous, (5.5) follows. □

Combination of the above lemma with (5.4) gives the *Itô–Föllmer formula*,

$$F(X_t) = F(X_0) + \int_0^t DF(X_s)dX + \frac{1}{2}\int_0^t D^2F(X_s)d[X, X]_t, \quad 0 \le t \le T$$
(5.6)

where the middle integral is given by the (now existent) limit of left-point Riemann-Stieltjes approximations

$$\lim_{n\to\infty} \sum_{[u,v]\in\mathcal{P}_n} DF(X_u), X_{u,v} =: \int_0^t DF(X)dX.$$

In fact, we encourage the reader to verify as an exercise that this formula is valid whenever $X : [0, T] \to \mathbf{R}^d$ is continuous, of finite quadratic variation, with $t \mapsto [X, X]_t^\pi$ continuous. Note, however, that Föllmer's notion of quadratic variation (and the above integral) can and will depend in general on the sequence $(\mathcal{P}_n)$.

5.4 Backward integration

Given a Brownian motion $B = B_t(\omega)$, one can define the *backward Itô-integral*

$$\int_0^T f_t \overleftarrow{dB_t} := \lim_n \sum_{[s,t]\in\mathcal{P}_n} f_t B_{s,t},$$

whenever $|\mathcal{P}_n| \to 0$ and this limit exists, in probability and uniformly on compact time intervals, and does not depend on the sequence of partitions $(\mathcal{P}_n)$ of $[0, T]$. For instance,

$$\int_0^T B_t \overleftarrow{dB_t} = \frac{1}{2}B_T^2 + \frac{T}{2}.$$

In many applications one encounters integrands $f = f_t(\omega)$ that are *backward adapted* in the sense that each f_t is measurable with respect to the σ-field $\mathcal{F}_t^T := \sigma(B_{u,v} : t \le u \le v \le T)$. For example,

$$\int_0^T (B_T - B_t) \overleftarrow{dB_t} = B_T^2 - \int_0^T B_t \overleftarrow{dB_t} = \frac{1}{2}B_T^2 - \frac{T}{2}$$

and we note (in contrast to the previous example) the zero mean property, which of course comes from a backward martingale structure. Indeed, $\hat{B}_t := B_T - B_{T-t}$ is a standard Brownian motion, adapted to $\hat{\mathcal{F}}_t := \mathcal{F}_{T-t}^T$ and so is $\hat{f}_t = f_{T-t}$. The backward integral can then be written as classical (forward) Itô integral

$$\int_0^T f_t \overleftarrow{dB}_t = \int_0^T \hat{f}_t \, d\hat{B}_t = \lim_n \sum_{[s,t] \in \mathcal{P}^n} \hat{f}_s \hat{B}_{s,t} \,. \tag{5.7}$$

Also, by analogy with its forward counterpart, the *backward Stratonovich integral* is defined as the backward Itô integral, minus $1/2$ times the quadratic variation of the integrand.

The purpose of this section is to understand backward integration as rough integration. To this end, recall that the "forward" rough integral of $(Y, Y') \in \mathscr{D}_X^{2\alpha}$ against $\mathbf{X} = (X, \mathbb{X})$ was given in Theorem 4.10 by

$$\int_0^T Y \, d\mathbf{X} = \lim_{|P| \downarrow 0} \sum_{[s,t] \in P} Y_s X_{s,t} + Y_s' \mathbb{X}_{s,t} \tag{5.8}$$

where P are partitions of $[0, T]$ with mesh-size $|P|$. Clearly, some sort of "left-point" evaluation has been hard-wired into our definition of rough integral. On the other hand, one can expect that feeding in explicit second order information makes this choice somewhat less important than in the case of classical stochastic integration. The next proposition, purely deterministic, answers the questions to what extent one can replace left-point by right-point evaluation. In fact, it provides the natural analogue of (5.7)[3] but without any need of "backward" rough integration: both rough integrals which appear in the following proposition are are "forward" in the sense of (5.8).

Proposition 5.12 (Backward representation of rough integral). *Given a rough path* $\mathbf{X} = (X, \mathbb{X}) \in \mathscr{C}^\alpha$ *with* $\alpha > 1/3$ *and* $(Y, Y') \in \mathscr{D}_X^{2\alpha}$ *we have, for all* $r \in [0, T]$,

$$\int_r^T (Y, Y') d\mathbf{X} = \lim_{|\mathcal{P}| \downarrow 0} \sum_{[s,t] \in \mathcal{P}} \left(Y_t X_{s,t} + Y_t' (\mathbb{X}_{s,t} - X_{s,t} \otimes X_{s,t}) \right) \tag{5.9}$$

$$= -\lim_{|\mathcal{P}| \downarrow 0} \sum_{[s,t] \in \mathcal{P}} \left(Y_t X_{t,s} + Y_t' \mathbb{X}_{t,s} \right) = -\int_0^{T-r} (\overleftarrow{Y}, \overleftarrow{Y}') d\overleftarrow{\mathbf{X}} \,.$$

with $\overleftarrow{\mathbf{X}}(t) = \mathbf{X}(T - t)$ *and similar for* Y *and* Y'.

Proof. It is clear from (5.8) the rough integral is given as (compensated) Riemann–Stieltjes limit

$$\int_r^T Y \, d\mathbf{X} = \lim_{|\mathcal{P}| \downarrow 0} \sum_{[s,t] \in \mathcal{P}} \left(Y_s X_{s,t} + Y_s' \mathbb{X}_{s,t} + (*)_{s,t} \right)$$

---

[3] With regard to (5.7), note that $d\hat{B} = -d\overleftarrow{B}$ where $\overleftarrow{B}_t = B_T - B_t$, not be mixed up with the backward Itô differential $\overleftarrow{dB}$.

whenever $(*)_{s,t} \approx 0$ in the sense that $(*)_{s,t} = O(|t-s|^{3\alpha}) = o(|t-s|)$, so that it does not contribute to the limit. (Recall (4.21) and Lemma 4.2.) But then

$$Y_s X_{s,t} + Y'_s \mathbb{X}_{s,t} = Y_t X_{s,t} - Y_{s,t} X_{s,t} + Y'_s \mathbb{X}_{s,t}$$
$$\approx Y_t X_{s,t} - Y'_s X_{s,t} \otimes X_{s,t} + Y'_s \mathbb{X}_{s,t}$$
$$\approx Y_t X_{s,t} + Y'_t (\mathbb{X}_{s,t} - X_{s,t} \otimes X_{s,t}) ,$$

which settles the first equality in (5.9). The second one follows from $X_{s,t} = -X_{t,s}$ and, from Chen's relation, $\mathbb{X}_{s,t} + \mathbb{X}_{t,s} + X_{s,t} \otimes X_{t,s} = \mathbb{X}_{s,s} = 0$. For the final equality, note that every partition $\mathcal{P}$ of $[r,T]$ induces a time-reversed partition $\overleftarrow{\mathcal{P}}$ of $[0, T-r]$, with each $[s,t]$ replaced by $[T-t, T-s]$. By Exercise 2.6, the (time T) time-reversal of $\mathbf{X}$ is again a rough path, $\overleftarrow{\mathbf{X}} \in \mathscr{C}^\alpha$, and since (easy to see) $(Y, Y') \in \mathscr{D}_X^{2\alpha}$ if and only if $(\overleftarrow{Y}, \overleftarrow{Y}') \in \mathscr{D}_{\overleftarrow{X}}^{2\alpha}$, we obtain the final equality. □

Remark 5.13 (Backward geometric integration). For $\mathbf{X} \in \mathscr{C}_g([0,T], \mathbf{R}^d)$, it was seen in Exercise 2.4) that $\mathbb{X}_{t,s} = \mathbb{X}_{s,t}^T$. It then follows from (5.9) that[4]

$$\int_0^T (Y, Y') d\mathbf{X} = \lim_{|\mathcal{P}| \downarrow 0} \sum_{[s,t] \in \mathcal{P}} \left(Y_t X_{s,t} - (Y'_t)^T \mathbb{X}_{s,t} \right) . \tag{5.10}$$

At this stage, one could rephrase the defining condition for $(Y, Y') \in \mathscr{D}_X^{2\alpha}$ in terms of a "backward" controlledness condition for $(\hat{Y}, \hat{Y}') := (Y, -(Y')^T)$, together with a "backward" rough integral given by[5]

$$\lim_{|\mathcal{P}| \downarrow 0} \sum_{[s,t] \in \mathcal{P}} \left(\hat{Y}_t X_{s,t} + \hat{Y}_t \mathbb{X}_{s,t} \right) =: \int_0^T (\hat{Y}, \hat{Y}') \overleftarrow{d\mathbf{X}} . \tag{5.11}$$

However, this is no different than the "forward" integral $\int (Y, Y') d\mathbf{X}$. Comparing (5.8) with (5.11), one changed left- to right-point evaluation, followed by twisting the meaning of controlled rough path, to make sure nothing happened!

As should be clear at this point, a naïve backward rough integral of $(Y, Y') \in \mathscr{D}_X^{2\alpha}$ against $\mathbf{X} \in \mathscr{C}([0,T], \mathbf{R}^d)$, with left- replaced by right-point evaluation in (5.8),

$$\lim_{|\mathcal{P}| \downarrow 0} \sum_{[s,t] \in \mathcal{P}} \left(Y_t X_{s,t} + Y'_t \mathbb{X}_{s,t} \right) ,$$

is, in general, not well-defined. In fact, in view of Proposition 5.12, existence of this limit is equivalent to existence of (either)

$$\lim_{|\mathcal{P}| \downarrow 0} \sum_{[s,t] \in \mathcal{P}} Y'_t (X_{s,t} \otimes X_{s,t}) = \lim_{|\mathcal{P}| \downarrow 0} \sum_{[s,t] \in \mathcal{P}} Y'_s (X_{s,t} \otimes X_{s,t}) .$$

---

[4] In coordinates: $(Y'\mathbb{X})^k = (Y')^k_{i,j} \mathbb{X}^{i,j}$ vs. $(Y')^T \mathbb{X} = (Y')^k_{j,i} \mathbb{X}^{i,j}$ with implicit summation over $i, j = 1, \ldots, d$.

[5] Not to be confused with a standard "forward" rough integral $\int (\ldots) d\overleftarrow{\mathbf{X}}$ seen in (5.9).

There is no reason why, for a general path $X \in C^\alpha$, the above limits should exist. On the other hand, we already considered such sums in the context of the Itô–Föllmer formula, cf. Lemma 5.11. The appropriate condition for X was seen to be "quadratic variation (in the sense of Föllmer, along some $(\mathcal{P}_n)$)". And under this assumption,

$$\sum_{[s,t]\in\mathcal{P}^n} Y_s'\left(X_{s,t} \otimes X_{s,t}\right) \to \int_0^T Y_s' d[X]_s^\pi . \tag{5.12}$$

Of course, with probability one, d-dimensional standard Brownian motion has quadratic variation in the sense of Föllmer, along dyadic partitions, for instance, with $[B,B]_t^\pi = t\mathrm{Id}$. These remarks are crucial for proving the following.

Theorem 5.14. *Define the random rough paths* $\mathbf{B}^{\mathrm{Strat}} = (B, \mathbb{B}^{\mathrm{Strat}})$ *and* $\mathbf{B}^{\mathrm{back}} \overset{def}{=} (B, \mathbb{B}^{\mathrm{back}})$ *by*

$$\mathbb{B}_{s,t}^{\mathrm{Strat}} \overset{def}{=} \int_s^t B_{s,r} \otimes \circ dB_r = \mathbb{B}_{s,t}^{\mathrm{Itô}} + \frac{1}{2}\mathrm{Id}(t-s) ,$$

$$\mathbb{B}_{s,t}^{\mathrm{back}} \overset{def}{=} \int_s^t B_{s,r} \otimes \overleftarrow{dB_r} = \mathbb{B}_{s,t}^{\mathrm{Itô}} + \mathrm{Id}(t-s) .$$

Then, the following statements hold.

i) Assume $(Y(\omega), Y'(\omega)) \in \mathscr{D}_{B(\omega)}^{2\alpha}$ *a.s. and* Y, Y' *are adapted as processes. Then, with probability one, for all* $t \in [0,T]$,

$$\int_0^t Y d\mathbf{B}^{\mathrm{Strat}} = \int_0^t Y_s dB_s + \frac{1}{2}\int_0^t Y_s'\mathrm{Id}\, ds = \int_0^t Y_s \circ dB_s ,$$

$$\int_0^t Y d\mathbf{B}^{\mathrm{back}} = \int_0^t Y_s dB_s + \int_0^t Y_s'\mathrm{Id}\, ds .$$

ii) Assume $(Y(\omega), Y'(\omega)) \in \mathscr{D}_{B(\omega)}^{2\alpha}$ *a.s. and* Y_t, Y_t' *are* $\mathcal{F}_t^T$*-measurable for all* $t < T$. *Then with probability one, for all* $r \in [0,T]$,

$$\int_r^T Y d\mathbf{B}^{\mathrm{Strat}} = \int_r^T Y_t \overleftarrow{dB_t} - \frac{1}{2}\int_r^T Y_t'\mathrm{Id}\, dt = \int_r^T Y_s \circ \overleftarrow{dB_s} ,$$

$$\int_r^T Y d\mathbf{B}^{\mathrm{back}} = \int_r^T Y_t \overleftarrow{dB_t} .$$

Proof. Regarding point i), it follows from the definition of the rough integral (see also Example 4.14) that

$$\int_0^t Y d\mathbf{B}^{\mathrm{back}} = \int_0^t Y d\mathbf{B}^{\mathrm{Itô}} + \int_0^t Y'\mathrm{Id}\, ds .$$

The claim then follows from Proposition 5.1. The Stratonovich case is similar, now using Corollary 5.2.

We now turn to point ii). Thanks to the backward presentation established in Proposition 5.12,

$$\int_r^T Y \, d\mathbf{B}^{\text{back}} = \lim_{n \to \infty} \sum_{[s,t] \in \mathcal{P}^n} Y_t B_{s,t} + Y_t' \left(\mathbb{B}_{s,t}^{\text{Itô}} + \text{Id}(t - s) - B_{s,t} \otimes B_{s,t} \right)$$

$$= \lim_{n \to \infty} \sum_{[s,t] \in \mathcal{P}^n} Y_t B_{s,t} + Y_t' \mathbb{B}_{s,t}^{\text{Itô}} - Y_s' \left(B_{s,t} \otimes B_{s,t} - \text{Id}(t - s) \right),$$

using $Y_{s,t}'(X_{s,t} \otimes X_{s,t}) \approx 0$ and $Y_{s,t}' \text{Id}(t - s) \approx 0$. (As before $(*)_{s,t} \approx 0$ means $(*)_{s,t} = \text{o}(|t - s|)$.) Now we know that with probability 1, $B(\omega)$ has finite quadratic variation $[B]_t^\pi = \text{Id} t$, in the sense of Föllmer along some sequence $\pi = (\mathcal{P}^n)$. As a purely deterministic consequence, cf. (5.12), on the same set of full measure,

$$\lim_{n \to \infty} \sum_{[s,t] \in \mathcal{P}^n} Y_s' B_{s,t} \otimes B_{s,t} = \int_0^T Y_s' d[B]_s^\pi = \lim_{n \to \infty} \sum_{[s,t] \in \mathcal{P}^n} Y_s' \text{Id}(t - s).$$

It follows at once that

$$\int_r^T Y \, d\mathbf{B}^{\text{back}}(\omega) = \lim_{n \to \infty} \sum_{[s,t] \in \mathcal{P}^n} Y_t B_{s,t} + Y_t' \mathbb{B}_{s,t}^{\text{Itô}}.$$

Since $\mathbb{B}_{s,t}^{\text{Itô}}$ is independent from $\mathcal{F}_t^T$ and Y_t, Y_t' are $\mathcal{F}_t^T$-measurable, a (backward) martingale argument shows that

$$\lim_{n \to \infty} \sum_{[s,t] \in \mathcal{P}^n} Y_t' \mathbb{B}_{s,t}^{\text{Itô}} = 0.$$

As a consequence, with probability one,

$$\int_r^T Y \, d\mathbf{B}^{\text{back}}(\omega) = \lim_{n \to \infty} \sum_{[s,t] \in \mathcal{P}^n} Y_t B_{s,t} = \int_r^T Y \, \overleftarrow{dB}.$$

The (backward) Stratonovich case is then treated as simple perturbation,

$$\int_r^T Y \, d\mathbf{B}^{\text{Strat}} = \lim_{n \to \infty} \sum_{[s,t] \in \mathcal{P}^n} \left(Y_t B_{s,t} + Y_t' \left(\mathbb{B}_{s,t}^{\text{Itô}} + \text{Id}(t - s) - B_{s,t} \otimes B_{s,t} \right) \right.$$

$$\left. - \frac{1}{2} Y_t' \text{Id}(t - s) \right)$$

$$= \int_0^T Y_t \, \overleftarrow{dB}_t - \frac{1}{2} \int_0^T Y_t' \text{Id} \, dt,$$

thus concluding the proof. □

5.5 Exercises

Exercise 5.1 *Complete the proof of Proposition 5.1 in the case of unbounded Y'.*

Solution. It suffices to show the convergence of (5.1) in probability; to this end, we introduce stopping times

$$\tau_M \overset{\text{def}}{=} \max\left\{ t \in \mathcal{P} : |Y_t'| < M \right\} \in [0, T] \cup \{+\infty\}$$

and note that $\lim_{M \to \infty} \tau_M = \infty$ almost surely. The stopped process S^{τ_M} is also a martingale, and we see as above that, for every fixed $M > 0$,

$$\left| \sum_{\substack{[u,v] \in \mathcal{P} \\ u \leq \tau_M}} Y_u' \mathbb{B}_{u,v} \right|^2_{L^2} = O(|\mathcal{P}|).$$

The proof is then easily finished by sending M to infinity.

Exercise 5.2 (Applications to statistics [DFM16]) *Let B be a d-dimensional Brownian motion. Consider a $d \times d$ matrix A, a non-degenerate volatility matrix σ of the same dimension and a sufficiently nice map $h : \mathbf{R}^d \to \mathbf{R}^d$ so that the Itô stochastic differential equation*

$$dY_t = A\, h(Y_t)dt + \sigma dB_t \tag{5.13}$$

has a unique solution, starting from any $Y_0 = y_0 \in \mathbf{R}^d$. (As a matter of fact, this SDE can be solved pathwise by considering the random ODE for $Z_t = Y_t - \sigma B_t$.) We are interested in the maximum likelihood estimation of the drift parameter A over a fixed time horizon $[0, T]$, given some observation path $Y = Y(\omega)$. Recall that this estimator, $\hat{A}_T(\omega)$, is based on the Radon–Nikodym density on pathspace, as given by Girsanov's theorem, relative to the drift free diffusion.

a) *Let $d = 1$, $h(y) = y$. Show that the estimator $\hat{A}$ can be "robustified" in the sense that $\hat{A}_T(\omega) = \tilde{A}_T(Y(\omega))$ where*

$$\tilde{A}_T(Y) = \frac{Y_T^2 - y_0^2 - \sigma^2 T}{2 \int_0^T Y_t^2 \, dt}. \tag{5.14}$$

is defined deterministically for any non-zero $Y \in \mathcal{C}([0, T], \mathbf{R}^d)$, and continuous with respect to uniform topology.

b) *Take again $h(x) = x$, but now in dimension $d > 1$. Show that $\hat{A}$ admits a robust representation on rough path space, i.e. one has $\hat{A}_T(\omega) = \tilde{A}_T(\mathbf{Y}(\omega))$ where $\tilde{A}_T = \tilde{A}_T(\mathbf{Y})$ is deterministically defined and continuous with respect to α-Hölder rough path topology for any fixed $\alpha \in (1/3, 1/2)$. Here, $\mathbf{Y}(\omega)$ is the geometric rough path constructed from Y by iterated Stratonovich integration. Explain why there cannot be a robust representation on path space (as was the case when $d = 1$). What about more general h?*

Exercise 5.3 (Rough vs. anticipating Skorokhod integration) *We have seen that Itô integration coincides with rough integration against* $\mathbf{B}^{\text{Itô}}(\omega)$, *subject to natural conditions (in particular: adaptedness of* (Y, Y') *which guarantees that both are well-defined). A well-known extension of the Itô integral to non-adapted integrands is given by the Skorokhod integral, details of which are found in any textbook on Malliavin calculus, see for example [Nua06].*

a) *Let* B *denote one-dimensional Brownian motion on* $[0, T]$. *Show that the Skorokhod integral of* B_T *against* B *over* $[0, T]$, *in symbols* $\int_0^T B_T \delta B_t$, *is given by* $B_T^2 - T$.

b) *Set* $Y_t(\omega) := B_T(\omega)$, *with (zero) increments (trivially) controlled by* B *with* $Y' := 0$. *(In view of true roughness of Brownian motion, cf. Section 6, there is no other choice for* Y'). *Show that the rough integral of* Y *against Brownian motion over* $[0, T]$ *equals* B_T^2. *Conclude that Skorokhod and rough integrals (against Itô enhanced Brownian motion) do not coincide beyond adapted integrals.*

Exercise 5.4 (Rough vs. anticipating Stratonovich integration [CFV07]) *In the spirit of Nualart–Pardoux [NP88], define the Stratonovich anticipating stochastic integral by*

$$\int_0^t u(s, \omega) \circ dB_s(\omega) \stackrel{def}{=} \lim_{n \to \infty} \int_0^t u(s, \omega) \frac{dB_s^n(\omega)}{ds} \, ds,$$

where B^n *is the dyadic piecewise linear approximation to a* (d-dimensinoal) *Brownian motion* B, *whenever this limit exists in probability and uniformly on compacts. Consider (possibly anticipating) random 1-forms,* $u(s, \omega) = F_\omega(B_s) \in \mathcal{C}_b^2$, *for a.e.* ω. *Show that with probability one,*

$$\int_0^{\cdot} F_\omega(B_s) d\mathbf{B}^{\text{Strat}}(\omega) \equiv \lim_{n \to \infty} \int_0^{\cdot} F_\omega(B_s) \frac{dB_s^n(\omega)}{ds} \, ds \ .$$

where the limit on the right-hand side exists in the almost sure sense. Conclude that in this case rough integration against $\mathbf{B}^{\text{Strat}}$ *coincides almost surely with Stratonovich anticipating stochastic integration, i.e.*

$$\int_0^{\cdot} F_\omega(B_s) d\mathbf{B}^{\text{Strat}}(\omega) \equiv \int_0^{\cdot} F_\omega(B_s) \circ dB_s(\omega).$$

Hint: *It is useful to consider the pair* $(\mathbf{B}^{\text{Strat}}, B^n)$, *canonically viewed as (geometric) rough paths over* $\mathbf{R}^{2d}$, *followed by its rough path convergence to the "doubled" rough path* $(\mathbf{B}^{\text{Strat}}, \mathbf{B}^{\text{Strat}})$ *(which needs to be defined rigorously).*

Remark. Nualart–Pardoux actually define their integral in terms of arbitrary deterministic (not necessarily dyadic) piecewise linear approximations and demand that the limit does not depend on the choice of the sequence of partitions. At the price of giving up the martingale argument, which made dyadic approximations easy (Proposition 3.6), everything can also be done in the general case; see Exercises 10.1 and 10.2 below.

Exercise 5.5 *Fix $t > 0$ and a sequence of dissections $(\mathcal{P}_n) \subset [0,t]$ with mesh $|\mathcal{P}_n| \to 0$. Consider the Itô–Föllmer integral given by*

$$\int_0^t DF(X)\,dX \stackrel{\text{def}}{=} \lim_{n\to\infty} \sum_{[u,v]\in\mathcal{P}_n} DF(X_u)\,X_{u,v}\,,$$

whenever this limit exists. Show that this limit does not exist, in general, when $X = B^H$, a d-dimensional fractional Brownian motion with Hurst parameter $H < 1/2$.

Hint: *Consider the simplest possible non-trival case, namely $d = 1$ and $F(x) = x^2$.*

Solution. Assume convergence in probability say along some $(\mathcal{P}_n)$ for the approximating (left-point) sum,

$$\sum_{[u,v]\in\mathcal{P}_n} X_u X_{u,v}.$$

We look for a contradiction. Elementary "calculus for sums" implies that the midpoint sum converges, i.e. where X_u above is replaced by $X_u + X_{u,v}/2$. It follows that convergence of the left-point sums is equivalent to to existence of quadratic variation, i.e. existence of

$$\lim_{n\to\infty} \sum_{[u,v]\in\mathcal{P}_n} |X_{u,v}|^2.$$

Note that $\mathbf{E}|X_{u,v}|^2 = (1/2^n)^{2H}$ so that the expectation of this sum equals $2^{n(1-2H)}$, which diverges when $H < 1/2$. In particular, quadratic variation does not exist as L^1 limit. But is also cannot exist as a limit in probability, for both types of convergence are equivalent on any finite Wiener–Itô chaos.

Exercise 5.6 *In Proposition 5.8, replace the assumption that $\mathbf{X} = (X, \mathbb{S}) \in \mathscr{C}_r^\alpha([0,T], V)$ with $\alpha > 1/3$, by a suitable p-variation assumption with $p < 3$. Show that $[\mathbf{X}]$ has finite p/2-variation and that $\int D^2 F(X)d[\mathbf{X}]$, as it appears in Itô's formula for reduced rough paths, remains a Young integral.*

♯ **Exercise 5.7** *Prove Proposition 1.1.*

Solution. Without loss of generality, we consider the problem on the interval $[0, 2\pi]$. Assume by contradiction that there is a space $\mathcal{B} \subset C([0, 2\pi])$ which carries the law μ of Brownian motion and such that $(f, g) \mapsto \int f\,dg$ is continuous on $\mathcal{B}$. By definition, the Cameron–Martin space of μ is $\mathcal{H} = W_0^{1,2}([0,1])$, which has an orthonormal basis $\{e_n\}_{n\in\mathbb{Z}}$ given by

$$e_0(t) = \frac{t}{\sqrt{2\pi}}\,, \quad e_k(t) = \frac{\sin kt}{k\sqrt{\pi}}\,, \quad e_{-k}(t) = \frac{1-\cos kt}{k\sqrt{\pi}}\,,$$

for $k > 0$. It follows from standard Gaussian measure theory [Bog98] that, given a sequence ξ_n of i.i.d. normal Gaussian random variables, the sequence $X_N =$

$\sum_{n=-N}^{N} e_n \xi_n$ converges almost surely in $\mathcal{B}$ to a limit X such that the law of X is μ. Write now $Y_N = \sum_{n=-N}^{N} \text{sign}(n) e_n \xi_n$, so that one also has $Y_N \to Y$ with law of Y given by μ.

This immediately leads to a contradiction: on the one hand, assuming that $(f, g) \mapsto \int f \, dg$ is continuous on $\mathcal{B}$, this implies that $\int_0^{2\pi} X_N(t) \, dY_N(t)$ converges to some finite (random) real number. On the other hand, an explicit calculation yields

$$\int_0^{2\pi} X_N(t) \, dY_N(t) = \frac{\xi_0^2}{2} + \sum_{n=1}^{N} \frac{\xi_n^2 + \xi_{-n}^2}{n} \, .$$

It is now straightforward to verify that this diverges logarithmically, thus concluding the proof.

5.6 Comments

Rough integrals of 1-forms against the Brownian rough path (and also continuous semimartingales enhanced to rough paths) are well known to coincide with stochastic integrals, see [LQ02, FV10b] and the references therein, [FS17, CF19] for the case of càdlàg semimartingales. Chouk and Tindel [TC15] discuss, from a rough path view, Skorohod and Stratonovich integration in the plane. Pathwise integration à la Föllmer is revisited and extended by Ananova, Cont and Perkowski [AC17, CP19].

Sharp rough path type p-variation and integrability estimates on martingale transforms (and then stochastic integrals against general càdlàg semimartingales) are given by Friz and Zorin-Kranich [FZK20], this extends and unifies the relevant parts of [Lep76, FV08a, KZK19], see also [DOP19] for the use of such an estimate. recently led to the notion of *rough semimartingale* [FZK20], which leads to a simultaneous development of (càdlàg) rough and stochastic integration. A parallel development [FHL20], in a Hölder setting, is based on stochastic sewing (Section 4.6), see also Exercise 4.15.

Chapter 6
Doob–Meyer type decomposition for rough paths

A deterministic Doob–Meyer type decomposition is established. It is closely related to the question to what extent Y' is determined by Y, given that $(Y, Y') \in \mathscr{D}_X^{2\alpha}$. The crucial property is true roughness of X, a deterministic property that guarantees that X varies in all directions, all the time.

6.1 Motivation from stochastic analysis

Consider a continuous semimartingale $(S_t : t \geq 0)$. By definition (e.g. [RY99, Ch. IV]) this means that $S = M + A$ where $M \in \mathcal{M}$, the space of continuous local martingales, and $A \in \mathcal{V}$, the space of continuous adapted process of finite variation. Then it is well known that the decomposition $S = M + A$ is unique in the following sense.

Proposition 6.1. *Assume* $M, \tilde{M} \in \mathcal{M}$, *vanishing at zero, and* $A, \tilde{A} \in \mathcal{V}$ *such that* $M + A \equiv \tilde{M} + \tilde{A}$ *(i.e. the respective processes are indistinguishable). Then*

$$M \equiv \tilde{M} \quad \text{and} \quad A \equiv \tilde{A} \,.$$

Furthermore, if $S = M + A \equiv 0$ *on some random interval* $[0, \tau)$ *where* τ *is a stopping time, then* $\langle M \rangle \equiv 0$ *on* $[0, \tau)$ *and* $A \equiv 0$ *on* $[0, \tau)$.

Proof. Assume $M + A \equiv \tilde{M} + \tilde{A}$. Then $M - \tilde{M} \in \mathcal{V}$, and null at zero. By a standard result in martingale theory, see for example [RY99, IV, Prop 1.2], this entails that $M - \tilde{M} \equiv 0$. But then $A \equiv \tilde{A}$ and the proof is complete.

Regarding the second statement, consider the stopped semimartingale, $S^\tau = M^\tau + A^\tau$ where $M_t^\tau = M_{t \wedge \tau}$ and similarly for A. By assumption $S^\tau \equiv 0$ and hence, by the first part, $M^\tau, A^\tau \equiv 0$. This also implies that the quadratic variation of M^τ, denoted by $\langle M^\tau \rangle$, vanishes. Since $\langle M^\tau \rangle = \langle M \rangle^\tau$ (see e.g. [RY99, Ch. IV]) it indeed follows that $\langle M \rangle \equiv 0$ on $[0, \tau)$. $\square$

© Springer Nature Switzerland AG 2020
P. K. Friz, M. Hairer, *A Course on Rough Paths*, Universitext,
https://doi.org/10.1007/978-3-030-41556-3_6

The above proposition applies in particular when M is given as multidimensional (say $\mathbf{R}^e$-valued) stochastic integral of a suitable $L(\mathbf{R}^d, \mathbf{R}^e)$-valued integrand Y (continuous and adapted will do) against d-dimensional Brownian motion B, while A is the indefinite integral of some suitable $\mathbf{R}^e$-valued process Z (again, continuous and adapted will do). We then have

Corollary 6.2. *Let B be a d-dimensional Brownian motion and let Y, Z, $\tilde{Y}$, $\tilde{Z}$ be continuous stochastic processes adapted to the filtration generated by B. Assume, in the sense of indistinguishability of left- and right-hand sides, that*

$$\int_0^{\cdot} Y \, dB + \int_0^{\cdot} Z \, dt \equiv \int_0^{\cdot} \tilde{Y} \, dB + \int_0^{\cdot} \tilde{Z} \, dt \qquad \text{on } [0, T]. \tag{6.1}$$

Then $Y \equiv \tilde{Y}$ and $Z \equiv \tilde{Z}$ on $[0, T]$.

Proof. We may take set the dimension to $e = 1$ by arguing componentwise. Also, by linearity, it suffices to consider the case $\tilde{Y} = 0$, $\tilde{Z} = 0$. By the second part of the previous proposition

$$\left\langle \int_0^{\cdot} Y \, dB \right\rangle \equiv \left\langle \sum_{k=1}^d \int_0^{\cdot} Y_k \, dB^k \right\rangle \equiv 0 \text{ on } [0, T].$$

On the other hand, since $\langle B^k, B^l \rangle_t = t$ if $k = l$, and zero otherwise,

$$\left\langle \sum_{k=1}^d \int_0^{\cdot} Y_k \, dB^k \right\rangle_{\cdot} \equiv \sum_{k,l=1}^d \int_0^{\cdot} Y_k Y_l \, d\langle B^k, B^l \rangle = \sum_{k=1}^d \int_0^{\cdot} Y_k^2 \, dt.$$

It follows that $Y \equiv 0$ as claimed. By differentiation, it then follows that also $Z \equiv 0$. $\square$

Clearly, the martingale and quadratic (co-)variation – i.e. *probabilistic* – properties of B play a key role in the proof of Corollary 6.2. It is worth noting that, with β a scalar Brownian motion and $B^1 = B^2 = \beta$ the conclusion fails; try non-zero $Y^1 \equiv -Y^2$, $Z \equiv 0$. It is crucial that d-dimensional standard Brownian motion "moves in all directions", captured through the non-degeneracy of the quadratic covariation matrix $\langle B^k, B^l \rangle_t$.

Surprisingly perhaps, one can formulate a purely *deterministic* decomposition of the form (6.1): the stochastic integrals will be replaced by rough integrals, the relevant probabilistic properties of B by certain conditions ("roughness from below[1], in all directions") on the sample path.

---

[1] As opposed to Hölder regularity which quantifies "roughness from above", in the sense of an upper estimate of the increment.

6.2 Uniqueness of the Gubinelli derivative and Doob–Meyer

Here and in the sequel of this section we fix $\alpha \in (\frac{1}{3}, \frac{1}{2}]$, a rough path $\mathbf{X} = (X, \mathbb{X}) \in \mathscr{C}^\alpha([0, T], V)$ and a controlled rough path $(Y, Y') \in \mathscr{D}_X^{2\alpha}$. We first address the question to what extent X and Y determine the Gubinelli derivative Y'. As it turns out, Y' is uniquely determined, provided that X is sufficiently "rough from below, in all directions". A Doob–Meyer type decomposition will then follow as a corollary.

Let us first consider the case when X is scalar, i.e. with values in $V = \mathbf{R}$. Assume that for some given $s \in [0, T)$, there exists a sequence of times $t_n \downarrow s$ such that $|X_{s,t_n}|/|t_n - s|^{2\alpha} \to \infty$, i.e.

$$\varlimsup_{t \downarrow s} \frac{|X_{s,t}|}{|t - s|^{2\alpha}} = +\infty.$$

Then Y_s' is uniquely determined from Y by (4.18) and the condition that $\|R^Y\|_{2\alpha} < \infty$. In fact, one necessarily has $X_{s,t_n} \in \mathbf{R} \setminus \{0\}$ for n large enough and so, from the very definition of R^Y,

$$Y_s' = \frac{Y_{s,t_n}}{X_{s,t_n}} - \frac{R_{s,t_n}^Y}{|t_n - s|^{2\alpha}} \frac{|t_n - s|^{2\alpha}}{X_{s,t_n}}$$

which implies that $\lim_{n \to \infty} Y_{s,t_n}/X_{s,t_n}$ exists and equals Y_s'. The multidimensional case is not that different, and the above consideration suggests the following definition.

Definition 6.3. For fixed $s \in [0, T)$ we call $X \in \mathcal{C}^\alpha([0, T], V)$ *"rough at time s"* if

$$\forall v^* \in V^* \setminus \{0\} : \quad \varlimsup_{t \downarrow s} \frac{|v^*(X_{s,t})|}{|t - s|^{2\alpha}} = \infty .$$

If X is rough on some dense set of $[0, T]$, we call it *truly rough*.

This definition is vindicated by the following result.

Proposition 6.4 (Uniqueness of Y'). *Let* $\mathbf{X} = (X, \mathbb{X}) \in \mathscr{C}^\alpha$, $(Y, Y') \in \mathscr{D}_X^{2\alpha}$, *so that the rough integral $\int Y \, d\mathbf{X}$ exists. Assume X is rough at some time $s \in [0, T)$. Then*

$$Y_{s,t} = \mathrm{O}\big(|t - s|^{2\alpha}\big) \quad \text{as } t \downarrow s \implies Y_s' = 0 . \tag{6.2}$$

As a consequence, if X is truly rough and $(Y, \tilde{Y}') \in \mathscr{D}_X^{2\alpha}$ is another controlled rough path (with respect to X) then $Y' \equiv \tilde{Y}'$.

Proof. From the definition of $(Y, \tilde{Y}') \in \mathscr{D}_X^{2\alpha}$, we have

$$Y_{s,t} = Y_s' X_{s,t} + \mathrm{O}\big(|t - s|^{2\alpha}\big) .$$

Hence, for $t \in (s, s + \varepsilon)$,

$$\frac{Y_s' X_{s,t}}{|t-s|^{2\alpha}} = \frac{Y_{s,t}}{|t-s|^{2\alpha}} + \mathrm{O}(1) = \mathrm{O}(1) \,,$$

where the second equality follows from the assumption made in (6.2). Now, $Y_s' X_{s,t}$ takes values in $\bar{W}$, the same Banach space in which Y takes its values. For every $w^* \in \bar{W}^*$, the map $V \ni v \mapsto w^*(Y_s' v)$ defines an element $v^* \in V^*$ so that

$$\frac{|v^*(X_{s,t})|}{|t-s|^{2\alpha}} = \left| \frac{w^*(Y_s' X_{s,t})}{|t-s|^{2\alpha}} \right| = \mathrm{O}(1) \text{ as } t \downarrow s;$$

Unless $v^* = 0$, the assumption that "X is rough at time s" implies that, along some sequence $t_n \downarrow s$, we have the divergent behaviour $|v^*(X_{s,t_n})|/|t_n - s|^{2\alpha} \to \infty$, which contradicts that the same expression is $\mathrm{O}(1)$ as $t_n \downarrow s$. We thus conclude that $v^* = 0$. In other words,

$$\forall w^* \in W^*, v \in V : \qquad w^*(Y_s' v) = 0 \,,$$

and this clearly implies $Y_s' = 0$. This finishes the proof of the implication stated in (6.2). □

The following result should be compared with Corollary 6.2.

Theorem 6.5 (Doob–Meyer for rough paths). *Assume that X is rough at some time $s \in [0, T)$ and let $(Y, Y') \in \mathscr{D}_X^{2\alpha}$. Then*

$$\int_s^t Y \, d\mathbf{X} = \mathrm{O}\big(|t-s|^{2\alpha}\big) \quad \text{as } t \downarrow s \implies Y_s = 0 \,. \tag{6.3}$$

As a consequence, if X is truly rough, $(\tilde{Y}, \tilde{Y}') \in \mathscr{D}_X^{2\alpha}$ and $Z, \tilde{Z} \in \mathcal{C}([0,T], W)$, then the identity

$$\int_0^{\cdot} Y \, d\mathbf{X} + \int_0^{\cdot} Z \, dt \equiv \int_0^{\cdot} \tilde{Y} \, d\mathbf{X} + \int_0^{\cdot} \tilde{Z} \, dt \tag{6.4}$$

on $[0, T]$ implies that $(Y, Y') \equiv (\tilde{Y}, \tilde{Y}')$ and $Z \equiv \tilde{Z}$ on $[0, T]$.

Proof. Recall from Theorem 4.10 that $(I, I') := \big(\int Y \, d\mathbf{X}, Y\big)$ is controlled by X, i.e. $(I, I') \in \mathscr{D}_X^{2\alpha}$. The statement (6.3) is then an immediate consequence of (6.2).

The claim is now straightforward. Pick any $s \in [0, T)$ such that X is rough at time s. From (6.4), and for all $0 \le s \le t \le T$,

$$\int_s^t \big(Y - \tilde{Y}\big) \, d\mathbf{X} = \int_s^t \big(Z_r - \tilde{Z}_r\big) \, dr = \mathrm{O}(|t-s|) = \mathrm{O}\big(|t-s|^{2\alpha}\big) \,,$$

where the last inequality is just the statement that $|t-s| = \mathrm{O}\big(|t-s|^{2\alpha}\big)$ as $t \downarrow s$, thanks to $\alpha \le 1/2$. We then conclude using (6.3) that $Y_s = \tilde{Y}_s$. If we now assume true roughness of X, this conclusion holds for a dense set of times s and hence, by continuity of Y and $\tilde{Y}$, we have $Y \equiv \tilde{Y}$. But then, by Proposition 6.4, we also have $Y' \equiv \tilde{Y}'$ and so

$$\int_0^{\cdot} Y \, d\mathbf{X} \equiv \int_0^{\cdot} \tilde{Y} \, d\mathbf{X} \, .$$

(Attention that the above notation "hides" the dependence on Y' resp. $\tilde{Y}'$.) But then (6.4) implies

$$\int_0^t Z_r \, dr \equiv \int_0^t \tilde{Z}_r \, dr \qquad \text{for } t \in [0, T],$$

and we conclude by differentiation with respect to t. $\square$

6.3 Brownian motion is truly rough

Recall that (say, d-dimensional standard) Brownian motion satisfies the so-called *(Khintchine) law of the iterated logarithm*, that is

$$\forall t \geq 0 : \quad \mathbf{P}\left(\overline{\lim_{h \downarrow 0}} \frac{|B_{t,t+h}|}{h^{\frac{1}{2}} (\ln \ln 1/h)^{1/2}} = \sqrt{2} \right) = 1. \tag{6.5}$$

See [McK69, p.18] or [RY99, Ch. II] for instance, typically proved with exponential martingales. Remark that it is enough to consider $t = 0$ since $(B_{t,t+h} : h \geq 0)$ is also a Brownian motion.

Theorem 6.6. *With probability one, Brownian motion on $V = \mathbf{R}^d$ is truly rough, relative to any Hölder exponent $\alpha \in [1/4, 1/2)$.*

Proof. It is enough to show that, for fixed time s, and any $\theta \in [1/2, 1)$,

$$\mathbf{P}\left(\forall v^* \in V^*, |v^*| = 1 : \overline{\lim_{t \downarrow s}} \frac{|v^*(B_{s,t})|}{|t-s|^\theta} = +\infty \right) = 1.$$

(Then take $s \in \mathbf{Q}$ and conclude that the above event holds true, simultaneously for all such s, with probability one.)

To this end, set $h^{\frac{1}{2}} (\ln \ln 1/h)^{1/2} \equiv \psi(h)$. We need the following two consequences of (6.5). There exists $c > 0$ (here $c = \sqrt{2}$) such that for every fixed unit dual vector $v^* \in V^* = \left(\mathbf{R}^d \right)^*$ and every fixed $s \in [0, T)$

$$\mathbf{P}\left(\overline{\lim_{t \downarrow s}} |v^*(B_{s,t})| / \psi(t-s) \geq c \right) = 1 \, ,$$

$$\mathbf{P}\left(\overline{\lim_{t \downarrow s}} \frac{|B_{s,t}|}{\psi(t-s)} < \infty \right) = 1 \, .$$

Take $K \subset V^*$ to be any dense, countable set of dual unit vectors. Since K is countable, the set on which the first condition holds simultaneously for all $v^* \in K$

has full measure,

$$\mathbb{P}\left(\forall v^* \in K : \overline{\lim_{t \downarrow s}} |v^*(B_{s,t})|/\psi(t-s) \geq c\right) = 1 \,.$$

On the other hand, every unit dual vector $v^* \in V^*$ is the limit of some $(v_n^*) \subset K$. Then

$$\frac{|v_n^*(B_{s,t})|}{\psi(t-s)} \leq \frac{|v^*(B_{s,t})|}{\psi(t-s)} + |v_n^* - v^*|_{V^*} \frac{|B_{s,t}|}{\psi(t-s)}$$

so that, using $\overline{\lim}\,(|a|+|b|) \leq \overline{\lim}\,(|a|) + \overline{\lim}\,(|b|)$, and restricting to the above set of full measure,

$$c \leq \overline{\lim_{t \downarrow s}} \frac{|v_n^*(B_{s,t})|}{\psi(t-s)} \leq \overline{\lim_{t \downarrow s}} \frac{|v^*(B_{s,t})|}{\psi(t-s)} + |v_n^* - v^*|_{V^*} \overline{\lim_{t \downarrow s}} \frac{|B_{s,t}|_V}{\psi(t-s)} \,.$$

Sending $n \to \infty$ gives, with probability one,

$$0 < c \leq \overline{\lim_{t \downarrow s}} \frac{|v^*(B_{s,t})|}{\psi(t-s)} \,.$$

Hence, for a.e. sample $B = B(\omega)$ we can pick a sequence (t_n) converging to s such that $|v^*(B_{s,t_n})|/\psi(t_n - s) \geq c - 1/n$. On the other hand, for any $\theta \geq 1/2$ we have

$$\frac{|v^*(B_{s,t_n}(\omega))|}{|t_n - s|^\theta} = \frac{|v^*(B_{s,t_n}(\omega))|}{\psi(t_n - s)} \frac{\psi(t_n - s)}{|t_n - s|^\theta}$$

$$\geq (c - 1/n)|t_n - s|^{\frac{1}{2}-\theta} L(t_n - s) \to \infty \,,$$

with $L(\tau) = (\ln \ln 1/\tau)^{1/2}$, where in the borderline case $\theta = 1/2$ (which corresponds to $\alpha = 1/4$) this divergence is only logarithmic. □

6.4 A deterministic Norris' lemma

We now turn our attention to a quantitative version of true roughness. In essence, we now replace 2α in Definition 6.3 by θ *and quantify the divergence, uniformly over all directions.*

Definition 6.7. A path $X : [0, T] \to V$ with values in a Banach space V is said to be θ-*Hölder rough* for $\theta \in (0, 1)$, on scale ⟨smaller than⟩ $\varepsilon_0 > 0$, if there exists a constant $L := L_\theta(X) := L(\theta, \varepsilon_0, T; X) > 0$ such that for every $v^* \in V^*$, $s \in [0, T]$ and $\varepsilon \in (0, \varepsilon_0]$ there exists $t \in [0, T]$ such that

$$|t - s| < \varepsilon \,, \quad \text{and} \quad |v^*(X_{s,t})| \geq L_\theta(X)\,\varepsilon^\theta |v^*| \,. \tag{6.6}$$

the largest such value of L is called the *modulus of θ-Hölder roughness* of X.

Observe that, indeed, any element in $\mathcal{C}^\alpha$ which is θ-Hölder rough for $\theta < 2\alpha$ is truly rough. (We shall see in the next section that multidimensional Brownian motion is θ-Hölder rough for any $\theta > 1/2$.) The following result can be viewed as quantitative version of Proposition 6.4.

Proposition 6.8. *Let $(X, \mathbb{X}) \in \mathscr{C}^\alpha([0, T], V)$ be such that X is θ-Hölder rough for some $\theta \in (0, 1]$. Then, for every controlled rough path $(Y, Y') \in \mathscr{D}_X^{2\alpha}([0, T], W)$ one has,*

$$\forall \varepsilon \in (0, \varepsilon_0] : L \varepsilon^\theta \|Y'\|_\infty \leq \mathrm{osc}(Y, \varepsilon) + \|R^Y\|_{2\alpha} \varepsilon^{2\alpha} . \qquad (6.7)$$

As immediate consequence, if $\theta < 2\alpha$, Y' is uniquely determined from Y, i.e. if (Y, Y') and $(\tilde{Y}, \tilde{Y}')$ both belong to $\mathscr{D}_X^{2\alpha}$ and $Y \equiv \tilde{Y}$, then $Y' \equiv \tilde{Y}'$.

Proof. Let us start with the consequence: apply estimate (6.7) with Y replaced by $Y - \tilde{Y} = 0$ and similarly Y' replaced by $Y' - \tilde{Y}'$. Thanks to $L > 0$ it follows that

$$\|Y' - \tilde{Y}'\|_\infty = \mathrm{O}(\varepsilon^{2\alpha - \theta})$$

and we send $\varepsilon \to 0$ to conclude $Y' = \tilde{Y}'$. The remainder of the proof is devoted to establish (6.7). Fix $s \in [0, T]$ and $\varepsilon \in (0, \varepsilon_0]$. From the definition of the remainder R^Y in (4.18), it then follows that

$$\sup_{|t-s|\leq\varepsilon} |Y_s' X_{s,t}| \leq \sup_{|t-s|\leq\varepsilon} \left(|Y_{s,t}| + |R_{s,t}^Y|\right) \leq \mathrm{osc}(Y, \varepsilon) + \|R^Y\|_{2\alpha} \varepsilon^{2\alpha} . \qquad (6.8)$$

Let now $w^* \in W^*$ be such that $|w^*| = 1$. Since X is θ-Hölder rough by assumption, there exists $u = u(w^*) \in [0, T]$ with $|u - s| < \varepsilon$ such that

$$\left|w^*(Y_s' X_{s,u})\right| = \left|\left((Y_s')^* w^*\right)(X_{s,u})\right| > L \varepsilon^\theta |(Y_s')^* w^*| . \qquad (6.9)$$

(Note that one has indeed $(Y_s')^* \colon W^* \to V^*$.) Combining both (6.8) and (6.9), we thus obtain that

$$L \varepsilon^\theta |(Y_s')^* w^*| \leq \mathrm{osc}(X, \varepsilon) + \|R^Y\|_{2\alpha} \varepsilon^{2\alpha} .$$

Taking the supremum over all such $w^* \in W^*$ of unit length and using the fact that the norm of a linear operator is equal to the norm of its dual, we obtain

$$L \varepsilon^\theta |Y_s'| \leq \mathrm{osc}(Y, \varepsilon) + \|R^Y\|_{2\alpha} \varepsilon^{2\alpha} .$$

Since s was also arbitrary, the stated bound follows at once. $\square$

Remark 6.9. Even though the argument presented above is independent of the dimension of V, we are not aware of any example where $L = L(\theta, X) > 0$ and $\dim V = \infty$. The reason why this definition works well only in the finite-dimensional case will be apparent in the proof of Proposition 6.11 below.

This leads us to the following quantitative version of our previous Doob–Meyer result for rough paths, Theorem 6.5. As usual, we assume that $\alpha \in (1/3, 1/2)$.

Theorem 6.10 (Norris lemma for rough paths). *Let* $\mathbf{X} = (X, \mathbb{X}) \in \mathscr{C}^{\alpha}([0, T], V)$ *be such that X is θ-Hölder rough with $\theta < 2\alpha$. Let $(Y, Y') \in \mathscr{D}_X^{2\alpha}([0, T], \mathcal{L}(V, W))$ and $Z \in \mathcal{C}^{\alpha}([0, T], W)$ and set*

$$I_t = \int_0^t Y_s d\mathbf{X}_s + \int_0^t Z_s ds.$$

Then there exist constants $r > 0$ and $q > 0$ such that, setting

$$\mathcal{R} := 1 + L_\theta(X)^{-1} + \|\mathbf{X}\|_\alpha + \|Y, Y'\|_{X;2\alpha} + |Y_0| + |Y_0'| + \|Z\|_\alpha + |Z_0|$$

one has the bound

$$\|Y\|_\infty + \|Z\|_\infty \le M\mathcal{R}^q \|I\|_\infty^r,$$

for a constant M depending only on α, θ, and the final time T.

Proof. We leave the details of the proof as an exercise, see [HP13], and only sketch its broad lines.

First, we conclude from Proposition 6.8 that I small in the supremum norm implies that $\|Y\|_\infty$ is also small. Then, we use interpolation to conclude from this that (Y, Y') is small when viewed as an element of $\mathscr{D}^{2\bar{\alpha}}$ for $\bar{\alpha} < \alpha$, thus implying that $\int Y \, d\mathbf{X}$ is necessarily small. This implies that $\int Z \, ds$ is itself small from which, using again interpolation, we finally conclude that Z itself must be small in the supremum norm. $\square$

6.5 Brownian motion is Hölder rough

We now turn to Hölder-roughness of Brownian motion. Our focus will be on the unit interval $T = 1$, and we consider scales up to $\varepsilon_0 = 1/2$ for the sake of argument.

Proposition 6.11. *Let B be a standard Brownian motion on $[0, 1]$ taking values in $\mathbf{R}^d$. Then, for every $\theta > \frac{1}{2}$, the sample paths of B are almost surely θ-Hölder rough. Moreover, with scale $\varepsilon_0 = 1/2$ and writing $L_\theta(B)$ for the modulus of θ-Hölder roughness, there exist constants M and c such that*

$$\mathbf{P}(L_\theta(B) < \varepsilon) \le M \exp\left(-c\varepsilon^{-2}\right),$$

for all $\varepsilon \in (0, 1)$.

The proof of Proposition 6.11 relies on the following variation of the standard small ball estimate for Brownian motion:

Lemma 6.12. *Let B be a d-dimensional standard Brownian motion. Then, there exist constants $c > 0$ and $C > 0$ such that*

$$\mathbf{P}\left(\inf_{|\varphi|=1} \sup_{t \in [0, \delta]} |\langle \varphi, B(t) \rangle| \le \varepsilon \right) \le C \exp(-c\delta\varepsilon^{-2}). \tag{6.10}$$

Proof. The standard small ball estimate for Brownian motion (see for example [LS01]) yields the bound

$$\sup_{|\varphi|=1} \mathbf{P}\left(\sup_{t\in[0,\delta]} |\langle \varphi, B(t)\rangle| \le \varepsilon \right) \le C \exp(-c\delta\varepsilon^{-2}) \,. \tag{6.11}$$

The required estimate then follows from a standard chaining argument, as in [Nor86, p. 127]: cover the sphere $|\varphi| = 1$ with $\varepsilon^{-2(d-1)}$ balls of radius ε^2, say, centred at φ_i. We then use the fact that, since the supremum of B has Gaussian tails, if $\sup_{t\in[0,\delta]} |\langle \varphi_i, B(t)\rangle| \le \varepsilon$, then the same bound, but with ε replaced by 2ε holds with probability exponentially close to 1 uniformly over all φ in the ball of radius ε^2 centred at φ_i. Since there are only polynomially many such balls required to cover the whole sphere, (6.10) follows. Note that this chaining argument uses in a crucial way that the number of balls or radius ε^2 required to cover the sphere $\|\varphi\| = 1$ grows only polynomially with ε^{-1}.

It is clear that bounds of the type (6.10) break down in infinite dimensions: if we consider a cylindrical Wiener process, then (6.11) still holds, but the unit sphere of a Hilbert space cannot be covered by a finite number of small balls anymore. If on the other hand, we consider a process with a non-trivial covariance, then we can get the chaining argument to work, but the bound (6.11) would break down due to the fact that $\langle \varphi, B(t)\rangle$ can then have arbitrarily small variance. $\qquad\square$

Proof (Proposition 6.11). With $T = 1, \varepsilon_0 = 1/2$, a different way of formulating Definition 6.7 is given by

$$L_\theta(X) = \inf \sup_{t:|t-s|\le \varepsilon} \frac{1}{\varepsilon^\theta} |\langle \varphi, X_{s,t}\rangle|.$$

where the inf is taken over $|\varphi| = 1$, $s \in [0,1]$ and $\varepsilon \in (0,1/2]$. We then define the "discrete analog" $D_\theta(X)$ of $L_\theta(X)$ to be given by

$$D_\theta(X) = \inf \sup_{s,t\in I_{k,n}} 2^{n\theta} |\langle \varphi, X_{s,t}\rangle| \,,$$

where $I_{k,n} = [\frac{k-1}{2^n}, \frac{k}{2^n}]$ and the inf is taken over $|\varphi| = 1$, $n \ge 1$ and $k \le 2^n$. We first claim that

$$L_\theta(X) \ge \frac{1}{2}\frac{1}{2^\theta} D_\theta(X). \tag{6.12}$$

To this end, fix a unit vector $\varphi \in V^*$, $s \in [0,1]$ and $\varepsilon \in (0,1/2]$. Pick $n \ge 1$: $\varepsilon/2 < 2^{-n} \le \varepsilon$. It follows that there exists some k such that $I_{k,n}$ is included in the set $\{t : |t - s| \le \varepsilon\}$. Then, by definition of D_θ, for any unit vector φ there exist two points $t_1, t_2 \in I_{k,n}$ such that

$$|\langle \varphi, X_{t_1,t_2}\rangle| \ge 2^{-n\theta} D_\theta(X).$$

Therefore, by the triangle inequality, we conclude that the magnitude of the difference between $\langle \varphi, X_s\rangle$ and one of the two terms $\langle \varphi, X_{t_i}\rangle, i = 1, 2$ (say t_1) is at least

$$|\langle \varphi, X_{s,t_1} \rangle| \geq \frac{1}{2} 2^{-n\theta} D_\theta(X)$$

and therefore

$$\frac{|\langle \varphi, X_{s,t_1} \rangle|}{\varepsilon^\theta} \geq \frac{1}{2} \frac{2^{-n\theta}}{\varepsilon^\theta} D_\theta(X) \geq \frac{1}{2} \frac{1}{2^\theta} D_\theta(X).$$

Since s, ε and φ were chosen arbitrarily, the claim (6.12) follows.

Applying this to Brownian sample paths, $X = B(\omega)$, it follows that it is sufficient to obtain the requested bound on $\mathbf{P}(D_\theta(B) < \varepsilon)$. We have the straightforward bound

$$\mathbf{P}(D_\theta(B) < \varepsilon) \leq \mathbf{P}\left(\inf_{\|\varphi\|=1} \inf_{n \geq 1} \inf_{k \leq 2^n} \sup_{s,t \in I_{k,n}} \frac{|\langle \varphi, B_{s,t} \rangle|}{2^{-n\theta}} < \varepsilon \right)$$

$$\leq \sum_{n=1}^{\infty} \sum_{k=1}^{2^n} \mathbf{P}\left(\inf_{\|\varphi\|=1} \sup_{s,t \in I_{k,n}} |\langle \varphi, B_{s,t} \rangle| < 2^{-n\theta}\varepsilon \right).$$

Trivially $\sup_{s,t \in I_{k,n}} |\langle \varphi, B_{s,t} \rangle| \geq \sup_{t \in I_{k,n}} |\langle \varphi, B_{r,t} \rangle|$, where r is the left boundary of the interval $I_{k,n}$, we can bound this by applying Lemma 6.12. Noting that the bound obtained in this way is independent of k, we conclude that

$$\mathbf{P}(D_\theta(B) < \varepsilon) \leq M \sum_{n=1}^{\infty} 2^n \exp\left(-c2^{(2\theta-1)n}\varepsilon^{-2}\right) \leq \tilde{M} \sum_{n=1}^{\infty} \exp\left(-\tilde{c}n\varepsilon^{-2}\right).$$

Here, we used the fact that as soon as $\theta > \frac{1}{2}$, we can find constants K and $\tilde{c}$ such that

$$n \log 2 - c2^{(2\theta-1)n}\varepsilon^{-2} \leq K - \tilde{c}n\varepsilon^{-2},$$

uniformly over all $\varepsilon < 1$ and all $n \geq 1$. (Consider separately the cases $\varepsilon^2 \in (0, 1/n)$ and $\varepsilon^2 \in [1/n, 1)$.) We deduce from this the bound

$$\mathbf{P}(D_\theta(B) < \varepsilon) \leq M\left(e^{-\tilde{c}\varepsilon^{-2}} + \int_1^\infty \exp\left(-\tilde{c}\varepsilon^{-2}x\right) dx \right),$$

which immediately implies the result.   □

Note that the proof given above is quite robust. In particular, we did not really make use of the fact that B has independent increments. In fact, it transpires that all that is required in order to prove the Hölder roughness of sample paths of a Gaussian process W with stationary increments is a small ball estimate of the type

$$\mathbf{P}\left(\sup_{t \in [0,\delta]} |W_t - W_0| \leq \varepsilon \right) \leq C \exp\left(-c\delta^\alpha \varepsilon^{-\beta}\right),$$

for some exponents $\alpha, \beta > 0$. These kinds of estimates are available for example for fractional Brownian motion with arbitrary Hurst parameter $H \in (0, 1)$.

6.6 Exercises

Exercise 6.1 *Show that the Q-Wiener process (as introduced in Exercise 3.4) is truly rough.*

Exercise 6.2 *Prove and state precisely: multidimensional fractional Brownian motion B^H, $H \in (1/3, 1/2]$, is truly rough.*

Exercise 6.3 *In (6.7), estimate $\mathrm{osc}(Z, \varepsilon)$ by $2\|Y\|_\infty$ (or alternatively by $\|Y\|_\alpha \varepsilon^\alpha$) and deduce the estimate*

$$\|Z'\|_\infty \leq \frac{1}{L} \inf_{\varepsilon \in (0, \varepsilon_0]} \left(2\varepsilon^{-\theta} \|Y\|_\infty + \|R^Z\|_{2\alpha} \varepsilon^{2\alpha - \theta} \right).$$

Carry out the elementary optimisation, e.g. when $\varepsilon_0 = T/2$, to see that

$$\|Z'\|_\infty \leq \frac{4\|Y\|_\infty}{L(\theta, X)} \left(\|R^Z\|_{2\alpha}^{\frac{\theta}{2\alpha}} \|Y\|_\infty^{-\frac{\theta}{2\alpha}} \vee T^{-\theta} \right).$$

∗ **Exercise 6.4 (Norris' lemma for rough paths; [HP13])** *Give a complete proof of Theorem 6.10.*

6.7 Comments

The notion of θ-roughness was first introduced in Hairer–Pillai [HP13], which also contains Proposition 6.8, although some of the ideas underlying the concepts presented here were already apparent in Baudoin–Hairer [BH07] and Hairer–Mattingly [HM11]. A version of this "Norris lemma" in the context of SDEs driven by fractional Brownian motion was proposed independently by Hu–Tindel [HT13]. The simplified condition of "true" roughness (which may be verified in infinite dimensions), targeted directly at a Doob–Meyer decomposition, is taken from Friz–Shekhar [FS13]; the quantitative "Norris lemma" is taken from Hairer–Pillai [HP13]. These results also hold in "rougher" situations, i.e. when $\alpha \leq 1/3$, see [FS13, CHLT15].

Chapter 7
Operations on controlled rough paths

At first sight, the notation $\int Y\, dX$ introduced in Chapter 4 is ambiguous since the resulting controlled rough path depends in general on the choices of both the second-order process $\mathbb{X}$ and the derivative process Y'. Fortunately, this "lack of completeness" in our notations is mitigated by the fact that in virtually all situations of interest, Y is constructed by using a small number of elementary operations described in this chapter. For all of these operations, it turns out to be intuitively rather clear how the corresponding derivative process is constructed.

7.1 Relation between rough paths and controlled rough paths

Consider $\mathbf{X} = (X, \mathbb{X}) \in \mathscr{C}^\alpha([0, T], V)$. It is easy to see that X itself can be interpreted as a path "controlled by X". Indeed, we can identify X with the element $(X, \mathrm{Id}) \in \mathscr{D}_X^{2\alpha}$, where Id is the identity matrix (more precisely: the constant path with value Id for all times).[1] Conversely, an element $(Y, Y') \in \mathscr{D}_X^{2\alpha}([0, T], W)$ can itself be interpreted as a rough path again, say $\mathbf{Y} = (Y, \mathbb{Y})$. Indeed, with the interpretation of the integral in the sense of (4.24), below fully spelled out for the reader's convenience, we can set

$$\mathbb{Y}_{s,t} = \int_s^t Y_{s,r} \otimes dY_r \stackrel{\text{def}}{=} \lim_{|\mathcal{P}| \to 0} \int_{\mathcal{P}} \Xi\,, \quad \Xi_{u,v} = Y_u \otimes Y_{u,v} + Y_u' \otimes Y_u' \mathbb{X}_{u,v}\,.$$

where $Y_u' \otimes Y_u' \in \mathcal{L}(V \otimes V, W \otimes W)$ is given by $(Y_u' \otimes Y_u')(v \otimes \tilde{v}) = (Y_u'(v)) \otimes (Y_u'(\tilde{v}))$. The fact that $\|\mathbb{Y}\|_{2\alpha}$ is finite is then a consequence of (4.25). On the other hand, the algebraic relations (2.1) already hold for the "Riemann sum" approximations to the three integrals, provided that the partition used for the approximation of $\mathbb{Y}_{s,t}$ is the union of the one used for the approximation of $\mathbb{Y}_{s,u}$ with the one used for $\mathbb{Y}_{u,t}$.

---

[1] It can also be useful to consider $t \mapsto \mathbb{X}_{0,t}$ as a path "controlled by X", resulting in the controlled rough path $(\mathbb{X}, X)$; cf. Exercise 4.6.

© Springer Nature Switzerland AG 2020
P. K. Friz, M. Hairer, *A Course on Rough Paths*, Universitext,
https://doi.org/10.1007/978-3-030-41556-3_7

We summarise the above consideration in saying that for every fixed $\mathbf{X} \in \mathscr{C}^\alpha([0, T], V)$, we have a continuous canonical injection

$$\mathscr{D}_X^{2\alpha}([0, T], W) \hookrightarrow \mathscr{C}^\alpha([0, T], W) .$$

Furthermore, this interpretation of elements of $\mathscr{D}_X^{2\alpha}$ as elements of $\mathscr{C}^\alpha$ is coherent in terms of the theory of integration constructed in the previous section, as can be seen by the following result:

Proposition 7.1. *Let $(X, \mathbb{X}) \in \mathscr{C}^\alpha$, let $(Y, Y') \in \mathscr{D}_X^{2\alpha}$, and let $\mathbf{Y} = (Y, \mathbb{Y}) \in \mathscr{C}^\alpha$ be the associated rough path constructed as above. If $(\tilde{Z}, \tilde{Z}') \in \mathscr{D}_Y^{2\alpha}$, then $(Z, Z') \in \mathscr{D}_X^{2\alpha}$, where $Z_t = \tilde{Z}_t$ and $Z_t' = \tilde{Z}_t' Y_t'$. Furthermore, one has the identity*

$$\int_0^t Z_s dY_s = \int_0^t \tilde{Z}_s d\mathbf{Y}_s . \tag{7.1}$$

Here, the left-hand side uses (4.24) to define the integral of two controlled rough paths against each other and the right-hand side uses the original definition (4.21) of the integral of a controlled rough path against its reference path.

Proof. By assumption, one has $Y_{s,t} = Y_s' X_{s,t} + \mathrm{O}(|t - s|^{2\alpha})$ and $\tilde{Z}_{s,t} = \tilde{Z}_s' Y_{s,t} + \mathrm{O}(|t - s|^{2\alpha})$. Combining these identities, it follows immediately that

$$Z_{s,t} = \tilde{Z}_s' Y_s' X_{s,t} + \mathrm{O}(|t - s|^{2\alpha}) = Z_s' X_{s,t} + \mathrm{O}(|t - s|^{2\alpha}) ,$$

so that $(Z, Z') \in \mathscr{D}_X^{2\alpha}$ as required. Now the left-hand side of (7.1) is given by $\mathcal{I}\Xi_{0,t}$ with $\Xi_{s,t} = Z_s Y_{s,t} + Z_s' Y_s' \mathbb{X}_{s,t}$, whereas the right-hand side is given by $\mathcal{I}\tilde{\Xi}_{0,t}$, where we set $\tilde{\Xi}_{s,t} = \tilde{Z}_s \tilde{Y}_{s,t} + \tilde{Z}_s' \mathbb{Y}_{s,t}$. Since $|\mathbb{Y}_{s,t} - Y_s' Y_s' \mathbb{X}_{s,t}| \le C|t - s|^{3\alpha}$ by (4.22), the claim now follows from Remark 4.13. $\square$

Remark 7.2. It is straightforward to see that if $\frac{1}{3} < \beta < \alpha$, then $\mathscr{C}^\alpha \hookrightarrow \mathscr{C}^\beta$ and, for every $X \in \mathscr{C}^\alpha$, we have a canonical embedding $\mathscr{D}_X^{2\alpha} \hookrightarrow \mathscr{D}_X^{2\beta}$. Furthermore, in view of the definition (4.10) of $\mathcal{I}$, the values of the integrals defined above do not depend on the interpretation of the integrand and integrator as elements of one or the other space.

7.2 Lifting of regular paths.

There is a canonical embedding $\iota \colon C^{2\alpha} \hookrightarrow \mathscr{D}_X^{2\alpha}$ given by $\iota Y = (Y, 0)$, since in this case $R_{s,t} = Y_{s,t}$ does indeed satisfy $\|R\|_{2\alpha} < \infty$. Recall that we are only interested in the case $\alpha \le \frac{1}{2}$. After all, if $Y_{s,t} = \mathrm{O}(|t - s|^{2\alpha})$ with $\alpha > \frac{1}{2}$, then Y has a vanishing derivative and must be constant.

7.3 Composition with regular functions.

Let W and $\bar{W}$ be two Banach spaces and let $\varphi \colon W \to \bar{W}$ be a function in C_b^2. Let furthermore $(Y, Y') \in \mathscr{D}_X^{2\alpha}([0, T], W)$ for some $X \in C^\alpha$. (In applications X will be part of some $\mathbf{X} = (X, \mathbb{X}) \in \mathscr{C}^\alpha$ but this is irrelevant here.) Then one can define a (candidate) controlled rough path $(\varphi(Y), \varphi(Y)') \in \mathscr{D}_X^{2\alpha}([0, T], \bar{W})$ by

$$\varphi(Y)_t = \varphi(Y_t), \qquad \varphi(Y)'_t = D\varphi(Y_t)Y'_t . \tag{7.2}$$

It is straightforward to check that the corresponding remainder term does indeed satisfy the required bound. It is also straightforward to check that, as a consequence of the chain rule, this definition is consistent in the sense that $(\varphi \circ \psi)(Y, Y') = \varphi(\psi(Y, Y'))$. We have the following result. Note that, since φ (and its derivatives) are only evaluated in a compact set (namely $Y([0, T]) \subset W$), there is no loss in generality in assuming φ (and its derivatives) bounded.

Lemma 7.3. *Let* $\varphi \in C_b^2$, $(Y, Y') \in \mathscr{D}_X^{2\alpha}([0, T], W)$ *for some* $X \in C^\alpha$ *with* $|Y'_0| + \|Y, Y'\|_{X,2\alpha} \le M \in [1, \infty)$. *Let* $(\varphi(Y), \varphi(Y)') \in \mathscr{D}_X^{2\alpha}([0, T], \bar{W})$ *be given by (7.2). Then, there exists a constant* C *depending only on* $T > 0$ *and* $\alpha > \frac{1}{3}$ *such that one has the bound*

$$\left\| \varphi(Y), \varphi(Y)' \right\|_{X,2\alpha} \le C_{\alpha,T} M \|\varphi\|_{C_b^2} (1 + \|X\|_\alpha)^2 \left(|Y'_0| + \|Y, Y'\|_{X,2\alpha} \right).$$

At last, C *can be chosen uniformly over* $T \in (0, 1]$.

Proof. We have $(\varphi(Y), \varphi(Y)') = (\varphi(Y_.), D\varphi(Y_.)Y'_.) \in \mathscr{D}_X^{2\alpha}$. Indeed,

$$\|\varphi(Y_.)\|_\alpha \le \|D\varphi\|_\infty \|Y_.\|_\alpha$$
$$\|\varphi(Y)'\|_\alpha \le \|D\varphi(Y_.)\|_\infty \|Y'_.\|_\alpha + \|Y'_.\|_\infty \|D\varphi(Y_.)\|_\alpha$$
$$\le \|D\varphi(Y_.)\|_\infty \|Y'_.\|_\alpha + \|Y'_.\|_\infty \|D^2\varphi(Y_.)\|_\infty \|Y_.\|_\alpha ,$$

which shows that $\varphi(Y), \varphi(Y)' \in C^\alpha$. Furthermore, $R^\varphi \equiv R^{\varphi(Y)}$ is given by

$$R_{s,t}^\varphi = \varphi(Y_t) - \varphi(Y_s) - D\varphi(Y_s)Y'_s X_{s,t}$$
$$= \varphi(Y_t) - \varphi(Y_s) - D\varphi(Y_s)Y_{s,t} + D\varphi(Y_s)R_{s,t}^Y$$

so that,

$$\|R^\varphi\|_{2\alpha} \le \frac{1}{2} |D^2\varphi|_\infty \|Y\|_\alpha^2 + |D\varphi|_\infty \|R^Y\|_{2\alpha}.$$

It follows that

$$\left\| \varphi(Y), \varphi(Y)' \right\|_{X,2\alpha} \le \|D\varphi(Y_.)\|_\infty \|Y'_.\|_\alpha + \|Y'_.\|_\infty \|D^2\varphi(Y_.)\|_\infty \|Y_.\|_\alpha$$
$$+ \frac{1}{2} |D^2\varphi|_\infty \|Y\|_\alpha^2 + |D\varphi|_\infty \|R^Y\|_{2\alpha}$$
$$\le \|\varphi\|_{C_b^2} \left(\|Y'_.\|_\alpha + \|Y'_.\|_\infty \|Y_.\|_\alpha + \|Y\|_\alpha^2 + \|R^Y\|_{2\alpha} \right)$$

$$\leq C_{\alpha,T}\|\varphi\|_{C_b^2}(1+\|X\|_\alpha)^2\left(1+|Y_0'|+\|Y,Y'\|_{X,2\alpha}\right)$$
$$\times\left(|Y_0'|+\|Y,Y'\|_{X,2\alpha}\right),$$

where we used in particular (4.20). $\square$

It follows immediately that one has the following "Leibniz rule", the proof of which is left to the reader:

Corollary 7.4. *Let* (Y,Y') *and* (Z,Z') *be two controlled paths in* $\mathscr{D}_X^{2\alpha}$ *for some* $X\in C^\alpha$. *Then the path* $U=YZ$, *with Gubinelli derivative* $U'=YZ'+ZY'$ *also belongs to* $\mathscr{D}_X^{2\alpha}$.

7.4 Stability II: Regular functions of controlled rough paths

In Lemma 7.3 we showed that controlled rough paths composed with (sufficiently) regular functions are again controlled rough paths. We shall be interested to quantify the continuity of this operation. As a useful warm-up, we start with the case of Hölder paths.

Lemma 7.5. *Assume* $\varphi\in C_b^2(W,\bar{W})$ *and* $T\leq 1$. *Then there exists a constant* $C_{\alpha,K}$ *such that for all* $X,Y\in C^\alpha([0,T],W)$ *with* $\|X\|_{\alpha;[0,T]},\|Y\|_{\alpha;[0,T]}\leq K\in[1,\infty)$,

$$\|\varphi(X)-\varphi(Y)\|_{\alpha;[0,T]}\leq C_{\alpha,K}\|\varphi\|_{C_b^2}\left(|X_0-Y_0|+\|X-Y\|_{\alpha;[0,T]}\right).$$

Proof. Consider the difference

$$\varphi(X)_{s,t}-\varphi(Y)_{s,t}=(\varphi(X_t)-\varphi(Y_t))-(\varphi(X_s)-\varphi(Y_s)).$$

The idea is to use a *division property* of sufficiently smooth functions. In the present context, this simply means that one has

$$\varphi(x)-\varphi(y)=g(x,y)(x-y)\quad\text{with}\quad g(x,y):=\int_0^1 D\varphi(tx+(1-t)y)\,dt\,,$$

where $g:W\times W\to\mathcal{L}(W,\bar{W})$ is obviously bounded by $\|D\varphi\|_\infty$ and in fact Lipschitz with $\|g\|_{Lip}\leq C\|D^2\varphi\|_\infty$ for some constant $C\geq 1$ relative to any product norm on $W\times W$, such as $|(x,y)|_{W\times W}=|x|+|y|$. It follows that

$$|(g(x,y)-g(\tilde{x},\tilde{y}))|\leq\|g\|_{Lip}|(x-\tilde{x},y-\tilde{y})|\leq C\|D^2\varphi\|_\infty(|x-\tilde{x}|+|y-\tilde{y}|).$$

Setting $\Delta_t=X_t-Y_t$ then allows to write

$$|\varphi(X)_{s,t}-\varphi(Y)_{s,t}|=|g(X_t,Y_t)\Delta_t-g(X_s,Y_s)\Delta_s|$$
$$=|g(X_t,Y_t)(\Delta_t-\Delta_s)+(g(X_t,Y_t)-g(X_s,Y_s))\Delta_s|$$

$$\leq \|g\|_\infty |X_{s,t} - Y_{s,t}| + \|g\|_{Lip} |(X_{s,t}, Y_{s,t})|_{W \times W} |X_s - Y_s|$$

$$\leq \|D\varphi\|_\infty |X_{s,t} - Y_{s,t}| + C\|D^2\varphi\|_\infty (|X_{s,t}| + |Y_{s,t}|)\|X - Y\|_{\infty;[0,T]}$$

$$\lesssim |t - s|^\alpha \left(\|D\varphi\|_\infty \|X - Y\|_\alpha + K\|D^2\varphi\|_\infty \|X - Y\|_{\infty;[0,T]} \right).$$

Since $T \leq 1$ we can also estimate $\|X - Y\|_{\infty;[0,T]} \leq |X_0 - Y_0| + \|X - Y\|_{\alpha;[0,T]}$ and the claimed estimate on $\varphi(X) - \varphi(Y)$ follows immediately. $\square$

We can now show the analogous statement for controlled rough paths, using notation previously introduced in Section 4.4.

Theorem 7.6 (Stability of composition). *Let* $X, \tilde{X} \in \mathcal{C}^\alpha([0,T])$ *with* $T \leq 1$, $(Y, Y') \in \mathscr{D}_X^{2\alpha}$, $(\tilde{Y}, \tilde{Y}') \in \mathscr{D}_{\tilde{X}}^{2\alpha}$. *For* $\varphi \in \mathcal{C}_b^3$ *define*

$$(Z, Z') := (\varphi(Y), D\varphi(Y)Y') \in \mathscr{D}_X^{2\alpha} \tag{7.3}$$

and similarly for $(\tilde{Z}, \tilde{Z}')$. *Then, one has the local Lipschitz estimates*

$$\|Z, Z'; \tilde{Z}, \tilde{Z}'\|_{X, \tilde{X}, 2\alpha} \leq C_M \Big(\|X - \tilde{X}\|_\alpha + |Y_0 - \tilde{Y}_0| + |Y_0' - \tilde{Y}_0'|$$
$$+ \|Y, Y'; \tilde{Y}, \tilde{Y}'\|_{X, \tilde{X}, 2\alpha} \Big), \tag{7.4}$$

as well as

$$\|Z - \tilde{Z}\|_\alpha \leq C_M \Big(\|X - \tilde{X}\|_\alpha + |Y_0 - \tilde{Y}_0| + |Y_0' - \tilde{Y}_0'| + \|Y, Y'; \tilde{Y}, \tilde{Y}'\|_{X, \tilde{X}, 2\alpha} \Big), \tag{7.5}$$

for a suitable constant $C_M = C(M, \alpha, \varphi)$.

Proof. (The reader is urged to revisit Lemma 7.3 where the composition (7.3) was seen to be well-defined for $\varphi \in \mathcal{C}_b^2$.) Similar as in the previous proof, noting that

$$|Z_0' - \tilde{Z}_0'| = |D\varphi(Y_0)Y_0' - D\varphi(\tilde{Y}_0)\tilde{Y}_0'| \leq C_M (|Y_0 - \tilde{Y}_0| + |Y_0' - \tilde{Y}_0'|)$$

it suffices to establish the first estimate, for (7.5) is an immediate consequence of (7.4) and (4.30). In order to establish the first estimate we need to bound

$$\|D\varphi(Y)Y' - D\varphi(\tilde{Y})\tilde{Y}'\|_\alpha + \|R^Z - R^{\tilde{Z}}\|_{2\alpha}.$$

Write $C_M(\varepsilon_X + \varepsilon_0 + \varepsilon_0' + \varepsilon)$ for the right-hand side of (7.4). Note that with this notation, from (4.30),

$$\|Y - \tilde{Y}\|_\alpha \lesssim \varepsilon_X + \varepsilon_0' + \varepsilon =: \varepsilon_Y,$$

and also $\|Y - \tilde{Y}\|_{\infty;[0,T]} \lesssim \varepsilon_0 + \varepsilon_Y$ (uniformly over $T \leq 1$). Since $D\varphi \in \mathcal{C}_b^2$, we know from Lemma 7.5 that

$$\|D\varphi(\tilde{Y}) - D\varphi(Y)\|_{\mathcal{C}^\alpha} = |D\varphi(\tilde{Y}_0) - D\varphi(Y_0)| + \|D\varphi(\tilde{Y}) - D\varphi(Y)\|_\alpha$$

$$\leq C(\varepsilon_0 + \varepsilon_Y)$$

where C depends on the $\mathcal{C}_b^3$-norm of φ. Also, $\left\| Y' - \tilde{Y}' \right\|_{\mathcal{C}^\alpha} \leq \varepsilon_0' + \varepsilon$. Clearly then ($\mathcal{C}^\alpha$ is a Banach algebra under pointwise multiplication), we have, for a constant C_M,

$$\left\| D\varphi(Y)Y' - D\varphi(\tilde{Y})\tilde{Y}' \right\|_\alpha \leq C_M(\varepsilon_0 + \varepsilon_Y + \varepsilon_0' + \varepsilon)$$
$$\lesssim C_M(\varepsilon_X + \varepsilon_0 + \varepsilon_0' + \varepsilon) \,.$$

To deal with $R^Z - R^{\tilde{Z}}$, write

$$R_{s,t}^Z = \varphi(Y_t) - \varphi(Y_s) - D\varphi(Y_s)Y_s'X_{s,t}$$
$$= \varphi(Y_t) - \varphi(Y_s) - D\varphi(Y_s)Y_{s,t} + D\varphi(Y_s)R_{s,t}^Y.$$

Taking the difference with $R^{\tilde{Z}}$ (replace Y, Y', R^Y above by $\tilde{Y}, \tilde{Y}', R^{\tilde{Y}}$) leads to the bound $\left| R_{s,t}^Z - R_{s,t}^{\tilde{Z}} \right| \leq T_1 + T_2$ where

$$T_1 := \varphi(Y_t) - \varphi(Y_s) - D\varphi(Y_s)Y_{s,t} - \left(\varphi(\tilde{Y}_t) - \varphi(\tilde{Y}_s) - D\varphi(\tilde{Y}_s)\tilde{Y}_{s,t} \right)$$
$$= \int_0^1 \left(D^2\varphi(Y_s + \theta Y_{s,t})(Y_{s,t}, Y_{s,t}) - D^2\varphi(\tilde{Y}_s + \theta \tilde{Y}_{s,t})\left(\tilde{Y}_{s,t}, \tilde{Y}_{s,t} \right) \right)(1 - \theta)d\theta$$
$$T_2 := D\varphi(Y_s)R_{s,t}^Y - D\varphi(\tilde{Y}_s)R_{s,t}^{\tilde{Y}} \,.$$

As for the second term, we know $R_{s,t}^Y - R_{s,t}^{\tilde{Y}} \leq (\varepsilon_0' + \varepsilon)|t - s|^{2\alpha}$, for all s, t, while

$$\left| D\varphi(\tilde{Y}_s) - D\varphi(Y_s) \right| \leq \left\| D^2\varphi \right\|_\infty |\tilde{Y}_s - Y_s| \leq \left\| D^2\varphi \right\|_\infty (\varepsilon_0 + \varepsilon_Y).$$

By elementary estimates of the form $\left| ab - \tilde{a}\tilde{b} \right| \leq |a| \left| b - \tilde{b} \right| + |a - \tilde{a}| \left| \tilde{b} \right|$ it then follows immediately that one has $T_2 \leq C(\varepsilon_X + \varepsilon_0 + \varepsilon_0' + \varepsilon)|t - s|^{2\alpha}$.

One argues similarly for the first term. This time, we consider the expression under the above integral $\int(\ldots)(1 - \theta)d\theta$ for fixed integration variable $\theta \in [0, 1]$. Using $Y^n \to Y$ in α-Hölder norm, we obtain

$$\left| D^2\varphi(\tilde{Y}_s + \theta\tilde{Y}_{s,t}) - D^2\varphi(Y_s + \theta Y_{s,t}) \right| \leq \left\| D^3\varphi \right\|_\infty \left(\left| \tilde{Y}_s - Y_s \right| + \left| \tilde{Y}_{s,t} - Y_{s,t} \right| \right)$$
$$\leq 3\left\| D^3\varphi \right\|_\infty \left\| \tilde{Y} - Y \right\|_\infty \lesssim \varepsilon_0 + \varepsilon_Y \,,$$

noting that this estimate is uniform in $s, t \in [0, T]$ and $\theta \in [0, 1]$. It then suffices to insert / subtract $D^2\varphi(Y_s + \theta Y_{s,t})\left(\tilde{Y}_{s,t}, \tilde{Y}_{s,t} \right)$ under the integral $\int \ldots (1 - \theta)d\theta$ appearing in the definition of T_1 and conclude with the triangle inequality and some simple estimates, keeping in mind that $\left\| Y - \tilde{Y} \right\|_\alpha \leq \varepsilon_Y$ and $\left\| Y \right\|_\alpha, \left\| \tilde{Y} \right\|_\alpha \lesssim C_M$.
□

7.5 Itô's formula revisited

Let $\mathbf{X} = (X, \mathbb{X}) \in \mathscr{C}^\alpha$, with $\alpha \in \left(\frac{1}{3}, \frac{1}{2}\right]$ as usual. In Proposition 5.8 we derived the following Itô formula

$$F(X_t) = F(X_0) + \int_0^t DF(X_s)d\mathbf{X}_s + \frac{1}{2}\int_0^t D^2F(X_s)d[\mathbf{X}]_s , \qquad (7.6)$$

and now ask for a similar formula for $F(Y_t)$, when $(Y, Y') \in \mathscr{D}_X^{2\alpha}$ is a controlled rough path. It turns out that we need to be more specific and assume

$$Y_t = Y_0 + \int_0^t Y_s' \, d\mathbf{X}_s + \Gamma_t , \qquad (7.7)$$

with $(Y', Y'') \in \mathscr{D}_X^{2\alpha}$, such as to have a well-defined rough integral; some flexibility is added in form of a "drift" term Γ, assumed regular in time. Such paths arise naturally as rough integrals of 1-forms, cf. Section 4.2, and also if Y is the solution to a rough differential equation driven by $\mathbf{X}$ to be discussed in Section 8.1. In analogy with similar Itô formulae from stochastic calculus, we expect

$$F(Y_t) = F(Y_0) + \int_0^t DF(Y_s)Y_s'd\mathbf{X}_s + \int_0^t DF(Y_s)\,d\Gamma_s$$
$$+ \frac{1}{2}\int_0^t D^2F(Y_s)\big(Y_s', Y_s'\big)\,d[\mathbf{X}]_s . \qquad (7.8)$$

Before going on, we note that the right-hand side above is indeed meaningful: the last two integrals are Young integrals and the first is a bona-fide rough integral. Indeed, by Lemma 7.3 and Corollary 7.4, the integrand $Z' := DF(Y)Y'$ is controlled by X, with Gubinelli derivative $Z'' = D^2F(Y)(Y', Y') + DF(Y)Y''$, so that the rough integral, following Theorem 4.10,

$$\int_0^t DF(Y_s)Y_s'd\mathbf{X}_s = \int_0^t Z_s'd\mathbf{X}_s = \lim_{|P|\to 0} \sum_{[u,v]\in P} \big(Z_u'X_{u,v} + Z_u''\mathbb{X}_{u,v}\big) , \quad (7.9)$$

is well-defined. (The extra structure $(Y', Y'') \in \mathscr{D}_X^{2\alpha}$ was crucially used.)

We note that, when $\mathbf{X} = \mathbf{B}^{\text{Itô}}(\omega)$, Itô enhanced Brownian motion, and Y, Y', Y'' are all adapted, then so is (Z', Z'') and the rough integral in (7.9) becomes, by Proposition 5.1, a classical Itô integral.

Theorem 7.7 (Itô formula II). *Let* $F : V \to W$ *in* C^3, $\mathbf{X} = (X, \mathbb{X}) \in \mathscr{C}^\alpha$ *and* $(Y, Y') \in \mathscr{D}_X^{2\alpha}$ *a controlled rough path of the form (7.7) for some controlled rough path* $(Y', Y'') \in \mathscr{D}_X^{2\alpha}$ *and some path* $\Gamma \in C^{2\alpha}$. *Then the the Itô formula (7.8) holds true.*

Proof. Assumption (7.7) implies that increments of Y are of the form

$$Y_{s,t} = Y_s' X_{s,t} + Y_s'' \mathbb{X}_{s,t} + \Gamma_{s,t} + \mathrm{o}(|t - s|) . \tag{7.10}$$

Thanks to (7.6), we know that $F(Y_t) - F(Y_0)$ equals

$$\lim_{|\mathcal{P}| \to 0} \sum_{[u,v] \in \mathcal{P}} \left(DF(Y_u) Y_{u,v} + D^2 F(Y_u) \mathbb{Y}_{u,v} \right) + \lim_{|\mathcal{P}| \to 0} \sum_{[u,v] \in \mathcal{P}} D^2 F(Y_u) [\mathbf{Y}]_{u,v} \tag{7.11}$$

where $\mathbb{Y}_{u,v} = \int_u^v Y_{u,\cdot} \otimes dY$ in the sense of Remark 4.12, noting that $\mathbb{Y}_{u,v} = Y_u' Y_u' \mathbb{X}_{u,v} + \mathrm{o}(|v - u|)$. Also,

$$\begin{aligned}
[\mathbf{Y}]_{u,v} &= Y_{u,v} \otimes Y_{u,v} - 2\,\mathrm{Sym}\,(\mathbb{Y}_{u,v}) \\
&= Y_u' Y_u' (X_{u,v} \otimes X_{u,v} - 2\,\mathrm{Sym}\,(\mathbb{X}_{u,v})) + \mathrm{o}(|v - u|) \\
&= Y_u' Y_u' [\mathbf{X}]_{u,v} + \mathrm{o}(|v - u|).
\end{aligned}$$

Let us also subtract / add $DF(Y_u) Y_u'' \mathbb{X}_{u,v}$ from (7.11). Then $F(Y_t) - F(Y_0)$ equals

$$\lim_{|\mathcal{P}| \to 0} \sum_{[u,v] \in \mathcal{P}} \left(DF(Y_u)(\mathbb{Y}_{u,v} - Y_u'' \mathbb{X}_{u,v}) + DF(Y_u) Y_u'' \mathbb{X}_{u,v} + D^2 F(Y_u) Y_u' Y_u' \mathbb{X}_{u,v} \right)$$

$$+ \lim_{|\mathcal{P}| \to 0} \sum_{[u,v] \in \mathcal{P}} D^2 F(Y_u) Y_u' Y_u' [\mathbf{X}]_{u,v}$$

$$= \lim_{|\mathcal{P}| \to 0} \sum_{[u,v] \in \mathcal{P}} DF(Y_u) Y_u' X_{u,v} + \left(DF(Y_u) Y_u'' + D^2 F(Y_u) Y_u' Y_u' \right) \mathbb{X}_{u,v}$$

$$+ \lim_{|\mathcal{P}| \to 0} \sum_{[u,v] \in \mathcal{P}} DF(Y_u) \Gamma_{u,v} + \int_0^t D^2 F(Y_u) Y_u' Y_u' d[\mathbf{X}]_u .$$

In view of (7.9), also noting the appearance of two Young integrals in the last line, the proof is complete.   □

It is worth having a different perspective on this Itô formula and take $\Gamma = 0$ for an unobstructed view. Then assumption 7.7 means exactly that $(Y, Y', Y'') \in \mathscr{D}_{\mathbf{X}}^{3\alpha}$ in the sense (cf. Definition 4.18)

$$\delta Y_{s,t} \overset{3\alpha}{\equiv} Y_s' X_{s,t} + Y_s'' \mathbb{X}_{s,t}, \qquad \delta Y_{s,t}' \overset{2\alpha}{\equiv} Y_s'' X_{s,t}, \qquad \delta Y_{s,t}'' \overset{\alpha}{\equiv} 0 . \tag{7.12}$$

If we furthermore restrict to $\mathbf{X}$ geometric, so that $[\mathbf{X}] \equiv 0$, Itô's formula takes the form of a classical chain rule,

$$F(Y_t) - F(Y_s) = Z_{s,t} = \int_s^t Z_r' d\mathbf{X}_r \overset{3\alpha}{\equiv} Z_s' X_{s,t} + Z_s'' \mathbb{X}_{s,t} .$$

On the other hand, $(Z', Z'') \in \mathscr{D}_{\mathbf{X}}^{2\alpha}$ means exactly $\delta Z_{s,t}' \overset{2\alpha}{\equiv} Z_s'' X_{s,t}$, $\delta Z_{s,t}'' \overset{\alpha}{\equiv} 0$. This discussion leads us to the following.

Proposition 7.8. *Let $F \in \mathcal{C}^3$ and $\mathbf{Y} = (Y, Y', Y'') \in \mathscr{D}_{\mathbf{X}}^{3\alpha}$ with geometric $\mathbf{X} = (X, \mathbb{X}) \in \mathscr{C}_g^\alpha$. Then*

$$Z = (Z, Z', Z'') := (F(Y), DF(Y)Y', DF(Y)Y'' + D^2F(Y)(Y', Y'))$$

is also an element in $\mathscr{D}_{\mathbf{X}}^{3\alpha}$. By abuse of notation, we write $\mathbf{Z} = F(\mathbf{Y})$.

Remark 7.9. The conclusion $\mathbf{Z} \in \mathscr{D}_{\mathbf{X}}^{3\alpha}$ can be "itemised", similar to (7.12). The $k\alpha$ estimates ($k = 1, 2, 3$) are then uniform over $F \in C_b^3$, in analogy with the estimate for elements in $\mathscr{D}_{\mathbf{X}}^{2\alpha}$, as was detailed in Lemma 7.3.

Proof. We give a direct proof, without intermediate use of rough integrals (and in fact no need for $\alpha > 1/3$) to emphasise the analogy with our previous Lemma 7.3 on composition of elements in $\mathscr{D}_X^{2\alpha}$ with regularity functions. By Taylor's theorem,

$$F(Y_t) \overset{3\alpha}{=} F(Y_s) + DF(Y_s)(Y_s' X_{s,t} + Y_s'' \mathbb{X}_{s,t}) + \frac{1}{2} D^2 F(Y_s)(Y_{s,t}, Y_{s,t}).$$

Note that $Y_{s,t} \otimes Y_{s,t} \overset{3\alpha}{=} (Y_s' X_{s,t})^{\otimes 2}$ and by geometricity $\frac{1}{2} X_{s,t}^{\otimes 2} = \mathbb{X}_{s,t}$, so that the second order term in the Taylor expansion can be replaced by

$$D^2 F(Y_s)(Y_s', Y_s') \mathbb{X}_{s,t} .$$

The remaining details are left to the reader.   □

As will be discussed in the next section, similar composition formulae can be obtained for arbitrary $\mathbf{Y} \in \mathscr{D}_{\mathbf{X}}^{\gamma}$ as long as $\gamma > 0$.

7.6 Controlled rough paths of low regularity

Let us conclude this section by showing how these canonical operations can be lifted to the case of controlled rough paths of low regularity, i.e. when $\alpha < \frac{1}{3}$. Recall from Section 4.5 that basis vectors in $T^{(N)}(\mathbf{R}^d)$ are of the form $e_w = e_1 \otimes \ldots \otimes e_k$, for words of the form $w = w_1 \cdots w_k$ with letters in $\{1, \ldots, d\}$, whereas we words themselves are identified via the dual dual basis of $T^{(N)}(\mathbf{R}^d)^*$,

$$w \leftrightarrow e_w^* .$$

Controlled rough paths $\mathbf{Y}$ are $T^{(N-1)}(\mathbf{R}^d)^*$-valued functions, which are controlled by increments of $\mathbf{X}$ in the sense of Definition 4.18.

This suggests that, in order to define the product of two controlled rough paths Y and $\bar{Y}$, we should first ask ourselves how a product of the type $\mathbf{X}_{s,t}^w \mathbf{X}_{s,t}^{\bar{w}}$ for two different words w a $\bar{w}$ can be rewritten as a linear combination of the increments of $\mathbf{X}$. It was seen in Section 2.4 that such a product is described by the *shuffle product* of words.

With this definition at hand, we saw that for any (weakly) geometric rough path $\mathbf{X}$ satisfies the identity

$$\mathbf{X}_{s,t}^w \mathbf{X}_{s,t}^{\bar{w}} = \mathbf{X}_{s,t}^{w \sqcup\!\sqcup \bar{w}} .$$

Also, $T^{(N)}(\mathbf{R}^d)^*$ becomes a commutative algebra, the *shuffle algbera*, via

$$e_w^* \star e_{\bar{w}}^* = e_{w \sqcup \bar{w}}^* \,.$$

This strongly suggests that the "correct" way of multiplying two controlled rough paths Y and $\bar{\text{Y}}$ is to define their product Z by

$$Z_t = Y_t \star \bar{Y}_t \,.$$

It is possible to check that Z is indeed again a controlled rough path. Similarly, if F is a (sufficiently) smooth function and Y is a controlled rough path, we abuse notation and define $F(\text{Y})$ by

$$F(\text{Y})_t \stackrel{\text{def}}{=} F(\text{Y}_t^\emptyset) + \sum_{k=1}^{N-1} \frac{1}{k!} F^{(k)}(\text{Y}_t^\emptyset) \, \tilde{\text{Y}}_t^{\star k} \,, \tag{7.13}$$

where $F^{(k)}$ denotes the kth derivative of F and $\tilde{\text{Y}}_t \stackrel{\text{def}}{=} \text{Y}_t - \text{Y}_t^\emptyset$ is the part describing the "local fluctuations". It is again possible to show that $F(\text{Y})$ is a controlled rough path if Y is a controlled rough path and F is sufficiently smooth (class $\mathcal{C}^p$ will do). This is nothing but the natural generalisation of the Itô formula in the formulation of Corollary 7.8. (The detailed verification of this is left to the reader in Exercise 7.5.)

Remark 7.10. A generalisation of (7.13) in the context of regularity structures is given in Proposition 14.8.

7.7 Exercises

♯ **Exercise 7.1** *Verify that* $\mathbb{X}_{s,t} = \int_s^t X_{s,r} \otimes dX_r$ *where the integral is to be interpreted in the sense of (4.24), taking* (Y, Y') *to be* (X, I). *In fact, check that this holds not only in the limit* $|\mathcal{P}| \to 0$ *but in fact for every fixed* $|\mathcal{P}|$, *i.e.* $\mathbb{X}_{s,t} = \int_{\mathcal{P}} \Xi$. *Compare this with formula (2.26), obtained in Exercise 2.4.*

Exercise 7.2 *Let* $\varphi \colon W \times [0,T] \to \bar{W}$ *be a function which is uniformly* $\mathcal{C}^2$ *in its first argument (i.e.* φ *is bounded and both* $D_y\varphi$ *and* $D_y^2\varphi$ *are bounded, where* D_y *denotes the Fréchet derivative with respect to the first argument) and uniformly* $\mathcal{C}^{2\alpha}$ *in its second argument. Let furthermore* $(Y, Y') \in \mathscr{D}_X^{2\alpha}([0,T], W)$. *Show that*

$$\varphi(Y)_t = \varphi(Y_t, t) \,, \qquad \varphi(Y)_t' = D_y\varphi(Y_t, t)Y_t' \,.$$

defines an element $(\varphi(Y), \varphi(Y)') \in \mathscr{D}_X^{2\alpha}([0,T], \bar{W})$. *In fact, show that there exists a constant* C, *depending only on* T, *such that one has the bound*

$$\|\varphi(Y)\|_{X,2\alpha} \le C\big(\|D_y^2\varphi\|_\infty + \|\varphi\|_\infty + \|\varphi\|_{2\alpha;t}\big)\big(1 + \|X\|_\alpha\big)^2\big(1 + \|Y\|_{X,2\alpha}\big)^2 \,,$$

where we denote by $\|\varphi\|_{2\alpha;t}$ *the supremum over* y *of the* 2α-*Hölder norm of* $\varphi(y, \cdot)$.

Exercise 7.3 (Composition with smooth functions; from [GH19]) *We return to the Hilbert/semigroup setting of Exercise 4.16. Let $\alpha \in \mathbf{R}$ and let $F \in C^2(H_\beta, H_\beta)$, consistently for every $\beta \geq \alpha$, with derivatives up to order 2 bounded. Let $\mathbf{X} = (X, \mathbb{X}) \in \mathscr{C}^\gamma([0, T], \mathbf{R}^d)$ for $\gamma \in (1/3, 1/2]$ and $(Y, Y') \in \mathscr{D}_{S,X}^{2\gamma}([0, T], H_\alpha)$. Moreover assume that in addition $Y \in L^\infty([0, T], H_{\alpha+2\gamma})$ and $Y' \in L^\infty([0, T], H_{\alpha+2\gamma})$. Show that $(Z_t, Z_t') := (F(Y_t), DF(Y_t) \circ Y_t')$ defines an element $(Z, Z') \in \mathscr{D}_{S,X}^{2\gamma}([0, T], H_\alpha)$ with the quantitative bound*

$$\|(Z, Z')\|_{X,2\gamma,\alpha} \lesssim (1 + |X|_\gamma)^2 (1 + \|Y\|_{\infty,\alpha+2\gamma} + \|Y'\|_{\infty,\alpha+2\gamma} + \|(Y, Y')\|_{X,2\gamma})^2.$$
$$(7.14)$$

The proportionality constant depends on the bounds on F and its derivatives. It also depends on time T, but is uniform over $T \in (0, 1]$.

Exercise 7.4 (Rough product formula) *Assume $Y = Y_0 + \int (Y', Y'')d\mathbf{X} + \Gamma$ as in Theorem 7.7, and similarly for $\bar{Y}$. Assume $\mathbf{X}$ is geometric, so that the bracket $[\mathbf{X}]$ vanishes. Then the following product formula holds*

$$Y_t \bar{Y}_t = Y_0 \bar{Y}_0 + \int_0^t (M, M')d\mathbf{X} + \int_0^t \left((d\Gamma_s)\bar{Y}_s + Y_s d\bar{\Gamma}_s\right)$$
$$\text{with} \quad M_s = Y_s' \bar{Y}_s + Y_s \bar{Y}_s', \qquad M_s' = Y_s'' \bar{Y}_s + 2Y_s' \bar{Y}_s' + Y_s \bar{Y}_s'' \,.$$

(If $Y, \bar{Y}$ take values in a Banach space V, the formula is understood as an identity in $V \otimes V$.)

Hint: *Apply Theorem 7.7 with $F(y, \bar{y}) = y\bar{y}$ (or $y \otimes \bar{y}$).*

Exercise 7.5 a) *Consider a controlled rough path $(Y, Y') \in \mathscr{D}_X^{2\alpha}$, with $X \in C^\alpha$, and verify that the composition formula (7.13), with $p = 2$, is consistent with Lemma 7.3.*

  b) *Consider then a controlled rough path $(Y, Y', Y'') \in \mathscr{D}_X^{3\alpha}$, with $\mathbf{X} = (X, \mathbb{X}) \in \mathscr{C}_g^\alpha$, and verify that the composition formula (7.13), with $p = 3$, is consistent with Corollary 7.8.*

7.8 Comments

Stability of controlled rough paths under composition with regular functions goes back in Gubinelli [Gub04], also in an α-Hölder setting $\alpha > \frac{1}{3}$, similar to our Sections 7.3 and 7.4. Extension to lower order regularity and then the "branched" setting are given by in [Gub10, HK15, FZ18], see also [BDFT20, Thm 2.11] for a concise proof in the geometric setting and connections to a multivariate Faà di Bruno formula.

  Our discussion of Itô's formula, Section 7.5, expands on a similar section of the first edition (2014), and makes more explicit the point that Itô's formula is really a composition formula for higher order controlled rough paths. Assuming $\alpha > \frac{1}{3}$ for the sake of argument,

Such formulae are sometimes directly given for RDE solutions, in which case the equation dictates a particular controlled structure, as seen spelled out directly in Davie's approach, Section 8.7. This is also a natural way to define manifold valued RDE solutions, similar to the definition of manifold valued semimartingales. See also comment Section 12.5 for some pointers to Itô formulae in the context of rough and stochastic PDEs.

Chapter 8
Solutions to rough differential equations

We show how to solve differential equations driven by rough paths by a simple Picard iteration argument. This yields a pathwise solution theory mimicking the standard solution theory for ordinary differential equations. We start with the simple case of differential equations driven by a signal that is sufficiently regular for Young's theory of integration to apply and then proceed to the case of more general rough signals.

8.1 Introduction

We now turn our attention to (rough) differential equations of the form

$$dY_t = f(Y_t)\, dX_t\,, \qquad Y_0 = \xi \in W\,. \tag{8.1}$$

Here, $X : [0, T] \to V$ is the driving or input signal, while $Y : [0, T] \to W$ is the output signal. As usual V and W are Banach spaces, and $f : W \to \mathcal{L}(V, W)$. When $\dim V = d < \infty$, one may think of f as a collection of vector fields $(f_1, \ldots, f_d)$ on W. As usual, the reader is welcome to think $V = \mathbf{R}^d$ and $W = \mathbf{R}^n$ but there is really no difference in the argument. Such equations are familiar from the theory of ODEs, and more specifically, control theory, where X is typically assumed to be absolutely continuous so that $dX_t = \dot{X}_t\, dt$. The case of SDEs, *stochastic differential equations*, with dX interpreted as Itô or Stratonovich differential of Brownian motion, is also well known. Both cases will be seen as special examples of RDEs, *rough differential equations*.

We may consider (8.1) on the unit time interval. Indeed, equation (8.1) is invariant under time-reparametrisation so that any (finite) time horizon may be rescaled to $[0, 1]$. Alternatively, global solutions on a larger time horizon are constructed successively, i.e. by concatenating $Y|_{[0,1]}$ (started at Y_0) with $Y|_{[1,2]}$ (started at Y_1) and so on. As a matter of fact, we shall construct solutions by a variation of the classical *Picard iteration* on intervals $[0, T]$, where $T \in (0, 1]$ will be chosen sufficiently small to guarantee invariance of suitable balls and the contraction property. Our key

© Springer Nature Switzerland AG 2020
P. K. Friz, M. Hairer, *A Course on Rough Paths*, Universitext,
https://doi.org/10.1007/978-3-030-41556-3_8

ingredients are estimates for rough integrals (cf. Theorem 4.10) and the composition of controlled paths with smooth maps (Lemma 7.3). Recall that, for rather trivial reasons (of the sort $|t - s|^{2\alpha} \leq |t - s|$, when $0 \leq s \leq t \leq T \leq 1$), all constants in these estimates were seen to be uniform in $T \in (0, 1]$.

8.2 Review of the Young case: a priori estimates

Let us postulate that there exists a solution to a differential equation in Young's sense and let us derive an a-priori estimate. (In finite dimension, this can actually be used to prove the existence of solutions. Note that the regularity requirement here is "one degree less" than what is needed for the corresponding uniqueness result.)

Proposition 8.1. *Assume* $X, Y \in \mathcal{C}^{\beta}([0, 1], V)$ *for some* $\beta \in (1/2, 1]$ *such that, given* $\xi \in W$, $f \in \mathcal{C}_b^1(W, \mathcal{L}(V, W))$, *we have*

$$dY_t = f(Y_t)dX_t, \qquad Y_0 = \xi,$$

in the sense of a Young integral equation. Then

$$\|Y\|_{\beta} \leq C\left[\left(\|f\|_{\mathcal{C}_b^1}\|X\|_{\beta}\right) \vee \left(\|f\|_{\mathcal{C}_b^1}\|X\|_{\beta}\right)^{1/\beta}\right].$$

Proof. By assumption, for $0 \leq s < t \leq 1$, $Y_{s,t} = \int_s^t f(Y_r)dX_r$. Using Young's inequality (4.3), with $C = C(\beta)$,

$$|Y_{s,t} - f(Y_s)X_{s,t}| = \left|\int_s^t (f(Y_r) - f(Y_s))dX_r\right|$$
$$\leq C\|Df\|_{\infty}\|Y\|_{\beta;[s,t]}\|X\|_{\beta;[s,t]}|t - s|^{2\beta}$$

so that

$$|Y_{s,t}|/|t - s|^{\beta} \leq \|f\|_{\infty}\|X\|_{\beta} + C\|Df\|_{\infty}\|Y\|_{\beta;[s,t]}\|X\|_{\beta;[s,t]}|t - s|^{\beta}.$$

Write $\|Y\|_{\beta;h} \equiv \sup |Y_{s,t}|/|t - s|^{\beta}$ where the sup is restricted to times $s, t \in [0, 1]$ for which $|t - s| \leq h$. Clearly then,

$$\|Y\|_{\beta;h} \leq \|f\|_{\infty}\|X\|_{\beta} + C\|Df\|_{\infty}\|Y\|_{\beta;h}\|X\|_{\beta}h^{\beta}$$

and upon taking h small enough, s.t. $\delta h^{\beta} \asymp 1$, with $\delta = \|X\|_{\beta}$, more precisely s.t.

$$C\|Df\|_{\infty}\|X\|_{\beta}h^{\beta} \leq C\left(1 + \|f\|_{\mathcal{C}_b^1}\right)\|X\|_{\beta}h^{\beta} \leq 1/2$$

(we will take h such that the second $\leq$ becomes an equality; adding 1 avoids trouble when $f \equiv 0$)

$$\frac{1}{2}\|Y\|_{\beta;h} \leq \|f\|_\infty \|X\|_\beta.$$

It then follows from Exercise 4.5 that, with $h \propto \|X\|_\beta^{-1/\beta}$,

$$\|Y\|_\beta \leq \|Y\|_{\beta;h}\left(1 \vee h^{-(1-\beta)}\right) \leq C\|X\|_\beta\left(1 \vee h^{-(1-\beta)}\right)$$
$$= C\left(\|X\|_\beta \vee \|X\|_\beta^{1/\beta}\right).$$

Here, we have absorbed the dependence on $f \in \mathcal{C}_b^1$ into the constants. By scaling (any non-zero f may be normalised to $\|f\|_{\mathcal{C}_b^1} = 1$ at the price of replacing X by $\|f\|_{\mathcal{C}_b^1} \times X$) we then get immediately the claimed estimate. □

8.3 Review of the Young case: Picard iteration

The reader may be helped by first reviewing the classical Picard argument in a Young setting, i.e. when $\beta \in (1/2, 1]$. Given $\xi \in W$, $f \in \mathcal{C}_b^2(W, \mathcal{L}(V, W))$, $X \in \mathcal{C}^\beta([0, 1], V)$ and $Y : [0, T] \to W$ of suitable Hölder regularity, $T \in (0, 1]$, one defines the map $\mathcal{M}_T$ by

$$\mathcal{M}_T(Y) := \left(\xi + \int_0^t f(Y_s)dX_s : t \in [0, T]\right).$$

Following a classical pattern of proof, we shall establish invariance of suitable balls, and then a contraction property upon taking $T = T_0$ small enough. The resulting unique fixed point is then obviously the unique solution to (8.1) on $[0, T_0]$. The unique solution on $[0, 1]$ is then constructed successively, i.e. by concatenating the solution Y on $[0, T_0]$, started at $Y_0 = \xi$, with the solution Y on $[T_0, 2T_0]$ started at Y_{T_0} and so on. Care is necessary to ensure that T_0 can be chosen uniformly; for instance, if f were only $\mathcal{C}^2$ (without the boundedness assumption) one can still obtain local existence on $[0, T_1]$, and then $[T_1, T_2]$, etc, but the resulting maximal solution (with respect to extension of solutions) may only be exist on $[0, \tau)$, for some $\lim_n T_n = \tau \leq T = 1$. In finite dimension, τ can be identified as explosion time, see also Exercise 8.4. (The situation here is completely analogous to the theory of Banach valued ODEs.)

We will need the Hölder norm of X over $[0, T]$ to tend to zero as $T \downarrow 0$. Now, as the example of the map $t \mapsto t$ and $\beta = 1$ shows, this may not be true relative to the β-Hölder norm; the (cheap) trick is to take $\alpha \in (1/2, \beta)$ and to view $\mathcal{M}_T$ as map from the Banach space $\mathcal{C}^\alpha([0, T], W)$, rather than $\mathcal{C}^\beta([0, T], W)$, into itself. Young's inequality is still applicable since all paths involved will be (at least) α-Hölder continuous with $\alpha > 1/2$. On the other hand,

$$\|X\|_{\alpha;[0,T]} \leq T^{\beta-\alpha}\|X\|_{\beta;[0,T]},$$

8 Solutions to rough differential equations

and so the α-Hölder norm of X has the desired behaviour. As previously, when no confusion is possible, we write $\|\cdot\|_\alpha \equiv \|\cdot\|_{\alpha;[0,T]}$.

To avoid norm versus seminorm considerations, it is convenient to work on the space of paths started at ξ, namely $\{Y \in C^\alpha([0,T], W) : Y_0 = \xi\}$. This affine subspace is a complete metric space under $(Y, \tilde{Y}) \mapsto \|Y - \tilde{Y}\|_\alpha$ and so is the closed unit ball

$$\mathcal{B}_T = \{Y \in C^\alpha([0,T], W) : Y_0 = \xi, \|Y\|_\alpha \leq 1\} .$$

Young's inequality (4.41) shows that there is a constant C which only depends on α (thanks to $T \leq 1$) such that for every $Y \in \mathcal{B}_T$,

$$\begin{aligned}
\|\mathcal{M}_T(Y)\|_\alpha &\leq C(|f(Y_0)| + \|f(Y)\|_\alpha)\|X\|_\alpha \\
&\leq C(|f(\xi)| + \|Df\|_\infty \|Y\|_\alpha)\|X\|_\alpha \\
&\leq C(|f|_\infty + \|Df\|_\infty)\|X\|_\alpha \leq C|f|_{C_b^1}\|X\|_\beta T^{\beta-\alpha} .
\end{aligned}$$

Similarly, for $Y, \tilde{Y} \in \mathcal{B}_T$, using Young, $f(Y_0) = f(\tilde{Y}_0)$ and Lemma 7.5 (with $K = 1$)

$$\begin{aligned}
\left\|\mathcal{M}_T(Y) - \mathcal{M}_T(\tilde{Y})\right\|_\alpha &= \left\|\int_0^\cdot f(Y_s)dX_s - \int_0^\cdot f(\tilde{Y}_s)dX_s\right\|_\alpha \\
&\leq C\left(|f(Y_0) - f(\tilde{Y}_0)| + \|f(Y) - f(\tilde{Y})\|_\alpha\right)\|X\|_\alpha \\
&\leq C\|f\|_{C_b^2}\|X\|_\beta T^{\beta-\alpha}\|Y - \tilde{Y}\|_\alpha .
\end{aligned}$$

It is clear from the previous estimates that a small enough $T_0 = T_0(f, \alpha, \beta, X) \leq 1$ can be found such that $\mathcal{M}_{T_0}(\mathcal{B}_{T_0}) \subset \mathcal{B}_{T_0}$ and, for all $Y, \tilde{Y} \in \mathcal{B}_{T_0}$,

$$\left\|\mathcal{M}_{T_0}(Y) - \mathcal{M}_{T_0}(\tilde{Y})\right\|_{\alpha;[0,T_0]} \leq \frac{1}{2}\|Y - \tilde{Y}\|_{\alpha;[0,T_0]} .$$

Therefore, $\mathcal{M}_{T_0}(\cdot)$ admits a unique fixed point $Y \in \mathcal{B}_{T_0}$ which is the unique solution Y to (8.1) on the (small) interval $[0, T_0]$. Noting that the choice $T_0 = T_0(f, \alpha, \beta, X)$ can indeed be done uniformly (in particular it does not change when the starting point ξ is replaced by Y_{T_0}), the unique solution on $[0, 1]$ is then constructed iteratively, as explained in the beginning.

8.4 Rough differential equations: a priori estimates

We now consider a priori estimates for rough differential equations, similar to Section 8.2. Recall that the homogeneous rough path norm $\|\mathbf{X}\|_\alpha$ was introduced in (2.4).

Proposition 8.2. *Let $\xi \in W$, $f \in C_b^3(W, \mathcal{L}(V, W))$ and a rough path $\mathbf{X} = (X, \mathbb{X}) \in \mathscr{C}^\alpha$ with $\alpha \in (1/3, 1/2]$ and assume that $(Y, Y') = (Y, f(Y)) \in \mathscr{D}_X^{2\alpha}$ is an RDE* *solution to* $dY = f(Y) \, d\mathbf{X}$ *started at* $Y_0 = \xi \in W$. *That is, for all* $t \in [0, T]$,

$$Y_t = \xi + \int_0^t f(Y_s) \, d\mathbf{X}_s \, , \tag{8.2}$$

with integral interpreted in the sense of Theorem 4.10 and $(f(Y), f(Y)') \in \mathscr{D}_X^{2\alpha}$ *built from* Y *by Lemma 7.3. (Thanks to* $\mathcal{C}_b^2$-*regularity of* f *and Lemma 7.3 the above rough integral equation (8.2) is well-defined.[1])*

Then the following (a priori) estimate holds true

$$\|Y\|_\alpha \le C\left[\left(\|f\|_{\mathcal{C}_b^2}\|\mathbf{X}\|_\alpha\right) \vee \left(\|f\|_{\mathcal{C}_b^2}\|\mathbf{X}\|_\alpha\right)^{1/\alpha}\right]$$

where $C = C(\alpha)$ *is a suitable constant.*

Proof. Consider an interval $I := [s, t]$ so that, using basic estimates for rough integrals (cf. Theorem 4.10),

$$
\begin{aligned}
\left|R_{s,t}^Y\right| &= |Y_{s,t} - f(Y_s)X_{s,t}| \\
&\le \left|\int_s^t f(Y)dX - f(Y_s)X_{s,t} - Df(Y_s)f(Y_s)\mathbb{X}_{s,t}\right| + |Df(Y_s)f(Y_s)\mathbb{X}_{s,t}| \\
&\lesssim \left(\|X\|_{\alpha;I}\left\|R^{f(Y)}\right\|_{2\alpha;I} + \|\mathbb{X}\|_{2\alpha;I}\|f(Y)\|_{\alpha;I}\right)|t - s|^{3\alpha} \\
&\quad + \|\mathbb{X}\|_{2\alpha;I}|t - s|^{2\alpha}.
\end{aligned}
\tag{8.3}
$$

Recall that $\|\cdot\|_\alpha$ is the usual Hölder seminorm over $[0, T]$, while $\|\cdot\|_{\alpha;I}$ denotes the same norm, but over $I \subset [0, T]$, so that trivially $\|X\|_{\alpha;I} \le \|X\|_\alpha$. Whenever notationally convenient, multiplicative constants depending on α and f are absorbed in $\lesssim$, at the very end we can use scaling to make the f dependence reappear. We will also write $\|\cdot\|_{\alpha;h}$ for the supremum of $\|\cdot\|_{\alpha;I}$ over all intervals $I \subset [0, T]$ with length $|I| \le h$. Again, one trivially has $\|X\|_{\alpha;I} \le \|X\|_{\alpha;h}$ whenever $|I| \le h$. Using this notation, we conclude from (8.3) that

$$\left\|R^Y\right\|_{2\alpha;h} \lesssim \|\mathbb{X}\|_{2\alpha;h} + \left(\|X\|_{\alpha;h}\left\|R^{f(Y)}\right\|_{2\alpha;h} + \|\mathbb{X}\|_{2\alpha;h}\|f(Y)\|_{\alpha;h}\right)h^\alpha.$$

We would now like to relate $R^{f(Y)}$ to R^Y. As in the proof of Lemma 7.3, we obtain the bound

$$
\begin{aligned}
R_{s,t}^{f(Y)} &= f(Y_t) - f(Y_s) - Df(Y_s)Y_s'X_{s,t} \\
&= f(Y_t) - f(Y_s) - Df(Y_s)Y_{s,t} + Df(Y_s)R_{s,t}^Y
\end{aligned}
$$

so that,

$$\left\|R^{f(Y)}\right\|_{2\alpha;h} \le \frac{1}{2}\left|D^2f\right|_\infty\|Y\|_{\alpha;h}^2 + |Df|_\infty\left\|R^Y\right\|_{2\alpha;h}$$

---

[1] Later we will establish existence and uniqueness under $\mathcal{C}_b^3$-regularity.

$$\lesssim \|Y\|_{\alpha;h}^2 + \|R^Y\|_{2\alpha;h}.$$

Hence, also using $\|f(Y)\|_{\alpha;h} \lesssim \|Y\|_{\alpha;h}$, there exists $c_1 > 0$, not dependent on $\mathbf{X}$ or Y, such that

$$\|R^Y\|_{2\alpha;h} \le c_1\|\mathbb{X}\|_{2\alpha;h} + c_1\|X\|_{\alpha;h}h^\alpha\|Y\|_{\alpha;h}^2 \tag{8.4}$$
$$+ c_1\|X\|_{\alpha;h}h^\alpha\|R^Y\|_{2\alpha;h} + c_1\|\mathbb{X}\|_{2\alpha;h}h^\alpha\|Y\|_{\alpha;h}.$$

We now restrict ourselves to h small enough so that $\|\mathbf{X}\|_\alpha h^\alpha \ll 1$. More precisely, we choose it such that

$$c_1\|X\|_\alpha h^\alpha \le \frac{1}{2}, \qquad c_1\|\mathbb{X}\|_{2\alpha}^{1/2}h^\alpha \le \frac{1}{2}.$$

Inserting this bound into (8.4), we conclude that

$$\|R^Y\|_{2\alpha;h} \le c_1\|\mathbb{X}\|_{2\alpha;h} + \frac{1}{2}\|Y\|_{\alpha;h}^2 + \frac{1}{2}\|R^Y\|_{2\alpha;h} + \|\mathbb{X}\|_{2\alpha;h}^{1/2}\|Y\|_{\alpha;h}.$$

This in turn yields the bound

$$\|R^Y\|_{2\alpha;h} \le 2c_1\|\mathbb{X}\|_{2\alpha;h} + \|Y\|_{\alpha;h}^2 + 2\|\mathbb{X}\|_{2\alpha;h}^{1/2}\|Y\|_{\alpha;h}$$
$$\le c_2\|\mathbb{X}\|_{2\alpha;h} + 2\|Y\|_{\alpha;h}^2, \tag{8.5}$$

with $c_2 = (2c_1 + 1)$. On the other hand, since $Y_{s,t} = f(Y_s)X_{s,t} - R_{s,t}^Y$ and f is bounded, we have the bound

$$\|Y\|_{\alpha;h} \lesssim \|X\|_\alpha + \|R^Y\|_{2\alpha;h}h^\alpha.$$

Combining this bound with (8.5) yields

$$\|Y\|_{\alpha;h} \le c_3\|X\|_\alpha + c_3\|\mathbb{X}\|_{2\alpha;h}h^\alpha + c_3\|Y\|_{\alpha;h}^2 h^\alpha$$
$$\le c_3\|X\|_\alpha + c_4\|\mathbb{X}\|_{2\alpha;h}^{1/2} + c_3\|Y\|_{\alpha;h}^2 h^\alpha,$$

for some constants c_3 and c_4. Multiplication with $c_3 h^\alpha$ then yields, with $\psi_h := c_3\|Y\|_{\alpha;h}h^\alpha$ and $\lambda_h := c_5\|\mathbf{X}\|_\alpha h^\alpha \to 0$ as $h \to 0$,

$$\psi_h \le \lambda_h + \psi_h^2.$$

Clearly, for all h small enough depending on Y (so that $\psi_h \le 1/2$) $\psi_h \le \lambda_h + \psi_h/2$ implies $\psi_h \le 2\lambda_h$ and so

$$\|Y\|_{\alpha;h} \le c_6\|\mathbf{X}\|_\alpha.$$

To see that this is true for all h small enough without dependence on Y, pick h_0 small enough so that $\lambda_{h_0} < 1/4$. It then follows that for each $h \le h_0$, one of the following two estimates must hold true

$$\psi_h \geq \psi_+ \equiv \frac{1}{2} + \sqrt{\frac{1}{4} - \lambda_h} \geq \frac{1}{2}$$

$$\psi_h \leq \psi_- \equiv \frac{1}{2} - \sqrt{\frac{1}{4} - \lambda_h} = \frac{1}{2}\left(1 - \sqrt{1 - 4\lambda_h}\right) \sim \lambda_h \text{ as } h \downarrow 0.$$

(In fact, for reasons that will become apparent shortly, we may decrease h_0 further to guarantee that for $h < h_0$ we have not only $\psi_h < 1/2$ but $\psi_h < 1/6$.) We already know that we are in the regime of the second estimate above as $h \downarrow 0$. Noting that $\psi_h(< 1/6) < 1/2$ in the second regime, the only reason that could prevent us from being in the second regime for all $h < h_0$ is an (upwards) jump of the (increasing) function $(0, h_0] \ni h \mapsto \psi_h$. But $\psi_h \leq 3 \lim_{g \uparrow h} \psi_g$, as seen from

$$\|Y\|_{\alpha;h} \leq 3\|Y\|_{\alpha;h/3} \leq 3 \lim_{g \uparrow h} \|Y\|_{\alpha;g} \ ,$$

(and similarly: $\lim_{g \downarrow h} \psi_g \leq 3\psi_h$) which rules out any jumps of relative jump size greater than 3. However, given that $\psi_h \geq 1/2$ in the first regime and $\psi_h < 1/6$ in the second, we can never jump from the second into the first regime, as h increases (from zero). And so, we indeed must be in the second regime for all $h \leq h_0$. Elementary estimates on ψ_-, as function of λ_h then show that

$$\|Y\|_{\alpha;h} \leq c_6 \|\mathbf{X}\|_\alpha \ ,$$

for all $h \leq h_0 \sim \|\mathbf{X}\|^{-1/\alpha}$. We conclude with Exercise 4.5, arguing exactly as in the Young case, Proposition 8.1. □

8.5 Rough differential equations

The aim of this section is to show that if f is regular enough and $(X, \mathbb{X}) \in \mathscr{C}^\beta$ with $\beta > \frac{1}{3}$, then we can solve differential equations driven by the rough path $\mathbf{X} = (X, \mathbb{X})$ of the type

$$dY = f(Y) \, d\mathbf{X} \ .$$

Such an equation will yield solutions in $\mathscr{D}_X^{2\alpha}$ and will be interpreted in the corresponding integral formulation, where the integral of $f(Y)$ against X is defined using Lemma 7.3 and Theorem 4.10. More precisely, one has the following local existence and uniqueness result. (The construction of a *maximal solution* is left as Exercise 8.4.)

Theorem 8.3. *Given $\xi \in W$, $f \in C^3(W, \mathcal{L}(V, W))$ and a rough path $\mathbf{X} = (X, \mathbb{X}) \in \mathscr{C}^\beta([0, T], V)$ with $\beta \in (\frac{1}{3}, \frac{1}{2})$, there exists $0 < T_0 \leq T$ and a unique element $(Y, Y') \in \mathscr{D}_X^{2\beta}([0, T_0], W)$, with $Y' = f(Y)$, such that, for all $0 \leq t \leq T_0$,*

$$Y_t = \xi + \int_0^t f(Y_s) \, d\mathbf{X}_s \ . \tag{8.6}$$

Here, the integral is interpreted in the sense of Theorem 4.10 and $f(Y) \in \mathscr{D}_X^{2\beta}$ is built from Y by Lemma 7.3. Moreover, if f is linear or $f \in C_b^3$, we may take $T_0 = T$, and thus global existence holds on $[0, T]$.

Remark 8.4. The condition $Y' = f(Y)$ (and then $f(Y)' = Df(Y)Y'$ by Lemma 7.3) is crucial for uniqueness. To see what can happen, consider the canonical lift of $X \in C^1$ to $\mathbf{X} = (X, \int X \otimes dX)$, in which case *any* choice of $f(Y)' \in C^\beta$ yields a pair $(f(Y), f(Y)') \in \mathscr{D}_X^{2\beta}$. (Indeed, thanks to $|X_{s,t}| \lesssim |t - s|$, the term $f(Y)'_s X_{s,t}$ can always be absorbed in the 2β-remainder.) On the other hand, regardless of the choice of Y', or $f(Y)'$, the rough integral in (8.6) here always agrees with the Riemann-Stieltjes integral $\int f(Y)dX$, so that (8.6) is satisfied whenever Y solves the ODE $\dot{Y} = f(Y)\dot{X}$, with $Y_0 = \xi$.

Proof. With $\mathbf{X} = (X, \mathbb{X}) \in \mathscr{C}^\beta \subset \mathscr{C}^\alpha$, $\frac{1}{3} < \alpha < \beta$ and $(Y, Y') \in \mathscr{D}_X^{2\alpha}$ we know from Lemma 7.3 that

$$(\Xi, \Xi') := \big(f(Y), f(Y)'\big) := (f(Y), Df(Y)Y') \in \mathscr{D}_X^{2\alpha} .$$

Restricting from $[0, 1]$ to $[0, T]$, any $T \leq 1$, Theorem 4.10 allows to define the map

$$\mathcal{M}_T(Y, Y') \overset{\text{def}}{=} \left(\xi + \int_0^\cdot \Xi_s d\mathbf{X}_s, \Xi \right) \in \mathscr{D}_X^{2\alpha} .$$

The RDE solution on $[0, T]$ we are looking for is a fixed point of this map. Strictly speaking, this would only yield a solution (Y, Y') in $\mathscr{D}_X^{2\alpha}$. But since $\mathbf{X} \in \mathscr{C}^\beta$, it turns out that this solution is automatically an element of $\mathscr{D}_X^{2\beta}$. Indeed, $|Y_{s,t}| \leq |Y'|_\infty |X_{s,t}| + \|R^Y\|_{2\alpha} |t - s|^{2\alpha}$, so that $Y \in C^\beta$. From the fixed point property it then follows that $Y' = f(Y) \in C^\beta$ and also $R^Y \in C_2^{2\beta}$, since $\mathbb{X} \in C_2^{2\beta}$ and

$$|R_{s,t}^Y| = |Y_{s,t} - Y_s'X_{s,t}| = \left| \int_s^t (f(Y_r) - f(Y_s))d\mathbf{X}_t \right|$$

$$\leq |Y'|_\infty |\mathbb{X}_{s,t}| + O\big(|t - s|^{3\alpha}\big) .$$

Note that if (Y, Y') is such that $(Y_0, Y_0') = (\xi, f(\xi))$, then the same is true for $\mathcal{M}_T(Y, Y')$. Therefore, $\mathcal{M}_T$ can be viewed as map on the space of controlled paths started at $(\xi, f(\xi))$, i.e.

$$\{(Y, Y') \in \mathscr{D}_X^{2\alpha}([0, T], W) : Y_0 = \xi, Y_0' = f(\xi)\} .$$

Since $\mathscr{D}_X^{2\alpha}$ is a Banach space (under the norm $(Y, Y') \mapsto |Y_0| + |Y_0'| + \|Y, Y'\|_{X,2\alpha}$) the above (affine) subspace is a complete metric space under the induced metric. This is also true for the (closed) unit ball $\mathcal{B}_T$ centred at, say

$$t \mapsto (\xi + f(\xi)X_{0,t}, f(\xi)).$$

(Note here that the apparently simpler choice $t \mapsto (\xi, f(\xi))$ does in general not belong to $\mathscr{D}_X^{2\alpha}$.) In other words, $\mathcal{B}_T$ is the set of all $(Y, Y') \in \mathscr{D}_X^{2\alpha}([0, T], W)$:

$Y_0 = \xi, Y_0' = f(\xi)$ and

$$|Y_0 - \xi| + |Y_0' - f(\xi)| + \|(Y - (\xi + f(\xi)X_{0,\cdot}), Y' - f(\xi))\|_{X,2\alpha}$$
$$= \|(Y - f(\xi)X_{0,\cdot}, Y' - f(\xi))\|_{X,2\alpha} \leq 1.$$

In fact, $\|(Y - f(\xi)X_{0,\cdot}, Y' - f(\xi))\|_{X,2\alpha} = \|Y, Y'\|_{X,2\alpha}$ as a consequence of the triangle inequality and $\|(f(\xi)X_{0,\cdot}, f(\xi))\|_{X,2\alpha} = \|f(\xi)\|_\alpha + \|0\|_{2\alpha} = 0$, so that

$$\mathcal{B}_T = \left\{ (Y, Y') \in \mathscr{D}_X^{2\alpha}([0,T], W) : Y_0 = \xi, Y_0' = f(\xi) : \|(Y, Y')\|_{X,2\alpha} \leq 1 \right\}.$$

Let us also note that, for all $(Y, Y') \in \mathcal{B}_T$, one has the bound

$$|Y_0'| + \|(Y, Y')\|_{X,2\alpha} \leq |f|_\infty + 1 =: M \in [1, \infty). \tag{8.7}$$

We now show that, for T small enough, $\mathcal{M}_T$ leaves $\mathcal{B}_T$ invariant and in fact is contracting. Constants below are denoted by C, may change from line to line and may depend on $\alpha, \beta, X, \mathbb{X}$ without special indication. They are, however, uniform in $T \in (0, 1]$ and we prefer to be explicit (enough) with respect to f such as to see where $\mathcal{C}_b^3$-regularity is used. With these conventions, we recall the following estimates, direct consequences from Lemma 7.3 and Theorem 4.10 , respectively,

$$\|\Xi, \Xi'\|_{X,2\alpha} \leq CM \|f\|_{\mathcal{C}_b^2} (|Y_0'| + \|Y, Y'\|_{X,2\alpha})$$

$$\left\| \int_0^{\cdot} \Xi_s d\mathbf{X}_s, \Xi \right\|_{X,2\alpha} \leq \|\Xi\|_\alpha + \|\Xi'\|_\infty \|\mathbb{X}\|_{2\alpha}$$
$$+ C(\|X\|_\alpha \|R^\Xi\|_{2\alpha} + \|\mathbb{X}\|_{2\alpha} \|\Xi'\|_\alpha)$$
$$\leq \|\Xi\|_\alpha + C(|\Xi_0'| + \|\Xi, \Xi'\|_{X,2\alpha})(\|X\|_\alpha + \|\mathbb{X}\|_{2\alpha})$$
$$\leq \|\Xi\|_\alpha + C(|\Xi_0'| + \|\Xi, \Xi'\|_{X,2\alpha})T^{\beta-\alpha}.$$

Invariance: For $(Y, Y') \in \mathcal{B}_T$, noting that $\|\Xi\|_\alpha = \|f(Y)\|_\alpha \leq \|f\|_{\mathcal{C}_b^1} \|Y\|_\alpha$ and that $|\Xi_0'| = |Df(Y_0)Y_0'| \leq \|f\|_{\mathcal{C}_b^1}^2$, we obtain the bound

$$\|\mathcal{M}_T(Y, Y')\|_{X,2\alpha} = \left\| \int_0^{\cdot} \Xi_s d\mathbf{X}_s, \Xi \right\|_{X,2\alpha}$$
$$\leq \|\Xi\|_\alpha + C(|\Xi_0'| + \|\Xi, \Xi'\|_{X,2\alpha})T^{\beta-\alpha}$$
$$\leq \|f\|_{\mathcal{C}_b^1} \|Y\|_\alpha + C\left(\|f\|_{\mathcal{C}_b^1}^2 + CM \|f\|_{\mathcal{C}_b^2}(|Y_0'| + \|Y, Y'\|_{X,2\alpha}) \right)T^{\beta-\alpha}$$
$$\leq \|f\|_{\mathcal{C}_b^1}(\|f\|_\infty + 1)T^{\beta-\alpha} + CM\left(\|f\|_{\mathcal{C}_b^1}^2 + \|f\|_{\mathcal{C}_b^2}(\|f\|_\infty + 1) \right)T^{\beta-\alpha},$$

where in the last step we used (8.7) and also $\|Y\|_{\alpha;[0,T]} \leq C_f T^{\beta-\alpha}$, seen from

$$|Y_{s,t}| \leq |Y'|_\infty |X_{s,t}| + \|R^Y\|_{2\alpha} |t - s|^{2\alpha}$$

$$\leq \left(|Y_0'| + \|Y'\|_\alpha\right)\|X\|_\beta |t-s|^\beta + \left\|R^Y\right\|_{2\alpha}|t-s|^{2\alpha} .$$

Then, using $T^\alpha \leq T^{\beta-\alpha}$ and $\left\|R^Y\right\|_{2\alpha} \leq \|Y,Y'\|_{X,2\alpha} \leq 1$, we obtain the bound

$$\|Y\|_{\alpha;[0,T]} \leq \left(|Y_0'| + \|Y,Y'\|_{X,2\alpha}\right)\|X\|_\beta T^{\beta-\alpha} + \left\|R^Y\right\|_{2\alpha}T^{\beta-\alpha} \qquad (8.8)$$
$$\leq \left((\|f\|_\infty + 1)\|X\|_\beta + 1\right)T^{\beta-\alpha} .$$

In other words, $\|\mathcal{M}_T(Y,Y')\|_{X,2\alpha} = \|\mathcal{M}_T(Y,Y')\|_{X,2\alpha;[0,T]} = \mathrm{O}(T^{\beta-\alpha})$ with constant only depending on $\alpha, \beta, \mathbf{X}$ and $f \in \mathcal{C}_b^2$. By choosing $T = T_0$ small enough, we obtain the bound $\|\mathcal{M}_{T_0}(Y,Y')\|_{X,2\alpha;[0,T_0]} \leq 1$ so that $\mathcal{M}_{T_0}$ leaves $\mathcal{B}_{T_0}$ invariant, as desired.

Contraction: Setting $\Delta_s = f(Y_s) - f(\tilde{Y}_s)$ as a shorthand, we have the bound

$$\left\|\mathcal{M}_T(Y,Y') - \mathcal{M}_T(\tilde{Y},\tilde{Y}')\right\|_{X,2\alpha} = \left\|\int_0^\cdot \Delta_s d\mathbf{X}_s, \Delta\right\|_{X,2\alpha}$$
$$\leq \|\Delta\|_\alpha + C\left(|\Delta_0'| + \|\Delta,\Delta'\|_{X,2\alpha}\right)T^{\beta-\alpha}$$
$$\leq C\|f\|_{\mathcal{C}_b^2}\left\|Y - \tilde{Y}\right\|_\alpha + C\|\Delta,\Delta'\|_{X,2\alpha}T^{\beta-\alpha} .$$

The contraction property is obvious, provided that we can establish the following two estimates:

$$\left\|Y - \tilde{Y}\right\|_\alpha \leq CT^{\beta-\alpha}\left\|Y - \tilde{Y}, Y' - \tilde{Y}'\right\|_{X,2\alpha} , \qquad (8.9)$$
$$\left\|\Delta,\Delta'\right\|_{X,2\alpha} \leq C\left\|Y - \tilde{Y}, Y' - \tilde{Y}'\right\|_{X,2\alpha} . \qquad (8.10)$$

To obtain (8.9), replace Y by $Y - \tilde{Y}$ in (8.8), noting $Y_0' - \tilde{Y}_0' = 0$, and this shows

$$\left\|Y - \tilde{Y}\right\|_\alpha \leq \left\|Y' - \tilde{Y}'\right\|_\alpha\|X\|_\beta T^{\beta-\alpha} + \left\|R^Y - R^{\tilde{Y}}\right\|_{2\alpha}T^{\beta-\alpha}$$
$$\leq CT^{\beta-\alpha}\left\|Y - \tilde{Y}, Y' - \tilde{Y}'\right\|_{X,2\alpha}.$$

We now turn to (8.10). Similar to the proof of Lemma 7.5, $f \in \mathcal{C}^3$ allows to write $\Delta_s = G_s H_s$ where

$$G_s := g\left(Y_s, \tilde{Y}_s\right) , \quad H_s := Y_s - \tilde{Y}_s ,$$

and $g \in \mathcal{C}_b^2$ with $\|g\|_{\mathcal{C}_b^2} \leq C\|f\|_{\mathcal{C}_b^3}$. Lemma 7.3 tells us that $(G,G') \in \mathscr{D}_X^{2\alpha}$ (with $G' = (D_Y g)Y' + (D_{\tilde{Y}}g)\tilde{Y}'$) and in fact immediately yields an estimate of the form

$$\|G,G'\|_{X,2\alpha} \leq C\|f\|_{\mathcal{C}_b^3} ,$$

uniformly over $(Y,Y'), (\tilde{Y},\tilde{Y}') \in \mathcal{B}_T$ and $T \leq 1$. On the other hand, $\mathscr{D}_X^{2\alpha}$ is an algebra in the sense that $(GH,(GH)') \in \mathscr{D}_X^{2\alpha}$ with $(GH)' = G'H + GH'$. In fact, we leave it as easy exercise to the reader to check that

$$\|GH, (GH)'\|_{X,2\alpha} \lesssim (|G_0| + |G_0'| + \|G, G'\|_{X,2\alpha})$$
$$\times (|H_0| + |H_0'| + \|H, H'\|_{X,2\alpha}) .$$

In our situation, $H_0 = Y_0 - \tilde{Y}_0 = \xi - \xi = 0$, and similarly $H_0' = 0$, so that, for all $(Y, Y'), (\tilde{Y}, \tilde{Y}') \in \mathcal{B}_T$, we have

$$\|\Delta, \Delta'\|_{X,2\alpha} \lesssim (|G_0| + |G_0'| + \|G, G'\|_{X,2\alpha})\|H, H'\|_{X,2\alpha}$$
$$\lesssim (\|g\|_\infty + \|g\|_{\mathcal{C}_b^1} (|Y_0'| + |\tilde{Y}_0'|) + C\|f\|_{\mathcal{C}_b^3})\|Y - \tilde{Y}, Y' - \tilde{Y}'\|_{X,2\alpha}$$
$$\lesssim \|Y - \tilde{Y}, Y' - \tilde{Y}'\|_{X,2\alpha} ,$$

where we made use of $\|g\|_\infty, \|g\|_{\mathcal{C}_b^1} \lesssim \|f\|_{\mathcal{C}_b^3}$ and $|Y_0'| = |\tilde{Y}_0'| = |f(\xi)| \le |f|_\infty$.

The argument from here on is identical to the Young case: the previous estimates allow for a small enough $T_0 \le 1$ such that $\mathcal{M}_{T_0}(\mathcal{B}_{T_0}) \subset \mathcal{B}_{T_0}$ and for all $(Y, Y'), (\tilde{Y}, \tilde{Y}') \in \mathcal{B}_{T_0}$:

$$\|\mathcal{M}_{T_0}(Y, Y') - \mathcal{M}_{T_0}(\tilde{Y}, \tilde{Y}')\|_{X,2\alpha} \le \frac{1}{2}\|Y - \tilde{Y}, Y' - \tilde{Y}'\|_{X,2\alpha}$$

and so $\mathcal{M}_{T_0}(\cdot)$ admits a unique fixed point $(Y, Y') \in \mathcal{B}_{T_0}$, which is then the unique solution Y to (8.1) on the (possibly rather small) interval $[0, T_0]$. Noting that the choice of T_0 can again be done uniformly in the starting point, the solution on $[0, 1]$ is then constructed iteratively as before.   □

In many situations, one is interested in solutions to an equation of the type

$$dY = f_0(Y, t) \, dt + f(Y, t) \, d\mathbf{X}_t , \qquad (8.11)$$

instead of (8.6). On the one hand, it is possible to recast (8.11) in the form (8.6) by writing it as an RDE for $\hat{Y}_t = (Y_t, t)$ driven by $\hat{\mathbf{X}}_t = (\hat{X}, \hat{\mathbb{X}})$ where $\hat{X} = (X_t, t)$ and $\hat{\mathbb{X}}$ is given by $\mathbb{X}$ and the "remaining cross integrals" of X_t and t, given by usual Riemann-Stieltjes integration. However, it is possible to exploit the structure of (8.11) to obtain somewhat better bounds on the solutions. See [FV10b, Ch. 12].

8.6 Stability III: Continuity of the Itô–Lyons map

We now obtain continuity of solutions to rough differential equations as function of their (rough) driving signals.

Theorem 8.5 (Rough path stability of the Itô–Lyons map). *Let $f \in \mathcal{C}_b^3$ and, for $\alpha \in \left(\frac{1}{3}, \frac{1}{2}\right]$, let $(Y, f(Y)) \in \mathscr{D}_X^{2\alpha}$ be the unique RDE solution given by Theorem 8.3 to*

$$dY = f(Y) \, d\mathbf{X}, \quad Y_0 = \xi \in W .$$

Similarly, let $(\tilde{Y}, f(\tilde{Y}))$ *be the RDE solution driven by* $\tilde{\mathbf{X}}$ *and started at* $\tilde{\xi}$ *where* $\mathbf{X}, \tilde{\mathbf{X}} \in \mathscr{C}^{\alpha}$. *Assuming*

$$\|\mathbf{X}\|_{\alpha}, \|\tilde{\mathbf{X}}\|_{\alpha} \leq M < \infty$$

we have the local Lipschitz estimates

$$d_{X, \tilde{X}, 2\alpha}\big(Y, f(Y); \tilde{Y}, f(\tilde{Y})\big) \leq C_M \big(|\xi - \tilde{\xi}| + \varrho_{\alpha}(\mathbf{X}, \tilde{\mathbf{X}})\big),$$

and also

$$\|Y - \tilde{Y}\|_{\alpha} \leq C_M \big(|\xi - \tilde{\xi}| + \varrho_{\alpha}(\mathbf{X}, \tilde{\mathbf{X}})\big),$$

where $C_M = C(M, \alpha, f)$ *is a suitable constant.*

Remark 8.6. The proof only uses the a priori information that RDE solutions remain bounded if the driving rough paths do, combined with basic stability properties of rough integration and composition.

Proof. Recall that, for given $\mathbf{X} \in \mathscr{C}^{\alpha}$, the RDE solution $(Y, f(Y)) \in \mathscr{D}_X^{2\alpha}$ is constructed as the unique fixed point of

$$\mathcal{M}_T(Y, Y') := (Z, Z') := \left(\xi + \int_0^{\cdot} f(Y_s) d\mathbf{X}_s, f(Y_{\cdot})\right) \in \mathscr{D}_X^{2\alpha},$$

and similarly for $\tilde{\mathcal{M}}_T\big(\tilde{Y}, f(\tilde{Y})\big) \in \mathcal{C}_{\tilde{X}}^{\alpha}$. Then, thanks to the fixed point property

$$(Y, f(Y)) = (Y, Y') = (Z, Z') = (Z, f(Y)),$$

(similarly with tilde) and the local Lipschitz estimate for rough integration, Theorem 4.17, and writing $(\Xi, \Xi') := \big(f(Y), f(Y)'\big)$ for the integrand, we obtain the bound

$$\begin{aligned}
d_{X, \tilde{X}, 2\alpha}\big(Y, Y'; \tilde{Y}, \tilde{Y}'\big) &= d_{X, \tilde{X}, 2\alpha}\big(Z, Z'; \tilde{Z}, \tilde{Z}'\big) \\
&\lesssim \varrho_{\alpha}(\mathbf{X}, \tilde{\mathbf{X}}) + |\xi - \tilde{\xi}| + T^{\alpha} d_{X, \tilde{X}, 2\alpha}\big(\Xi, \Xi'; \tilde{\Xi}, \tilde{\Xi}'\big),
\end{aligned}$$

Thanks to the local Lipschitz estimate for composition, Theorem 7.6, uniform in $T \leq 1$,

$$d_{X, \tilde{X}, 2\alpha}\big(\Xi, \Xi'; \tilde{\Xi}, \tilde{\Xi}'\big) \lesssim \varrho_{\alpha}(\mathbf{X}, \tilde{\mathbf{X}}) + |\xi - \tilde{\xi}| + d_{X, \tilde{X}, 2\alpha}\big(Y, f(Y); \tilde{Y}, f(\tilde{Y})\big).$$

In summary, for some constant $C = C(\alpha, f, M)$, we have the bound

$$\begin{aligned}
d_{X, \tilde{X}, 2\alpha}\big(Y, f(Y); \tilde{Y}, f(\tilde{Y})\big) \leq C\big(\varrho_{\alpha}(\mathbf{X}, \tilde{\mathbf{X}}) + |\xi - \tilde{\xi}| \\
+ T^{\alpha} d_{X, \tilde{X}, 2\alpha}\big(Y, f(Y); \tilde{Y}, f(\tilde{Y})\big)\big).
\end{aligned}$$

By taking $T = T_0(M, \alpha, f)$ smaller, if necessary, we may assume that $CT^{\alpha} \leq 1/2$, from which it follows that

$$d_{X, \tilde{X}, 2\alpha}\big(Y, f(Y); \tilde{Y}, f(\tilde{Y})\big) \leq 2C\big(\varrho_{\alpha}(\mathbf{X}, \tilde{\mathbf{X}}) + |\xi - \tilde{\xi}|\big),$$

which is precisely the required bound. The bound on $\left\|Y - \tilde{Y}\right\|_{\alpha}$ then follows as in (4.32), and these bounds can be iterated to cover a time interval of arbitrary (fixed) length. □

8.7 Davie's definition and numerical schemes

Fix $f \in C_b^2(W, \mathcal{L}(V, W))$ and $\mathbf{X} = (X, \mathbb{X}) \in \mathscr{C}^\beta([0, T], V)$ with $\beta > \frac{1}{3}$. Under these assumptions, the rough differential equation $dY = f(Y)d\mathbf{X}$ makes sense as well-defined integral equation. (In Theorem 8.3 we used additional regularity, namely C_b^3, to establish existence of a *unique* solution on $[0, T]$.) By the very definition of an RDE solution, unique or not, $(Y, f(Y)) \in \mathscr{D}_X^{2\beta}$, i.e.

$$Y_{s,t} = f(Y_s)X_{s,t} + \mathrm{O}\left(|t - s|^{2\beta}\right) ,$$

and we recognise a step of *first-order Euler* approximation, $Y_{s,t} \approx f(Y_s)X_{s,t}$, started from Y_s. Clearly $\mathrm{O}\left(|t - s|^{2\beta}\right) = \mathrm{o}(|t - s|)$ if and only if $\beta > 1/2$ and one can show that iteration of such steps along a partition $\mathcal{P}$ of $[0, T]$ yields a convergent "Euler" scheme as $|\mathcal{P}| \downarrow 0$, see [Dav08] or [FV10b].

In the case $\beta \in \left(\frac{1}{3}, \frac{1}{2}\right]$ we have to exploit that we know more than just $(Y, f(Y)) \in \mathscr{D}_X^{2\beta}$. Indeed, since $Y_{s,t} = \int_s^t f(Y)d\mathbf{X}$, estimate (4.22) for rough integrals tells us that, for all pairs s, t

$$Y_{s,t} = f(Y_s)X_{s,t} + (f(Y))'_s \mathbb{X}_{s,t} + \mathrm{O}\left(|t - s|^{3\beta}\right) . \tag{8.12}$$

Using the identity $f(Y)' = Df(Y)Y' = Df(Y)f(Y)$, this can be spelled out further to

$$Y_{s,t} = f(Y_s)X_{s,t} + Df(Y_s)f(Y_s)\mathbb{X}_{s,t} + \mathrm{o}(|t - s|) \tag{8.13}$$

and, omitting the small remainder term, we recognise a step of a *second-order Euler* or *Milstein* approximation. Again, one can show that iteration of such steps along a partition $\mathcal{P}$ of $[0, T]$ yields a convergent "Euler" scheme as $|\mathcal{P}| \downarrow 0$; see [Dav08] or [FV10b].

Remark 8.7. This schemes can be understood from simple Taylor expansions based on the differential equation $dY = f(Y)dX$, at least when X is smooth (enough), or via Itô's formula in a semimartingale setting. With focus on the smooth case, the Euler approximation is obtained by a "left-point freezing" approximation $f(Y_\cdot) \approx f(Y_s)$ over $[s, t]$ in the integral equation,

$$Y_{s,t} = \int_s^t f(Y_r)dX_r \approx f(Y_s)X_{s,t}$$

whereas the Milstein scheme, with $\mathbb{X}_{s,t} = \int_s^t X_{s,r}dX_r$ for smooth paths, is obtained from the next-best approximation

$$f(Y_r) \approx f(Y_s) + Df(Y_s)Y_{s,r}$$
$$\approx f(Y_s) + Df(Y_s)f(Y_s)X_{s,r} \; .$$

It turns out that the description (8.13) is actually a formulation that is equivalent to the RDE solution built previously in the following sense.

Proposition 8.8. *The following two statements are equivalent*

i) $(Y, f(Y))$ is a RDE solution to (8.6), as constructed in Theorem 8.3.
ii) $Y \in C([0,T], W)$ is an "RDE solution in the sense of Davie", i.e. in the sense of (8.13).

Proof. We already discussed how (8.13) is obtained from an RDE solution to (8.6). Conversely, (8.13) implies immediately $Y_{s,t} = f(Y_s)X_{s,t} + \mathrm{O}(|t-s|^{2\beta})$ which shows that $Y \in C^{\beta}$ and also $Y' := f(Y) \in C^{\beta}$, thanks to $f \in C_b^2$, so that $(Y, f(Y)) \in \mathscr{D}_X^{2\beta}$. It remains to see, in the notation of the proof of Theorem 4.10, that $Y_{s,t} = (\mathcal{I}\Xi)_{s,t}$ with

$$\Xi_{s,t} = f(Y_s)X_{s,t} + (f(Y))'_s \mathbb{X}_{s,t} = f(Y_s)X_{s,t} + Df(Y_s)f(Y_s)\mathbb{X}_{s,t} \; .$$

To see this, we note that trivially $Y_{s,t} = (\mathcal{I}\tilde{\Xi})_{s,t}$ with $\tilde{\Xi}_{s,t} := Y_{s,t}$. But $\tilde{\Xi}_{s,t} = \Xi_{s,t} + \mathrm{o}(|t-s|)$ and one sees as in Remark 4.13 that $\mathcal{I}\tilde{\Xi} = \mathcal{I}\Xi$. □

8.8 Lyons' original definition

A slightly different notion of solution was originally introduced in [Lyo98] by Lyons.[2] This notion only uses the spaces $\mathscr{C}^{\alpha}$, without ever requiring the use of the spaces $\mathscr{D}_X^{2\alpha}$ of "controlled rough paths". Indeed, for $\mathbf{X} = (X, \mathbb{X}) \in \mathscr{C}^{\alpha}([0,T], V)$ and $F \in C_b^2(V, \mathcal{L}(V, W))$ we can define an element $\mathbf{Z} = (Z, \mathbb{Z}) = I_F(X) \in \mathscr{C}^{\alpha}([0,T], W)$ directly by

$$Z_t \stackrel{\text{def}}{=} (\mathcal{I}\Xi)_{0,t} \; , \qquad \Xi_{s,t} = F(X_s)X_{s,t} + DF(X_s)\mathbb{X}_{s,t} \; ,$$
$$\mathbb{Z}_{s,t} \stackrel{\text{def}}{=} (\mathcal{I}\bar{\Xi}^s)_{s,t} \; , \qquad \bar{\Xi}_{u,v}^s = Z_{s,u}Z_{u,v} + (F(X_u) \otimes F(X_u))\mathbb{X}_{u,v} \; .$$

It is possible to check that $\bar{\Xi}^s \in C_2^{\alpha, 3\alpha}$ for every fixed s (see the proof of Theorem 4.10) so that the second line makes sense. It is also straightforward to check that $(Z, \mathbb{Z})$ satisfies (2.1), so that it does indeed belong to $\mathscr{C}^{\alpha}$. Actually, one can see that

$$Z_t = \int_0^t F(X_s)\,d\mathbf{X}_s \; , \qquad \mathbb{Z}_{s,t} = \int_s^t Z_{s,r} \otimes dZ_r \; ,$$

---

[2] As always, we only consider the step-2 α-Hölder case, i.e. $\alpha > \frac{1}{3}$, whereas Lyons' theory is valid for every Hölder-exponent $\alpha \in (0,1]$ (or: variation parameter $p \geq 1$) at the complication of heaving to deal with $\lfloor p \rfloor$ levels.

where the integrals are defined as in the previous sections, where $F(X) \in \mathscr{D}_X^{2\alpha}$ as in Section 7.3.

We can now define solutions to (8.6) in the following way.

Definition 8.9. A rough path $\mathbf{Y} = (Y, \mathbb{Y}) \in \mathscr{C}^\alpha([0, T], W)$ is a solution in the sense of Lyons to (8.6) if there exists $\mathbf{Z} = (Z, \mathbb{Z}) \in \mathscr{C}^\alpha(V \oplus W)$ such that the projection of $(Z, \mathbb{Z})$ onto $\mathscr{C}^\alpha(V)$ is equal to $(X, \mathbb{X})$, the projection onto $\mathscr{C}^\alpha(W)$ is equal to $(Y, \mathbb{Y})$, and $Z = I_F(Z)$ where

$$F(x, y) = \begin{pmatrix} I & 0 \\ f(y) & 0 \end{pmatrix}.$$

It is straightforward to see that if $(Y, Y') \in \mathscr{D}_X^{2\alpha}(W)$ is a solution to (8.6) in the sense of the previous section, then the path $Z = (X, Y) \in V \oplus W$ is controlled by X. As seen in Section 7.1, it can therefore be interpreted as an element of $\mathscr{C}^\alpha$. It follows immediately from the definitions that it is then also a solution in the sense of Lyons. Conversely, if $(Y, \mathbb{Y})$ is a solution in the sense of Lyons, then one can check that one necessarily has $(Y, f(Y)) \in \mathscr{D}_X^{2\alpha}(W)$ and that this is a solution in the sense of the previous section. We leave the verification of this fact as an exercise to the reader.

8.9 Linear rough differential equations

Let $X \in \mathcal{C}^1([0, 1], V)$, $A \in \mathcal{L}(W, \mathcal{L}(V, W))$ with finite operator norm $\|A\|_{\text{op}} = a \in [0, \infty)$, and consider the linear differential equation $dY = AY\,dX$, with initial data $Y_0 \in W$, written in integral form as

$$Y_t = Y_0 + \int_0^t AY_s\,dX_s.$$

Clearly $|Y_t| \le |Y_0| + a \int_0^t |Y_s| d|X|_s$ in terms of the Lipschitz path $|X|_t := \int_0^t |\dot{X}_s| ds$, and the classical Gronwall lemma gives

$$\|Y\|_{\infty;[0,1]} \le |Y_0| \exp(a\|X\|_{1;[0,1]}),$$

with $\|X\|_{1;[0,1]} = \sup_{0 \le s < t \le 1} \frac{|X_{s,t}|}{|t-s|} = \sup_{0 \le s \le 1} |\dot{X}_s|$. Alternatively, one can extract from the integral formulation the estimate, valid for all $0 \le s < t \le 1$,

$$|Y_{s,t}| \le a\|X\|_{1;[0,1]}\|Y\|_{\infty;[s,t]}|t - s|.$$

The following lemma, applied with $\alpha = 1$, then leads to a similar conclusion. More importantly, it will be seen to be applicable in rough situations with $\alpha < 1$.

Lemma 8.10. *(Rough Gronwall) Assume* $Y \in C([0, 1])$, $\alpha \in (0, 1]$, *and*

$$|Y_{s,t}| \leq M\|Y\|_{\infty;[s,t]}|t-s|^{\alpha}$$

whenever $0 \leq s < t \leq 1$. *Then there exists* $c = c_{\alpha} < \infty$ *such that*

$$\|Y\|_{\infty;[0,1]} \leq c\exp(cM^{1/\alpha})|Y_0|.$$

Remark 8.11. Since $|Y_{s,t}| \leq 2\|Y\|_{\infty;[s,t]}$ the assumption is trivially satisfied for "distant" times s, t such that $M|t-s|^{\alpha} \geq 2$. It then suffices to check the assumption for "nearby" times with $M|t-s|^{\alpha} \leq \theta$ with $\theta = 2$, and in fact any $\theta > 0$, at the price of replacing M by $\frac{2M}{\theta \wedge 2}$.

Proof. For any $\xi \in [s,t]$ have $|Y_{\xi}| \leq |Y_s| + |Y_{s,\xi}| \leq |Y_s| + M\|Y\|_{\infty;[s,t]}|t-s|^{\alpha}$, and so

$$\|Y\|_{\infty;[s,t]}(1 - M|t-s|^{\alpha}) \leq |Y_s| .$$

Since $e^{-2x} \leq 1 - x$ for $x \in [0, 1/2]$, we have, for $M|t-s|^{\alpha} \in [0, 1/2]$,

$$\|Y\|_{\infty;[s,t]} \leq |Y_s|e^{2M|t-s|^{\alpha}} \leq e|Y_s| .$$

This induces a greedy partition of $[0, 1]$, of mesh-size $(2M)^{-1/\alpha}$ and hence no more than $(2M)^{1/\alpha} + 1$ intervals. The final estimate is then

$$\|Y\|_{\infty;[0,1]} \leq e^{1+(2M)^{1/\alpha}}|Y_0| ,$$

so that the claimed estimate holds with $c = e \vee 2^{1/\alpha}$. □

We now apply this to linear (Young and rough) differential equations, without loss of generality posed on $[0, 1]$. By general theory, Theorem 8.3, we have a (non-explosive) solution.

Proposition 8.12. *Let* Y *solve the linear Young differential equation* $dY = AY\,dX$, *started from* Y_0 *and driven by* $X \in C^{\alpha}([0,1])$, $\alpha > 1/2$, *with* A *of finite operator norm* a. *Then there exists* $c = c(\alpha) \in (0, \infty)$ *so that*

$$\|Y\|_{\infty;[0,1]} \leq c\exp\left(c(a\|X\|_{\alpha;[0,1]})^{1/\alpha}\right)|Y_0|.$$

Proof. By scaling A, we can and will assume $\|X\|_{\alpha;[0,1]} = 1$. Young's inequality gives, with $a = |A|$ and $c = c(\alpha)$,

$$|Y_{s,t}| \leq |AY_sX_{s,t}| + \left|\int_s^t A(Y_r - Y_s)dX_r\right| \leq a|Y_s||t-s|^{\alpha} + ca\|Y\|_{\alpha;[s,t]}|t-s|^{2\alpha}$$

and so $\frac{1}{2}\|Y\|_{\alpha;[s,t]} \leq a|Y|_{\infty;[s,t]}$ whenever $ca|t-s|^{\alpha} \leq 1/2$. Re-insert the estimate on $\|Y\|_{\alpha;[s,t]}$ (and also use $ca|t-s|^{\alpha} \leq 1/2$) above to obtain precisely

$$|Y_{s,t}| \leq a|Y_s||t-s|^{\alpha} + a|Y|_{\infty;[s,t]}|t-s|^{\alpha} \leq 2a\|Y\|_{\infty;[s,t]}|t-s|^{\alpha}.$$

This holds whenever $ca|t - s|^\alpha \leq 1/2$ and so we can conclude with the rough Gronwall lemma (and the remark after it). The constant c is allowed to change of course, but remains $c = c(\alpha)$.   □

A similar result also holds in the rough case.

Proposition 8.13. *Let Y solve the linear rough differential equation $dY = AY\,d\mathbf{X}$, started from Y_0 and driven by $\mathbf{X} \in \mathscr{C}^\alpha([0,1])$, $\alpha > 1/3$, with A of finite operator norm a. Then there exists $c = c(\alpha) \in (0, \infty)$ so that*

$$\|Y\|_{\infty;[0,1]} \leq c \exp\left(c(a\|\mathbf{X}\|_{\alpha;[0,1]})^{1/\alpha}\right)|Y_0|.$$

Proof. By scaling A, we can again assume unit (homogeneous) rough path norm for $\mathbf{X}$. By a basic estimate for rough integrals it then holds, with $c = c(\alpha) \in [1, \infty)$ and $a = |A|$,

$$|Y^\natural_{s,t}| \leq c\|AY, A^2Y\|_{2\alpha,X}|t - s|^{3\alpha} \leq ca\|Y, AY\|_{2\alpha,X}|t - s|^{3\alpha}$$
$$= ca(\|AY\|_\alpha + \|Y^\#\|_{2\alpha})|t - s|^{3\alpha},$$

using musical notation $Y_{s,t} \equiv AY_s X_{s,t} + Y^\#_{s,t} \equiv AY_s X_{s,t} + A^2 Y_s \mathbb{X}_{s,t} + Y^\natural_{s,t}$. This entails

$$|Y^\#_{s,t}| \leq |A^2 Y_s \mathbb{X}_{s,t}| + |Y^\natural_{s,t}| \leq a^2 |Y_s||t - s|^{2\alpha} + (a\|Y\|_\alpha + \|Y^\#\|_{2\alpha})ca|t - s|^{3\alpha}$$

and so for all $s < t$ with $ca|t - s|^\alpha \leq 1/2$ we obtain

$$\frac{1}{2}\|Y^\#\|_{2\alpha;[s,t]} \leq a^2\|Y\|_{\infty;[s,t]} + \frac{a}{2}\|Y\|_\alpha.$$

Similarly, $|Y_{s,t}| \leq |AY_s X_{s,t}| + |Y^\#_{s,t}| \leq a|Y_s||t-s|^\alpha + (2a^2\|Y\|_{\infty;[s,t]} + a\|Y\|_\alpha)|t - s|^{2\alpha}$ and so

$$\frac{1}{2}\|Y\|_{\alpha;[s,t]} \leq a\|Y\|_{\infty;[s,t]} + 2a^2\|Y\|_{\infty;[s,t]}|t - s|^\alpha \leq 3a\|Y\|_{\infty;[s,t]}.$$

for all $s < t$ with $a|t - s|^\alpha \leq 1/2$. Re-inserting this and the bound for $\|Y\|_\alpha = \|Y\|_{\alpha;[s,t]}$ in the above estimate for $|Y_{s,t}|$, we obtain

$$|Y_{s,t}| \leq a|Y_s||t - s|^\alpha + 8a^2\|Y\|_{\infty;[s,t]}|t - s|^{2\alpha} \leq 5a\|Y\|_{\infty;[s,t]}|t - s|^\alpha.$$

We conclude with the rough Gronwall lemma, just as in the Young case.   □

Remark 8.14. All this can be vector-valued. Assuming X takes values in some space V and Y takes values in W, we should view A as a linear map $A \colon W \otimes V \to W$. The operator $A^2 \colon W \otimes V \otimes V \to W$ should then be interpreted as $A \circ (A \otimes \mathrm{Id})$.

8.10 Stability IV: Flows

We briefly state, without proof, a result concerning regularity of flows associated to rough differential equations, as well as local Lipschitz estimates of the Itô–Lyons maps on the level of such flows. More precisely, given a geometric rough path $\mathbf{X} \in \mathscr{C}_g^\alpha([0, T], \mathbf{R}^d)$, we saw in Theorem 8.3 that, for $\mathcal{C}_b^3$ vector fields $f = (f_1, \ldots, f_d)$ on $\mathbf{R}^e$, there is a unique global solution to the rough integral equation

$$Y_t = y + \int_0^t f(Y_s) \, d\mathbf{X}_s, \qquad t \geq 0. \tag{8.14}$$

Write $\pi_{(f)}(0, y; \mathbf{X}) = Y$ for this solution. Note that the inverse flow exists trivially, by following the RDE driven by $\mathbf{X}(t - .)$,

$$\pi_{(f)}(0, \cdot; \mathbf{X})_t^{-1} = \pi_{(f)}(0, \cdot; \mathbf{X}(t - .))_t.$$

We call the map $y \mapsto \pi_{(f)}(0, y; \mathbf{X})$ the flow associated to the above RDE. Moreover, if X^ϵ is a smooth approximation to $\mathbf{X}$ (in rough path metric), then the corresponding ODE solution Y^ϵ is close to Y, with a local Lipschitz estimate as given in Section 8.6.

It is natural to ask if the flow depends smoothly on y. Given a multi-index $k = (k_1, \ldots, k_e) \in \mathbf{N}^e$, write D^k for the partial derivative with respect to $y^1, \ldots, y^e$. The proof of the following statement is an easy consequence of [FV10b, Chapter 12].

Theorem 8.15. *Let* $\alpha \in (1/3, 1/2]$ *and* $\mathbf{X}, \tilde{\mathbf{X}} \in \mathscr{C}_g^\alpha$. *Assume* $f \in \mathcal{C}_b^{3+n}$ *for some integer* n. *Then the associated flow is of regularity* $\mathcal{C}^{n+1}$ *in* y, *as is its inverse flow. The resulting family of partial derivatives,* $\{D^k \pi_{(f)}(0, \xi; \mathbf{X}), |k| \leq n\}$ *satisfies the RDE obtained by formally differentiating* $dY = f(Y)d\mathbf{X}$.

At last, for every $M > 0$ *there exist* C, K *depending on* M *and the norm of* f *such that, whenever* $\|\mathbf{X}\|_\alpha, \|\tilde{\mathbf{X}}\|_\alpha \leq M < \infty$ *and* $|k| \leq n$,

$$\sup_{\xi \in \mathbf{R}^e} \left| D^k \pi_{(f)}(0, \xi; \mathbf{X}) - D^k \pi_{(f)}(0, \xi; \tilde{\mathbf{X}}) \right|_{\alpha;[0,t]} \leq C \varrho_\alpha(\mathbf{X}, \tilde{\mathbf{X}}),$$

$$\sup_{\xi \in \mathbf{R}^e} \left| D^k \pi_{(f)}(0, \xi; \mathbf{X})^{-1} - D^k \pi_{(f)}(0, \xi; \tilde{\mathbf{X}})^{-1} \right|_{\alpha;[0,t]} \leq C \varrho_\alpha(\mathbf{X}, \tilde{\mathbf{X}}),$$

$$\sup_{\xi \in \mathbf{R}^e} \left| D^k \pi_{(f)}(0, \xi; \mathbf{X}) \right|_{\alpha;[0,t]} \leq K,$$

$$\sup_{\xi \in \mathbf{R}^e} \left| D^k \pi_{(f)}(0, \xi; \mathbf{X})^{-1} \right|_{\alpha;[0,t]} \leq K.$$

8.11 Exercises

Exercise 8.1 *a) Consider the case of a smooth, one-dimensional driving signal* X : $[0, T] \to \mathbf{R}$. *Show that the solution map to the (ordinary) differential equation* $dY = f(Y)dX$, *for sufficiently nice* f *(say bounded with bounded derivatives)* *and started at some fixed point $Y_0 = \xi$, is locally Lipschitz continuous with respect to the driving signal in the supremum norm on $[0, T]$. Conclude that it admits a unique continuous extension to every continuous driving signal X.*

  b) *Show by an example that no such continuous extension is possible, in general, in a multi-dimensional situation, with vector fields $f = (f_1, \ldots, f_d)$ driven by a d-dimensional signal $X : [0, T] \to \mathbf{R}^d$, with $d > 2$.*

♯ c) *Show that a continuous extension is possible for commuting vector fields, in the sense that all Lie bracket $[f_i, f_j]$, $1 \leq i, j \leq d$, vanish or, equivalently, their flows commute.*

Exercise 8.2 (Explicit solution, Chen–Strichartz formula) *View*

$$f = (f_1, \ldots, f_d) \in \mathcal{C}_b^\infty \left(\mathbf{R}^e, \mathcal{L}\left(\mathbf{R}^d, \mathbf{R}^e\right)\right),$$

as a collection of d (smooth, bounded with bounded derivatives of all orders) vector fields on $\mathbf{R}^e$. Assume that f is step-2 nilpotent in the sense that $[f_i, [f_j, f_k]] \equiv 0$ for all $i, j, k \in \{1, \ldots, d\}$. Here, $[\cdot, \cdot]$ denotes the Lie bracket between two vector fields. Let $(Y, f(Y))$ be the RDE solution to $dY = f(Y)d\mathbf{X}$ started at some $\xi \in \mathbf{R}^e$ and assume that the rough path $\mathbf{X}$ is geometric. Give an explicit formula of the type $Y_t = \exp(\ldots)\xi$ where $\exp$ denotes the unit time solution flow along a vector field $(\ldots)$ which you should write down explicitly.

∗ **Exercise 8.3 (Explosion along linear-growth vector fields)** *Give an example of smooth f with linear growth, and $\mathbf{X} \in \mathcal{C}^\alpha$ so that $dY = f(Y)d\mathbf{X}$ started at some ξ fails to have a global solution.*

♯ **Exercise 8.4 (Maximal RDE solution)** *We are in the setting of the local existence and uniqueness Theorem 8.3, with $\mathcal{C}^3$-regular coefficients, $f \in \mathcal{C}^3(W, \mathcal{L}(V, W))$, and local solution Y to (8.6) with values in the Banach space W.*

  a) *Show that Y can either be extended to a global solution on the whole interval $[0, T]$ or only on a subinterval $[0, \tau)$ which is maximal with respect to extension of solutions.*
  b) *Show that $\tau = \tau(\mathbf{X})$ is a lower semicontinuous function of the driving rough path, i.e. $\underline{\lim}_{n\to\infty} \tau(\mathbf{X}^n) \geq \tau(\mathbf{X})$ whenever $\mathbf{X}^n \to \mathbf{X} \in \mathcal{C}^\alpha$.*
  c) *Assume f is $\mathcal{C}^3$-bounded on bounded sets. (This is always the case for $f \in \mathcal{C}^3$ with W, V finite-dimensional.) If a solution only exists on $[0, \tau)$, then $\overline{\lim}_{t\uparrow\tau} |Y_t| = +\infty$ and we call $\tau \in (0, T]$ explosion time.*

  Remark: *In infinite dimensions, there are examples of Banach-valued ODEs with smooth coefficients, where global existence fails but the solution does not explode. In essence, this is possible because a smooth vector field need not map bounded sets into bounded sets.*

Exercise 8.5 *Let $T > 0, \alpha \in (1/3, 1/2]$ and $\mathbf{X}, \tilde{\mathbf{X}} \in \mathcal{C}^\alpha([0, T], \mathbf{R}^d)$. Establish existence, continuity and stability for rough differential equations with drift (cf. (8.6)),*

$$dY_t = f_0(Y_t)\, dt + f(Y_t)\, d\mathbf{X}_t \,. \tag{8.15}$$

a) First assume f_0 to have the same regularity as f, in which case you may solve $dY = \bar{f}(Y)\bar{\mathbf{X}}$ with $\bar{f} = (f, f_0)$ and $\bar{\mathbf{X}}$ as (canonical) space-time rough path extension of $\mathbf{X}$. (The missing integrals $\int X^i dt, \int t dX^i, i = 1, \ldots, d$ are canonically defined as Riemann–Stieltjes integrals.)

b) Give a direct analysis for $f_0 \in C_b^1$ (or in fact f_0 Lipschitz continuous, without boundedness assumption).

Exercise 8.6 *Let $f \in C_b^2$ and assume $(Y, f(Y))$ is a RDE solution to (8.6), as constructed in Theorem 8.3. Show that the o-term in Davie's definition, (8.13), can be bounded uniformly over $(X, \mathbb{X}) \in B_R$, any $R < \infty$, where*

$$B_R := \left\{ (X, \mathbb{X}) \in \mathscr{C}^\beta : \|X\|_\beta + \|\mathbb{X}\|_{2\beta} \leq R \right\}, \text{ any } R < \infty.$$

Show also that RDE solutions are β-Hölder, uniformly over $(X, \mathbb{X}) \in B_R$, any $R < \infty$.

Exercise 8.7 *Show that $\|Y, f(Y); Y^n, f(Y^n)\|_{X, X^n, 2\alpha} \to 0$, together with $\mathbf{X} \to \mathbf{X}^n$ in $\mathscr{C}^\beta$ implies that also $(Y^n, \mathbb{Y}^n) \to (Y, \mathbb{Y})$ in $\mathscr{C}^\alpha$. Since, at the price of replacing f by F, cf. Definition 8.9, there is no loss of generality in solving for the controlled rough path $Z = (X, Y)$, conclude that continuity of the RDE solution map (Itô–Lyons map) also holds with Lyons' definition of a solution.*

Exercise 8.8 *Show that $\|Y, f(Y); Y^n, f(Y^n)\|_{X, X^n, 2\alpha} \to 0$, together with $\mathbf{X} \to \mathbf{X}^n$ in $\mathscr{C}^\beta$ implies that also $(Y^n, \mathbb{Y}^n) \to (Y, \mathbb{Y})$ in $\mathscr{C}^\alpha$. Since, at the price of replacing f by F, cf. Definition 8.9, there is no loss of generality in solving for the controlled rough path $Z = (X, Y)$, conclude that continuity of the RDE solution map (Itô–Lyons map) also holds with Lyons' definition of a solution.*

Exercise 8.9 (Lyons extension theorem revisited) *Let $\alpha \in (\frac{1}{3}, \frac{1}{2}]$ and consider $\mathbf{X} = (X, \mathbb{X}) \in \mathscr{C}^\alpha([0, T], V)$. Show that $\bar{\mathbf{X}} = (1, \mathbf{X}^{(1)}, \mathbf{X}^{(2)}, \mathbf{X}^{(3)}, \ldots, \mathbf{X}^{(N)})$, the (level-N) Lyons lift of $\mathbf{X}$ from Exercise 4.6, solves a linear RDE. Use this and a scaling argument for another proof of the estimate, $0 \leq s < t \leq T$, $n = 1, \ldots, N$,*

$$|\mathbf{X}_{s,t}^{(n)}|^{\frac{1}{n}} \lesssim \|\mathbf{X}\|_\alpha |t - s|^\alpha .$$

8.12 Comments

ODEs driven by not too rough paths, i.e. paths that are α-Hölder continuous for some $\alpha > 1/2$ or of finite p-variation with $p < 2$, understood in the (Young) integral sense were first studied by Lyons in [Lyo94]; nonetheless, the terminology Young-ODEs is now widely used. Existence and uniqueness for such equations via Picard iterations is by now classical, our discussion in Section 8.3 is a mild variation of [LCL07, p.22] where also the *division property* (cf. proof of Lemma 7.5) is emphasised. Existence and uniqueness of solutions to RDEs via Picard iteration in the (Banach!) space of controlled rough paths originates in [Gub04] for regularity $\alpha \in (\frac{1}{3}, \frac{1}{2}]$. This approach also allows to treat arbitrary regularities, [Gub10, HK15]. In case of driving rough paths with jumps, one has to distinguish between forward (think Itô or branched) and geometric (think Marcus canonical) sense, this was started in [Wil01], and the general forward resp. geometric case completed by Chevyrev, Friz and Zhang in [FZ18, CF19], see also comment Section 9.6.

The continuity result of Theorem 8.5 is due to T. Lyons; proofs of uniform continuity on bounded sets were given in [Lyo98, LQ02, LCL07]. Local Lipschitz estimates were pointed out subsequently and in different settings by various authors including Lyons–Qian [LQ02], Gubinelli [Gub04], Friz–Victoir [FV10b], Inahama [Ina10], Deya et al. [DNT12]; Bailleul [Bai15a, Bai14] and Bailleul–Riedel [BR19] take a flow perspective, initially studied in [LQ98]. Smoothness of the Lyons–Itô map is discussed in [LL06, FV10b, Bai15b, CL18], see also comment Section 11.5.

The name *universal limit theorem* was suggested by P. Malliavin, meaning continuity of the Itô–Lyons map in rough path metrics. As we tried to emphasise, the stability in rough path metrics is seen at all levels of the theory.

Lyons' original argument (for arbitrary regularity) also involves a Picard iteration, see e.g. [LCL07, p.88]. In his p-variation setting, vector fields are assumed Lip$^\gamma$, $\gamma > p$, which agrees with our C_b^γ in finite dimensions, cf. Sections 1.4 and 1.5, with the usual disclaimer $\gamma \notin \mathbf{N}$ (Lipschitz vs continuously. differentiable). In finite dimensions, existence results are given for $\gamma > p - 1$, see [Dav08, FV10b] for $p < 3$ and general p respectively. In infinite dimensions, due to lack of compactness, extra assumptions on the vector fields are necessary; a *Peano existence theorem*, as in the case of Banach valued RDEs is shown by Caruana [Car10]. On the other hand, under local C^γ regularity one has a unique (in infinite dimensions: not necessarily exploding) maximal solution, cf. Exercise 8.4. In finite dimensions, global existence is guaranteed by non-explosion, discussed in [Dav08, FV10b, Lej12, RS17].

For regularity $1/p = \alpha > 1/3$, Davie [Dav08] establishes existence and uniqueness for Young resp. rough differential equations via discrete Euler resp. Milstein approximations. Step-N Euler schemes. with $\lfloor p \rfloor \le N$, are studied in [FV08b] via sub-Riemannian geodesics in $G^{(N)}(\mathbf{R}^d)$, Boutaib et al. [BGLY14] establish similar estimates in the Banach setting, Boedihardjo, Lyons and Yang [BLY15] study $N \to \infty$.

Our regularity assumption as stated in Theorem 8.3, namely C^3 for a unique (local) solution is not sharp; it is straightforward to push this to C^γ any $\gamma > 1/\alpha$ for $\alpha \in \left(\frac{1}{3}, \frac{1}{2}\right]$ (due to our level-2 exposition) in agreement with [Lyo98, Dav08]. It is less straightforward [Dav08, FV10b] to show that uniqueness also holds for $\gamma = 1/\alpha$ and this is optimal, with counter-examples constructed in [Dav08]. Local existence results on the other hand are available for $\gamma > (1/\alpha) - 1$. Setting $\alpha = 1$, this is consistent with the theory of ODEs where it is well known that, at least modulo possible logarithmic divergencies and in finite dimensions, Lipschitz continuity of the coefficients is required for the uniqueness of local solutions, but continuity is sufficient for their existence.

Theorem 8.3 gives global existence for $f \in C_b^3$ or (affine) linear f. Linear rough differential equations are important (Jacobian of the flow, equations for Malliavin type derivatives, etc) and studied e.g. in [Lyo98, FV10b, CL14], see also [HH10] for related analysis. Solutions can be estimated by the *rough Gronwall lemma* [DGHT19b, Hof18], in a sense a real-analysis abstraction of previously used arguments for linear RDE solutions, [HN07, FV10b].

The existence and uniqueness results for rough differential equations have seen many variations over recents years. Gubinelli, Imkeller and Perkowski apply their theory of paracontrolled distributions to (level-2) RDEs with Hölder drivers [GIP15, Sec.3], extended to Besov drivers by Prömel–Trabs [PT16], revisited with "classical" rough path tools in [FP18].

Rough/stochastic Volterra equations are discussed from a rough path point of view in [DT09, HT19, Com19], from a paracontrolled point of view in [PT18] and in a regularity structure context in [BFG+19, Sec.5]. Bailleul–Diehl then study the inverse problem for rough differential equations [BD15]. For a "joint development" of RDEs and SDEs with stochastic sewing, by a fixed point argument in a space of stochastic controlled rough paths, see [FHL20]. Rough partial differential equations are discussed in Chapter 12.

Last not least, we note that the point of view to construct RDE solutions by fixed point arguments in the (linear) space of controlled rough paths, where the rough path figures as parameter of the fixed point problem, extends naturally to the framework of regularity structures developed in [Hai14b], cf. Chapter 13 onwards. In that context, solutions (to singular SPDEs, say) are found by similar fixed point arguments in a linear space of "modelled distributions"), with enhanced noise ("the model") again as parameter of the fixed point problem. (The question of renormalisation is a priori disconnected from the construction of a solution and only concerns the model / rough path. However, one would like to understand the equation driven by renormalised noise, at least when the latter is smooth. In the setting of rough differential equations such effects have been observed in [FO09], a systematic study in case of branched RDE is found in Bruned et al. [BCFP19], see also [BCEF20].)

Chapter 9
Stochastic differential equations

We identify the solution to a rough differential equation driven by the Itô or Stratonovich lift of Brownian motion with the solution to the corresponding stochastic differential equation. In combination with continuity of the Itô–Lyons maps, a quick proof of the Wong–Zakai theorem is given. Applications to Stroock–Varadhan support theory and Freidlin–Wentzell large deviations are briefly discussed.

9.1 Itô and Stratonovich equations

We saw in Section 3 that d-dimensional Brownian motion lifts in an essentially canonical way to $\mathbf{B} = (B, \mathbb{B}) \in \mathscr{C}^\alpha([0, T], \mathbf{R}^d)$ almost surely, for any $\alpha \in \left(\frac{1}{3}, \frac{1}{2} \right)$. In particular, we may use almost every realisation of $(B, \mathbb{B})$ as the driving signal of a rough differential equation. This RDE is then solved "pathwise" i.e. for a *fixed* realisation of $(B(\omega), \mathbb{B}(\omega))$. Recall that the choice of $\mathbb{B}$ is never unique: two important choices are the Itô and the Stratonovich lift, we write $\mathbf{B}^{\text{Itô}}$ and $\mathbf{B}^{\text{Strat}}$, where $\mathbb{B}$ is defined as $\int B \otimes dB$ and $\int B \otimes \circ dB$ respectively. We now discuss the interplay with classical stochastic differential equations (SDEs).

Theorem 9.1. *Let $f \in C_b^3\left(\mathbf{R}^e, \mathcal{L}\left(\mathbf{R}^d, \mathbf{R}^e\right)\right)$, let $f_0 : \mathbf{R}^e \to \mathbf{R}^e$ be Lipschitz continuous, and let $\xi \in \mathbf{R}^e$. Then,*

i) With probability one, $\mathbf{B}^{\text{Itô}}(\omega) \in \mathscr{C}^\alpha$, any $\alpha \in (1/3, 1/2)$ and there is a unique RDE solution $(Y(\omega), f(Y(\omega))) \in \mathscr{D}_{B(\omega)}^{2\alpha}$ to

$$dY = f_0(Y)dt + f(Y)\, d\mathbf{B}^{\text{Itô}}, \quad Y_0 = \xi.$$

Moreover, $Y = (Y_t(\omega))$ is a strong solution to the Itô SDE $dY = f_0(Y)dt + f(Y)dB$ started at $Y_0 = \xi$.

ii) Similarly, the RDE solution driven by $\mathbf{B}^{\text{Strat}}$ yields a strong solution to the Stratonovich SDE $dY = f_0(Y)dt + f(Y) \circ dB$ started at $Y_0 = \xi$.

Proof. We assume zero drift f_0, but see Exercise 8.5. The map

© Springer Nature Switzerland AG 2020
P. K. Friz, M. Hairer, *A Course on Rough Paths*, Universitext,
https://doi.org/10.1007/978-3-030-41556-3_9

$$B|_{[0,t]} \mapsto (B, \mathbb{B}^{\text{Strat}})|_{[0,t]} \in \mathscr{C}_g^{0,\alpha}([0,t], \mathbf{R}^d)$$

is measurable, where $\mathscr{C}_g^{0,\alpha}$ denotes the (separable, hence Polish) subspace of $\mathscr{C}^\alpha$ obtained by taking the closure, in α-Hölder rough path metric, of piecewise smooth paths. This follows, for instance, from Proposition 3.6. By the continuity of the Itô–Lyons map (adding a drift vector field is left as an easy exercise) the RDE solution $Y_t \in \mathbf{R}^e$ is the continuous image of the driving signal $(B, \mathbb{B}^{\text{Strat}})|_{[0,t]} \in \mathscr{C}_g^{0,\alpha}([0,t], \mathbf{R}^d)$. It follows that Y_t is adapted to

$$\sigma\{B_{r,s}, \mathbb{B}_{r,s} : 0 \le r \le s \le t\} = \sigma\{B_s : 0 \le s \le t\},$$

and it suffices to apply Corollary 5.2. Since $\mathbb{B}_{s,t}^{\text{Itô}} = \mathbb{B}_{s,t}^{\text{Strat}} - \frac{1}{2}(t-s)I$, measurability is also guaranteed and we conclude with the same argument, using Proposition 5.1.
□

Remark 9.2. In contrast to standard SDE theory, the present solution constructed via RDEs is immediately well-defined as a *flow*, i.e. for all ξ on a common set of probability one. The price to pay is that of C^3 regularity of f, as opposed to the mere Lipschitz regularity required for the standard theory.

9.2 The Wong–Zakai theorem

A classical result (e.g. [IW89, p.392]) asserts that SDE approximations based on piecewise linear approximations to the driving Brownian motions converge to the solution of the Stratonovich equation. Using the machinery built in the previous sections, we can now give a simple proof of this by combining Proposition 3.6, Theorem 8.5 and the understanding that RDEs driven by $\mathbf{B}^{\text{Strat}}$ yield solutions to the Stratonovich equation (Theorem 9.1).

Theorem 9.3 (Wong–Zakai, Clark, Stroock–Varadhan). *Let f, f_0, ξ be as in Theorem 9.1 above. Let $\alpha < 1/2$. Consider dyadic piecewise linear approximations (B^n) to B on $[0, T]$, as defined in Proposition 3.6. Write Y^n for the (random) ODE solutions to $dY^n = f_0(Y^n)dt + f(Y^n)dB^n$ and Y for the Stratonovich SDE solution to $dY = f_0(Y)dt + f(Y) \circ dB$, all started at ξ. Then the Wong–Zakai approximations converge a.s. to the Stratonovich solution. More precisely, with probability one,*

$$\|Y - Y^n\|_{\alpha;[0,T]} \to 0.$$

The only reason for *dyadic* piecewise linear approximations in the above statement is the formulation of the martingale-based Proposition 3.6. In Section 10 we shall present a direct analysis (going far beyond the setting of Brownian drivers) which easily entails quantitative convergence (in probability and L^q, any $q < \infty$) for all piecewise linear approximations towards a (Gaussian) rough path.

In the forthcoming Exercise 10.2 it will be seen that (non-dyadic) piecewise linear approximations of mesh size $\sim 1/n$, viewed canonically as rough paths, converge a.s.

in $\mathscr{C}^\alpha$ with rate anything less than $1/2 - \alpha$. As long as $\alpha > 1/3$, it then follows from (local) Lipschitzness of the Itô–Lyons map that Wong–Zakai approximations also converge with rate $(1/2 - \alpha)^-$. Note that the "best" rate one obtains in this way is $(1/2 - 1/3)^- = 1/6^-$; the reason being that rate is measured in some Hölder space with exponent $1/3^+$, rather than the uniform norm. The well-known almost sure "strong" rate $1/2^-$ can be obtained from rough path theory at the price of working in rough path spaces of much lower regularity, see [FR14].

9.3 Support theorem and large deviations

We briefly discuss two fundamental results in diffusion theory and explain how the theory of rough paths provides elegant proofs, reducing a question for general diffusion to one for Brownian motion and its Lévy area.

The results discussed in this section were among the very first applications of rough path theory to stochastic analysis, see Ledoux et al. [LQZ02]. Much more on these topics is found in [FV10b], so we shall be brief. The first result, due to Stroock–Varadhan [SV72] concerns the support of diffusion processes.

Theorem 9.4 (Stroock–Varadhan support theorem). *Let f, f_0, ξ be as in Theorem 9.1 above. Let $\alpha < 1/2$, B be a d-dimensional Brownian motion and consider the unique Stratonovich SDE solution Y on $[0, T]$ to*

$$dY = f_0(Y)dt + \sum_{i=1}^{d} f_i(Y) \circ dB^i \tag{9.1}$$

started at $Y_0 = \xi \in \mathbf{R}^e$. Write y^h for the ODE solution obtained by replacing $\circ dB$ with $dh \equiv \dot{h}\, dt$, whenever $h \in \mathcal{H} = W_0^{1,2}$, i.e. absolutely continuous, $h(0) = 0$ and $\dot{h} \in L^2([0, T], \mathbf{R}^d)$. Then, for every $\delta > 0$,

$$\lim_{\varepsilon \to 0} \mathbf{P}\left(\|Y - Y^h\|_{\alpha;[0,T]} < \delta \mid \|B - h\|_{\infty;[0,T]} < \varepsilon \right) = 1 \tag{9.2}$$

(where Euclidean norm is used for the conditioning $\|B - h\|_{\infty,[0,T]} < \varepsilon$). As a consequence, the support of the law of Y, viewed as measure on the pathspace $\mathcal{C}^{0,\alpha}([0, T], \mathbf{R}^e)$, is precisely the α-Hölder closure of $\{y^h : \dot{h} \in L^2([0, T], \mathbf{R}^d)\}$.

Proof. Using Theorem 9.1 we can and will take Y as RDE solution driven by $\mathbf{B}^{\text{Strat}}(\omega)$. For $h \in \mathcal{H}$ and some fixed $\alpha \in (\frac{1}{3}, \frac{1}{2})$, we furthermore denote by $S^{(2)}(h) = (h, \int h \otimes dh) \in \mathscr{C}_g^{0,\alpha}$ the canonical lift given by computing the iterated integrals using usual Riemann–Stieltjes integration. It was then shown in [FLS06][1] that for every $\delta > 0$,

---

[1] Strictly speaking, this was shown for $h \in \mathcal{C}^2$; the extension to $h \in \mathcal{H}$ is non-trivial and found in [FV10b].

$$\lim_{\varepsilon \to 0} \mathbf{P}\Big(\varrho_{\alpha;[0,T]}\big(\mathbf{B}^{\text{Strat}}, S^{(2)}(h)\big) < \delta \;\Big|\; \|B - h\|_{\infty;[0,T]} < \varepsilon\Big) = 1. \tag{9.3}$$

The conditional statement then follows easily from continuity of the Itô–Lyons map and so yields the "difficult" support inclusion: every y^h is in the support of Y. The easy inclusion, support of Y contained in the closure of $\{y^h\}$, follows from the Wong–Zakai theorem, Theorem 9.3. If one is only interested in the support statement, but without the conditional statement (9.2), there are "softer" proofs; see Exercise 9.1 below. □

The second result to be discussed here, due to Freidlin–Wentzell, concerns the behaviour of diffusion in the singular ($\varepsilon \to 0$) limit when B is replaced by εB. We assume the reader is familar with large deviation theory.

Theorem 9.5 (Freidlin–Wentzell large deviations). *Let f, f_0, ξ be as in Theorem 9.1 above. Let $\alpha < 1/2$, B be a d-dimensional Brownian motion and consider the unique Stratonovich SDE solution $Y = Y^\varepsilon$ on $[0, T]$ to*

$$dY = f_0(Y)dt + \sum_{i=1}^d f_i(Y) \circ \varepsilon dB^i \tag{9.4}$$

started at $Y_0 = \xi \in \mathbf{R}^e$. Write Y^h for the ODE solution obtained by replacing $\circ \varepsilon dB$ with dh where $h \in \mathcal{H} = W_0^{1,2}$. Then $(Y_t^\varepsilon : 0 \le t \le T)$ satisfies a large deviation principle (in α-Hölder topology) with good rate function on pathspace given by

$$J(y) = \inf\big\{I(h) : Y^h = y\big\}.$$

Here I is Schilder's rate function for Brownian motion, i.e. $I(h) = \frac{1}{2}\|\dot{h}\|^2_{L^2([0,T],\mathbf{R}^d)}$ for $h \in \mathcal{H}$ and $I(h) = +\infty$ otherwise.

Proof. The key remark is that large deviation principles are robust under continuous maps, a simple fact known as *contraction principle*. The problem is then reduced to establishing a suitable large deviation principle for the Stratonovich lift of εB (which is exacly $\delta_\varepsilon \mathbf{B}^{\text{Strat}}$) in the α-Hölder rough path topology. Readers familiar with general facts of large deviation theory, in particular the *inverse* and *generalised* contraction principles, are invited to complete the proof along Exercise 9.2 below. □

9.4 Laplace method

We have seen that $(Y_t^\varepsilon : 0 \le t \le T)$, given as continuous images of the rescaled Brownian rough path, $Y^\varepsilon = \Phi(\delta_\varepsilon \mathbf{B}^{\text{Strat}})$, satisfies a large deviations principle (in Hölder and hence also in uniform topology) with rate function

$$J(y) = \inf\{I(h) : \Phi(h) = y, h \in \mathcal{H}\} \tag{9.5}$$

with the mild abuse of notation $\Phi(h) \equiv \Phi(\mathbf{h})$ where $\mathbf{h} = (h, \int h \otimes dh)$ is the canonical lift of $h \in \mathcal{H}$. A standard fact of large deviation theory, *Varadhan's lemma*, implies the following Laplace principle: for bounded continuous $F : C([0,T], \mathbf{R}^e) \to \mathbf{R}$,

$$\lim_{\varepsilon \to 0} \varepsilon^2 \log \mathbf{E}\left[\exp\left(-F(Y^\varepsilon)/\varepsilon^2\right)\right] = -\inf\{F_\Lambda(h) : h \in \mathcal{H}\},$$

where we set $F_\Lambda = F \circ \Phi + I$, for I as in Theorem 9.5. We are interested in precise asymptotics, hence the following collection of hypotheses.

(H1) The function F is bounded continuous on $C([0,T], \mathbf{R}^e)$.
(H2) The function F_Λ attains its unique minimum at $\gamma \in \mathcal{H}$.
(H3) The function F is C^3 in the Fréchet sense at $\varphi := \Phi(\gamma)$.
(H4) The element γ is a non-degenerate minimum of F_Λ restricted to $\mathcal{H}$ namely, for all $h \in \mathcal{H}\backslash\{0\}$,

$$D^2 F_\Lambda(\gamma)(h,h) = D^2(F \circ \Phi)\big|_\gamma(h,h) + \|h\|_{\mathcal{H}}^2 > 0$$

Theorem 9.6. *Let Y^ε be the unique Stratonovich SDE solution on $[0,T]$ in the small noise regime from Theorem 9.5. Under conditions (H1-H4), the following precise Laplace asymptotic holds*

$$\mathbf{E}\left[\exp\left(-F(Y^\varepsilon)/\varepsilon^2\right)\right] = \exp\left(-\frac{F_\Lambda(\gamma)}{\varepsilon^2}\right)(c_0 + o(1)) \qquad \text{as } \varepsilon \downarrow 0, \qquad (9.6)$$

for some constant $c_0 \in (0, \infty)$.

Proof. (i) Localisation around the minimiser. We regard $\mathbf{B} = \mathbf{B}^{\text{Strat}}$ (and its ε-dilations) as random variables in the (Polish) rough path space $\mathscr{C} := \mathscr{C}_g^{0,\alpha}([0,T])$. Write $\boldsymbol{\gamma} := (\gamma, \int \gamma \otimes d\gamma) \in \mathscr{C}$ for the canonical lift of the minimiser $\gamma \in \mathcal{H}$. Take now an arbitrary neighbourhood O of $\boldsymbol{\gamma} \in \mathscr{C}$ and decompose

$$\mathbf{E}\left[\exp\left(-F(Y^\varepsilon)/\varepsilon^2\right)\right] = \mathbf{E}\left[\exp\left(-F \circ \Phi(\delta_\varepsilon \mathbf{B})/\varepsilon^2\right)\right]$$
$$= \mathbf{E}[\ldots; \{\delta_\varepsilon \mathbf{B} \in O\}] + \mathbf{E}[\ldots; \{\delta_\varepsilon \mathbf{B} \in O\}^c].$$

Since $(\delta_\varepsilon \mathbf{B})$ satisfies an LDP with *good* rate function, (H1) implies that there exists $d > a := F_\Lambda(\gamma)$ and $\varepsilon_0 > 0$ such that for all $\varepsilon \in (0, \varepsilon_0)$

$$\mathbf{E}\left[\exp\left(-F \circ \Phi(\delta_\varepsilon \mathbf{B})/\varepsilon^2\right); \{\delta_\varepsilon \mathbf{B} \in O\}^c\right] \leq \exp\left(-d/\varepsilon^2\right). \qquad (9.7)$$

Hence this term does not contribute to the asymptotics (9.6). In the sequel, we shall take, for some $\varrho > 0$,

$$O := O_\varrho := \{T_\gamma \mathbf{X} : \mathbf{X} \in \mathscr{C}, \|\mathbf{X}\| < \varrho\} = \{\mathbf{X} \in \mathscr{C} : \|T_{-\gamma} \mathbf{X}\| < \varrho\}.$$

(By continuity of the translation operator, this is indeed an open neighbourhood of $T_\gamma 0 = \boldsymbol{\gamma}$.) We are thus left to analyse

$$J_\varrho := \mathbf{E}\left[\exp\left(-F \circ \Phi(\delta_\varepsilon \mathbf{B})/\varepsilon^2\right); \|T_{-\gamma} \delta_\varepsilon \mathbf{B}\| < \varrho\right].$$

(ii) Cameron–Martin shift. It is easy to see that, for Wiener a.e. ω, one has $\mathbf{B}(\omega + h) = T_h \mathbf{B}(\omega)$. In particular, the Cameron–Martin shift $\varepsilon B \rightsquigarrow \varepsilon B + \gamma$ (or $\omega \rightsquigarrow \omega + \gamma/\varepsilon$) induces a translation of $\delta_\varepsilon \mathbf{B}$ in the sense that

$$\delta_\varepsilon \mathbf{B} = \left(\varepsilon B, \int \varepsilon B \otimes d(\varepsilon B) \right) \rightsquigarrow \left(\varepsilon B + \gamma, \int (\varepsilon B + \gamma) \otimes d(\varepsilon B + \gamma) \right) = T_\gamma \delta_\varepsilon \mathbf{B} \,.$$

From the Cameron–Martin theorem, with all integrals below understood over $[0, T]$,

$$J_\varrho(\varepsilon) = \mathbf{E}\left(\exp\left(-\frac{\|\gamma\|_\mathcal{H}^2}{2\varepsilon^2} - \frac{\int \dot{\gamma} d(\varepsilon B)}{\varepsilon^2} \right) \exp\left(-\frac{F \circ \Phi(T_\gamma \delta_\varepsilon \mathbf{B})}{\varepsilon^2} \right); \|\delta_\varepsilon \mathbf{B}\| < \varrho \right)$$

$$= \exp\left(-\frac{\|\gamma\|_\mathcal{H}^2 + F \circ \Phi(\gamma)}{2\varepsilon^2} \right) \mathbf{E}\left(\exp\left(-\frac{(*)}{\varepsilon^2} \right); \|\delta_\varepsilon \mathbf{B}\| < \varrho \right);$$

where we recognise $F_\Lambda(\gamma)$ in the first exponential and also set

$$(*) = F \circ \Phi(T_\gamma \delta_\varepsilon \mathbf{B}) - F \circ \Phi(\gamma) + \varepsilon \int \dot{\gamma} dB \,.$$

(iii) Local analysis around the minimiser. We argue on a fixed rough path realisation $\mathbf{X} := \mathbf{B}(\omega)$. One checks that $\varepsilon \mapsto \Phi(T_\gamma \delta_\varepsilon \mathbf{X})$ is sufficiently smooth so that

$$\Phi(T_\gamma \delta_\varepsilon \mathbf{X}) = \Phi(\gamma) + \varepsilon G^1(\mathbf{X}) + \frac{\varepsilon^2}{2} G^2(\mathbf{X}) + \varepsilon^3 R_\varepsilon(\mathbf{X})$$

with remainder $R_\varepsilon(\mathbf{X})$, uniformly bounded in $\varepsilon \in (0, 1]$. We now use (H3) to obtain the expansion

$$(F \circ \Phi)(T_\gamma \delta_\varepsilon \mathbf{X}) = (F \circ \Phi)(\gamma) + \varepsilon \, DF|_\varphi \left(G^1(\mathbf{X}) \right)$$

$$+ \frac{\varepsilon^2}{2} \underbrace{\left[DF|_\varphi \left(G^2(\mathbf{X}) \right) + D^2 F|_\varphi \left(G^1(\mathbf{X}), G^1(\mathbf{X}) \right) \right]}_{=:Q(\mathbf{X})} + \varepsilon^3 R_\varepsilon^F(\mathbf{X}) \,,$$

where (H3) requires us to take ε less than some $\varepsilon_1(\mathbf{X})$, with remainder $R_\varepsilon^F(\mathbf{X})$, uniformly bounded in $\varepsilon \in (0, \varepsilon_1)$. Write $G^1 = G^1(h)$, and similar for G^2, Q, when evaluated at the canonical lift of an element $h \in \mathcal{H}$. We note for later

$$Q(h) = \frac{\partial^2}{\partial \varepsilon^2}\bigg|_{\varepsilon=0} (F \circ \Phi)(\gamma + \varepsilon h) \,.$$

Since γ minimises $F_\Lambda = F \circ \Phi + I$, first order optimality leads precisely to

$$DF|_\varphi \left(G^1(h) \right) + \int \dot{\gamma} dh = 0 \,, \tag{9.8}$$

for any $h \in \mathcal{H}$. By continuous extension we have $DF|_\varphi \left(G^1(\mathbf{B}(\omega)) \right) + \int \dot{\gamma} dB = 0$, see Exercise 9.3 (ii), and so

$$J_\varrho(\varepsilon) = \exp\left(-\frac{F_\Lambda(\gamma)}{2\varepsilon^2}\right)\mathbf{E}\left(\exp\left(-Q(\mathbf{B})/2 + \varepsilon R_\varepsilon^F(\mathbf{B})\right); \|\delta_\varepsilon\mathbf{B}\| < \varrho\right).$$

We claim that, as one would expect from exchanging $\varepsilon \to 0$ with expectation,

$$\lim_{\varepsilon\to 0}\mathbf{E}\left[\exp\left(-Q(\mathbf{B})/2 + \varepsilon R_\varepsilon^F(\mathbf{B})\right); \|\delta_\varepsilon\mathbf{B}\| < \varrho\right] = \mathbf{E}[\exp(-Q(\mathbf{B}))/2] < \infty.$$

To see why this is so, we first show integrability and even $\exp\left[-Q(\mathbf{B})/2\right] \in L^{1+\beta}$, for some $\beta > 0$, as consequence of the non-degeneracy assumption on the minimizer. The claimed integrability follows from the tail estimate $\mathbf{P}(-Q(\mathbf{B})/2 \geq r) \leq e^{-Cr}$, with $C > 1$ and for sufficiently large r. Now Q is "quadratic" in the precise sense $Q(\delta_\lambda\mathbf{X}) = \lambda^2 Q(\mathbf{X})$, $\lambda > 0$, so that upon setting $r \equiv 1/\varepsilon^2$, we are left to show

$$\mathbf{P}(-Q(\delta_\varepsilon\mathbf{B}) \geq 2) \leq e^{-C/\varepsilon^2}.$$

Since Q is seen to be continuous on rough path space, we have a good Large Deviations Principle for $\{-Q(\delta_\varepsilon\mathbf{B}) : \varepsilon > 0\}$, and using the upper LDP bound

$$\mathbf{P}(-Q(\delta_\varepsilon\mathbf{B})/2 \geq 1) \leq e^{-(C^* + o(1))/\varepsilon^2},$$

it remains to see $1 < C^*$, where, using goodness of the rate function,

$$C^* = \inf\left\{\tfrac{1}{2}\|h\|_{\mathcal{H}}^2 : h \in \mathcal{H}, -Q(h)/2 \geq 1\right\} = \tfrac{1}{2}\|h^*\|_{\mathcal{H}}^2 \quad \text{for some } h^* \in \mathcal{H}.$$

But this follows exactly from "$D^2(F \circ \Phi + I)(\gamma) > 0$" in direction h^*,

$$1 \leq -Q(h^*)/2 = \frac{1}{2}\frac{\partial^2}{\partial\varepsilon^2}\Big|_{\varepsilon=0}(-F \circ \Phi)(\gamma + \varepsilon h^*) < \frac{1}{2}\|h^*\|_{\mathcal{H}}^2.$$

This establishes $\exp\left[-Q(\mathbf{B})/2\right] \in L^{1+\beta}$. This additional amount of integrability, $\beta > 0$, is now used to give a uniform L^1-bound on $\exp\left(-Q(\mathbf{B})/2 + \varepsilon R_\varepsilon^F(\mathbf{B})\right)$ over $\|\delta_\varepsilon\mathbf{B}\| < \varrho$, after which one can conclude by dominated convergence. To this end, we revert to a pathwise consideration, $\mathbf{X} := \mathbf{B}(\omega)$. We need the remainder estimate, Exercise 9.4,

$$\sup_{\varepsilon\in(0,\varepsilon_1]}\left|R_\varepsilon^F(\mathbf{X})\right| \lesssim 1 + \|\mathbf{X}\|^3, \tag{9.9}$$

valid whenever $\varepsilon\|\mathbf{X}\| = \|\delta_\varepsilon\mathbf{X}\|$ remains bounded. It follows that, on $\|\delta_\varepsilon\mathbf{B}\| < \varrho$, we have the (uniform in small ε) estimate

$$\varepsilon\left|R_\varepsilon^F(\mathbf{B})\right| \lesssim 1 + \varepsilon\|\mathbf{B}\|^3 \lesssim 1 + \varrho\|\mathbf{B}\|^2 \tag{9.10}$$

and this estimate is uniform over $\varepsilon \in (0, 1]$. By Fernique's estimate for the (homogeneous!) rough path norm $\|\mathbf{B}\|$ of $\mathbf{B} = \mathbf{B}(\omega)$ and by choosing $\varrho = \varrho(\beta)$ small enough, we can guarantee that

$$e^{\varepsilon R_\varepsilon^F(\mathbf{B})}1_{\{\|\delta_\varepsilon\mathbf{B}\|<\varrho\}} \lesssim \exp\left(C\varrho\|\mathbf{B}\|^2\right) \in L^{\beta'},$$

where $\beta' < \infty$ is the Hölder conjugate of $\beta > 1$. Hence $\exp[-Q(\mathbf{B})/2 + \varrho\|\mathbf{B}\|^2] \in L^1$ serves as the uniform L^1-bound we were looking for and the proof is complete.
□

9.5 Exercises

Exercise 9.1 (Support of Brownian rough path [FV10b]) *Fix* $\alpha \in (\frac{1}{3}, \frac{1}{2})$ *and view the law* μ *of* $\mathbf{B}^{\text{Strat}}$ *as probability measure on the Polish space* $\mathscr{C}_{g,0}^{0,\alpha}$, *the (closed) subspace of* $\mathscr{C}_g^{0,\alpha}$ *of rough paths* $\mathbf{X}$ *started at* $X_0 = 0$. *Show that* $\mathbf{B}^{\text{Strat}}$ *has full support. The "easy" inclusion,* $\text{supp}\,\mu \subset \mathscr{C}_g^{0,\alpha}$ *is clear from Proposition 3.6. For the other inclusion, recall the translation operator from Exercise 2.15 and follow the steps below.*

a) **(Cameron–Martin theorem for Brownian rough path)** *Let* $h \in [0,T] \in \mathcal{H} = W_0^{1,2}$. *Show that* $\mathbf{X} \in \text{supp}\,\mu$ *implies* $T_h(\mathbf{X}) \in \text{supp}\,\mu$.

b) *Show that the support of* μ *contains at least one point, say* $\hat{\mathbf{X}} \in \mathscr{C}_g^{0,\alpha}$ *with the property that there exists a sequence of Lipschitz paths* $(h^{(n)})$ *so that* $T_{h^{(n)}}(\hat{\mathbf{X}}) \to (0,0)$ *in* α-*Hölder rough path metric.*

Hint: *Almost every realisation of* $\mathbf{B}^{\text{Strat}}(\omega)$ *will do, with* $-h^{(n)} = B^{(n)}$, *the dyadic piecewise linear approximations from Proposition 3.6.*

c) *Conclude that* $(0,0) = \lim_{n\to\infty} T_{h^{(n)}}(\hat{\mathbf{X}}) \in \text{supp}\,\mu$.

d) *As a consequence, any* $(h, \int h \otimes dh) = T_h(0,0) \in \text{supp}\,\mu$, *for any* $h \in \mathcal{H}$ *and taking the closure yields the "difficult" inclusion.*

e) *Appeal to continuity of the Itô–Lyons map to obtain the "difficult" support inclusion ("every* y^h *is in the support of* Y *") in the context of Theorem 9.4.*

Exercise 9.2 ("Schilder" large deviations, see [FV10b]) *Fix* $\alpha \in (\frac{1}{3}, \frac{1}{2})$ *and consider*

$$\delta_\varepsilon \mathbf{B}^{\text{Strat}} = (\varepsilon B, \varepsilon^2 \mathbb{B}^{\text{Strat}}),$$

the laws of which are viewed as probability measures μ^ε *on the Polish space* $\mathscr{C}_{g,0}^{0,\alpha}$. *Show that* $(\mu^\varepsilon) : \varepsilon > 0$ *satisfies a large deviation principle in* α-*Hölder rough path topology with good rate function*

$$J(\mathbf{X}) = I(X),$$

where $\mathbf{X} = (X, \mathbb{X})$ *and* I *is Schilder's rate function for Brownian motion, i.e.* $I(h) = \frac{1}{2}\|\dot{h}\|_{L^2([0,T],\mathbf{R}^d)}^2$ *for* $h \in \mathcal{H} = W_0^{1,2}$ *and* $I(h) = +\infty$ *otherwise.*

Hint: *Thanks to Gaussian integrability for the homogeneous rough paths norm of* $\mathbf{B}^{\text{Strat}}$ *it is actually enough to establish a large deviation principle for* $(\delta_\varepsilon \mathbf{B}^{\text{Strat}} : \varepsilon > 0)$ *in the (much coarser) uniform topology, which is not very hard to do "by hand", cf. [FV10b].*

Exercise 9.3 *In the context of Laplace asymptotics given in Theorem 9.6:*

a) *Detail the localisation estimate (9.7).*
b) *Derive the first order optimality condition (9.8) and justify its "continuous extension", i.e. replacing h by $B(\omega)$.*
c) *Show that $G_2 = G_2(\mathbf{X})$ is continuous in rough path sense. Conclude that the same holds for $Q = Q(\mathbf{X})$.*

Remark: *Related results appear in [BA88] (on path space) and [Ina06, Lemma 8.2].*

Exercise 9.4 (Stochastic Taylor-like rough path expansion) *We aim to show the remainder estimate (9.9).*

a) *As a warmup, consider $\Phi : \mathcal{C}([0,1], \mathbf{R}^d) \to \mathbf{R}$ so that $\Phi(X) = \varphi(X_1)$, for some $\varphi \in \mathcal{C}^3(\mathbf{R}^d)$. Fix $\gamma \in \mathcal{C}([0,1], \mathbf{R}^d)$ and establish the expansion*

$$\Phi(\gamma + \varepsilon X) \equiv g_0 + \varepsilon g_1(X) + \varepsilon^2 g_2(X) + \varepsilon^3 r_\varepsilon(X) \,,$$

such that $|r_\varepsilon(X)| \lesssim |X_1|^3$, uniformly in $\varepsilon \in (0,1]$, provided $|\varepsilon X_1|$ remains bounded.
b) *Show that an extra ε-dependent drift, say εX replaced by $\varepsilon X + \varepsilon \mu$ for some fixed $\mu \in \mathcal{C}([0,1], \mathbf{R}^d)$, alters the remainder estimate to $|r_\varepsilon(X)| \lesssim 1 + |X_1|^3$.*
c) *Generalise a) and b) to the situation when Φ is $\mathcal{C}^3$-regular in Fréchet sense. (This trivially covers the case $F \circ \Phi$, with another $F \in \mathcal{C}^3$.)*
d) *Prove the real thing, i.e. the remainder estimate (9.9) based on the expansion of $\varepsilon \mapsto F \circ \Phi(T_\gamma \delta_\varepsilon \mathbf{X})$ where Φ is the Itô–Lyons map. (See e.g. [IK07, Thm 5.1] and the references therein. For a similar estimate in a slightly different setting, see also [FGP18].)*

9.6 Comments

The rough path approach to solving stochastic differential equations (SDEs) driven by d-dimensional noise, can be seen as far-reaching extension of the works of Doss and Sussmann [Dos77, Sus78], and the Wong–Zakai approximation result [WZ65] ($d = 1$) and Clark [Cla66], Stroock-Varadhan [SV72] for $d > 1$. Lyons [Lyo98] used the Wong–Zakai theorem in conjunction with his continuity result to deduce the fact that RDE solutions (driven by the Brownian rough path $\mathbf{B}^{\text{Strat}}$) coincide with solution to (Stratonovich) stochastic differential equations. Similar to Friz–Victoir [FV10b], the logic is reversed in our presentation: thanks to an a priori identification of $\int f(Y) \, d\mathbf{B}^{\text{Strat}}$ as a Stratonovich stochastic integral, the Wong–Zakai results is obtained. Ikeda–Watanabe [IW89] present "twisted" Wong–Zakai approximation, based on McShane [McS72], in which case an additional limiting drift vector field appears; see also [Sus91, FO09]. Wong-Zakai type results for SPDEs (with finite-dimensional noise) is a straight-forward consequence of continuity statements for rough partial differential equations, as discussed in Sections 12.1 and 12.2. A version of the Wong–Zakai theorem for a singular SPDEs with space-time white noise via regularity structures is established by Hairer–Pardoux [HP15].

Almost sure rates for Wong–Zakai approximations in Brownian (and then more general Gaussian) rough path situations, were studied by Hu–Nualart [HN09], Deya–Neuenkirch–Tindel [DNT12] and Friz–Riedel [FR14]; see also Riedel–Xu [RX13]. Let us also note that L^q-rates for the convergence of approximations are not easy to obtain with rough path techniques (in contrast to Itô calculus which is ideally suited for moment calculations). Nonetheless, such rates can be obtained by Gaussian techniques, as discussed in Section 11.2.3 below; applications include multi-level Monte Carlo for SDEs and more generally Gaussian RDEs [BFRS16]. The rough path approach to SDEs (and more generally Gaussian RDEs) leads naturally to random dynamical systems, cf. comment Section 10.5.

The rough path approach to the *Stroock-Varadhan support theorem* [SV72] in Section 9.3 goes back to Ledoux–Qian–Zhang [LQZ02] in p-variation and Friz [Fri05] in Hölder topology, simplified and extended with Victoir in [FV05, FV07, FV10b]; the conditional estimate (9.3) is due to Friz, Lyons and Stroock [FLS06]. We note that this strategy of proof applies whenever one has rough path stability, which includes many stochastic partial differential equations (with finite-dimensional noise) discussed in Chapter 12. In the case of infinite-dimensional noise, a general support theorem for singular SPDEs was obtained via regularity structures by Hairer–Schönbauer [HS19] and extends the paracontrolled work of Chouk–Friz [CF18], as well as classical results such as the work of Bally, Millet and Sanz-Sole [BMSS95].

The rough path approach to *Freidlin–Wentzell* (small noise) *large deviations* in Section 9.3 goes also back to Ledoux, Qian and Zhang [LQZ02]; in p-variation, strengthened to Hölder topology in [FV05]; Inahama studies large deviations for pinned diffusions [Ina15], see also [Ina16a]. Once more, the strategy of proof applies whenever one has rough path stability, and thus applies to many stochastic partial differential equations as discussed in Chapter 12. Large deviations for Banach valued Wiener–Itô chaos proved useful in extensions to Gaussian rough paths and then Gaussian models (in the sense of regularity structures), see [FV07] and [HW15], where Hairer–Weber establish small noise large deviations for large classes of singular SPDEs.

Theorem 9.6 is an elegant application of rough paths, due to Aida [Aid07], to the classical theme of *Laplace method on Wiener space*, in a setting close to Ben Arous [BA88]; see also Inahama [Ina06], his work with Kawabi [IK07] and [Ina13]. Our presentation borrows from Friz, Gassiat and Pigato [FGP18]. See Friz–Klose [FK20] for a recent extension of these works to singular SPDEs via regularity structures. Recent applications to heat kernel expansions include [IT17].

The pathwise approach has also been useful to study *mean field* or *McKean–Vlasov* stochastic differential equations. This goes back to Tanaka [Tan84], with pathwise analysis of additive noise, revisited and extended by Coghi et al. [CDFM18]. The rough path case was pioneered by Cass–Lyons [CL15], with measure dependent drift, followed by Bailleul, Catellier and Delarue [BCD20, BCD19] to a setup that includes the important case of measure dependent noise vector fields. Dawson–Gärtner type large deviations from the McKean-Vlasov limit of weakly interacting diffusions is studied in by [Tan84, CDFM18], and also in Deuschel et al. [DFMS18] via rough paths, always under additive noise. Coghi–Nilssen [CN19] study, from a rough path point of view, McKean-Vlasov diffusion with "common" noise.

The Lions–Sznitman theory of reflecting SDEs [LS84] was revisited from a purely analytic rough path perspective by Aida [Aid15] and Deya et al. [DGHT19a] (existence) Gassiat [Gas20] shows non-uniqueness.

Homogenisation has also seen much impetus from rough path theory. After early works by Lejay–Lyons [LL03], we mention Bailleul–Catellier [BC17] and Kelly–Melbourn [KM16, KM17], who pioneered applications to deterministic homogenisation for fast-slow systems with chaotic noise, work continued by Chevyrev et al. [CFK+19b, CFK+19a, CFKM19].

Stochastic differential equations with jumps, driven by Lévy or general semimartingale noise, noise are well-known [KPP95, Pro05, App09] to require a careful interpretation: forward vs. geometric (a.k.a. Marcus canonical) sense. The pathwise interpretation of such differential equations was started by Williams [Wil01] and essentially completed by Chevyrev, Friz, Shekhar and Zhang [FS17, FZ18, CF19], consistency with the corresponding stochastic theories is also shown.

Rough analysis is "strong" by nature, yet has also proven a powerful tool for "weak" (or martingale) problems. This was pioneered by Delarue–Diehl [DD16], using rough paths to study a one-dimensional *SDE with distributional drift*, with applications to *polymer measures*. The extension to higher dimensions was carried out with paracontrolled methods by Cannizzaro–Chouk [CC18a].

Bruned et al. [BCF18] construct examples of renormalised SDE solutions, partially based on the "Hoff" process [Hof06, FHL16], related to Itô SDE solutions as averaging Stratonovich solutions [LY16].

Chapter 10
Gaussian rough paths

We investigate when multidimensional stochastic processes can be viewed – in a "canonical" fashion – as random rough paths. Gaussianity only enters through equivalence of moments. A simple criterion is given which applies in particular to fractional Brownian motion with suitable Hurst parameter.

10.1 A simple criterion for Hölder regularity

We now consider a driving signal modelled by a continuous, centred Gaussian process with values in $V = \mathbf{R}^d$. We thus have continuous sample paths

$$X(\omega) : [0, T] \to \mathbf{R}^d$$

and may take the underlying probability space as $\mathcal{C}([0, T], \mathbf{R}^d)$, equipped with a Gaussian measure μ so that $X_t(\omega) = \omega(t)$. Recall that μ, the law of X, is fully determined by its covariance function

$$R : [0, T]^2 \to \mathbf{R}^{d \times d}$$
$$(s, t) \mapsto \mathbf{E}[X_s \otimes X_t] .$$

In this section, a major role will be played by the *rectangular increments* of the covariance, namely

$$R\begin{pmatrix} s, t \\ s', t' \end{pmatrix} \overset{\text{def}}{=} \mathbf{E}[X_{s,t} \otimes X_{s',t'}] .$$

As far as the Hölder regularity of sample paths is concerned, we have the following classical result, which is nothing but a special case of Kolmogorov's continuity criterion:

Proposition 10.1. *Assume there exists positive ϱ and M such that for every $0 \le s \le t \le T$,*

© Springer Nature Switzerland AG 2020
P. K. Friz, M. Hairer, *A Course on Rough Paths*, Universitext,
https://doi.org/10.1007/978-3-030-41556-3_10

$$\left| R\begin{pmatrix} s,t \\ s,t \end{pmatrix} \right| \le M |t-s|^{1/\varrho}. \tag{10.1}$$

Then, for every $\alpha < 1/(2\varrho)$ *there exists* $K_\alpha \in L^q$, *for all* $q < \infty$, *such that*

$$|X_{s,t}(\omega)| \le K_\alpha(\omega)|t-s|^\alpha.$$

Proof. We may argue componentwise and thus take $d = 1$ without loss of generality. Since

$$|X_{s,t}|_{L^2} = (\mathbf{E}[X_{s,t}X_{s,t}])^{1/2} \le \left| R\begin{pmatrix} s,t \\ s,t \end{pmatrix} \right|^{1/2} \le M^{1/2}|t-s|^{\frac{1}{2\varrho}}$$

and $|X_{s,t}|_{L^q} \le c_q |X_{s,t}|_{L^2}$ by Gaussianity, we conclude immediately with an application of the Kolmogorov criterion. $\square$

Whenever the above proposition applies with $\varrho < 1$, the resulting sample paths can be taken with Hölder exponent $\alpha \in (\frac{1}{2}, \frac{1}{2\varrho})$; differential equations driven by X can then be handled with Young's theory, cf. Section 8.3. Therefore, our focus will be on Gaussian processes which satisfy a suitable modification of condition (10.1) with $\varrho \ge 1$ such that the process X allows for a *probabilistic* construction of a suitable second order process[1]

$$\mathbb{X}(\omega) : [0,T]^2 \to \mathbf{R}^{d \times d},$$

which is tantamount to making sense of the "formal" stochastic integrals

$$\int_s^t X_{s,r}^i dX_r^j \qquad \text{for} \qquad 0 \le s < t \le T, \quad 1 \le i,j \le d, \tag{10.2}$$

such that almost every realisation $\mathbb{X}(\omega)$ satisfies the algebraic and analytical properties of Section 2, notably (2.1) and (2.3) for some $\alpha \in (\frac{1}{3}, \frac{1}{2}]$. We shall also look for $(X, \mathbb{X})$ as (random) *geometric* rough path; thanks to (2.6), only the case $i < j$ in (10.2) then needs to be considered.

At the risk of being repetitive, the reader should keep in mind the following three points: (i) the sample paths $X(\omega)$ will not have, in general, enough regularity to define (10.2) as Young integrals; (ii) the process X will not be, in general, a semimartingale, so (10.2) cannot be defined using classical stochastic integrals; (iii) a lift of the process X to $(X, \mathbb{X}) \in \mathscr{C}_g^\alpha$ for some $\alpha \in (\frac{1}{3}, \frac{1}{2}]$, if at all possible, will never be unique (as discussed in Chapter 2, one can always perturb the area, i.e. Anti$(\mathbb{X})$ by the increments of a 2α-Hölder path). But there might still be one distinguished *canonical* choice for $\mathbb{X}$, in the same way as $\mathbb{B}^{\text{Strat}}$ is canonically obtained as limit (in probability) of $\int B^n \otimes dB^n$, for many natural approximations B^n of Brownian motion B.

---

[1] Despite the two parameters (s,t) one should not think of a random field here: as was noted in Exercise 2.4, $(X, \mathbb{X})$ is really a *path*.

10.2 Stochastic integration and variation regularity of the covariance

Our standing assumption from here on is independence of the d components of X, which is tantamount to saying that the covariance takes values in the diagonal matrices. Basic examples to have in mind are d-dimensional standard Brownian motion B with

$$R(s,t) = (s \wedge t)\mathrm{Id} \in \mathbf{R}^{d \times d}$$

(here Id denotes the identity matrix in $\mathbf{R}^{d \times d}$) or fractional Brownian motion B^H, with

$$R_H(s,t) = \frac{1}{2}\left[s^{2H} + t^{2H} - |t - s|^{2H}\right]\mathrm{Id} \in \mathbf{R}^{d \times d}$$

where $H \in (0,1)$; note the implication $\mathbf{E}\left[\left(B_t^H - B_s^H\right)^2\right] = |t - s|^{2H}$. The reader should observe that Proposition 10.1 above applies with $\varrho = 1/(2H)$; the focus on $\varrho \geq 1$ (to avoid trivial situations covered by Young theory) translates to $H \leq 1/2$.

We return to the task of making sense of (10.2), componentwise for fixed $i < j$, and it will be enough to do so for the unit interval; the interval $[s,t]$ is handled by considering $\left(X_{s+\tau(t-s)} : 0 \leq \tau \leq 1\right)$. Writing $\left(X, \tilde{X}\right)$, rather than $\left(X^i, X^j\right)$, we attempt a definition of the form

$$\int_0^1 X_{0,u}\, d\tilde{X}_u \overset{\text{def}}{=} \lim_{|\mathcal{P}| \downarrow 0} \sum_{[s,t] \in \mathcal{P}} X_{0,\xi} \tilde{X}_{s,t} \quad \text{with} \quad \xi \in [s,t], \tag{10.3}$$

where the limit is understood in probability, say. Classical stochastic analysis (e.g. [RY99, p144]) tells us that care is necessary: if $X, \tilde{X}$ are semimartingales, the choice $\xi = s$ ("left-point evaluation") leads to the Itô integral; $\xi = t$ ("right-point evaluation") to the backward Itô – and $\xi = (s+t)/2$ to the Stratonovich integral. On the other hand, all these integrals only differ by a bracket term $\langle X, \tilde{X} \rangle$ which vanishes if $X, \tilde{X}$ are independent. While we do not assume a semimartingale structure here, we do have the standing assumption of componentwise independence. This suggests a Riemann sum approximation of (10.2) in which we expect the precise point of evaluation to play no rôle; we thus consider left-point evaluation (but mid- or rightpoint evaluation would lead to the same result; cf. Exercise 10.5, (ii) below). Given partitions $\mathcal{P}, \mathcal{P}'$ of $[0,1]$ we set

$$\int_{\mathcal{P}} X_{0,s}\, d\tilde{X}_s := \sum_{[s,t] \in \mathcal{P}} X_{0,s} \tilde{X}_{s,t},$$

so that under the *assumption that X and $\tilde{X}$ are independent*, we have

$$\mathbf{E}\left[\int_{\mathcal{P}} X_{0,s}\, d\tilde{X}_s \int_{\mathcal{P}'} X_{0,s}\, d\tilde{X}_s\right] = \sum_{\substack{[s,t] \in \mathcal{P} \\ [s',t'] \in \mathcal{P}'}} R\begin{pmatrix} 0,s \\ 0,s' \end{pmatrix} \tilde{R}\begin{pmatrix} s,t \\ s',t' \end{pmatrix}. \tag{10.4}$$

On the right-hand side we recognise a 2D Riemann-Stieltjes sum and set

$$\int_{\mathcal{P} \times \mathcal{P}'} R \, d\tilde{R} := \sum_{\substack{[s,t] \in \mathcal{P} \\ [s',t'] \in \mathcal{P}'}} R\binom{0,s}{0,s'} \tilde{R}\binom{s,t}{s',t'}.$$

Let us now assume that R has finite ϱ-variation in the sense $\|R\|_{\varrho;[0,1]^2} < \infty$ where the ϱ-variation on a rectangle $I \times I'$ is given by

$$\|R\|_{\varrho;I \times I'} := \left(\sup_{\substack{\mathcal{P} \subset I, \\ \mathcal{P}' \subset I'}} \sum_{\substack{[s,t] \in \mathcal{P} \\ [s',t'] \in \mathcal{P}'}} \left| R\binom{s,t}{s',t'} \right|^{\varrho} \right)^{1/\varrho} < \infty, \qquad (10.5)$$

and similarly for $\tilde{R}$, with $\theta = 1/\varrho + 1/\tilde{\varrho} > 1$. A generalisation of Young's maximal inequality due to Towghi [Tow02] states that [2]

$$\sup_{\substack{\mathcal{P} \subset I, \\ \mathcal{P}' \subset I'}} \left| \int_{\mathcal{P} \times \mathcal{P}'} R \, d\tilde{R} \right| \leq C(\theta) \|R\|_{\varrho;I \times I'} \|\tilde{R}\|_{\tilde{\varrho};I \times I'}.$$

In particular, if the covariance of $\tilde{X}$ has similar variation regularity as X, the condition simplifies to $\varrho < 2$ and we obtain the following L^2-*maximal inequality.*

Lemma 10.2. *Let* $X, \tilde{X}$ *be independent, continuous, centred Gaussian processes with respective covariances* $R, \tilde{R}$ *of finite* ϱ-*variation, some* $\varrho < 2$. *Then*

$$\sup_{\mathcal{P} \subset [0,1]} \mathbf{E}\left[\left(\int_{\mathcal{P}} X_{0,r} \, d\tilde{X}_r \right)^2 \right] \leq C \|R\|_{\varrho;[0,1]^2} \|\tilde{R}\|_{\varrho;[0,1]^2},$$

where the constant C *depends on* ϱ.

We can now show existence of (10.3) as L^2-limit.

Proposition 10.3. *Under the assumptions of the previous lemma,*

$$\lim_{\varepsilon \to 0} \sup_{\substack{\mathcal{P}, \mathcal{P}' \subset [0,1]: \\ |\mathcal{P}| \vee |\mathcal{P}'| < \varepsilon,}} \left| \int_{\mathcal{P}} X_{0,r} d\tilde{X}_r - \int_{\mathcal{P}'} X_{0,r} d\tilde{X}_r \right|_{L^2} = 0. \qquad (10.6)$$

Hence, $\int_0^1 X_{0,r} d\tilde{X}_r$ *exists as the* L^2-*limit of* $\int_{\mathcal{P}} X_{0,r} d\tilde{X}_r$ *as* $|\mathcal{P}| \downarrow 0$ *and*

$$\mathbf{E}\left[\left(\int_0^1 X_{0,r} d\tilde{X}_r \right)^2 \right] \leq C \|R\|_{\varrho;[0,1]^2} \|\tilde{R}\|_{\varrho;[0,1]^2} \qquad (10.7)$$

with a constant $C = C(\varrho)$.

---

[2] This holds more generally if R is evaluated at $[0, \xi] \times [0, \xi']$ where $\xi \in [s, t]$, $\xi' \in [s', t']$.

Proof. At first glance, the situation looks similar to Young's part in the proof of Theorem 4.10 where we deduce (4.14) from Young's maximal inequality. However, the same argument fails if re-run with $\Xi_{s,t} = X_{0,s}\tilde{X}_{s,t}$ and $|\cdot|$ replaced by $|\cdot|_{L^2}$; in effect, the triangle inequality is too crude and does not exploit probabilistic cancellations present here. We now present two arguments for the key estimate (10.6). **First argument**: at the price of adding / subtracting $\mathcal{P} \cap \mathcal{P}'$, we may assume without loss of generality that $\mathcal{P}'$ refines $\mathcal{P}$. This allows to write

$$\int_{\mathcal{P}'} X_{0,r}\, d\tilde{X}_r - \int_{\mathcal{P}} X_{0,r}\, d\tilde{X}_r = \sum_{[u,v]\in\mathcal{P}} \int_{\mathcal{P}'\cap[u,v]} X_{u,r}\, d\tilde{X}_r \overset{\text{def}}{=} \mathcal{I},$$

and we need to show convergence of $\mathcal{I}$ to zero in L^2 as $|\mathcal{P}| = |\mathcal{P}| \vee |\mathcal{P}'| \to 0$. To see this, we rewrite the square of the expectation of this quantity as

$$\mathbf{E}\mathcal{I}^2 = \sum_{[u,v]\in\mathcal{P}} \sum_{[u',v']\in\mathcal{P}} \mathbf{E}\left(\int_{\mathcal{P}'\cap[u,v]} X_{u,r}\, d\tilde{X}_r \int_{\mathcal{P}'\cap[u',v']} X_{u',r'}\, d\tilde{X}_{r'} \right)$$

$$= \sum_{[u,v]\in\mathcal{P}} \sum_{[u',v']\in\mathcal{P}} \int_{\mathcal{P}'\cap[u,v]\times\mathcal{P}'\cap[u',v']} R\, d\tilde{R}.$$

Thanks to Towghi's maximal inequality, the absolute value of this term is bounded from above by a constant $C = C(\varrho)$ times

$$\sum_{[u,v]\in\mathcal{P}} \sum_{[u',v']\in\mathcal{P}} \|R\|_{\varrho;[u,v]\times[u',v']} \|\tilde{R}\|_{\varrho;[u,v]\times[u',v']}$$

$$\leq \sum_{[u,v]\in\mathcal{P}} \sum_{[u',v']\in\mathcal{P}} \omega([u,v]\times[u',v'])^{\frac{1}{\varrho}} \tilde{\omega}([u,v]\times[u',v'])^{\frac{1}{\varrho}},$$

where $\omega = \omega([s,t]\times[s',t'])$ (and similarly for $\tilde{\omega}$) is a so-called 2D control [FV11]: super-additive, continuous and zero when $s = t$ or $s' = t'$. A possible choice, if finite, is

$$\omega([s,t]\times[s',t']) \overset{\text{def}}{=} \sup_{\mathcal{Q}\subset[s,t]\times[s',t']} \sum_{[u,v]\times[u',v']\in\mathcal{Q}} \left| R\begin{pmatrix} u,v \\ u',v' \end{pmatrix} \right|^{\varrho}. \tag{10.8}$$

The difference to (10.5) is that the sup is taken over all (finite) partitions $\mathcal{Q}$ of $[s,t]\times[s',t']$ into rectangles; not just "grid-like" partitions induced by $\mathcal{P}\times\mathcal{P}'$. At this stage it looks like one should the change assumption "covariance of finite ϱ-variation" to "finite *controlled* ϱ-variation", which by definition means $\omega([0,1]^2) < \infty$. But in fact there is little difference [FV11]: finite controlled ϱ-variation trivially implies finite ϱ-variation; conversely, finite ϱ-variation implies finite controlled ϱ'-variation, any $\varrho' > \varrho$. Since (10.6) does not depend on ϱ, we may as well (at the price of replacing ϱ by ϱ') assume finite controlled ϱ-variation. The Cauchy–Schwarz inequality for finite sums shows that $\bar{\omega} := \omega^{1/2}\tilde{\omega}^{1/2}$ is again a 2D control; the above estimates can then be continued to

$$
\mathbf{E}\mathcal{I}^2 \le C \sum_{[u,v]\in\mathcal{P}} \sum_{[u',v']\in\mathcal{P}} \bar{\omega}([u,v]\times[u',v'])^{2/\varrho}
$$

$$
\le C \max_{\substack{[u,v]\in\mathcal{P}\\[u',v']\in\mathcal{P}}} \bar{\omega}([u,v]\times[u',v'])^{\frac{2-\varrho}{\varrho}} \times \sum_{[u,v]\in\mathcal{P}} \sum_{[u',v']\in\mathcal{P}} \bar{\omega}([u,v]\times[u',v'])
$$

$$
\le \mathrm{o}(1) \times \bar{\omega}([0,1]\times[0,1]) \, ,
$$

where we used the facts that $|\mathcal{P}| \downarrow 0$, $\varrho < 2$ and super-additivity of $\bar{\omega}$ to obtain the last inequality. This is precisely the required bound. The **second argument** makes use of Riemann-Stieltjes theory, applicable after mollification of $\tilde{X}$, and a uniformity property of ϱ-variation upon mollification. Let thus denote $\tilde{X}^n := \tilde{X} * f_n$ the convolution of $t \mapsto \tilde{X}_t$ with (f_n), a family of smooth, compactly supported probability density functions, weakly convergent to a Dirac at 0. Writing $\tilde{R}^n_{s,t} := \mathbf{E}\big(\tilde{X}^n_s \tilde{X}^n_t\big)$ for the covariance of $\tilde{X}^n$, and also $\tilde{S}^n_{s,t} := \mathbf{E}\big(\tilde{X}_s \tilde{X}^n_t\big)$ for the "mixed" covariance, we leave the fact that

$$
\sup_n \big\|\tilde{R}^n\big\|_{\varrho;[0,1]^2}, \sup_n \big\|\tilde{S}^n\big\|_{\varrho;[0,1]^2} \le \big\|\tilde{R}\big\|_{\varrho;[0,1]^2} \, , \tag{10.9}
$$

as and easy exercise for the reader. (**Hint:** Note $\tilde{R}^n = \tilde{R} * (f_n \otimes f_n)$, $\tilde{S}^n = \tilde{R} * (\delta \otimes f_n)$; estimate then the rectangular increments of $\tilde{R}_n$, respectively $\tilde{S}^n$, to the power ϱ with Jensen's inequality.)

Since $\tilde{X}^n$ has finite variation sample paths, basic Riemann-Stieltjes theory implies

$$
\int_{\mathcal{P}} X_{0,r}\, d\tilde{X}^n_r \to \int X_{0,r}\, d\tilde{X}^n_r \quad \text{as} \quad |\mathcal{P}| \to 0. \tag{10.10}
$$

In fact, this convergence (n fixed) takes also place in L^2 which may be seen as consequence of Lemma 10.2. On the other hand, pick $\varrho' \in (\varrho, 2)$ and apply Lemma 10.2 to obtain[3]

$$
\sup_{\mathcal{P}} \left| \int_{\mathcal{P}} X_{0,r}\, d\tilde{X}_r - \int_{\mathcal{P}} X_{0,r}\, d\tilde{X}^n_r \right|^2_{L^2} \le C\|R_X\|_{\varrho';[0,1]^2} \|R_{\tilde{X}-\tilde{X}^n}\|_{\varrho';[0,1]^2}
$$

$$
\le C\|R_X\|_{\varrho';[0,1]^2} \|R_{\tilde{X}-\tilde{X}^n}\|_{\varrho;[0,1]^2}^{\varrho/\varrho'} \|R_{\tilde{X}-\tilde{X}^n}\|_{\infty;[0,1]^2}^{1-\varrho/\varrho'} \, , \tag{10.11}
$$

where $C = C(\varrho)$. Now $\varrho' > \varrho$ implies $\|R_X\|_{\varrho';[0,1]^2} \le \|R_X\|_{\varrho;[0,1]^2}$ (immediate consequence of $|x|_{\varrho'} \le |x|_\varrho \equiv (\sum_{i=1}^m |x_i|^\varrho)^{1/\varrho}$ on $\mathbf{R}^m$) and thanks to (10.9) we also have the (uniform in n) estimate

$$
\|R_{\tilde{X}-\tilde{X}^n}\|_{\varrho;[0,1]^2} \le C_\varrho \Big(\|R_{\tilde{X}}\|_{\varrho;[0,1]^2} + 2\|\tilde{S}^n\|_{\varrho;[0,1]^2} + \|R_{\tilde{X}^n}\|_{\varrho;[0,1]^2}\Big)
$$

$$
\le 4C_\varrho \|\tilde{R}\|_{\varrho;[0,1]^2}.
$$

---

[3] Define $|f|_{\infty;[0,1]^2} = \sup \left| f\begin{pmatrix} u,v \\ u',v' \end{pmatrix} \right|$ where the sup is taken over all $[u,v], [u',v'] \subset [0,1]$.

Since $\tilde{X}^n$ converges to $\tilde{X}$ uniformly and in L^2, it is not hard to see that $R_{\tilde{X}-\tilde{X}^n} \to 0$ uniformly on $[0,1]^2$. We then see that (10.11) tends to zero as $n \to \infty$. It is now an elementary exercise to combine this with (10.10) to conclude the (second) proof of (10.6).

At last, the L^2-estimate is an immediate corollary of the maximal inequality given in Lemma 10.2 and L^2-convergence of the approximating Riemann–Stieltjes sums. $\square$

Note that there was nothing special about the time horizon $[0,1]$ in the above discussion. Indeed, given any time horizon $[s,t]$ of interest, it suffices to apply the same argument to the process $(X_{s+\tau(t-s)} : 0 \le \tau \le 1)$. Since variation norms are conveniently invariant under reparametrisation, (10.7) translates immediately to an estimate of the form

$$\mathbf{E}\left[\left(\int_s^t X_{s,r}\, d\tilde{X}_r\right)^2\right] \le C \|R\|_{\varrho;[s,t]^2} \|\tilde{R}\|_{\varrho;[s,t]^2}, \qquad (10.12)$$

first for the approximating Riemann–Stieltjes sums and then for their L^2-limits.

Theorem 10.4. *Let $(X_t : 0 \le t \le T)$ be a d-dimensional, continuous, centred Gaussian process with independent components and covariance R such that there exists $\varrho \in [1,2)$ and $M < \infty$ such that for every $i \in \{1,\dots,d\}$ and $0 \le s \le t \le T$,*

$$\|R_{X^i}\|_{\varrho;[s,t]^2} \le M|t-s|^{1/\varrho}. \qquad (10.13)$$

Define, for $1 \le i < j \le d$ and $0 \le s \le t \le T$, in L^2-sense (cf. Proposition 10.3),

$$\mathbb{X}_{s,t}^{i,j} := \lim_{|\mathcal{P}|\to 0} \int_{\mathcal{P}} (X_r^i - X_s^i)\, dX_r^j,$$

and then also (the algebraic conditions (2.1) and (2.6) leave no other choice!)

$$\mathbb{X}_{s,t}^{i,i} := \frac{1}{2}(X_{s,t}^i)^2 \quad and \quad \mathbb{X}_{s,t}^{j,i} := -\mathbb{X}_{s,t}^{i,j} + X_{s,t}^i X_{s,t}^j. \qquad (10.14)$$

Then, the following properties hold:

a) *For every $q \in [1,\infty)$ there exists $C_1 = C_1(q,\varrho,d,T)$ such that for all $0 \le s \le t \le T$,*

$$\mathbf{E}\left(|X_{s,t}|^{2q} + |\mathbb{X}_{s,t}|^q\right) \le C_1 M^q |t-s|^{q/\varrho}. \qquad (10.15)$$

b) *There exists a continuous modification of $\mathbb{X}$, denoted by the same letter from here on. Moreover, for any $\alpha < 1/(2\varrho)$ and $q \in [1,\infty)$ there exists $C_2 = C_2(q,\varrho,d,\alpha)$ such that*

$$\mathbf{E}\left(\|X\|_\alpha^{2q} + \|\mathbb{X}\|_{2\alpha}^q\right) \le C_2 M^q. \qquad (10.16)$$

c) *For any $\alpha < \frac{1}{2\varrho}$, with probability one, the pair $(X, \mathbb{X})$ satisfies conditions (2.1), (2.3) and (2.6). In particular, for $\varrho \in [1, \frac{3}{2})$ and any $\alpha \in (\frac{1}{3}, \frac{1}{2\varrho})$ we have $(X, \mathbb{X}) \in \mathscr{C}_g^\alpha$ almost surely.*

Proof. By scaling, we can take $M = 1$ without loss of generality. Regarding the first property, the "first level" estimates are contained in Proposition 10.1. Thus, in view of (10.14), in order to establish (10.15) only $\mathbf{E}(|\mathbb{X}_{s,t}^{i,j}|^q)$ for $i < j$ needs to be considered. For $q = 2$ this is an immediate consequence of (10.12) and our assumption (10.13). The case of general q follows from the well-known equivalence of L^q- and L^2-norm on the second Wiener–Itô chaos (e.g. [FV10b, Appendix D]).

Regarding the remaining two properties, almost sure validity of the algebraic constraint (2.1) for any fixed pair of times is an easy consequence of algebraic identities for Riemann sums. The construction of a continuous modification of $(s, t) \mapsto \mathbb{X}_{s,t}$ under the assumed bound is then standard (in fact, the proof of Theorem 3.1 shows this for dyadic times and the unique continuous extension is the desired modification). At last, Theorem 3.1 yields $K_\alpha, \mathbb{K}_\alpha$, with moments of all orders, such that

$$|X_{s,t}| \le K_\alpha(\omega)|t - s|^\alpha , \qquad |\mathbb{X}_{s,t}| \le \mathbb{K}_\alpha(\omega)|t - s|^{2\alpha} .$$

The dependence of the moments of K_α and $\mathbb{K}_\alpha$ on M finally follows by simple rescaling. $\square$

Theorem 10.5. *Let $(X, Y) = (X^1, Y^1, \ldots, X^d, Y^d)$ be a centred continuous Gaussian process on $[0, T]$ such that (X^i, Y^i) is independent of (X^j, Y^j) when $i \ne j$. Assume that there exists $\varrho \in [1, 2)$ and $M \in (0, \infty)$ such that the bounds*

$$\|R_{X^i}\|_{\varrho;[s,t]^2} \le M|t - s|^{1/\varrho} , \qquad \|R_{Y^i}\|_{\varrho;[s,t]^2} \le M|t - s|^{1/\varrho} ,$$

$$\|R_{X^i - Y^i}\|_{\varrho;[s,t]^2} \le \varepsilon^2 M|t - s|^{1/\varrho} , \tag{10.17}$$

hold for all $i \in \{1, \ldots, d\}$ and all $0 \le s \le t \le T$. Then

a) *For every $q \in [1, \infty)$, the bounds*

$$\mathbf{E}(|Y_{s,t} - X_{s,t}|^q)^{\frac{1}{q}} \lesssim \varepsilon \sqrt{M}|t - s|^{\frac{1}{2\varrho}} ,$$

$$\mathbf{E}(|\mathbb{Y}_{s,t} - \mathbb{X}_{s,t}|^q)^{\frac{1}{q}} \lesssim \varepsilon M|t - s|^{\frac{1}{\varrho}} ,$$

hold for all $0 \le s \le t \le T$.

b) *For any $\alpha < 1/(2\varrho)$ and $q \in [1, \infty)$, one has*

$$|\mathbf{E}(\|Y - X\|_\alpha^q)|^{\frac{1}{q}} \lesssim \varepsilon \sqrt{M} ,$$

$$|\mathbf{E}(\|\mathbb{Y} - \mathbb{X}\|_{2\alpha}^q)|^{\frac{1}{q}} \lesssim \varepsilon M .$$

c) *For $\varrho \in [1, \frac{3}{2})$ and any $\alpha \in (\frac{1}{3}, \frac{1}{2\varrho})$, $q < \infty$, one has*

$$|\varrho_\alpha(\mathbf{X}, \mathbf{Y})|_{L^q} \lesssim \varepsilon .$$

(Here, $\varrho_\alpha(\mathbf{X}, \mathbf{Y})$ denotes the α-Hölder rough path distance between $\mathbf{X} = (X, \mathbb{X})$ and $\mathbf{Y} = (Y, \mathbb{X})$ in $\mathscr{C}_g^\alpha$.)

Proof. By scaling we may without loss of generality assume $M = 1$. As for a) we note (again) that equivalence of L^q- and L^2-norm on Wiener–Itô chaos allow to reduce our discussion to $q = 2$. The first level estimate being easy, we focus on the second level estimate; to this end fix $i \neq j$. Since L^2-convergence implies a.s. convergence along a subsequence there exists $(\mathcal{P}_n)$, with mesh tending to zero, so that we can use Fatou's lemma to estimate

$$\mathbf{E}\big(|\mathbb{Y}_{s,t}^{i,j} - \mathbb{X}_{s,t}^{i,j}|^2\big) = \mathbf{E}\Big(\lim_{n \to \infty} \Big| \int_{\mathcal{P}_n} Y_{s,r}^i dY_r^j - X_{s,r}^i dX_r^j \Big|^2 \Big)$$

$$\leq \liminf_n \mathbf{E}\Big(\Big| \int_{\mathcal{P}_n} Y_{s,r}^i dY_r^j - X_{s,r}^i dX_r^j \Big|^2 \Big)$$

$$\leq \sup_{\mathcal{P}} \mathbf{E}\Big(\Big| \int_{\mathcal{P}} Y_{s,r}^i dY_r^j - X_{s,r}^i dX_r^j \Big|^2 \Big) .$$

The result now follows from the bound

$$\Big| \int_{\mathcal{P}} Y_{s,r}^i dY_r^j - X_{s,r}^i dX_r^j \Big| \leq \Big| \int_{\mathcal{P}} Y_{s,r}^i d(Y - X)_r^j \Big| + \Big| \int_{\mathcal{P}} (Y - X)_{s,r}^i dX_r^j \Big| ,$$

where we estimate the second moment of each term on the right-hand side by the respective variation norms of the covariances; e.g.

$$\mathbf{E}\Big(\Big| \int_{\mathcal{P}} Y_{s,r}^i d(Y - X)_r^j \Big|^2 \Big) \leq C \|R_{Y^i}\|_{\varrho;[s,t]^2} \|R_{Y^j - X^j}\|_{\varrho;[s,t]^2}$$

$$\leq C \varepsilon^2 |t - s|^{\frac{2}{\varrho}} .$$

The case $i = j$ is easier: it suffices to note that

$$\mathbf{E}\big(|\mathbb{Y}_{s,t}^{i,i} - \mathbb{X}_{s,t}^{i,i}|^2\big) = \frac{1}{4} \mathbf{E}\big((Y_{s,t}^i)^2 - (X_{s,t}^i)^2\big)$$

$$= \frac{1}{4} \big| \mathbf{E}\big((Y_{s,t}^i - X_{s,t}^i)(Y_{s,t}^i + X_{s,t}^i)\big) \big| ,$$

then conclude with Cauchy–Schwarz.

Regarding b), given the pointwise L^q-estimates as stated in a), the L^q-estimates for $\|X - Y\|_\alpha$ and $\|\mathbb{Y} - \mathbb{X}\|_{2\alpha}$ are obtained from Theorem 3.3. The last statement is then an immediate consequence of the definition of ϱ_α. $\square$

Corollary 10.6. *As above, let $(X, Y) = (X^1, Y^1, \ldots, X^d, Y^d)$ be a centred continuous Gaussian process such that (X^i, Y^i) is independent of (X^j, Y^j) when $i \neq j$. Assume that there exists $\varrho \in [1, \frac{3}{2})$ and $M \in (0, \infty)$ such that*

$$\|R_{(X,Y)}\|_{\varrho;[s,t]^2} \leq M|t - s|^{1/\varrho} \ \forall 0 \leq s \leq t \leq T. \tag{10.18}$$

Then, for every $\alpha \in (\frac{1}{3}, \frac{1}{2\varrho})$, *every* $\theta \in (0, \frac{1}{2} - \varrho\alpha)$ *and* $q < \infty$, *there exists a constant* C *such that*

$$|\varrho_\alpha(\mathbf{X}, \mathbf{Y})|_{L^q} \le C \sup_{s,t \in [0,T]} \left[\mathbf{E}|X_{s,t} - Y_{s,t}|^2 \right]^\theta. \tag{10.19}$$

Proof. At the price of replacing (X, Y) by the rescaled process $M^{-1/2}(X, Y)$ we may take $M = 1$. (The concluding L^q-estimate on $\varrho_\alpha(M^{-1/2}X, M^{-1/2}Y)$ is then readily translated into an estimate on $\varrho_\alpha(X, Y)$, given that we allow the final constant to depend on M.) Assumption (10.18) then spells out precisely to

$$\|R_{X^i}\|_{\varrho;[s,t]^2} \le |t - s|^{1/\varrho}, \|R_{Y^i}\|_{\varrho;[s,t]^2} \le |t - s|^{1/\varrho}$$

and (not present in the assumptions of the previous theorem!)

$$\|R_{(X^i, Y^i)}\|_{\varrho;[s,t]^2} \le |t - s|^{1/\varrho}$$

where $R_{(X^i, Y^i)}(u, v) = \mathbf{E}(X_u^i Y_v^i)$. Thanks to this assumption we have

$$\|R_{X^i - Y^i}\|_{\varrho;[s,t]^2} \le C_\varrho \left(\|R_{X^i}\|_{\varrho;[s,t]^2} + 2\|R_{(X^i, Y^i)}\|_{\varrho;[s,t]^2} + \|R_{Y^i}\|_{\varrho;[s,t]^2} \right)$$
$$\le 4C_\varrho |t - s|^{1/\varrho},$$

which is handy in the following interpolation argument. Set

$$\eta := \max\{\|R_{X^i - Y^i}\|_{\infty;[0,T]^2} : 1 \le i \le d\}$$

and note that, for any $\varrho' > \varrho$,

$$\|R_{X^i - Y^i}\|_{\varrho';[s,t]^2} \le \|R_{X^i - Y^i}\|_{\infty;[s,t]^2}^{1 - \varrho/\varrho'} \|R_{X^i - Y^i}\|_{\varrho;[s,t]^2}^{\varrho/\varrho'}$$
$$\le (4C_\varrho)^{\varrho/\varrho'} \eta^{1 - \varrho/\varrho'} |t - s|^{1/\varrho'}.$$

Also, with $\tilde{M} = 1 \vee T^{1/\varrho - 1/\varrho'}$, and then similar for R_{Y^i},

$$\|R_{X^i}\|_{\varrho';[s,t]^2} \le \|R_{X^i}\|_{\varrho;[s,t]^2} \le |t - s|^{1/\varrho} \le \tilde{M}|t - s|^{1/\varrho'}$$

and so, picking $\varrho' = \frac{\varrho}{1 - 2\theta}$ the previous theorem (with $\varrho' \leftarrow \varrho$ and $\varepsilon^2 \leftarrow \eta^{1 - \varrho/\varrho'}$, $M \leftarrow \tilde{M} \vee (4C_\varrho)^{\varrho/\varrho'}$) yields

$$|\varrho_\alpha(X, Y)|_{L^q} \le C\varepsilon = C\eta^{\frac{1}{2} - \varrho\frac{1}{2\varrho'}} = C\eta^\theta.$$

for any given $\theta \in (0, \frac{1}{2} - \varrho\alpha)$. At last, take $i_* \in \{1, \ldots, d\}$ as the arg max in the definition of η and set $\Delta = X^{i_*} - Y^{i_*}$. Then, by Cauchy–Schwarz,

$$\eta = \|R_\Delta\|_{\infty;[0,T]^2} = \sup_{\substack{0 \le s \le t \le T \\ 0 \le s' \le t' \le T}} \mathbf{E}(\Delta_{s,t}\Delta_{s',t'}) \le \sup_{0 \le s \le t \le T} \mathbf{E}\Delta_{s,t}^2$$

and the proof is finished. $\square$

Remark 10.7. Corollary 10.6 suggests an alternative route to the construction of a rough path lift $\mathbf{X} = (X, \mathbb{X})$ for some Gaussian process X as in Theorem 10.4. The idea is to establish the crucial estimate (10.19) only for processes with regular sample paths, in which case $\mathbb{X}$ is canonically given by iterated Riemann–Stieltjes integration. Apply this to piecewise linear (or mollifier) approximations X^n, X^m to see that $(X^n, \mathbb{X}^n)$ is Cauchy, in probability and rough path metric in the space $\mathscr{C}_g^{0,\alpha}$. The resulting limiting (random) rough path $\mathbf{X}$ is easily seen to be indistinguishable from the one constructed in Theorem 10.4. All estimates are then seen to remain valid in the limit. (This is the approach taken in [FV10a, FV10b].)

10.3 Fractional Brownian motion and beyond

We remarked in the beginning of Section 10.2 that (d-dimensional) fractional Brownian motion B^H, with Hurst parameter $H \in (0,1)$, determined through its covariance

$$R_H(s,t) = \frac{1}{2}\Big[s^{2H} + t^{2H} - |t-s|^{2H}\Big]\mathrm{Id} \in \mathbf{R}^{d \times d}$$

has α-Hölder sample paths for any $\alpha < H$. For $H > 1/2$, there is little need for rough path analysis - after all, Young's theory is applicable. For $H = 1/2$, one deals with d-dimensional standard Brownian motion which, of course, renders the classical martingale based stochastic analysis applicable. For $H < 1/2$, however, all these theories fail but rough path analysis works. In the remainder of this section we detail the construction of a fractional Brownian rough path.

In fact, we shall consider centred, continuous Gaussian processes with independent components $X = (X^1, \ldots, X^d)$ and *stationary increments*. The construction of a (geometric) rough path associated to X then naturally passes through an understanding of the two-dimensional ϱ-variation of $R = R_X$, the covariance of X; cf. Theorem 10.4. To this end, it is enough to focus on one component and we may take X to be scalar until further notice. The law of such a process is fully determined by

$$\sigma^2(u) := \mathbf{E}\big[X_{t,t+u}^2\big] = R\binom{t,t+u}{t,t+u}.$$

Lemma 10.8. *Assume that $\sigma^2(\cdot)$ is concave on $[0,h]$ for some $h > 0$. Then, one has non-positive correlation of non-overlapping increments in the sense that, for $0 \le s \le t \le u \le v \le h$,*

$$\mathbf{E}[X_{s,t}X_{u,v}] = R\binom{s,t}{u,v} \le 0.$$

If in addition $\sigma^2(\cdot)$ restricted to $[0,h]$ is non-decreasing (which is always the case for some possibly smaller h), then for $0 \leq s \leq u \leq v \leq t \leq h$,

$$0 \leq \mathbf{E}[X_{s,t}X_{u,v}] = |\mathbf{E}[X_{s,t}X_{u,v}]| \leq \mathbf{E}[X_{u,v}^2] = \sigma^2(v-u) \ .$$

Proof. Using the identity $2ac = (a+b+c)^2 + b^2 - (b+c)^2 - (a+b)^2$ with $a = X_{s,t}, b = X_{t,u}$ and $c = X_{u,v}$, we see that

$$2\mathbf{E}[X_{s,t}X_{u,v}] = \mathbf{E}[X_{s,v}^2] + \mathbf{E}[X_{t,u}^2] - \mathbf{E}[X_{t,v}^2] - \mathbf{E}[X_{s,u}^2]$$
$$= \sigma^2(v-s) + \sigma^2(u-t) - \sigma^2(v-t) - \sigma^2(u-s).$$

The first claim now easily follows from concavity, cf. [MR06, Lemma 7.2.7].

To show the second bound, note that $X_{s,t}X_{u,v} = (a+b+c)b$ where $a = X_{s,u}$, $b = X_{u,v}$, and $c = X_{v,t}$. Applying the algebraic identity

$$2(a+b+c)b = (a+b)^2 - a^2 + (c+b)^2 - c^2$$

and taking expectations yields

$$2\mathbf{E}[X_{s,t}X_{u,v}] = \mathbf{E}[X_{s,v}^2] - \mathbf{E}[X_{s,u}^2] + \mathbf{E}[X_{u,t}^2] - \mathbf{E}[X_{v,t}^2]$$
$$= (\sigma^2(v-s) - \sigma^2(u-s)) + (\sigma^2(t-u) - \sigma^2(t-v)) \geq 0 \ ,$$

where we used that $\sigma^2(\cdot)$ is non-decreasing. On the other hand, using $(a+b+c)b = b^2 + ab + cb$ and the non-positive correlation of non-overlapping increments, we have

$$\mathbf{E}[X_{s,t}X_{u,v}] = \mathbf{E}[X_{u,v}^2] + \mathbf{E}[X_{s,u}X_{u,v}] + \mathbf{E}[X_{v,t}X_{u,v}] \leq \mathbf{E}[X_{u,v}^2] \ ,$$

thus concluding the proof. $\square$

Theorem 10.9. *Let X be a real-valued Gaussian process with stationary increments and $\sigma^2(\cdot)$ concave and non-decreasing on $[0,h]$, some $h > 0$. Assume also, for constants $L, \varrho \geq 1$, and all $\tau \in [0,h]$,*

$$|\sigma^2(\tau)| \leq L|\tau|^{1/\varrho} \ .$$

Then the covariance of X has finite ϱ-variation. More precisely

$$\|R_X\|_{\varrho\text{-var};[s,t]^2} \leq M|t-s|^{1/\varrho} \tag{10.20}$$

for all intervals $[s,t]$ with length $|t-s| \leq h$ and some $M = M(\varrho, L) > 0$.

Proof. Consider some interval $[s,t]$ with length $|t-s| \leq h$. The proof relies on separating "diagonal" and "off-diagonal" contributions. Let $\mathcal{D} = \{t_i\}, \mathcal{D}' = \{t_j'\}$ be two dissections of $[s,t]$. For fixed i, we have

$$3^{1-\varrho} \sum_{t'_j \in \mathcal{D}'} \left| \mathbf{E}\left(X_{t_i,t_{i+1}} X_{t'_j,t'_{j+1}}\right) \right|^{\varrho} \leq 3^{1-\varrho} \left\| \mathbf{E}X_{t_i,t_{i+1}} X. \right\|^{\varrho}_{\varrho\text{-var};[s,t]} \quad (10.21)$$

$$\leq \left\| \mathbf{E}X_{t_i,t_{i+1}} X. \right\|^{\varrho}_{\varrho\text{-var};[s,t_i]} + \left\| \mathbf{E}X_{t_i,t_{i+1}} X. \right\|^{\varrho}_{\varrho\text{-var};[t_i,t_{i+1}]}$$

$$+ \left\| \mathbf{E}X_{t_i,t_{i+1}} X. \right\|^{\varrho}_{\varrho\text{-var};[t_{i+1},t]}.$$

By Lemma 10.8 above, we have

$$\left\| \mathbf{E}X_{t_i,t_{i+1}} X. \right\|_{\varrho\text{-var};[s,t_i]} \leq \left| \mathbf{E}X_{t_i,t_{i+1}} X_{s,t_i} \right| \leq \left| \mathbf{E}X_{t_i,t_{i+1}} X_{s,t_{i+1}} \right| + \left| \mathbf{E}X^2_{t_i,t_{i+1}} \right|$$

$$\leq 2\sigma^2(t_{i+1} - t_i).$$

The third term is bounded analogously. For the middle term in (10.21) we estimate

$$\left\| \mathbf{E}X_{t_i,t_{i+1}} X. \right\|^{\varrho}_{\varrho\text{-var};[t_i,t_{i+1}]} = \sup_{\mathcal{D}'} \sum_{t'_j \in \mathcal{D}'} \left| \mathbf{E}X_{t_i,t_{i+1}} X_{t'_j,t'_{j+1}} \right|^{\varrho}$$

$$\leq \sup_{\mathcal{D}'} \sum_{t'_j \in \mathcal{D}'} \left| \sigma^2(t'_{j+1} - t'_j) \right|^{\varrho} \leq L|t_{i+1} - t_i|,$$

where we used the second estimate of Lemma 10.8 for the penultimate bound and the assumption on σ^2 for the last bound. Using these estimates in (10.21) yields

$$\sum_{t'_j \in \mathcal{D}'} \left| \mathbf{E}X_{t_i,t_{i+1}} X_{t'_j,t'_{j+1}} \right|^{\varrho} \leq C|t_{i+1} - t_i|,$$

and (10.20) follows by summing over t_i and taking the supremum over all dissections of $[s,t]$. $\square$

Corollary 10.10. *Let $X = (X^1, \ldots, X^d)$ be a centred continuous Gaussian process with independent components such that each X^i satisfies the assumption of the previous theorem, with common values of h, L and $\varrho \in [1, 3/2)$. Then X, restricted to any interval $[0,T]$, lifts to $\mathbf{X} = (X, \mathbb{X}) \in \mathscr{C}^{\alpha}_g([0,T], \mathbf{R}^d)$.*

Proof. Set $I_n = [(n-1)h, nh]$ so that $[0,T] \subset I_1 \cup I_2 \cup \ldots \cup I_{[T/h]+1}$. On each interval I_n, we may apply Theorem 10.4 to lift $X_n := X|_{I_n}$ to a (random) rough path $\mathbf{X}_n \in \mathscr{C}^{\alpha}_g(I_n, \mathbf{R}^d)$. The concatenation of $\mathbf{X}_1, \mathbf{X}_2, \ldots$ then yields the desired rough path lift on $[0,T]$. $\square$

Example 10.11 (Fractional Brownian motion). Clearly, d-dimensional fractional Brownian motion B^H with Hurst parameter $H \in (\frac{1}{3}, \frac{1}{2}]$ satisfies the assumptions of the above theorem / corollary for all components with

$$\sigma(u) = u^{2H},$$

obviously non-decreasing and concave for $H \leq \frac{1}{2}$ and on any time interval $[0,T]$. This also identifies

$$\varrho = \frac{1}{2H}$$

and $\varrho < \frac{3}{2}$ translates to $H > \frac{1}{3}$ in which case we obtain a canonical geometric rough path $\mathbf{B}^H = (B^H, \mathbb{B}^H)$ associated to fBm. In fact, a canonical "level-3" rough path $\mathbf{B}^H$ can be constructed as long as $\varrho < \varrho^* = 2$, corresponding to $H > 1/4$ but this requires level-3 considerations which we do not discuss here (see [FV10b, Ch.15]).

Example 10.12 (Ornstein-Uhlenbeck process). Consider the d-dimensional (stationary) OU process, consisting of i.i.d. copies of a scalar Gaussian process X with covariance

$$\mathbf{E}[X_s X_t] = K(|t - s|), \qquad K(u) = \exp(-cu),$$

where $c > 0$ is fixed. Note that $\sigma^2(u) = \mathbf{E}X_{t,t+u}^2 = \mathbf{E}X_{t+u}^2 + \mathbf{E}X_t^2 - 2\mathbf{E}X_{t,t+u} = 2[K(0) - K(u)] = 1 - \exp(-cu)$, so that $\sigma^2(u)$ is indeed increasing and concave:

$$\partial_u[\sigma^2(u)] = c \exp(-cu) > 0$$
$$\partial_u^2[\sigma^2(u)] = -c^2 \exp(-cu) < 0.$$

One also has the bound $\sigma^2(u) = 1 - \exp(-cu) \le cu$, which shows that the assumptions of the above corollary are satisfied with $\varrho = 1$, $L = c$ and arbitrary $h > 0$.

10.4 Exercises

Exercise 10.1 *Let $X^{\mathcal{D}}$ be a piecewise linear approximation to X. Show that $(\mathbb{X}_{s,t})$ as constructed in Theorem 10.4 is the limit, in probability and uniformly on $\{(s,t) : 0 \le s \le t \le T\}$ say, of $\int_s^t X_{s,u}^{\mathcal{D}} \otimes dX_u^{\mathcal{D}}$ as $|\mathcal{D}| \to 0$. (In particular, any algebraic relations which hold for (piecewise) smooth paths and their iterated integrals then hold true in the limit. This yields an alternative proof that $(X, \mathbb{X})$ satisfies conditions (2.1) and (2.6).)*

Exercise 10.2 (Convergence to Brownian rough path [HN09, FR11]) *Let $X = B$ and $Y = B^n$ be a d-dimensional Brownian motion and its piecewise linear approximation with mesh size $1/n$, respectively. Show that the covariance of (B, B^n) has finite 1-variation, uniformly in n. Show also that*

$$\sup_{s,t \in [0,T]} \left[\mathbf{E}(|B_{s,t} - B_{s,t}^n|^2) \right] = \mathrm{O}\left(\frac{1}{n}\right).$$

Conclude that, for any $\theta < 1/2 - \alpha$

$$\left| \|B - B^n\|_\alpha + \sqrt{\|\mathbb{B} - \mathbb{B}^n\|_{2\alpha}} \right|_{L^q} = \mathrm{O}\left(\frac{1}{n^\theta}\right).$$

Use a Borel–Cantelli argument to show that, also for any $\theta < 1/2 - \alpha$,

$$\|B - B^n\|_\alpha + \|\mathbb{B} - \mathbb{B}^n\|_{2\alpha} \le C(\omega)\frac{1}{n^\theta}.$$

When $\alpha \in \left(\frac{1}{3}, \frac{1}{2}\right)$, we can conclude convergence in α-Hölder rough path metric, i.e.

$$\varrho_\alpha((B, \mathbb{B}), (B^n, \mathbb{B}^n)) \to 0,$$

almost surely with rate $1/2 - \alpha - \varepsilon$ for every $\varepsilon > 0$.

Exercise 10.3 *Let $(B, \tilde{B})$ be a 2-dimensional standard Brownian motion. The (Gaussian) process given by*

$$X = (B_t, B_t + \tilde{B}_t)$$

fails to have independent components and yet lifts to a Gaussian rough path. Explain how and detail the construction.

Exercise 10.4 *Assume $R(s, t) = K(|t - s|)$ for some C^2-function K. (This was exactly the situation in the above Ornstein–Uhlenbeck case, Example 10.12.) Give a direct proof that R has finite 2-dimensional 1-variation, more precisely,*

$$\|R\|_{1\text{-var};[s,t]^2} \le C|t - s|, \qquad \forall \, 0 \le s \le t \le T,$$

for a constant C which depends on T and K.

Solution. If $(s, t) \mapsto R(s, t) := \mathbf{E}[X_s X_t]$ is smooth, the 2-dimensional 1-variation is given by

$$\|R\|_{1\text{-var};[0,T]^2} = \int_{[0,T]^2} \left|\partial_{s,t}^2 R(s, t)\right| ds \, dt$$

This remains true when the mixed derivative is a signed measure, which in turn is the case when $R(s, t) = K(|t - s|)$ for some C^2-function K. Indeed, write H and 2δ for the distributional derivatives of $|\cdot|$. Formal application of the chain-rule gives $\partial_t R = K'(|t - s|)H(t - s)$ and then, using $|H| \le 1$ a.s.,

$$\left|\partial_{s,t}^2 R(s, t)\right| \le |K''(|t - s|)| + 2|K'(|t - s|)|\delta(t - s).$$

Integration again over $[s, t]^2 \subset [0, T]^2$ yields

$$\|R\|_{1\text{-var};[s,t]^2} = \int_{[s,t]^2} \left|\partial_{u,v}^2 R(u, v)\right| du \, dv \le (T|K''|_\infty + 2|K'(0)|)|t - s|.$$

This is easily made rigorous by replacing $|\cdot|$ (and then $H, 2\delta$) by a mollified version, say $|\cdot|_\varepsilon$ (and $H_\varepsilon, 2\delta_\varepsilon$), noting that variation norms are lower semicontinuous fashion under pointwise limits; that is

$$\|R\|_{1\text{-var};[s,t]^2} \le \liminf_{\varepsilon \to 0} \|R_\varepsilon\|_{1\text{-var};[s,t]^2}$$

whenever $R^\varepsilon \to R$ pointwise. To see this, it suffices to take arbitrary dissections $\mathcal{D} = (t_i)$ and $\mathcal{D}' = (t_j')$ of $[u, v]$ and note that

$$\sum_{i,j} \left|R\begin{pmatrix} t_{i-1}, t_i \\ t_{j-1}', t_j' \end{pmatrix}\right| = \lim_{\varepsilon \to 0} \sum_{i,j} \left|R^\varepsilon \begin{pmatrix} t_{i-1}, t_i \\ t_{j-1}', t_j' \end{pmatrix}\right| \le \liminf_{\varepsilon \to 0} \|R_\varepsilon\|_{1\text{-var};[u,v]^2}.$$

Exercise 10.5 *Assume $X = (X^1, \ldots, X^d)$ is a centred, continuous Gaussian process with independent components.*

(i) *Assume covariance of finite ϱ-variation with $\varrho < 2$. Show that each component $X = X^i$, for $i = 1, \ldots, d$, has almost surely vanishing compensated quadratic variation on $[0, T]$ by which we mean*

$$\lim_{n \to \infty} \sum_{[s,t] \in \mathcal{P}_n} \left(X_{s,t}^2 - \mathbf{E}(X_{s,t}^2) \right) = 0 ,$$

in probability (and L^q, any $q < \infty$) for any sequence of partitions $(\mathcal{P}_n)$ of $[0, T]$ with mesh $|\mathcal{P}_n| \to 0$.

(ii) *Under the assumptions of (i), show that there exists $(\mathcal{P}_n)$ with $|\mathcal{P}_n| \to 0$ so that, with probability one, the quadratic (co)variation $[X^i, X^j]$, in the sense of Definition 5.10, vanishes, for any $i \neq j$, with $i, j \in \{1, \ldots, d\}$.*
Conclude that, with regard to Theorem 10.4, the off-diagonal elements $\mathbb{X}_{s,t}^{i,j}$, defined as the L^2 limit of left-point Riemann–Stieltjes sums, could have been equivalently defined via mid- or right-point Riemann sums.

(iii) *Assume $\varrho = 1$. Show that, for all $i = 1, \ldots, d$, there exists a sequence $(\mathcal{P}_n)$ with mesh $|\mathcal{P}_n| \to 0$ so that, with probability one, the quadratic variation $[X^i, X^i]$, in the sense of Definition 5.10, exists and equals*

$$\left[X^i \right]_t := \lim_{\varepsilon \to 0} \sup_{|\mathcal{P}| < \varepsilon} \sum_{\substack{[u,v] \in \mathcal{P} \\ u < t}} \mathbf{E}\left(X_{u,v}^i \right)^2 .$$

Discuss the possibility of lifting X to a (random) non-geometric rough path, similar to the Itô-lift of Brownian motion.

(iv) *Consider the case of a zero-mean, stationary Gaussian process on $[0, 2\pi]$ with i.i.d. components, each specified by*

$$\mathbf{E}(X_{s,t}^2) = \cosh(-\pi) - \cosh(|t - s| - \pi).$$

Verify that $\varrho = 1$ and compute $[X]$. (This example is related to the stochastic heat equation, where s, t should be thought of as spatial variables, cf. Lemma 12.30)

Solution. (i) Using Wick's formula for the expectation of products of centred Gaussians, namely

$$\mathbf{E}(ABCD) = \mathbf{E}(AB)\mathbf{E}(CD) + \mathbf{E}(AC)\mathbf{E}(BD) + \mathbf{E}(AD)\mathbf{E}(BC) ,$$

we obtain the identity

$$\mathbf{E}\left| \sum_{[s,t] \in \mathcal{P}_n} X_{s,t}^2 - \mathbf{E}(X_{s,t}^2) \right|^2$$
$$= \sum_{[s,t] \in \mathcal{P}_n} \sum_{[s',t'] \in \mathcal{P}_n} \left(\mathbf{E}([X_{s,t}^2 X_{s',t'}^2) - \mathbf{E}(X_{s,t}^2)\mathbf{E}(X_{s',t'}^2) \right)$$

$$= \sum_{[s,t]\in\mathcal{P}_n} \sum_{[s',t']\in\mathcal{P}_n} 2\mathbf{E}(X_{s,t}X_{s',t'})\mathbf{E}(X_{s,t}X_{s',t'})$$

$$= 2 \sum_{[s,t]\in\mathcal{P}_n} \sum_{[s',t']\in\mathcal{P}_n} \left| R\begin{pmatrix} s,t \\ s',t' \end{pmatrix} \right|^2$$

$$\leq \sup_{\substack{t-s\leq|\mathcal{P}_n| \\ t'-s'\leq|\mathcal{P}_n|}} \left| R\begin{pmatrix} s,t \\ s',t' \end{pmatrix} \right|^{2-\varrho} \|R\|^{\varrho}_{\varrho\text{-var};[0,T]^2} .$$

This term on the other hand converges to 0 as $|\mathcal{P}_n| \to 0$. This gives L^2-convergence and hence convergence in probability. Convergence in L^q for any $q < \infty$ follows from general facts on Wiener–Itô chaos.

(ii) Left to the reader.

(iii) We fix i and drop the index. We easily see that (i) holds uniformly on compacts, say, in the sense that

$$\sup_{t\in[0,T]} \sum_{\substack{[u,v]\in D_n \\ u<t}} \left(X^2_{u,v} - \mathbf{E}(X^2_{u,v}) \right) \to 0 \text{ as } n \to \infty$$

in probability whenever $|\mathcal{P}_n| \to 0$. On the other hand,

$$\sup_{|\mathcal{P}|<\varepsilon} \sum_{\substack{[u,v]\in\mathcal{P} \\ u<t}} \mathbf{E}(X_{u,v})^2 < \infty$$

thanks to finite 1-variation of the covariance. By monotonicity, the limit as $\varepsilon = 1/n \to 0$ exists, and we call it $[[X]]_t$. Then, along a suitable sequence $(\tilde{\mathcal{P}}_n)$,

$$[[X]]_t = \lim_n \sum_{\substack{[u,v]\in\tilde{\mathcal{P}}_n \\ u<t}} \mathbf{E}(X_{u,v})^2 .$$

On the other hand, at the price of passing to another subsequence also denoted by $(\tilde{\mathcal{P}}_n)$, we have

$$\sup_{t\in[0,T]} \sum_{\substack{[u,v]\in\tilde{\mathcal{P}}_n \\ u<t}} \left(X^2_{u,v} - \mathbf{E}(X^2_{u,v}) \right) \to 0 \qquad \text{almost surely,}$$

and so with probability one, and uniformly in $t \in [0,T]$,

$$\sum_{\substack{[u,v]\in\tilde{\mathcal{P}}_n \\ u<t}} X^2_{u,v} \to [[X]]_t .$$

(iv) One has $\mathbf{E}(X^2_{s,t}) = \cosh(-\pi) - \cosh(|t-s| - \pi) = \sinh(\pi)|t-s| + \mathrm{o}(|t-s|)$ and so $[X]_t = t \sinh(\pi)$.

Exercise 10.6 *Assume finite 1-variation of the covariance (as e.g. defined in (10.5)) of a zero-mean Gaussian process X and $\mathbf{E}[X_{t,t+h}^2] = f(t)h + \mathrm{o}(h)$ as $h \downarrow 0$, for some $f \in C([0,T], \mathbf{R})$. Show that, for every smooth test function φ,*

$$\int_0^T \varphi(t) \frac{X_{t,t+h}^2}{h}\, dt \;\to\; \int_0^T \varphi(t) f(t)\, dt \qquad \text{as } h \to 0,$$

where the convergence takes place in L^q for any $q < \infty$ (and hence also in probability).

Solution. Since all types of L^q-convergence are equivalent on the finite Wiener–Itô chaos (here we only need the chaos up to level 2), it suffices to consider $q = 2$. A dissection (t_k) of $[0,T]$ is given by $t_k = kh \wedge T$. We have

$$\sum_k \frac{1}{h} \int_{t_k}^{t_{k+1}} \varphi(t) X_{t,t+h}^2\, dt = \int_0^1 d\theta \sum_k \varphi(t_k + \theta h) X_{t_k+\theta h, t_k+\theta h+h}^2$$

$$\equiv \int_0^1 \langle \varphi, \mu_{\theta,h} \rangle\, d\theta \,,$$

where the random measure $\mu_{\theta,h} := \sum_k \delta_{t_k+\theta h} X_{t_k+\theta h, t_k+\theta h+h}^2$ acts on test functions by integration. It obviously suffices to establish $\langle \varphi, \mu_{\theta,h} \rangle \to \langle \varphi, f \rangle$ in L^2, uniformly in $\theta \in [0,1]$. Define the (random) distribution function of $\mu_{\theta,h}$

$$F(t) := \mu_{\theta,h}([0,t]) = \sum_{k:t_k+\theta h \leq t} X_{t_k+\theta h, t_k+\theta h+h}^2 \,,$$

and also $\bar{F}(t) = \mathbf{E}F(t)$. Note that,

$$\bar{F}(t) = \sum_{k:t_k+\theta h \leq t} f(t_k + \theta h)h + \mathrm{o}(h) \sim \int_0^t f(s)ds \quad \text{as } h \downarrow 0,$$

uniformly in $\theta \in [0,1], t \in [0,T]$. On the other hand, the Gaussian (or Wick) identity $\mathbf{E}(A^2 B^2) - \mathbf{E}[A^2]\mathbf{E}(B^2) = 2(\mathbf{E}(AB))^2$, applied with $A = X_{t_k+\theta h, t_k+\theta h+h}$ and $B = X_{t_j+\theta h, t_j+\theta h+h}$, gives

$$\mathbf{E}\big(F(t) - \bar{F}(t)\big)^2 = \mathbf{E}\big(F^2(t)\big) - \bar{F}^2(t)$$

$$= 2 \sum_{\substack{k:t_k+\theta h \leq t \\ j:t_j+\theta h \leq t}} R_X \begin{pmatrix} t_k + \theta h, t_k + \theta h + h \\ t_j + \theta h, t_j + \theta h + h \end{pmatrix}^2$$

$$\lesssim \operatorname{osc}\big(R^{2-\varrho}; h\big) \to 0 \quad \text{as } h \to 0 \,,$$

uniformly in $\theta \in [0,1], t \in [0,T]$. It follows that

$$F(t) = \mu_{\theta,h}([0,t]) \to \int_0^t f(s)ds$$

in L^2, again uniformly in t and θ. Now, for fixed smooth φ, one has the bound

$$\left| \int \varphi(t)\mu_{\theta,h}(dt) - \int \varphi(t)f(t)dt \right|^2 = \left| \int \left(\int_0^t f(s)ds - \mu_{\theta,h}([0,t]) \right) \dot{\varphi}(t)dt \right|^2$$

$$\lesssim \int_0^1 \left(\int_0^t f(s)ds - \mu_{\theta,h}([0,t]) \right)^2 dt$$

and so

$$\mathbf{E}\left| \int \varphi(t)\mu_{\theta,h}(dt) - \int \varphi(t)f(t)dt \right|^2 \lesssim \int_0^1 \mathbf{E}\left(\int_0^t f(s)ds - \mu_{\theta,h}([0,t]) \right)^2 dt \ .$$

This expression converges to 0 as $h \to 0$, uniformly in θ, thus completing the proof.

10.5 Comments

Classes of Gaussian processes which admit (canonical) lifts to random rough paths were first studied by Coutin–Qian [CQ02], with focus on fBm with Hurst parameter $H > 1/4$. Ledoux, Qian and Zhang [LQZ02] used Gaussian techniques to establish large deviation and support for the Brownian rough paths, extensions to fractional Brownian motions were investigated by Millet–Sanz-Solé [MSS06], Feyel and de la Pradelle [FdLP06], Friz–Victoir [FV07, FV06a]. When $H \leq 1/4$, there is no canonical rough path lift: as noted in [CQ02], the L^2-norm of the area associated to piecewise linear approximations to fBm diverges. See however the works of Unterberger and then Nualart–Tindel [Unt10, NT11]. Parameter estimation for fractional SDEs via rough paths is studied in Papavasiliou–Ladroue [PL11], see also [DFM16]. The notion of two-dimensional ϱ-variation of the covariance, as adopted in this chapter, is due to Friz–Victoir, [FV10a], [FV10b, Ch.15], [FV11], and allows for an elegant and general construction of Gaussian rough paths. It also leads naturally to useful Cameron–Martin embeddings, see Section 11.1. If restricted to the "diagonal", ϱ-variation of the covariance relates to a classical criterion of Jain–Monrad [JM83]. The question remains how one checks finite ϱ-variation when faced with a non-trivial (and even non-explicit, e.g. given as Fourier series) covariance function. A general criterion based on a certain covariance measure structure (reminiscent of Kruk, Russo and Tudor [KRT07]) was recently given by Friz, Gess, Gulisashvili and Riedel [FGGR16], a special case of which is the "concavity criterion" of Theorem 10.9. Cass-Lim establish a Stratonovich-Skorohod integral formula for Gaussian rough paths. Multi-level Monte Carlo for Gaussian RDEs is analysed by Bayer et al. [BFRS16]. Bailleul, Riedel and Scheutzow [BRS17] show that random RDEs driven by suitable Gaussian rough paths constitute random dynamical system. It is interesting to note that many key results for Gaussian rough paths (tail estimate, support, densities, ...) can be shown with different tools to hold in a Markovian setting [CO17, CO18], using the framework of Markovian rough paths [FV08c, FV10b].

Chapter 11
Cameron–Martin regularity and applications

A continuous Gaussian process gives rise to a Gaussian measure on path-space. Thanks to variation regularity properties of Cameron–Martin paths, powerful tools from the analysis on Gaussian spaces become available. A general Fernique type theorem leads us to integrability properties of rough integrals with Gaussian integrator akin to those of classical stochastic integrals. We then discuss Malliavin calculus for differential equations driven by Gaussian rough paths. As application a version of Hörmander's theorem in this non-Markovian setting is established.

11.1 Complementary Young regularity

Although we have chosen to introduce (rough) paths subject to α-Hölder regularity, the arguments are not difficult to adapt to continuous paths with finite p-variation with $p = 1/\alpha \in [1, \infty)$. Recall that $C^{p\text{-var}}([0, T], \mathbf{R}^d)$ is the space of continuous paths $X : [0, T] \to \mathbf{R}^d$ so that

$$\|X\|_{p\text{-var};[0,T]} \overset{\text{def}}{=} \left(\sup_{\mathcal{P}} \sum_{[s,t] \in \mathcal{P}} |X_{s,t}|^p \right)^{\frac{1}{p}} < \infty , \tag{11.1}$$

with supremum taken over all partitions of $[0, T]$ and this constitutes a seminorm on $C^{p\text{-var}}$. The 1-variation ($p = 1$) of such a path is of course nothing but its length, possibly $+\infty$. Hölder implies variation regularity, one has the immediate estimate

$$\|X\|_{p\text{-var};[0,T]} \leq T^\alpha \|X\|_{\alpha;[0,T]}.$$

Conversely, a time-change renders p-variation paths Hölder continuous with exponent $\alpha = 1/p$. Given two paths $X \in C^{p\text{-var}}([0, T], \mathbf{R}^d)$, $h \in C^{q\text{-var}}([0, T], \mathbf{R}^d)$ let us say that they enjoy *complementary Young regularity* if Young's condition

$$\frac{1}{p} + \frac{1}{q} > 1 , \tag{11.2}$$

© Springer Nature Switzerland AG 2020
P. K. Friz, M. Hairer, *A Course on Rough Paths*, Universitext,
https://doi.org/10.1007/978-3-030-41556-3_11

is satisfied.

We are now interested in the regularity of Cameron–Martin paths. As in the last section, X is an $\mathbf{R}^d$-valued, continuous and centred Gaussian process on $[0,T]$, realised as $X(\omega) = \omega \in \mathcal{C}([0,T], \mathbf{R}^d)$, a Banach space under the uniform norm, equipped with a Gaussian measure. General principles of Gaussian measures on (separable) Banach spaces thus apply, see e.g. [Led96]. Specialising to the situation at hand, the associated *Cameron–Martin space* $\mathcal{H} \subset \mathcal{C}([0,T], \mathbf{R}^d)$ consists of paths $t \mapsto h_t = \mathbf{E}(ZX_t)$ where $Z \in \mathcal{W}^1$ is an element in the so-called *first Wiener chaos*, the L^2-closure of span$\{X_t^i : t \in [0,T], 1 \leq i \leq d\}$, consisting of Gaussian random variables. We recall that if $h' = \mathbf{E}(Z'X.)$ denotes another element in $\mathcal{H}$, the inner product $\langle h, h' \rangle_{\mathcal{H}} = \mathbf{E}(ZZ')$ makes $\mathcal{H}$ a Hilbert space; $Z \mapsto h$ is an isometry between $\mathcal{W}^1$ and $\mathcal{H}$.

Example 11.1. (Brownian motion). Let B be a d-dimensional Brownian motion, let $g \in L^2([0,T], \mathbf{R}^d)$, and set

$$Z = \sum_{i=1}^{d} \int_0^T g_s^i dB_s^i \equiv \int_0^T \langle g, dB \rangle .$$

By Itô's isometry, $h_t^i := \mathbf{E}(ZB_t^i) = \int_0^t g_s^i ds$ so that $\dot{h} = g$ and $\|h\|_{\mathcal{H}}^2 := \mathbf{E}(Z^2) = \int_0^T |g_s|^2 ds = \|\dot{h}\|_{L^2}^2$ where $|\cdot|$ denotes Euclidean norm on $\mathbf{R}^d$. Clearly, h is of finite 1-variation, and its length is given by $\|\dot{h}\|_{L^1}$. On the other hand, Cauchy–Schwarz shows any $h \in \mathcal{H}$ is $1/2$-Hölder which, in general, "only" implies 2-variation.

The proposition below applies to Brownian motion with $\varrho = 1$, also recalling that $\|R\|_{1;[s,t]^2} = |t - s|$ in the Brownian motion case.

Proposition 11.2. *Assume the covariance* $R : (s,t) \mapsto \mathbf{E}(X_s \otimes X_t)$ *is of finite ϱ-variation (in 2D sense) for $\varrho \in [1,\infty)$. Then $\mathcal{H}$ is continuously embedded in the space of continuous paths of finite ϱ-variation. More, precisely, for all $h \in \mathcal{H}$ and all $s < t$ in $[0,T]$,*

$$\|h\|_{\varrho\text{-var};[s,t]} \leq \|h\|_{\mathcal{H}} \sqrt{\|R\|_{\varrho\text{-var};[s,t]^2}} .$$

Proof. We assume X, h to be scalar, the extension to d-dimensional X is straightforward (and even trivial when X has independent components, which will always be the case for us). Setting $h = \mathbf{E}(ZX.)$, we may assume without loss of generality (by scaling), that $\|h\|_{\mathcal{H}}^2 := \mathbf{E}(Z^2) = 1$. Let (t_j) be a dissection of $[s,t]$. Let ϱ' be the Hölder conjugate of ϱ. Using duality for l^{ϱ}-spaces, we have[1]

$$\left(\sum_j |h_{t_j,t_{j+1}}|^{\varrho} \right)^{1/\varrho} = \sup_{\beta, |\beta|_{l^{\varrho'}} \leq 1} \sum_j \langle \beta_j, h_{t_j,t_{j+1}} \rangle$$

$$= \sup_{\beta, |\beta|_{l^{\varrho'}} \leq 1} \mathbf{E}\left(Z \sum_j \langle \beta_j, X_{t_j,t_{j+1}} \rangle \right)$$

---

[1] The case $\varrho = 1$ may be seen directly by taking $\beta_j = \text{sgn}(h_{t_j,t_{j+1}})$.

$$\leq \sup_{\beta,|\beta|_{l\varrho'} \leq 1} \sqrt{\sum_{j,k} \langle \beta_j \otimes \beta_k, \mathbf{E}(X_{t_j,t_{j+1}} \otimes X_{t_k,t_{k+1}}) \rangle}$$

$$\leq \sup_{\beta,|\beta|_{l\varrho'} \leq 1} \sqrt{\left(\sum_{j,k} |\beta_j|^{\varrho'} |\beta_k|^{\varrho'}\right)^{\frac{1}{\varrho'}} \left(\sum_{j,k} |\mathbf{E}(X_{t_j,t_{j+1}} \otimes X_{t_k,t_{k+1}})|^{\varrho}\right)^{\frac{1}{\varrho}}}$$

$$\leq \left(\sum_{j,k} |\mathbf{E}(X_{t_j,t_{j+1}} \otimes X_{t_k,t_{k+1}})|^{\varrho}\right)^{1/(2\varrho)} \leq \sqrt{\|R\|_{\varrho\text{-var};[s,t]^2}}.$$

The proof is then completed by taking the supremum over all dissections (t_j) of $[0,t]$.
□

Remark 11.3. It is typical (e.g. for Brownian or fractional Brownian motion, with $\varrho = 1/(2H) \geq 1$) that

$$\forall s < t \text{ in } [0,T]: \qquad \|R\|_{\varrho\text{-var};[s,t]^2} \leq M|t-s|^{1/\varrho}.$$

In such a situation, Proposition 11.2 implies that

$$|h_{s,t}| \leq \|h\|_{\varrho\text{-var};[s,t]} \leq \|h\|_{\mathcal{H}} M^{1/2} |t-s|^{1/(2\varrho)},$$

which tells us that $\mathcal{H}$ is continuously embedded in the space of $1/(2\varrho)$-Hölder continuous paths (which can also be seen directly from $h_{s,t} = \mathbf{E}(ZX_{s,t})$ and Cauchy–Schwarz). The point is that $1/(2\varrho)$-Hölder only implies 2ϱ-variation regularity, in contrast to the sharper result of Proposition 11.2.

In part i) of the following lemma we allow $\mathbf{X} = (X, \mathbb{X})$ to be a (continuous) rough path of finite p-variation rather than of α-Hölder regularity. More formally, we write $\mathbf{X} \in \mathscr{C}^{p\text{-var}}([0,T], \mathbf{R}^d)$ when $p \in [2,3)$ and the analytic Hölder type condition (2.3) in the definition of a rough path is replaced by $\|X\|_{p\text{-var};[0,T]} < \infty$ and the second order regularity condition

$$\|\mathbb{X}\|_{p/2\text{-var};[0,T]} \stackrel{\text{def}}{=} \left(\sup_{\mathcal{P}} \sum_{[s,t]\in\mathcal{P}} |\mathbb{X}_{s,t}|^{p/2}\right)^{2/p} < \infty. \tag{11.3}$$

(As before, we shall drop $[0,T]$ from our notation whenever the time horizon is fixed.) The homogeneous p-variation rough path norm (over $[0,T]$) is then given by

$$\|\mathbf{X}\|_{p\text{-var};[0,T]} = \|\mathbf{X}\|_{p\text{-var}} \stackrel{\text{def}}{=} \|X\|_{p\text{-var}} + \sqrt{\|\mathbb{X}\|_{p/2\text{-var}}}. \tag{11.4}$$

Of course, a geometric rough path of finite p-variation, $\mathbf{X} \in \mathscr{C}_g^{p\text{-var}}$ is one for which the "first order calculus" condition (2.6) holds.

The following results will prove crucial in Section 11.2 where we will derive, based on the Gaussian isoperimetric inequality, good probabilistic estimates on Gaussian rough path objects. They are equally crucial for developing the Malliavin calculus for (Gaussian) rough differential equations in Section 11.3.

Recall from Exercise 2.15 that the translation of a rough path $\mathbf{X} = (X, \mathbb{X})$ in direction h is given by

$$T_h(\mathbf{X}) \stackrel{\text{def}}{=} \left(X^h, \mathbb{X}^h\right) \tag{11.5}$$

where $X^h := X + h$ and

$$\mathbb{X}^h_{s,t} := \mathbb{X}_{s,t} + \int_s^t h_{s,r} \otimes dX_r + \int_s^t X_{s,r} \otimes dh_r + \int_s^t h_{s,r} \otimes dh_r , \tag{11.6}$$

provided that h is sufficienly regular to make the final three integrals above well-defined.

Lemma 11.4. *i) Let $\mathbf{X} \in \mathscr{C}_g^{p\text{-var}}([0, T], \mathbf{R}^d)$, with $p \in [2, 3)$ and consider a function $h \in C^{q\text{-var}}([0, T], \mathbf{R}^d)$ with complementary Young regularity in the sense that*

$$1/p + 1/q > 1 .$$

Then the translation of $\mathbf{X}$ in direction h is well-defined in the sense that the integrals appearing in (11.6) are well-defined Young integrals and $T_h : \mathbf{X} \mapsto T_h(\mathbf{X})$ maps $\mathscr{C}_g^{p\text{-var}}([0, T], \mathbf{R}^d)$ into itself. Moreover, one has the estimate, for some constant $C = C(p, q)$,

$$\|T_h(\mathbf{X})\|_{p\text{-var}} \leq C\left(\|\mathbf{X}\|_{p\text{-var}} + \|h\|_{q\text{-var}}\right) .$$

ii) Similarly, let $\alpha = 1/p \in (\frac{1}{3}, \frac{1}{2}]$, $\mathbf{X} \in \mathscr{C}_g^\alpha([0, T], \mathbf{R}^d)$ and $h : [0, T] \to \mathbf{R}^d$ again of complementary Young regularity, but now "respectful" of α-Hölder regularity in the sense that [2]

$$\|h\|_{q\text{-var};[s,t]} \leq K|t - s|^\alpha, \tag{11.7}$$

uniformly in $0 \leq s < t \leq T$. Write $\|h\|_{q,\alpha}$ for the smallest constant K in the bound (11.7). Then again T_h is well-defined and now maps $\mathscr{C}_g^\alpha([0, T], \mathbf{R}^d)$ into itself. Moreover, one has the estimate, again with $C = C(p, q)$,

$$\|T_h(\mathbf{X})\|_\alpha \leq C(\|\mathbf{X}\|_\alpha + \|h\|_{q,\alpha}) .$$

Proof. This is essentially a consequence of Young's inequality which gives

$$\left| \int_s^t h_{s,r} \otimes dX_r \right| \leq C\|h\|_{q\text{-var};[s,t]} \|X\|_{p\text{-var};[s,t]} ,$$

and then similar estimates for the other (Young) integrals appearing in the definition of $\mathbb{X}^h$. One then uses elementary estimates of the form $\sqrt{ab} \leq a + b$ (for non-negative reals a, b), in view of the definition of *homogeneous* norm (which involves $\mathbb{X}^h$ with a square root). Details are left to the reader. □

By combining the Cameron–Martin regularity established in Proposition 11.2, see also Remark 11.3, with the previous lemma we obtain the following result.

---

[2] From Remark 11.3, $\|h\|_{\varrho,\alpha} \lesssim \|h\|_{\mathcal{H}}$ for all $\alpha \leq \frac{1}{2\varrho}$.

Theorem 11.5. *Assume* $(X_t : 0 \le t \le T)$ *is a continuous d-dimensional, centred Gaussian process with independent components and covariance R such that there exists* $\varrho \in [1, \frac{3}{2})$ *and* $M < \infty$ *such that for every* $i \in \{1, \dots, d\}$ *and* $0 \le s \le t \le T$,

$$\|R_{X^i}\|_{\varrho\text{-var};[s,t]^2} \le M|t - s|^{1/\varrho}.$$

Let $\alpha \in (\frac{1}{3}, \frac{1}{2\varrho}]$ *and* $\mathbf{X} = (X, \mathbb{X}) \in \mathscr{C}^\alpha([0, T], \mathbf{R}^d)$ *a.s. be the random Gaussian rough path constructed in Theorem 10.4. Then there exists a null set* N *such that for every* $\omega \in N^c$ *and every* $h \in \mathcal{H}$,

$$T_h(\mathbf{X}(\omega)) = \mathbf{X}(\omega + h).$$

Proof. Note that complementary Young regularity holds, with $p = \frac{1}{\alpha} < 3$ and $q = \varrho < \frac{3}{2}$, as is seen from $\frac{1}{p} + \frac{1}{q} > \frac{1}{3} + \frac{2}{3} = 1$. As a consequence of Lemma 11.4, the translation $T_h(\mathbf{X}(\omega))$ is well-defined whenever $\mathbf{X}(\omega) \in \mathscr{C}^\alpha$. The proof requires a close look at the precise construction of $\mathbf{X}(\omega) = (X(\omega), \mathbb{X}(\omega))$ in Theorem 10.4, using Kolmogorov's criterion to build a suitable (continuous, and then Hölder) modification from $\mathbf{X}$ restricted to dyadic times. We recall that $X(\omega) = \omega \in \mathcal{C}([0, T], \mathbf{R}^d)$. Let N_1 be the null set of ω where $X(\omega)$ fails to be of α-Hölder (or p-variation) regularity. Note that $\omega \in N_1^c$ implies $\omega + h \in N_1^c$ for all $h \in \mathcal{H}$. By the very construction of $\mathbb{X}_{s,t}$ as an L^2-limit, for fixed s, t there exists a sequence of partitions $(\mathcal{P}^m)$ of $[s, t]$ such that $\mathbb{X}_{s,t}(\omega) = \lim_m \int_{\mathcal{P}^m} X \otimes dX$ exists for a.e. ω, and we write $N_{2;s,t}$ for the null set on which this fails. The intersections of all these, for dyadic times s, t, is again a null set, denoted by N_2. Now take $\omega \in N_1^c \cap N_2^c$. For fixed dyadic s, t, consider the aforementioned partitions $(\mathcal{P}^m)$ and note

$$\int_{\mathcal{P}^m} X(\omega + h) \otimes dX(\omega + h)$$

$$= \int_{\mathcal{P}^m} X(\omega) \otimes dX(\omega) + \int_{\mathcal{P}^m} h \otimes dX + \int_{\mathcal{P}^m} X \otimes dh + \int_{\mathcal{P}^m} h \otimes dh.$$

Thanks to $\omega \in N_1^c$ and Proposition 11.2, $X(\omega)$ and h have complementary Young regularities, which guarantees convergence of the last three integrals to their respective Young integrals. On the other hand, $\omega \in N_2^c$ guarantees that $\int_{\mathcal{P}^m} X(\omega) \otimes dX(\omega) \to \mathbb{X}_{s,t}(\omega)$. This shows that the left-hand side converges, the limit being by definition $\mathbb{X}(\omega + h)$. In other words, for all $\omega \in N_1^c \cap N_2^c$, $h \in \mathcal{H}$ and dyadic times s, t,

$$T_h(\mathbf{X}(\omega))_{s,t} = \mathbf{X}(\omega + h)_{s,t}.$$

The construction of $\mathbf{X}_{s,t}$ for non-dyadic times was obtained by continuity (see Theorem 10.4) and the above almost sure identity remains valid. $\square$

11.2 Concentration of measure

11.2.1 Borell's inequality

Let us first recall a remarkable isoperimetric inequality for Gaussian measures. Following [Led96], we state it in the form due to C. Borell [Bor75], but an essentially equivalent result was obtained independently by Sudakov and Tsirelson [ST78]. In order to state things in their natural generality, we consider in this section an abstract Wiener-space $(E, \mathcal{H}, \mu)$. The reader may have in mind the Banach space $E = \mathcal{C}([0, T], \mathbf{R}^d)$, equipped with norm $\|x\|_E := \sup_{0 \leq t \leq T} |x_t|$ and a Gaussian measure μ, the law of a d-dimensional, continuous centred Gaussian process X. In this example, the Cameron–Martin space is given by $\mathcal{H} = \{\mathbf{E}(X.Z) : Z \in \mathcal{W}^1\}$ with $\|h\|_{\mathcal{H}} = \mathbf{E}(Z^2)^{1/2}$ for $h = \mathbf{E}(X.Z)$. Let us write

$$\Phi(y) = \frac{1}{\sqrt{2\pi}} \int_{-\infty}^{y} e^{-x^2/2} dx$$

for the cumulative distribution function of a standard Gaussian, noting the elementary tail estimate

$$\bar{\Phi}(y) := 1 - \Phi(y) \leq \exp\left(-y^2/2\right), \quad y \geq 0.$$

Theorem 11.6 (Borell's inequality). *Let $(E, \mathcal{H}, \mu)$ be an abstract Wiener space and $A \subset E$ a measurable Borel set with $\mu(A) > 0$ so that*

$$\hat{a} := \Phi^{-1}(\mu(A)) \in (-\infty, \infty]$$

Then, if $\mathcal{K}$ denotes the unit ball in $\mathcal{H}$, for every $r \geq 0$,

$$\mu((A + r\mathcal{K})^c) \leq \bar{\Phi}(\hat{a} + r).$$

where $A + r\mathcal{K} = \{x + rh : x \in A, h \in \mathcal{K}\}$ is the so-called Minkowski sum.[3]

Theorem 11.7 (Generalised Fernique Theorem). *Let $a, \sigma \in (0, \infty)$ and consider measurable maps $f, g : E \to [0, \infty]$ such that*

$$A_a = \{x : g(x) \leq a\}$$

has (strictly) positive μ measure[4] and set

$$\hat{a} := \Phi^{-1}(\mu(A_a)) \in (-\infty, \infty].$$

Assume furthermore that there exists a null-set N such that for all $x \in N^c, h \in \mathcal{H}$:

$$f(x) \leq g(x - h) + \sigma\|h\|_{\mathcal{H}}. \tag{11.8}$$

---

[3] Measurability is a delicate matter but circumventable by reading μ as outer measure; [Led96].
[4] Unless $g = +\infty$ almost surely, this holds true for sufficiently large a.

Then f has a Gaussian tail. More precisely, for all $r > a$ and with $\bar{a} := \hat{a} - a/\sigma$,

$$\mu(\{x : f(x) > r\}) \le \bar{\Phi}(\bar{a} + r/\sigma).$$

Proof. Note that $\mu(A_a) > 0$ implies $\hat{a} = \Phi^{-1}(\mu(A_a)) > -\infty$. We have for all $x \notin N$ and arbitrary $r, M > 0$ and $h \in r\mathcal{K}$,

$$\begin{aligned}
\{x : f(x) \le M\} &\supset \{x : g(x-h) + \sigma\|h\|_{\mathcal{H}} \le M\} \\
&\supset \{x : g(x-h) + \sigma r \le M\} \\
&= \{x + h : g(x) \le M - \sigma r\}.
\end{aligned}$$

Since $h \in r\mathcal{K}$ was arbitrary, this immediately implies the inclusion

$$\begin{aligned}
\{x : f(x) \le M\} &\supset \bigcup_{h \in r\mathcal{K}} \{x + h : g(x) \le M - \sigma r\} \\
&= \{x : g(x) \le M - \sigma r\} + r\mathcal{K},
\end{aligned}$$

and we see that

$$\mu(f(x) \le M) \ge \mu(\{x : g(x) \le M - \sigma r\} + r\mathcal{K}).$$

Setting $M = \sigma r + a$ and $A := \{x : g(x) \le a\}$, it then follows from Borell's inequality that

$$\mu(f(x) > \sigma r + a) \le \mu((A + r\mathcal{K})^c) \le \bar{\Phi}(\hat{a} + r).$$

It then suffices to rewrite the estimate in terms of $\tilde{r} := \sigma r + a > a$, noting that $\hat{a} + r = \bar{a} + \tilde{r}/\sigma$. $\quad\square$

Example 11.8 (Classical Fernique estimate). Take $f(x) = g(x) = \|x\|_E$. Then the assumptions of the generalised Fernique Theorem are satisfied with σ equal to the operator norm of the continuous embedding $\mathcal{H} \hookrightarrow E$. This applies in particular to Wiener measure on $\mathcal{C}([0, T], \mathbf{R}^d)$.

11.2.2 Fernique theorem for Gaussian rough paths

Theorem 11.9. *Let $(X_t : 0 \le t \le T)$ be a d-dimensional, centred Gaussian process with independent components and covariance R such that there exists $\varrho \in [1, \frac{3}{2})$ and $M < \infty$ such that for every $i \in \{1, \dots, d\}$ and $0 \le s \le t \le T$,*

$$\|R_{X^i}\|_{\varrho\text{-var};[s,t]^2} \le M|t - s|^{1/\varrho}.$$

Then, for any $\alpha \in (\frac{1}{3}, \frac{1}{2\varrho})$, the associated rough path $\mathbf{X} = (X, \mathbb{X}) \in \mathscr{C}_g^\alpha$ built in Theorem 10.4 is such that there exists $\eta = \eta(M, T, \alpha, \varrho)$ with

$$\mathbf{E}\exp\left(\eta\|\mathbf{X}\|_\alpha^2\right) < \infty .\tag{11.9}$$

Remark 11.10. Recall that the homogeneous "norm" $\|\mathbf{X}\|_\alpha$ was defined in (2.4) as the sum of $\|X\|_\alpha$ and $\sqrt{\|\mathbb{X}\|_{2\alpha}}$. Since $\mathbb{X}$ is "quadratic" in X (more precisely: in the second Wiener–Itô chaos), the square root is crucial for the Gaussian estimate (11.9) to hold.

Proof. Combining Theorem 11.5 with Lemma 11.4 and Proposition 11.2 shows that for a.e. ω and all $h \in \mathcal{H}$

$$\|\mathbf{X}(\omega)\|_\alpha \le C\left(\|(\mathbf{X}(\omega - h))\|_\alpha + M^{1/2}\|h\|_\mathcal{H}\right) .$$

We can thus apply the generalised Fernique Theorem with $f(\omega) = \|\mathbf{X}\|_\alpha(\omega)$ and $g(\omega) = Cf(\omega)$, noting that $\|\mathbf{X}\|_\alpha(\omega) < \infty$ almost surely implies that

$$A_a \overset{\text{def}}{=} \{x : g(x) \le a\}$$

has positive probability for a large enough (and in fact, any $a > 0$ thanks to a support theorem for Gaussian rough paths, [FV10b]). Gaussian integrability of the homogeneous rough path norm, for a fixed Gaussian rough path $\mathbf{X}$ is thus established. The claimed uniformity, $\eta = \eta(M, T, \alpha, \varrho)$ and not depending on the particular $\mathbf{X}$ under consideration requires an additional argument. We need to make sure that $\mu(A_a)$ is uniformly positive over all $\mathbf{X}$ with given bounds on the parameters (in particular M, ϱ, a, d); but this is easy, using (10.16),

$$\mu(\|\mathbf{X}\|_\alpha \le a) \ge 1 - \frac{1}{a^2}\mathbf{E}\|\mathbf{X}\|_\alpha^2 \ge 1 - \frac{1}{a^2}C ,$$

where $C = C(M, \varrho, \alpha, d)$ and so, say, $a = \sqrt{2C}$ would do. $\qquad\square$

11.2.3 Integrability of rough integrals and related topics

The price of a pathwise integration / SDE theory is that all estimates (have to) deal with the worst possible scenario. To wit, given $\mathbf{X} = (X, \mathbb{X}) \in \mathscr{C}_g^\alpha$ and a nice 1-form, $F \in \mathcal{C}_b^2$ say, we had the estimate

$$\left|\int_0^T F(X)d\mathbf{X}\right| \le C\left(\|\mathbf{X}\|_{\alpha;[0,T]} \vee \|\mathbf{X}\|_{\alpha;[0,T]}^{1/\alpha}\right) ,$$

where C may depend on F, T and $\alpha \in \left(\frac{1}{3}, \frac{1}{2}\right]$. In terms of p-variation, $p = 1/\alpha$, one can show similarly, with $\|\mathbf{X}\|_{p\text{-var};[0,T]}$ as introduced earlier, cf. (11.4),

$$\left|\int_0^T F(X)d\mathbf{X}\right| \le C\left(\|\mathbf{X}\|_{p\text{-var};[0,T]} \vee \|\mathbf{X}\|_{p\text{-var};[0,T]}^p\right) ,\tag{11.10}$$

where C depends on F and $\alpha \in \left(\frac{1}{3}, \frac{1}{2}\right]$ but not on T, thanks to invariance under reparametrisation. For the same reason, the integration domain $[0, T]$ in (11.10) may be replaced by any other interval.

Example 11.11. The estimate (11.10) is sharp, at least when $p = 1/\alpha = 2$, in the following sense. Consider the ("pure-area") rough path given by

$$ t \mapsto (0, At) , \quad A = \begin{pmatrix} 0 & c \\ -c & 0 \end{pmatrix} , $$

for some $c > 0$. The homogeneous (p-variation, or α-Hölder) rough path norm here scales with $c^{1/2}$. Hence, the right-hand side of (11.10) scales like c (for c large), as does the left-hand side which in fact is given by $T|DF(0)A|$.

The "trouble", in Brownian ($\varrho = 1$) or worse ($\varrho > 1$) regimes of Gaussian rough paths is that, despite Gaussian tails of the random variable $\|\mathbf{X}(\omega)\|_\alpha$, established in Theorem 11.9, the above estimate (11.10) fails to deliver Gaussian, or even exponential, integrability of the "random" rough integral

$$ Z(\omega) \overset{\text{def}}{=} \int_0^T F(X(\omega))d\mathbf{X}(\omega) , $$

something which is rather straightforward in the context of (Itô or Stratonovich) stochastic integration against Brownian motion.

As we shall now see, Borell's inequality, in the manifestation of our generalised Fernique estimate, allows to fully close this "gap" between integrability properties. The key idea, due to Cass–Litterer–Lyons [CLL13] is to define, for a fixed rough path $\mathbf{X}$ of finite homogeneous p-variation in the sense of (11.4), a tailor-made partition[5] of $[0, T]$, say

$$ \mathcal{P} = \{[\tau_i, \tau_{i+1}] : i = 0, \ldots, N\} $$

with the property that for all $i < N$

$$ \|\mathbf{X}\|_{p\text{-var};[\tau_i, \tau_{i+1}]} = 1, $$

i.e. for all but the very last interval for which one has $\|\mathbf{X}\|_{p\text{-var};[\tau_N, \tau_{N+1}]} \leq 1$. One can then exploit rough path estimates such as (11.10) on (small) intervals $[\tau_i, \tau_{i+1}]$ on which estimates are linear in $\|\mathbf{X}\|_{p\text{-var}} \sim 1$. The problem of estimating rough integrals is thus reduced to estimating $N = N(\mathbf{X})$ and it was a key technical result in [CLL13] to use Borell's inequality to establish good (probabilistic) estimates on N when $\mathbf{X} = \mathbf{X}(\omega)$ is a Gaussian rough path. (Our proof below is different from [CLL13] and makes good use of the generalised Fernique estimate.)

To formalise this construction, we fixed a (1D) control function $w = w(s, t)$, i.e. a continuous map on $\{0 \leq s \leq t \leq T\}$, super-additive, continuous and zero on the

---

[5] The construction is purely deterministic. Of course, when $\mathbf{X} = \mathbf{X}(\omega)$ is random, then so is the partition.

diagonal.[6] The canonical example of a control in this context is[7]

$$w_{\mathbf{X}}(s,t) = \|\mathbf{X}\|_{p\text{-var};[s,t]}^p.$$

Thanks to continuity of $w = w_{\mathbf{X}}$ we can then define a partition tailor-made for $\mathbf{X}$ based on eating up unit ($\beta = 1$ below) pieces of p-variation as follows. Set

$$\tau_0 = 0, \quad \tau_{i+1} = \inf\{t : w(\tau_i, t) \geq \beta, \ \tau_i < t \leq T\} \wedge T, \qquad (11.11)$$

so that $w(\tau_i, \tau_{i+1}) = \beta$ for all $i < N$, while $w(\tau_N, \tau_{N+1}) \leq \beta$, where N is given by

$$N(w) \equiv N_\beta(w; [0,T]) := \sup\{i \geq 0 : \tau_i < T\}.$$

As immediate consequence of super-additivity of controls,

$$\beta N_\beta(w; [0,T]) = \sum_{i=0}^{N-1} w(\tau_i, \tau_{i+1}) \leq w(0, \tau_N) \leq w(0, \tau_{N+1}) = w(0, T).$$

Note also that N is monotone in w, i.e. $w \leq \tilde{w}$ implies $N(w) \leq N(\tilde{w})$. At last, let us set $N(\mathbf{X}) = N(w_{\mathbf{X}})$. The following (purely deterministic) lemma is most naturally stated in variation regularity.

Lemma 11.12. *Assume $\mathbf{X} \in \mathscr{C}_g^{p\text{-var}}$, $p \in [2,3)$, and $h \in C^{q\text{-var}}$, $q \geq 1$, of complementary Young regularity in the sense that $\frac{1}{p} + \frac{1}{q} > 1$. Then there exists $C = C(p,q)$ so that*

$$N_1(\mathbf{X}; [0,T])^{\frac{1}{q}} \leq C\left(\|T_{-h}(\mathbf{X})\|_{p\text{-var};[0,T]}^{\frac{p}{q}} + \|h\|_{q\text{-var};[0,T]}\right). \qquad (11.12)$$

Proof. (Riedel) It is easy to see that all $N_\beta, N_{\beta'}$, with $\beta, \beta' > 0$ are comparable, it is therefore enough to prove the lemma for some fixed $\beta > 0$.

Given $h \in C^{q\text{-var}}$, $w_h(s,t) = \|h\|_{q\text{-var};[s,t]}^q$ is a control and so is w_h^θ whenever $\theta \geq 1$. (Noting $1 \leq q \leq p$, we shall use this fact with $\theta = p/q$.) From Lemma 11.4 we have, for any interval I

$$\|T_h \mathbf{X}\|_{p\text{-var};I} \lesssim \|\mathbf{X}\|_{p\text{-var};I} + \|h\|_{q\text{-var};I}.$$

Raise everything to the pth power to see that

$$(s,t) \mapsto \|T_h \mathbf{X}\|_{p\text{-var};[s,t]}^p \leq C\left(\|\mathbf{X}\|_{p\text{-var};[s,t]}^p + \|h\|_{q\text{-var};[s,t]}^p\right) =: C\tilde{w}(s,t).$$

where $C = C(p,q)$ and $\tilde{w}$ is a control. Choose $\beta = C$. By monotonicity of N_β in the control,

---

[6] Do not confuse a control w with "randomness" ω.

[7] Super-additivity, i.e. $\omega(s,t) + \omega(t,u) \leq \omega(s,u)$ whenever $s \leq t \leq u$ is immediate, but continuity is non-trivial see e.g. [FV10b, Prop. 5.8])

$$N_\beta(T_h\mathbf{X}; [0,T]) \leq N_\beta(C\tilde{w}; [0,T]) = N_1(\tilde{\omega}; [0,T]).$$

By definition, $\tilde{N} := N_1(\tilde{\omega}; [0,T])$ is the number of consecutive intervals $[\tau_i, \tau_{i+1}]$ for which

$$1 = \tilde{\omega}(\tau_i, \tau_{i+1}) = \|\mathbf{X}\|^p_{p\text{-var};[\tau_i,\tau_{i+1}]} + \|h\|^p_{q\text{-var};[\tau_i,\tau_{i+1}]}.$$

Using the manifest estimate $\|h\|^p_{q\text{-var};[\tau_i,\tau_{i+1}]} \leq 1$ and $q/p \leq 1$ we have

$$1 \leq \|\mathbf{X}\|^p_{p\text{-var};[\tau_i,\tau_{i+1}]} + \|h\|^q_{q\text{-var};[\tau_i,\tau_{i+1}]} = w_\mathbf{X}(\tau_i, \tau_{i+1}) + w_h(\tau_i, \tau_{i+1})$$

for $0 \leq i < \tilde{N}$. Summation over i yields

$$\tilde{N} \leq w_\mathbf{X}(0, \tau_{\tilde{N}}) + w_h(0, \tau_{\tilde{N}}) \leq \|\mathbf{X}\|^p_{p\text{-var};[0,T]} + \|h\|^q_{q\text{-var};[0,T]}.$$

Combination of these estimate hence shows that

$$N_\beta(T_h\mathbf{X}; [0,T]) \leq \|\mathbf{X}\|^p_{p\text{-var};[0,T]} + \|h\|^q_{q\text{-var};[0,T]}.$$

Replace $\mathbf{X} = T_hT_{-h}\mathbf{X}$ by $T_{-h}\mathbf{X}$ and then use elementary estimates of the type $(a+b)^{1/q} \leq (a^{1/q} + b^{1/q})$ for non-negative reals a, b, to obtain the claimed estimate (11.12). $\quad\square$

The previous lemma, combined with variation regularity of Cameron–Martin paths (Proposition 11.2) and the generalised Fernique Theorem 11.7 then gives immediately

Theorem 11.13 (Cass–Litterer–Lyons). *Let* $\mathbf{X} = (X, \mathbb{X}) \in \mathscr{C}_g^\alpha$ *a.s. be a Gaussian rough path, as in Theorem 11.9. (In particular, the covariance is assumed to have finite 2D ϱ-variation.) Then the integer-valued random variable*

$$N(\omega) := N_1(\mathbf{X}(\omega); [0,T])$$

has a Weibull tail with shape parameter $2/\varrho$ (by which we mean that $N^{1/\varrho}$ has a Gaussian tail).

Let us quickly illustrate how to use the above estimate.

Corollary 11.14. *Let* $\mathbf{X}$ *be as in the previous theorem and assume $F \in C_b^2$. Then the random rough integral*

$$Z(\omega) \stackrel{def}{=} \int_0^T F(X(\omega))d\mathbf{X}(\omega)$$

has a Weibull tail with shape parameter $2/\varrho$ by which we mean that $|Z|^{1/\varrho}$ has a Gaussian tail.

Proof. Let (τ_i) be the (random) partition associated to the p-variation of $\mathbf{X}(\omega)$ as defined in (11.11), with $\beta = 1$ and $w = w_\mathbf{X}$. Thanks to (11.10) we may estimate

$$\left| \int_0^T F(X(\omega)) d\mathbf{X}(\omega) \right| \le \sum_{[\tau_i, \tau_{i+1}] \in \mathcal{P}} \left| \int_{\tau_i}^{\tau_{i+1}} F(X(\omega)) d\mathbf{X}(\omega) \right|$$

$$\lesssim (N(\omega) + 1) \sup_i \left(\|\mathbf{X}\|_{p\text{-var};[\tau_i, \tau_{i+1}]} \vee \|\mathbf{X}\|_{p\text{-var};[\tau_i, \tau_{i+1}]}^p \right)$$

$$= (N(\omega) + 1) ,$$

where the proportionality constant may depend on F, T and $\alpha \in \left(\frac{1}{3}, \frac{1}{2\varrho} \right]$. $\quad\square$

11.3 Malliavin calculus for rough differential equations

In this section, we assume that the reader is already familiar with the basics of Malliavin calculus as exposed for example in the monographs [Mal97, Nua06].

11.3.1 Bouleau–Hirsch criterion and Hörmander's theorem

Consider some abstract Wiener space $(W, \mathcal{H}, \mu)$ and a Wiener functional of the form $F : W \to \mathbf{R}^e$. In the context of stochastic – or rough – differential equations driven by Gaussian signals, the Banach space W is of the form $\mathcal{C}([0, T], \mathbf{R}^d)$ where μ describes the statistics of the driving noise. If F denotes the solution to a stochastic differential equation at some time $t \in (0, T]$, then, in general, F is not a continuous, let alone Fréchet regular, function of the driving path. However, as we will see in this section, it can be the case that for μ-almost every ω, the map $\mathcal{H} \ni h \mapsto F(\omega + h)$, i.e. $F(\omega + \cdot)$ restricted to the Cameron-Martin space $(\mathcal{H}, \langle \cdot, \cdot \rangle)$ is Fréchet differentiable. (This implies $\mathbb{D}_{\text{loc}}^{1,p}$-regularity, based on the commonly used Shigekawa Sobolev space $\mathbb{D}^{1,p}$; our notation here follows [Mal97] or [Nua06, Sec. 1.2, 1.3.4].) More precisely, we introduce the following notion, see for example [Nua06, Sec. 4.1.3]:

Definition 11.15. Given an abstract Wiener space $(W, \mathcal{H}, \mu)$, a random variable $F \colon W \to \mathbf{R}$ is said to be continuously $\mathcal{H}$-differentiable, in symbols $F \in \mathcal{C}_{\mathcal{H}}^1$, if for μ-almost every ω, the map

$$\mathcal{H} \ni h \mapsto F(\omega + h)$$

is continuously Fréchet differentiable. A vector-valued random variable is said to be in $\mathcal{C}_{\mathcal{H}}^1$ if this is the case for each of its components. In particular, μ-almost surely, $DF(\omega) = \left(DF^1(\omega), \dots, DF^e(\omega) \right)$ is a linear bounded map from $\mathcal{H}$ to $\mathbf{R}^e$.

Given an $\mathbf{R}^e$-valued random variable F in $\mathcal{C}_{\mathcal{H}}^1$, we define the *Malliavin covariance matrix*

$$\mathcal{M}_{ij}(\omega) \overset{\text{def}}{=} \left\langle DF^i(\omega), DF^j(\omega) \right\rangle . \tag{11.13}$$

The following well-known criterion of Bouleau–Hirsch, see [BH91, Thm 5.2.2] and [Nua06, Sec. 1.2, 1.3.4] then provides a condition under which the law of F has a density with respect to Lebesgue measure:

Theorem 11.16. *Let $(W, \mathcal{H}, \mu)$ be an abstract Wiener space and let F be an $\mathbf{R}^e$-valued random variable F in $C^1_{\mathcal{H}}$. If the associated Malliavin matrix $\mathcal{M}$ is invertible μ-almost surely, then the law of F is has a density with respect to Lebesgue measure on $\mathbf{R}^e$.*

Remark 11.17. Higher order differentiability, together with control of inverse moments of $\mathcal{M}$ allow to strengthen this result to obtain smoothness of this density.

As beautifully explained in his own book [Mal97], Malliavin realised that the strong solution to the stochastic differential equation

$$dY_t = \sum_{i=1}^{d} V_i(Y_t) \circ dB_t^i , \qquad (11.14)$$

started at $Y_0 = y_0 \in \mathbf{R}^e$ and driven along C^∞-bounded vector fields V_i on $\mathbf{R}^e$, gives rise to a non-degenerate Wiener functional $F = Y_T$, admitting a density with respect to Lebesgue measure, provided that the vector fields satisfy Hörmander's famous "bracket condition" at the starting point y_0:

$$\text{Lie} \{V_1, \dots, V_d\} \big|_{y_0} = \mathbf{R}^e . \qquad \text{(H)}$$

(Here, Lie $\mathcal{V}$ denotes the Lie algebra generated by a collection $\mathcal{V}$ of smooth vector fields.) There are many variations on this theme, one can include a drift vector field (which gives rise to a modified Hörmander condition) and under the same assumptions one can show that Y_T admits a smooth density. This result can also (and was originally, see [Hör67, Koh78]) be obtained by using purely functional analytic techniques, exploiting the fact that the density solves Kolmogorov's forward equation. On the other hand, Malliavin's approach is purely stochastic and allows to go beyond the Markovian / PDE setting. In particular, we will see that it is possible to replace B by a somewhat generic sufficiently non-degenerate Gaussian process, with the interpretation of (11.14) as a random RDE driven by some Gaussian rough path $\mathbf{X}$ rather than Brownian motion.

11.3.2 Calculus of variations for ODEs and RDEs

Throughout, we assume that $V = (V_1, \dots, V_d)$ is a given set of smooth vector fields, bounded and with bounded derivatives of all orders. In particular, there is a unique solution flow to the RDE

$$dY = V(Y) \, d\mathbf{X} , \qquad (11.15)$$

for any α-Hölder geometric driving rough path $\mathbf{X} = (X, \mathbb{X}) \in \mathscr{C}_g^{0,\alpha}$, which may be obtained as limit of smooth, or piecewise smooth, paths in α-Hölder rough path metric. Set $p = 1/\alpha$. Recall that, thanks to continuity of the Itô–Lyons maps, RDE solutions are limits of the corresponding ODE solutions.

The unique RDE solution (11.15) passing through $Y_{t_0} = y_0$ gives rise to the solution flow $y_0 \mapsto U^{\mathbf{X}}_{t \leftarrow t_0}(y_0) = Y_t$. We call the derivative of the flow with respect to the starting point the *Jacobian* and denote it by $J^{\mathbf{X}}_{t \leftarrow t_0}$, so that

$$J^{\mathbf{X}}_{t \leftarrow t_0} a = \frac{d}{d\varepsilon} U^{\mathbf{X}}_{t \leftarrow t_0}(y_0 + \varepsilon a)\Big|_{\varepsilon = 0}.$$

We also consider the directional derivative

$$D_h U^{\mathbf{X}}_{t \leftarrow 0} = \frac{d}{d\varepsilon} U^{T_{\varepsilon h} \mathbf{X}}_{t \leftarrow 0}\Big|_{\varepsilon = 0},$$

for any sufficiently smooth path $h \colon \mathbf{R}_+ \to \mathbf{R}^e$. Recall that the translation operator T_h was defined in (11.5). In particular, we have seen in Lemma 11.4 that, if $\mathbf{X}$ arises from a smooth path X together with its iterated integrals, then the translated rough path $T_h \mathbf{X}$ is nothing but $X + h$ together with its iterated integrals. In the general case, given $h \in C^{q\text{-var}}$ of complementary Young regularity, i.e. with $1/p + 1/q > 1$, the translation $T_h \mathbf{X}$ can be written in terms of $\mathbf{X}$ and cross-integrals between X and h.

Suppose for a moment that the rough path $\mathbf{X}$ is the canonical lift of a smooth $\mathbf{R}^d$-valued path X. Then, it is classical to prove that $J^{\mathbf{X}}_{t \leftarrow t_0} = J^X_{t \leftarrow t_0}$, where $J^X_{t \leftarrow t_0}$ solves the linear ODE

$$dJ^X_{t \leftarrow t_0} = \sum_{i=1}^d DV_i(Y_t) J^X_{t \leftarrow t_0} \, dX^i_t, \tag{11.16}$$

and satisfies $J^X_{t_2 \leftarrow t_0} = J^X_{t_2 \leftarrow t_1} \cdot J^X_{t_1 \leftarrow t_0}$. Furthermore, the variation of constants formula leads to

$$D_h U^X_{t \leftarrow 0} = \int_0^t \sum_{i=1}^d J^X_{t \leftarrow s} V_i(Y_s) \, dh^i_s. \tag{11.17}$$

Similarly, given any smooth vector field W, a straightforward application of the chain rule yields

$$d\big(J^X_{0 \leftarrow t} W(Y_t)\big) = \sum_{i=1}^d J^X_{0 \leftarrow t} [V_i, W](Y_t) \, dX^i_t, \tag{11.18}$$

where $[V, W]$ denotes the Lie bracket between the vector fields V and W. All this extends to the rough path limit without difficulties. For instance, (11.16) can be interpreted as a linear equation driven by the rough path $\mathbf{X}$, using the fact that $DV(Y)$ is controlled by X to give meaning to the equation. It is then still the case that $J^{\mathbf{X}}_{t \leftarrow t_0}$ is the derivative of the flow associated to (11.15) with respect to its initial condition.

Proposition 11.18. *Let* $\mathbf{X} \in \mathscr{C}_g^{0,\alpha}([0,T], \mathbf{R}^d)$ *and* $h \in C^{q\text{-var}}([0,T], \mathbf{R}^d)$ *with* $\alpha \in (\frac{1}{3}, \frac{1}{2}]$ *and complementary Young regularity in the sense that* $\alpha + \frac{1}{q} > 1$. *Then*

$$D_h U_{t\leftarrow 0}^{\mathbf{X}}(y_0) = \int_0^t \sum_{i=1}^d J_{t\leftarrow s}^{\mathbf{X}}\big(V_i\big(U_{s\leftarrow 0}^{\mathbf{X}}\big)\big) dh_s^i \qquad (11.19)$$

where the right-hand side is well-defined as Young integral.

Proof. Both $J_{t\leftarrow 0}^{\mathbf{X}}$ and $D_h U_{t\leftarrow 0}^{\mathbf{X}}$ satisfy (jointly with $U_{t\leftarrow 0}^{\mathbf{X}}$) an RDE driven by $\mathbf{X}$. This is well known in the ODE case, i.e. when both X, h are smooth, (Duhamel's principle, variation of constant formula, ...) and remains valid in the geometric rough path limit by appealing to continuity of the Itô–Lyons and continuity properties of the Young integral. A little care is needed since the resulting vector fields are not bounded anymore. It suffices to rule out explosion so that the problem can be localised. The required remark is that that $J_{t\leftarrow 0}^{\mathbf{X}}$ also satisfies a linear RDE of form

$$dJ_{t\leftarrow 0}^{\mathbf{X}} = d\mathbf{M}^{\mathbf{X}} \cdot J_{t\leftarrow 0}^{\mathbf{X}}(y_0)$$

and linear RDEs do not explode. □

Consider now an RDE driven by a Gaussian rough path $\mathbf{X} = \mathbf{X}(\omega)$. We now show that the $\mathbf{R}^e$-valued random variable obtained from solving this random RDE enjoys $\mathcal{C}_{\mathcal{H}}^1$-regularity.

Proposition 11.19. *With* $\varrho \in [1, \frac{3}{2})$ *and* $\alpha \in (\frac{1}{3}, \frac{1}{2\varrho})$, *let* $\mathbf{X} = (X, \mathbb{X}) \in \mathscr{C}_g^\alpha$ *be a Gaussian rough path as constructed in Theorem 10.4. For fixed* $t \geq 0$, *the* $\mathbf{R}^e$-*valued random variable*

$$\omega \mapsto U_{t\leftarrow 0}^{\mathbf{X}(\omega)}(y_0)$$

is continuously $\mathcal{H}$-differentiable.

Proof. Recall $h \in \mathcal{H} \subset C^{\varrho\text{-var}}$ so that a.e. $\mathbf{X}(\omega)$ and h enjoy complementary Young regularity. As a consequence, we saw that the event

$$\{\omega : \mathbf{X}(\omega + h) \equiv T_h \mathbf{X}(\omega) \text{ for all } h \in \mathcal{H}\} \qquad (11.20)$$

has full measure. We show that $h \in \mathcal{H} \mapsto U_{t\leftarrow 0}^{\mathbf{X}(\omega+h)}(y_0)$ is continuously Fréchet differentiable for every ω in the above set of full measure. By basic facts of Fréchet theory, it is sufficient to show (a) Gâteaux differentiability and (b) continuity of the Gâteaux differential.

Ad (a): Using $\mathbf{X}(\omega + g + h) \equiv T_g T_h \mathbf{X}(\omega)$ for $g, h \in \mathcal{H}$ it suffices to show Gâteaux differentiability of $U_{t\leftarrow 0}^{\mathbf{X}(\omega+\cdot)}(y_0)$ at $0 \in \mathcal{H}$. For fixed t, define

$$Z_{i,s} \equiv J_{t\leftarrow s}^{\mathbf{X}}\big(V_i\big(U_{s\leftarrow 0}^{\mathbf{X}}\big)\big).$$

Note that $s \mapsto Z_{i,s}$ is of finite p-variation, with $p = 1/\alpha$. We have, with implicit summation over i,

$$\left| D_h U^{\mathbf{X}}_{t\leftarrow 0}(y_0) \right| = \left| \int_0^t J^{\mathbf{X}}_{t\leftarrow s}\left(V_i\left(U^{\mathbf{X}}_{s\leftarrow 0}\right)\right) dh^i_s \right| = \left| \int_0^t Z_i dh^i \right|$$
$$\lesssim \left(\|Z\|_{p\text{-var}} + |Z(0)| \right) \times \|h\|_{\varrho\text{-var}}$$
$$\lesssim \left(\|Z\|_{p\text{-var}} + |Z(0)| \right) \times \|h\|_{\mathcal{H}}.$$

Hence, the linear map $DU^{\mathbf{X}}_{t\leftarrow 0}(y_0) : h \mapsto D_h U^{\mathbf{X}}_{t\leftarrow 0}(y_0) \in \mathbf{R}^e$ is bounded and each component is an element of $\mathcal{H}^*$. We just showed that

$$h \mapsto \left. \frac{d}{d\varepsilon} U^{T_{\varepsilon h}\mathbf{X}(\omega)}_{t\leftarrow 0}(y_0) \right|_{\varepsilon=0} = \left\langle DU^{\mathbf{X}(\omega)}_{t\leftarrow 0}(y_0), h \right\rangle_{\mathcal{H}}$$

and hence

$$h \mapsto \left. \frac{d}{d\varepsilon} U^{\mathbf{X}(\omega+\varepsilon h)}_{t\leftarrow 0}(y_0) \right|_{\varepsilon=0} = \left\langle DU^{\mathbf{X}(\omega)}_{t\leftarrow 0}(y_0), h \right\rangle_{\mathcal{H}}$$

emphasizing again that $\mathbf{X}(\omega + h) \equiv T_h \mathbf{X}(\omega)$ almost surely for all $h \in \mathcal{H}$ simultaneously. Repeating the argument with $T_g \mathbf{X}(\omega) = \mathbf{X}(\omega + g)$ shows that the Gâteaux differential of $U^{\mathbf{X}(\omega+\cdot)}_{t\leftarrow 0}$ at $g \in \mathcal{H}$ is given by

$$DU^{\mathbf{X}(\omega+g)}_{t\leftarrow 0} = DU^{T_g\mathbf{X}(\omega)}_{t\leftarrow 0}.$$

(b) It remains to be seen that $g \in \mathcal{H} \mapsto DU^{T_g\mathbf{X}(\omega)}_{t\leftarrow 0} \in \mathcal{L}(\mathcal{H}, \mathbf{R}^e)$, the space of linear bounded maps equipped with operator norm, is continuous. We leave this as exercise to the reader, cf. Exercise 11.4 below. $\quad\square$

11.3.3 Hörmander's theorem for Gaussian RDEs

Recall that $\varrho \in [1, \frac{3}{2}), \alpha \in (\frac{1}{3}, \frac{1}{2\varrho})$ and $\mathbf{X} = (X, \mathbb{X}) \in \mathscr{C}^\alpha_g$ a.s. is the Gaussian rough path constructed in Theorem 10.4. Any $h \in \mathcal{H} \subset C^{\varrho\text{-var}}$ and a.e. $\mathbf{X}(\omega)$ enjoy complementary Young regularity. We now present the remaining conditions on X, followed by some commentary on each of the conditions, explaining their significance in the context of the problem and verifying them for some explicit examples of Gaussian processes.

Condition 1 *Fix $T > 0$. For every $t \in (0, T]$ we assume non-degeneracy of the law of X on $[0, t]$ in the following sense. Given $f \in C^\alpha([0, t], \mathbf{R}^d)$, if $\sum_{j=1}^d \int_0^t f_j dh^j = 0$ for all $h \in \mathcal{H}$, then one has $f = 0$.*

Note that, thanks to complementary Young regularity, the integral $\int_0^t f_j dh^j$ makes sense as a Young integral. Some assumption along the lines of Condition 1 is certainly necessary: just consider the trivial rough differential equation $dY = dX$, starting at $Y_0 = 0$, with driving process $X = X(\omega)$ given by a Brownian bridge which returns to the origin at time T (i.e. $X_t = B_t - \frac{t}{T} B_T$ in terms of a standard Brownian motion B). Clearly $Y_T = X_T = 0$ and so Y_T does not admit a density, despite the equation

$dY = dX$ being even "elliptic". However, it is straightforward to verify that in this example $\int_0^T dh = 0$ for every h belonging to the Cameron–Martin space of the Brownian bridge, so that Condition 1 is violated by taking for f a non-zero constant function.

Condition 2 *With probability one, sample paths of X are truly rough, at least in a right-neighbourhood of 0.*

These conditions obviously hold for d-dimensional Brownian motion: the first condition is satisfied because 0 is the only (continuous) function orthogonal to all of $L^2([0,T], \mathbf{R}^d)$; the second condition was already verified in Section 6.3. More interestingly, these conditions are very robust and also hold for the Ornstein–Uhlenbeck process, a Brownian bridge which returns to the origin at a time strictly greater than T, and some non-semimartingale examples such as fractional Brownian motion, including the rough regime of Hurst parameter less than $1/2$. We now show that under these conditions the process admits a density at strictly positive times. Note that the aforementioned situations are not at all covered by the "usual" Hörmander theorem.

Theorem 11.20. *With $\varrho \in [1, \frac{3}{2})$ and $\alpha \in (\frac{1}{3}, \frac{1}{2\varrho})$, let $\mathbf{X} = (X, \mathbb{X}) \in \mathscr{C}_g^\alpha$ be a Gaussian rough path as constructed in Theorem 10.4. Assume that the Gaussian process X satisfies Conditions 1 and 2. Let $V = (V_1, \ldots, V_d)$ be a collection of C^∞-bounded vector fields on $\mathbf{R}^e$, which satisfies Hörmander's condition (H) at some point $y_0 \in \mathbf{R}^e$. Then the law of the RDE solution*

$$dY_t = V(Y_t) \, d\mathbf{X}_t \,, \qquad Y(0) = y_0 \,,$$

admits a density with respect to Lebesgue measure on $\mathbf{R}^e$ for all $t \in (0, T]$.

Proof. Thanks to Proposition 11.19 and in view of the Bouleau–Hirsch criterion, Theorem 11.16 we only need to show almost sure invertibility of the Malliavin matrix associated to the solution map. As a consequence of (11.13) and (11.19), we have for every $z \in \mathbf{R}^e$ the identity

$$z^\mathsf{T} \mathcal{M}_t z = \sum_{j=1}^d \left\| z^\mathsf{T} J_{t \leftarrow \cdot}^{\mathbf{X}} V_j(Y_\cdot) \right\|_t^2 \,,$$

where we wrote $\| \cdot \|_t$ for the norm given by

$$\| f \|_t = \sup_{h \in \mathcal{H} : \|h\| = 1} \int_0^t f(s) \, dh(s) \,.$$

Before we proceed we note that, by the multiplicative property of $J_{t \leftarrow s}^{\mathbf{X}}$, see the remark following (11.16), one has

$$\mathcal{M}_t = J_{t \leftarrow 0}^{\mathbf{X}} \tilde{\mathcal{M}}_t \big(J_{t \leftarrow 0}^{\mathbf{X}} \big)^\mathsf{T} \,,$$

where $\tilde{\mathcal{M}}_t$ is given by

$$z^\mathsf{T} \tilde{\mathcal{M}}_t z = \sum_{j=1}^{d} \left\| z^\mathsf{T} J_{0\leftarrow\cdot}^{\mathbf{X}} V_j(Y_\cdot) \right\|_t^2 .$$

Since we know that the Jacobian is invertible, invertibility of $\mathcal{M}_t$ is equivalent to that of $\tilde{\mathcal{M}}_t$, and it is the invertibility of the latter that we are going to show.

Assume now by contradiction that $\tilde{\mathcal{M}}_t$ is not almost surely invertible. This implies that there exists a random unit vector $z \in \mathbf{R}^e$ such that $z^\mathsf{T} \tilde{\mathcal{M}}_t z = 0$ with non-zero probability. It follows immediately from Condition 1 that, with non-zero probability, the functions $s \mapsto z^\mathsf{T} J_{0\leftarrow s}^{\mathbf{X}(\omega)} V_j(Y_s)$ vanish identically on $[0, t]$ for every $j \in \{1, \ldots, d\}$. By (11.18), this is equivalent to

$$\sum_{i=1}^{d} \int_0^\cdot z^\mathsf{T} J_{0\leftarrow s}^{\mathbf{X}} [V_i, V_j](Y_s) \, d\mathbf{X}^i(s) \equiv 0$$

on $[0, t]$. Thanks to Condition 2, true roughness of X, we can apply Theorem 6.5 to conclude that one has

$$z^\mathsf{T} J_{0\leftarrow\cdot}^{\mathbf{X}} [V_i, V_j](Y_\cdot) \equiv 0 ,$$

for every $i, j \in \{1, \ldots, d\}$. Iterating this argument shows that, with non-zero probability, the processes $s \mapsto z^\mathsf{T} J_{0\leftarrow s}^{\mathbf{X}} W(Y_s)$ vanish identically for every vector field W obtained as a Lie bracket of the vector fields V_i. In particular, this is the case for $s = 0$, which implies that with positive probability, z is orthogonal to $W(z_0)$ for all such vector fields. Since Hörmander's condition (H) asserts precisely that these vector fields span the tangent space at the starting point y_0, we conclude that $z = 0$ with positive probability, which is in contradiction with the fact that z is a random unit vector and thus concludes the proof. $\quad\square$

11.4 Exercises

Exercise 11.1 (Improved Cameron–Martin regularity, [FGGR16]) *A combination of Theorem 10.9 with the Cameron–Martin embedding, Proposition 11.2, shows that every Cameron–Martin path associated to a Gaussian process enjoys finite q-variation regularity with $q = \varrho$. Show that, under the assumptions of Theorem 10.9, this can be improved to*

$$q = \frac{1}{\frac{1}{2} + \frac{1}{2\varrho}} . \tag{11.21}$$

As a consequence, "complementary Young regularity", now holds for all $\varrho < 2$. In the fBm setting, this covers every Hurst parameter $H > 1/4$. (To exploit this in the newly covered regime $H \in (1/4, 1/3]$, one would need to work in a "level-3" rough path setting.)

Exercise 11.2 *Formulate a quantitative version of Theorem 11.14. Show in particular that the Gaussian tail of $|Z|^{1/\varrho}$ is uniform over rough integrals against Gaussian rough paths, provided that $\|F\|_{\mathcal{C}_b^2}$ and the ϱ-variation of the covariance, say in the form of the constant M in Theorem 11.9, are uniformly bounded.*

Exercise 11.3 (Noise doubling, from [Ina14, Sch18]) *Let X be a d-dimensional Gaussian process as considered in Theorem 10.4 and $\mathbf{X} = (X, \mathbb{X})$ the random α-Hölder rough path over $\mathbf{R}^d$ constructed therein. Recall that any $h \in \mathcal{H}$, with $\mathcal{H}$ the associated Cameron–Martin space, is given by $h_t = \mathbf{E}(\Xi X_t) = \bar{\mathbf{E}}(\bar{\Xi} \bar{X}_t) \in \mathbf{R}^d$ where $\bar{X} = \bar{X}(\bar{\omega})$ is an IID copy of $X = X(\omega)$ and $\bar{\Xi}, \Xi$ are elements in their respective first Wiener chaoses with L^2-norm equal to $\|h\|_{\mathcal{H}}$.*

a) *Apply Theorem 10.4 to construct the "doubled" rough path associated to the 2d-dimensional process $(X, \bar{X})$ and use this to show that $Z^h := (X, h)$ can be extended canonically to a random rough path $\mathbf{Z}^h = (Z^h, \mathbb{Z}^h)$ over $\mathbf{R}^{2d}$.*

Hint: *Formally, in case $d = 1$ for notational simplicity,*

$$\mathbb{Z}^h = \begin{pmatrix} \int X \, dX & \bar{\mathbf{E}}\left(\bar{\Xi} \int X \, d\bar{X}\right) \\ \bar{\mathbf{E}}\left(\bar{\Xi} \int \bar{X} \, dX\right) & \bar{\mathbf{E}}\left(\bar{\Xi}\bar{\Xi} \int \bar{X} \, d\bar{X}\right) \end{pmatrix},$$

where $\bar{\mathbf{E}} = \bar{\mathbf{E}}^{\bar{\omega}}$ only averages over $\bar{\omega}$.

b) *Show further that*

$$\mathbf{E}\left(\|\mathbb{Z}^h - \mathbb{Z}^k\|_{2\alpha}^2\right) \lesssim \|h - k\|_{\mathcal{H}}^2.$$

(Since $\|Z^h - Z^k\|_\alpha = \|h - k\|_\alpha \lesssim \|h - k\|_{\mathcal{H}}$ this shows that the construction of the joint lift of (X, h) as a random rough path is continuous in $h \in \mathcal{H}$.)

Exercise 11.4 *Finish the proof of part (b) of Proposition 11.19.*

Solution. In the notation of the (proof of) this Proposition, we have to show that $g \in \mathcal{H} \mapsto DU_{t \leftarrow 0}^{T_g \mathbf{X}(\omega)} \in \mathcal{L}(\mathcal{H}, \mathbf{R}^e)$ is continuous. To this end, assume $g_n \to g$ in $\mathcal{H}$ (and hence in $\mathcal{C}^{\varrho\text{-var}}$). Continuity properties of the Young integral imply continuity of the translation operator viewed as map $h \in \mathcal{C}^{\varrho\text{-var}} \mapsto T_h \mathbf{X}(\omega)$ and so

$$T_{g_n}\mathbf{X}(\omega) \to T_g \mathbf{X}(\omega)$$

in p-variation rough path metric. The point here is that

$$\mathbf{x} \mapsto J_{t \leftarrow \cdot}^{\mathbf{x}} \text{ and } J_{t \leftarrow \cdot}^{\mathbf{x}}(V_i(U_{\cdot \leftarrow 0}^{\mathbf{x}})) \in \mathcal{C}^{p\text{-var}}$$

depends continuously on $\mathbf{x}$ with respect to p-variation rough path metric: using the fact that $J_{t \leftarrow \cdot}^{\mathbf{x}}$ and $U_{\cdot \leftarrow 0}^{\mathbf{x}}$ both satisfy rough differential equations driven by $\mathbf{x}$ this is just a consequence of Lyons' limit theorem (the *universal limit theorem* of rough path theory). We apply this with $\mathbf{x} = \mathbf{X}(\omega)$ where ω remains a fixed element in (11.20). It follows that

$$\left\|DU_{t \leftarrow 0}^{T_{g_n}\mathbf{X}(\omega)} - DU_{t \leftarrow 0}^{T_g \mathbf{X}(\omega)}\right\|_{op} = \sup_{h: \|h\|_{\mathcal{H}} = 1} \left|D_h U_{t \leftarrow 0}^{T_{g_n}\mathbf{X}(\omega)} - D_h U_{t \leftarrow 0}^{T_g \mathbf{X}(\omega)}\right|$$

and defining $Z_i^g(s) \equiv J_{t \leftarrow s}^{T_g \mathbf{X}(\omega)} \big(V_i \big(U_{s \leftarrow 0}^{T_g \mathbf{X}(\omega)} \big) \big)$, and similarly $Z_i^{g_n}(s)$, the same reasoning as in part (a) leads to the estimate

$$\left\| DU_{t \leftarrow 0}^{T_{g_n} \mathbf{X}(\omega)} - DU_{t \leftarrow 0}^{T_g \mathbf{X}(\omega)} \right\|_{op} \leq c \big(|Z^{g_n} - Z^g|_{p\text{-var}} + |Z^{g_n}(0) - Z^g(0)| \big).$$

From the explanations just given this tends to zero as $n \to \infty$ which establishes continuity of the Gâteaux differential, as required, and the proof is finished.

Exercise 11.5 *Prove Theorem 11.20 in presence of a drift vector field V_0. In particular, show that in this case condition* (H) *can be weakened to*

$$\text{Lie} \{ V_1, \dots, V_d, [V_0, V_1], \dots, [V_0, V_d] \} \big|_{y_0} = \mathbf{R}^e . \tag{11.22}$$

11.5 Comments

Section 11.1: Regularity of Cameron–Martin paths (q-variation, with $q = \varrho$) under the assumption of finite ϱ-variation of the covariance was established in Friz–Victoir, [FV10a], see also [FV10b, Ch.15]. In the context of Gaussian rough paths, this leads to complementary Young regularity (CYR) whenever $\varrho < \frac{3}{2}$ which covers general "level-2" Gaussian rough paths as discussed in Chapter 10. On the other hand, "level-3" Gaussian rough paths can be constructed for any $\varrho < 2$ which includes fBm with $H = \frac{1}{2\varrho} > \frac{1}{4}$). A sharper Cameron regularity result specific to fBm follows from a Besov–variation embedding theorem [FV06b], thereby leading to CYR for all $H > \frac{1}{4}$. The general case was understood in [FGGR16]: one can take q as in (11.21), provided one makes the slightly stronger assumption of finite "mixed" $(1, \varrho)$-variation of the covariance. The conclusion concerning ϱ-variation of Theorem 10.9 can in fact be strengthened to finite mixed $(1, \varrho)$-variation at no extra cost and indeed this theorem is only a special case of a general criterion given in [FGGR16].

Section 11.2: Theorem 11.9 was originally obtained by careful tracking of constants via the Garsia–Rodemich–Rumsey Lemma, see [FV10b]. The generalised Fernique estimate is taken from Friz–Oberhauser and then Diehl, Oberhauser and Riedel [FO10, DOR15]; Riedel [Rie17] establishes a further generalisation in form of a transportation cost inequality in the spirit of Talagrand. This yields an elegant proof of Theorem 11.13 with which Cass, Litterer, and Lyons [CLL13] have overcome the longstanding problem of obtaining moment bounds for the Jacobian of the flow of a rough differential equation driven by Gaussian rough paths, thereby paving the way for the proof of the Hörmander-type results, see below. As was illustrated, this above methodology can be adapted to many other situations of interest, a number of which are discussed in [FR13]. See also [CO17] for Fernique type estimate in a Markovian context.

Section 11.3: Baudoin–Hairer [BH07] proved a Hörmander theorem for differential equations driven by fBm in the regular regime of Hurst parameter $H > 1/2$ in a framework of Young differential equations. The Brownian case $H = 1/2$

of course classical, see the monographs [Nua06, Mal97] or the original articles [Mal78, KS84, KS85, KS87, Bis81b, Bis81a, Nor86], a short self-contained proof can be found in [Hai11a]. In the case of rough differential equations driven by less regular Gaussian rough path (including the case of fBm with $H > 1/4$), the relevance of complementary Young regularity of Cameron–Martin paths to Malliavin regularity or (Gaussian) RDE solutions was first recognised by Cass, Friz and Victoir [CFV09]. Existence of a density under Hörmander's condition for such RDEs was obtained by Cass–Friz [CF10], see also [FV10b, Ch.20], but with a Stroock-Varadhan support type argument instead of true roughness (already commented on at the end of Chapter 6.) Smoothness of densities was subsequently established by Hairer–Pillai [HP13] in the case of fBm and then Cass, Hairer, Litterer and Tindel [CHLT15] in the general Gaussian setting of Chapter 10, making crucial use of the integrability estimates discussed in Section 11.2. Indeed, combined with known estimates for the Jacobian of RDE flows (Friz–Victoir, [FV10b, Thm 10.16]) one readily obtains finite moments of the Jacobian of the inverse flow. This is a key ingredient in the smoothness proof via Malliavin calculus, as is the higher-order Malliavin differentiability of Gaussian RDE solutions established by Inahama [Ina14]. Several authors have studied the resulting density, see e.g. [BNOT16, Ina16b, GOT19, IN19] and the references therein.

We note that existence of densities via Malliavin calculus for singular SPDEs, in the framework of regularity structures, has been studied by Cannizzaro, Friz and Gassiat [CFG17], Gassiat–Labbé [GL20] and in great generality by Schönbauer [Sch18].

Chapter 12
Stochastic partial differential equations

Second order stochastic partial differential equations are discussed from a rough path point of view. In the linear and finite-dimensional noise case we follow a Feynman–Kac approach which makes good use of concentration of measure results, as those obtained in Section 11.2. Alternatively, one can proceed by flow decomposition and this approach also works in a number of nonlinear situations. Secondly, now motivated by some semilinear SPDEs of Burgers' type with infinite-dimension noise, we study the stochastic heat equation (in space dimension 1) as evolution in Gaussian rough path space *relative to the spatial variable*, in the sense of Chapter 10.

12.1 First order rough partial differential equations

12.1.1 Rough transport equation

As a prototypical linear first order PDE with noise we consider the transport equation, posed (without loss of generality) as a terminal value problem. This is,

$$-\partial_t u(t,x) = \sum_{i=1}^{d} f_i(x) \cdot D_x u(t,x)\dot{W}_t^i \equiv \Gamma u_t(x)\dot{W}_t , \quad u(T,\cdot) = g , \quad (12.1)$$

where $u : [0,T] \times \mathbf{R}^n \to \mathbf{R}$, with vector fields $f = (f_1,\ldots,f_d)$ driven by a $\mathcal{C}^1$ driving signal $W = (W^1,\ldots,W^d)$, and we write indifferently $u(t,x) = u_t(x)$. The canonical pairing of $Du = D_x u = (\partial_{x^1} u,\ldots,\partial_{x^n} u)$ with a vector field is indicated by a dot, and we already used the operator / vector notation

$$\Gamma_i = f_i(x) \cdot D_x, \quad \Gamma = (\Gamma_1,\ldots,\Gamma_d). \quad (12.2)$$

By the methods of characteristics, the unique (classical) $\mathcal{C}^{1,1}$-solution $u : [0,T] \times \mathbf{R}^n \to \mathbf{R}$, is given explicitly by

© Springer Nature Switzerland AG 2020
P. K. Friz, M. Hairer, *A Course on Rough Paths*, Universitext,
https://doi.org/10.1007/978-3-030-41556-3_12

$$u(s,x) = u(s,x;W) := g(X_T^{s,x})\,,\tag{12.3}$$

provided $g \in \mathcal{C}^1$ and the vector fields $f_1,\ldots,f_d$ are nice enough ($\mathcal{C}_b^1$ will do) to ensure a $\mathcal{C}^1$ solution flow for the ODE $\dot{X} = \sum_{i=1}^d f_i(X)\dot{W}^i \equiv f(X)\dot{W}$; here $X^{s,x}$ denotes the unique solution started from $X_s = x$.

We start with a rough path stability result for the transport equation, the proof of which is an immediate consequence of our results on flow stability of RDEs.

Proposition 12.1. *Let $g \in \mathcal{C}(\mathbf{R}^m)$ and $W^\varepsilon \in \mathcal{C}^1([0,T],\mathbf{R}^d)$, with geometric rough path limit $\mathbf{W} \in \mathscr{C}_g^{0,\alpha}$, $\alpha > 1/3$. Write $u^\varepsilon(s,x) := u(s,x;W^\varepsilon)$, defined as in (12.3) with W replaced by W^ε. Let $f \in \mathcal{C}_b^3$. Then u^ε converges locally uniformly to*

$$u(s,x;\mathbf{W}) := g(X_T^{s,x})\tag{12.4}$$

where $X^{s,x}$ denotes the (unique) RDE solution to $dX = f(X)d\mathbf{W}$, started from $X_s = x$. (In particular, the limit depends on $\mathbf{W}$ but not on the approximating sequence.)

It is instructive to consider the case of Brownian motion $B = B(t,\omega)$ with Stratonovich lift as prototypical example of a (random) geometric rough path. The RDE solution X is then equivalently described by a Stratonovich SDE and $u(t,x;\omega) = g(X_T^{t,x}(\omega))$ is $\mathcal{F}_t^T$-measurable. The so-defined random field should then constitute a (backward adapted) solution to the Stratonovich backward stochastic partial differential equation

$$-du_t(x) = \Gamma u_t(x) \circ \overleftarrow{dB_t}\,,\quad u(T,\cdot) = g\,,\tag{12.5}$$

where $\overleftarrow{dB}$ stands for backward Stratonovich integration (cf. Section 5.4) provided g (und then Γu_t) are sufficiently regular to make this Stratonovich integral meaningful. If rewritten in Itô-form, a matrix valued second order $\Gamma^2 = (\Gamma_i\Gamma_j)_{1\le i,j\le d}$ appears, which of course must not change the hyperbolic nature of the stochastic transport equation. (In classical SPDE theory on has the *stochastic parabolicity* condition, which in the transport case is fully degenerate.)

All this strongly suggests that rough transport noise must be geometric (i.e. $\mathbf{W} \in \mathscr{C}_g^\alpha$). We now prepare the definition of (regular, backward) solution to the rough transport equation. Since we are in the fortunate position to have an explicit solution (candidate) we derive a graded set of rough path estimates that provide a natural generalisation of the classical the transport differential equation. In what follows we abbreviate estimates of the form $|(a) - (b)| \lesssim |t - s|^\gamma$ by writing $(a) \overset{\gamma}{=} (b)$. (Both sides may depend on s,t and the multiplicative constant hidden in $\lesssim$ is assumed uniform over bounded intervals).

Proposition 12.2. *Consider vector fields $f = (f_1,\ldots,f_d) \in \mathcal{C}_b^5$, with associated first order differential operators $\Gamma = (\Gamma_1,\ldots,\Gamma_d)$. There is a unique $\mathcal{C}^3$ solution flow for the RDE $dX = f(X)d\mathbf{W}$ with $\mathbf{W} \in \mathscr{C}_g^\alpha$, $\alpha > 1/3$. Let $g \in \mathcal{C}^3$ and define $u(s,x;\mathbf{W}) := g(X_T^{s,x})$ as in (12.4). Then $u = u(s,x) \in \mathcal{C}^{\alpha,3}$, $u_T = g$, and we have the estimates, with Einstein summation,*

$$u_s(x) \overset{3\alpha}{\equiv} u_t(x) + \Gamma_i u_t(x) W^i_{s,t} + \Gamma_i \Gamma_j u_t(x) \mathbb{W}^{i,j}_{s,t}$$
$$\Gamma_i u_s(x) \overset{2\alpha}{\equiv} \Gamma_i u_t(x) + \Gamma_i \Gamma_j u_t(x) W^j_{s,t} ,$$
$$\Gamma_i \Gamma_j u_s(x) \overset{\alpha}{\equiv} \Gamma_i \Gamma_j u_t(x) ,$$

with $0 \leqslant s < t \leqslant T$, $i, j = 1, \dots, d$, locally uniformly in x, and, as consequence,

$$u_s(x) - g(x) = u_s(x) - u_T(x) = \int_s^T \Gamma u_t(x) \, d\mathbf{W}_t .$$

Remark 12.3. The first 3α estimate is nothing but Davie's definition of solution for a linear RDE, here of the form $-du = \Gamma u \, d\mathbf{W}$. In finite dimensions, a linear map Γ is necessarily bounded (equivalently: continuous) as linear operator, so that the cascade of lower order $(2\alpha, \alpha)$ estimates are a trivial consequence of the first. This is different in the present situation, where u_t takes values in a function space where each application of Γ amounts to take one derivative. These estimates then have the interpretation that time regularity of u, in the stated ("$k\alpha$") controlled sense, can be traded against space regularity.

Remark 12.4. The rough integral formulation needs explanation. Indeed, while it is clear from $\delta \Xi \overset{3\alpha}{\equiv} \delta u(x) = 0$ that $\Xi_{s,t} = \Gamma_i u_t(x) W^i_{s,t} + \Gamma_i \Gamma_j u_t(x) \mathbb{W}^{i,j}_{s,t}$ has a sewing limit, the right-point evalution requires attention, cf. Proposition 5.12 and the subsequent discussion about the subtleties of "right-point" rough integrals. Fortunately, one checks that $(\Gamma u, -(\Gamma^2 u)^T) \in \mathscr{D}^{2\alpha}_X$ so that, thanks to (5.10), Remark 5.13, this sewing limit, over all partitions of $[0, T]$ say, is exactly identified as

$$\lim_{|\mathcal{P}| \downarrow 0} \sum_{[s,t] \in \mathcal{P}} \left(\Gamma u_t X_{s,t} - (\Gamma^2 u_t)^T \mathbb{X}_{s,t} \right) = \int_0^T (\Gamma u, -\Gamma^2 u^T) d\mathbf{X} ,$$

where we omitted x for better readability. (Since the matrix $\Gamma^2 u_t = (\Gamma_i \Gamma_j u_t)_{1 \leq i,j \leq d}$ is in general not symmetric, a careful check of the controlledness condition is best spelled out in coordinates.)

Notwithstanding the elegance of the rough integral formulation, additional quantifiers, such as local uniformity in x, are better formulated at the level of the detailed estimates which brings us to

Definition 12.5. Any $C^{\alpha,3}$-function $u : [0, T] \times \mathbf{R}^n \to \mathbf{R}$, for which the (locally uniform) estimates in Proposition 12.2 hold is called a *regular solution* to the *rough backward transport equation*

$$-du = \Gamma u d\mathbf{W}.$$

Proof (Proposition 12.2). Consider a solution $X = X^{s,x}$ to $dX = f(X)d\mathbf{W}$, started from $X_s = x$ so that

$$X_t \overset{3\alpha}{\equiv} x + f(x)W_{s,t} + f'f(x)\mathbb{W}_{s,t}.$$

Fix times $s < t < T$. By uniqueness of RDE flow, $X_T^{t,y} = X_T^{s,x}$ whenever $y = X_t^{s,x}$. From $u(s,x) := g(X_T^{s,x})$ and uniqueness of the RDE flow it is clear that, for all such t,

$$u(s,x) = u(t, X_t^{s,x}).$$

Note that $u_t = u(t, \cdot) \in \mathcal{C}^3$ follows from $g \in \mathcal{C}^3, f \in \mathcal{C}^5$; the claimed $\mathcal{C}^{\alpha,3}$ regularity is then easy to see. We can expand

$$u_t(X_t^{s,x}) \overset{3\alpha}{=} u_t(x) + Du_t(x)(f(x)W_{s,t} + (Df)f(x)\mathbb{W}_{s,t}) + \frac{1}{2}D^2u_t(x)(f(x)W_{s,t})^2$$

where the final term is really the contraction $\partial_{ij}u_t f_k^i f_l^j(\frac{1}{2}W_{s,t} \otimes W_{s,t})^{k,l}$ with summation over all repeated indices. Using geometricity of $\mathbf{X}$ and symmetry of $D^2u_t(x)(f,f)$ the right-hand side becomes

$$u_t(x) + Du_t(x)f(x)W_{s,t} + \{Du_t(x)(Df)f(x) + D^2u_t(x)(f,f)(x)\}\mathbb{W}_{s,t}.$$

(We essentially repeated the proof of Itô's formula here, cf. Section 7.5.) In terms of the first order differential operators Γ_i associated to the vector fields f_i this can be written elegantly as

$$u_s(x) \overset{3\alpha}{=} u(t,x) + \Gamma u_t(x)W_{s,t} + \Gamma^2 u_t(x)\mathbb{W}_{s,t}.$$

This relation actually implies that with $\Xi_{s,t} := \Gamma_i u_t(x)W_{s,t}^i + \Gamma_i\Gamma_j u_t(x)\mathbb{W}_{s,t}^{i,j}$ we have $|(\delta\Xi)_{r,s,t}| = O(|t-r|^{3\alpha})$ and hence (after a line of algebra) $(\Gamma_i u_{s,t} - \Gamma_i\Gamma_j u_t W_{s,t}^j)W_{r,s}^i \overset{3\alpha}{=} 0$ which strongly suggests validity of the desired 2α-estimate, for all $i = 1, \ldots, d$,

$$\Gamma_i u_s(x) \overset{2\alpha}{=} \Gamma_i u_t(x) + \Gamma_i\Gamma_j u_t(x)W_{s,t}^j.$$

Since no true roughness condition on W is imposed (W could be zero!), one has to check this by hand from $u(s,x) = g(X_T^{s,x})$, left to the reader. Similarly, the previous relation gives $(\Gamma^2 u_t - \Gamma^2 u_s)W_{s,t} \overset{2\alpha}{=} 0$ so that the same argument suggests $\Gamma^2 u_s(x) - \Gamma^2 u_t(x) \overset{\alpha}{=} 0$. Here again, a direct verification is not hard (and amounts to check α-Hölder regularity of $s \mapsto \Gamma^2 g(X_T^{s,x})$, with $g \in \mathcal{C}^3$.) □

We can now show that solutions in the sense of Definition 12.5 are unique.

Theorem 12.6. *Consider vector fields $f = (f_1, \ldots, f_d) \in \mathcal{C}_b^5$, with associated first order differential operators $\Gamma = (\Gamma_1, \ldots, \Gamma_d)$ and $\mathbf{W} \in \mathscr{C}_g^\alpha([0,T], \mathbf{R}^d)$ with $\alpha > 1/3$. For $g \in \mathcal{C}^3$, there exists a unique regular solution $u : [0,T] \times \mathbf{R}^n \to \mathbf{R}$ of $\mathcal{C}^{\alpha,3}$ regularity to the rough backward transport equation*

$$-du = \Gamma u d\mathbf{W}, \quad u(T, \cdot) = g.$$

Proof. Existence is clear, since Proposition 12.2 exactly says that $(s,x) \mapsto g(X_T^{s,x})$ gives a regular solution. Let now u be any solution with $u_T = g$. We show that, whenever X solves $dX = f(X)d\mathbf{W}$,

$$u(t, X_t) - u(s, X_s) \overset{3\alpha}{=} 0.$$

Since $3\alpha > 1$ this entails that $t \mapsto u(t, X_t)$ is constant, and so $u(s, x) = u(T, X_T^{s,x}) = g(X_T^{s,x})$. In fact, we show for $k = 1, 2, 3$

$$\Gamma^{3-k} u_t(X_t) \overset{k\alpha}{=} \Gamma^{3-k} u_s(X_s).$$

(Case $k = 1$.) Write

$$\Gamma^2 u_t(X_t) - \Gamma^2 u_s(X_s) = \Gamma^2 u_t(X_t) - \Gamma^2 u_s(X_t) + \Gamma^2 u(s, X_t) - \Gamma^2 u(s, X_s).$$

From the (third) defining property of a solution, the first difference on the right-hand side of order α. Since solutions are $\mathcal{C}^3$ in space, hence $\Gamma^2 u(s, \cdot) \in \mathcal{C}^1$, always uniformly in $s \in [0, T]$ the final difference is also of order α, as required.
(Case $k = 2$.) Write

$$\Gamma u_t(X_t) - \Gamma u_s(X_s) = \Gamma u_t(X_t) - \Gamma u_s(X_t) + \Gamma u_s(X_t) - \Gamma u_s(X_s).$$

By the second defining property of a solution, the first difference on the right-hand side equals $-\Gamma^2 u_t(X_t) W_{s,t}$ (up to order 2α). On the other hand, $\Gamma u_s \in \mathcal{C}^2$ so that the final difference can be replaced by

$$D\Gamma u_s(X_s)(X_t - X_s) \overset{2\alpha}{=} D\Gamma u_s(X_s) f(X_s) W_{s,t} = \Gamma^2 u_s(X_s) W_{s,t}.$$

Put together we have $\Gamma u_t(X_t) - \Gamma u_s(X_s) = (\Gamma^2 u_s(X_s) - \Gamma^2 u_t(X_t)) W_{s,t}$. We see that this is of (desired) order 2α, thanks to the case $k = 1$ and $W_{s,t} \overset{\alpha}{=} 0$.
(Case $k = 3$.) We write

$$u(t, X_t) - u(s, X_s) = u(t, X_t) - u(s, X_t) + u(s, X_t) - u(s, X_s).$$

By the (first) defining property of a solution, the the first difference on the right-hand side equals $-\Gamma u_t(X_t) W_{s,t} - \Gamma^2 u_t(X_t) \mathbb{W}_{s,t}$ (up to order 3α). On the other hand, $u(s, \cdot) \in \mathcal{C}^3$ so that the final difference can be replaced, using a second order Taylor expansion, exactly as in the proof of Proposition 7.8, by

$$Du_s(X_s)(f(X_s) W_{s,t} + f'f(X_s) \mathbb{W}_{s,t}) + \frac{1}{2} D^2 u_s(f(X_s), f(X_s)) W_{s,t} \otimes W_{s,t}$$
$$= \Gamma u_s(X_s) W_{s,t} + \Gamma^2 u_s(X_s) \mathbb{W}_{s,t}$$

Put together (and using the cases $k = 1, 2$) gives the desired estimate. $\square$

12.1.2 Continuity equation and analytically weak formulation

Given a finite measure $\varrho \in \mathcal{M}(\mathbf{R}^n)$ and a continuous bounded function $\varphi \in \mathcal{C}_b(\mathbf{R}^n)$, we write $\varrho(\varphi) = \int \varphi(x) \varrho(dx)$ for the natural pairing. We are interested in measure-valued (forward) solutions to the continuity equation

$$\partial_t \varrho = - \sum_{i=1}^{d} \mathrm{div}_x (f_i(x)\varrho_t) \dot{W}_t^i \equiv \Gamma^* \varrho_t \dot{W}_t$$

when W becomes a (geometric) rough path. As before, $\Gamma_i = f_i(x) \cdot D_x$, with formal adjoint $\Gamma_i^* = - \mathrm{div}_x(f_i \cdot)$.

Definition 12.7. We say that $\varrho : [0, T] \to \mathcal{M}(\mathbf{R}^n)$ is a measure-valued forward RPDE solution to the rough continuity equation

$$d\varrho_t + \mathrm{div}_x(f(x)\varrho_t)d\mathbf{W}_t = 0 \tag{12.6}$$

if, uniformly over φ bounded in $\mathcal{C}_b^3$,

$$\varrho_t(\varphi) \overset{3\alpha}{=} \varrho_s(\varphi) + \varrho_s(\Gamma\varphi)W_{s,t} + \varrho_s(\Gamma^2\varphi)\mathbb{W}_{s,t}$$
$$\varrho_t(\Gamma\varphi) \overset{2\alpha}{=} \varrho_s(\Gamma\varphi) + \varrho_s(\Gamma^2\varphi)W_{s,t}$$
$$\varrho_t(\Gamma^2\varphi) \overset{\alpha}{=} \varrho_s(\Gamma^2\varphi).$$

(Note $\Gamma\varphi, \Gamma^2\varphi \in \mathcal{C}_b$ so all pairings are well-defined. Formally, the second and third estimate follow from the first with φ replaces by $\Gamma\varphi$ and $\Gamma^2\varphi$), however doing so would require test functions up to $\Gamma^4\varphi \notin \mathcal{C}_b$. Itemizing the estimates allows us to keep track of the correct regularity of φ.)

These estimates imply immediately the following (analytically) weak formulation

$$\forall \varphi \in \mathcal{C}_b^3 : \varrho_t(\varphi) - \varrho_0(\varphi) = \int_0^t (\varrho_s(\Gamma\varphi), \varrho_s(\Gamma^2\varphi))d\mathbf{W}_s ,$$

but the finer information, as put foward in the definition, is crucial for uniqueness. (Remark 12.9 below comments on time-dependent test functions.)

Theorem 12.8. *Consider vector fields* $f = (f_1, \ldots, f_d) \in \mathcal{C}_b^5$, *with associated first order differential operators* $\Gamma = (\Gamma_1, \ldots, \Gamma_d)$ *and* $\mathbf{W} \in \mathscr{C}_g^\alpha([0, T], \mathbf{R}^d)$ *with* $\alpha > 1/3$. *For every measure* $\nu \in \mathcal{M}(\mathbf{R}^n)$, *there exists a unique measure-valued solution to the rough continuity equation*

$$d\varrho_t + \mathrm{div}_x(f(x)\varrho_t)d\mathbf{W}_t , \quad \varrho_0 = \nu , \tag{12.7}$$

with explicit representation, for $\varphi \in \mathcal{C}_b^3$, *given by*

$$\varrho_t(\varphi) = \int \varphi(X_t^{0,x})\nu(dx) .$$

Proof. (Existence) Let $X = X^{0,x}$ be a solution to the RDE $dX = f(X)d\mathbf{W}$, started at $X_0 = x$. By Proposition 7.8, a form of Itô's formula for controlled rough paths,

$$\varphi(X_t) \overset{3\alpha}{=} \varphi(X_s) + \varphi(X_s)X_s' W_{s,t} + (D\varphi(X_s)X_s'' + D^2\varphi(X_s)(X_s', X_s'))\mathbb{W}_{s,t} ,$$

uniformly in $\varphi \in C_b^3$. Taking into account $X' = f(X), X'' = (Df)f$ gives

$$\varphi(X_t) \overset{3\alpha}{=} \varphi(X_s) + \Gamma\varphi(X_s)W_{s,t} + (\Gamma^2\varphi)(X_s)\mathbb{W}_{s,t} .$$

Combining this with $\varrho_t(\varphi) := \varphi(X_t)$ yields the claimed 3α-estimate. Similar, but now using standard facts on composition of controlled rough paths with regular functions, we obtain

$$\varphi(X_t) \overset{2\alpha}{=} \varphi(X_s) + \Gamma\varphi(X_s)W_{s,t},$$

uniformly over φ bounded in C_b^2. At last, the third estimate comes from α-Hölder regularity of $t \mapsto \varrho_t(\Gamma^2\varphi) = \Gamma^2\varphi(X_t)$, itself a manifest consequence of $\Gamma^2\varphi \in C_b^1$ and α-Hölder regularity of X.

We are not yet done, because until now, we have only handled the case of Dirac initial data $\varrho_0 = \delta_x$. (Since $\varrho_0(\varphi) = \varphi(X_0^{0,x}) = \varphi(x)$.) Fortunately, we are in a linear situation so that, given $\varrho_0 = \nu \in \mathcal{M}$, it suffices to generalise our construction and define

$$\varrho_t(\varphi) := \int \varphi(X_t^{0,x})\nu(dx).$$

It remains to see that such an integration in x respects all graded $3\alpha, 2\alpha, \alpha$ estimates. This is indeed the case, because all required estimates are uniform in $X_0 = x$. (A pleasant consequence of dealing with bounded vector fields so that all quantitative bounds are invariant under shift.)

(Uniqueness) Given any $g = u_T \in C_b^3$, there exists a regular backward RPDE solution, $u_t = u(t, \cdot) \in C_b^3$, with

$$u_s - u_t \overset{3\alpha}{=} u_t'W_{s,t} + u_t''\mathbb{W}_{s,t}$$

(and then $u' = \Gamma u \in C_b^2$ etc). Write $u_{s,t} = u_t - u_s$ and similarly for ϱ. Then

$$\varrho_t(u_t) - \varrho_s(u_s) = \varrho_{s,t}(u_t) + \varrho_s(u_{s,t}) .$$

The first summand on the right-hand side expands, using the very definition of weak solution (applied with $\varphi = u_t \in C_b^3$, uniformly in $t \in [0,T]$),

$$\varrho_{s,t}(u_t) \overset{3\alpha}{=} \varrho_s(\Gamma u_t)W_{s,t} + \varrho_s(\Gamma^2 u_t)\mathbb{W}_{s,t} .$$

The second summand on the other hand expands, using the defining property of regular backward equation,

$$\varrho_s(u_{s,t}) = -\varrho_s(u_s - u_t) \overset{3\alpha}{=} -\varrho_s(\Gamma u_t)W_{s,t} - \varrho_s(\Gamma^2 u_t)\mathbb{W}_{s,t} .$$

(Here one needs to argue that the 3α-bound on $u_{s,t}(x) + \Gamma u_t(x)W_{s,t} + \Gamma^2 u_t(x)\mathbb{W}_{s,t}$ is uniform in x, for $u_T \in C_b^3$, and thus the same 3α-estimate holds after integrating against $\varrho_s(dx)$.) Taken together we see a perfect cancellation so that $\varrho_t(u_t) - \varrho_s(u_s) \overset{3\alpha}{=} 0$. By a familiar argument (using $3\alpha > 1$) this implies that $t \mapsto \varrho_t(u_t)$ is constant and thus

$$\varrho_T(g) = \varrho_T(u_T) = \varrho_0(u_0) = \nu(u_0)$$

where u is a regular backward RPDE solution (with terminal data $g = u_T \in \mathcal{C}_b^3$). (Uniqueness of the regular backward RPDE solutions is not used here.) Hence, with given initial data $\varrho = \nu \in \mathcal{M}$ we see that $\varrho_T(g)$ is determined for all $g \in \mathcal{C}_b^3$ and this (uniquely) determines the measure $\varrho_T \in \mathcal{M}$. □

Remark 12.9. The uniqueness part of the proof actually shows that analytically weak solutions to the rough PDE (12.6) can be tested in space-time with test functions $\varphi = \varphi(t, x)$ that have a precise controlled structure, starting with

$$\varphi_s - \varphi_t \overset{3\alpha}{=} \varphi_t' W_{s,t} + \varphi_t'' \mathbb{W}_{s,t}$$

(and then 2α, resp. α expansions for φ' and φ''). This space of test functions is tailored to the realisation of the noise $\mathbf{W}$.

12.2 Second order rough partial differential equations

12.2.1 Linear theory: Feynman–Kac

As motivation, consider the second order stochastic partial differential equation with d-dimensional Brownian noise in (backward) Stratonovich form, posed as terminal value problem,

$$-du = L[u]dt + \Gamma[u] \circ \overleftarrow{dB} , \qquad u(T, \cdot) = g , \qquad (12.8)$$

for $u = u(\omega) : [0, T] \times \mathbf{R}^n \to \mathbf{R}$, with differential operators L and $\Gamma = (\Gamma_1, \ldots, \Gamma_d)$ given by

$$L[u] \overset{\text{def}}{=} \frac{1}{2} \mathrm{Tr}\big(\sigma(x)\sigma^T(x)D^2 u\big) + b(x) \cdot Du + c(x)u , \qquad (12.9)$$

$$\Gamma_i[u] \overset{\text{def}}{=} \beta_i(x) \cdot Du + \gamma_i(x)u .$$

The coefficients $\sigma = (\sigma_1, \ldots, \sigma_m)$, b and $\beta = (\beta_1, \ldots, \beta_d)$ are viewed as vector fields on $\mathbf{R}^n$, while $c, \gamma_1, \ldots, \gamma_d$ are scalar functions. For simplicity only, all coefficients are assumed to be bounded with bounded derivatives of all orders (but see Remark 12.12). We assume $g \in \mathcal{BC}(\mathbf{R}^n)$, that is bounded and continuous.[1] As in the previous section, we are interested in replacing W by a genuine (geometric) rough path $\mathbf{W}$, such as to solve the *rough partial differential equation* (RPDE)

$$-du = L[u]dt + \Gamma[u]d\mathbf{W} , \qquad u(T, \cdot) = g . \qquad (12.10)$$

---

[1] In contrast to the space $\mathcal{C}_b$ we shall equip $\mathcal{BC}$ with the topology of locally uniform convergence.

We have already treated the fully degenerate case $L = 0$, with pure transport noise, $\Gamma_i = \beta_i(x) \cdot D_x$, in Section 12.1.1. Since geometric rough paths are limits of smooth paths, we start with the case when $\mathbf{W}$ is replaced by $\dot{W} dt$, for $W \in C^1([0,T], \mathbf{R}^d)$. It is a basic exercise in Itô calculus that any bounded $C^{1,2}$ solution to

$$-\partial_t u = L[u] + \sum_{i=1}^d \Gamma_i[u] \dot{W}_t^i, \qquad u(T, \cdot) = g, \qquad (12.11)$$

is given by the classical Feynman–Kac formula (and hence also unique),

$$u(s,x) = \mathbf{E}^{s,x}\left[g(X_T) \exp\left(\int_s^T c(X_t) dt + \int_s^T \gamma(X_t) \dot{W}_t dt \right) \right] \quad (12.12)$$

$$=: \mathcal{S}[W; g](s,x), \qquad (12.13)$$

where X is the (unique) strong solution to

$$dX_t = \sigma(X_t) dB(\omega) + b(X_t) dt + \beta(X_t) \dot{W}_t dt, \qquad (12.14)$$

where B is a m-dimensional standard Brownian motion. When $\sigma \equiv 0$, this is nothing but the method of characteristics, previously encountered for the transport equation in (12.3). (For the moment, we keep $W \in C^1$, but will soon encounter *rough stochastic characteristics*.)

Remark 12.10. The natural form of the Feynman–Kac formula is the reason for considering terminal value problems here, rather than Cauchy problems of the form $\partial_t u = L[u] + \Gamma[u]\dot{W}$ with given initial data $u(0, \cdot)$. Of course, a change of the time variable $t \mapsto T - t$ allows to switch between these problems.

Clearly, there are situations when solutions cannot be expected to be $C^{1,2}$, notably when $g \notin C^2$ and L fails to provide smoothing as is the case, for example, in "transport" equations where L is of first order. In such a case, formula (12.12) is a perfectly good way to define a generalised solution to (12.11). Such a solution need not be $C^{1,2}$ although it is bounded and continuous on $[0,T] \times \mathbf{R}^n$, as one can see directly from (12.12). As a matter of fact, (12.12) yields a (analytically) weak PDE solution (cf. Exercise 12.1). It is also a stochastic representation of the unique (bounded) viscosity solution [CIL92, FS06] to (12.11) although this will play no role for us in the present section. The main result here is the following rough path stability for linear second order RPDEs.

Theorem 12.11. Let $\alpha \in (\frac{1}{3}, \frac{1}{2}]$. Given a geometric rough path $\mathbf{W} = (W, \mathbb{W}) \in \mathscr{C}_g^{0,\alpha}([0,T], \mathbf{R}^d)$, pick $W^\varepsilon \in C^1([0,T], \mathbf{R}^d)$ so that

$$(W^\varepsilon, \mathbb{W}^\varepsilon) := \left(W^\varepsilon, \int_0^\cdot W_{0,t}^\varepsilon \otimes dW_t^\varepsilon \right) \to \mathbf{W}$$

in α-Hölder rough path metric. Then there exists $u = u(t,x) \in \mathcal{BC}([0,T] \times \mathbf{R}^n)$, not dependent on the approximating (W^ε) but only on $\mathbf{W} \in \mathscr{C}_g^{0,\alpha}([0,T], \mathbf{R}^d)$, so *that, for $g \in \mathcal{BC}(\mathbf{R}^n)$,*

$$u^\varepsilon = \mathcal{S}[W^\varepsilon; g] \to u =: \mathcal{S}[\mathbf{W}; g]$$

as $\varepsilon \to 0$ in the sense of locally uniform convergence. Moreover, the resulting solution map

$$\mathcal{S} : \mathscr{C}_g^{0,\alpha}([0,T], \mathbf{R}^d) \times \mathcal{BC}(\mathbf{R}^n) \to \mathcal{BC}([0,T] \times \mathbf{R}^n)$$

is continuous. We say that u satisfies the RPDE (12.10).

Proof. **Step 1:** Write $X = X^W$ for the solution to (12.14) whenever $W \in \mathcal{C}^1$. The first step is to make sense of the *stochastic RDS*

$$dX_t = \sigma(X_t)dB_t + b(X_t)dt + \beta(X_t)d\mathbf{W}_t. \tag{12.15}$$

This is clearly not an equation that can be solved by Itô theory alone. But is also not immediately well-posed as rough differential equation since for this we would need to understand B and $\mathbf{W} = (W, \mathbb{W})$ jointly as a rough path. In view of the Itô-differential dB in (12.15), we take $(B, \mathbb{B}^{\text{Itô}})$, as constructed in Section 3.2), and are basically short of the cross-integrals between B and W. (For simplicity of notation only, pretend over the next few lines W, B to be scalar.) We can define $\int W dB(\omega)$ as Wiener integral (Itô with deterministic integrand), and then $\int B dW = WB - \int W dB$ by imposing integration by parts. We then easily get the estimate

$$\mathbf{E}\left(\int_s^t W_{s,r} dB_r \right)^2 \lesssim \|W\|_\alpha^2 |t-s|^{2\alpha+1},$$

also when switching the roles of W, B, thanks to the integration by parts formula. It follows from Kolmogorov's criterion that $\mathbf{Z}^W(\omega) := \mathbf{Z} = (Z, \mathbb{Z}) \in \mathscr{C}^{\alpha'}$ a.s. for any $\alpha' \in (1/3, \alpha)$ where

$$Z_t = \begin{pmatrix} B_t(\omega) \\ W_t \end{pmatrix}, \qquad \mathbb{Z}_{s,t} = \begin{pmatrix} \mathbb{B}_{s,t}^{\text{Itô}}(\omega) & \int_s^t W_{s,r} \otimes dB_r(\omega) \\ \int_s^t B_{s,r} \otimes dW_r(\omega) & \mathbb{W}_{s,t} \end{pmatrix}$$

where we reverted to tensor notation reflecting the multidimensional nature of B, W. It is easy to deduce from Theorem 3.3 that, for any $q < \infty$,

$$\left| \varrho_{\alpha'}\left(\mathbf{Z}^\mathbf{W}, \mathbf{Z}^{\tilde{\mathbf{W}}} \right) \right|_{L^q} \lesssim \varrho_\alpha\left(\mathbf{W}, \tilde{\mathbf{W}} \right). \tag{12.16}$$

We are hence able to say that a solution $X = X(\omega)$ of (12.15) is, by definition, a solution to the genuine (random) rough differential equation

$$dX = (\sigma, \beta)(X)d\mathbf{Z}^\mathbf{W}(\omega) + b(X)dt \tag{12.17}$$

driven by the random rough path $\mathbf{Z} = \mathbf{Z}^\mathbf{W}(\omega)$. Moreover, as an immediate consequence of (12.16) and continuity of the Itô–Lyons map, we see that X is really the limit, e.g. in probability and uniformly on $[0, T]$, of classical Itô SDE solutions X^ε, obtained by replacing $d\mathbf{W}_t$ by the $\dot{W}^\varepsilon_t dt$ in (12.15).

Step 2: Given (s, x) we have a solution $(X_t : s \leq t \leq T)$ to the hybrid equation (12.15), started at $X_s = x$. In fact $(X, X') \in \mathscr{D}_Z^{2\alpha'}$ with $X' = (\sigma, \beta)(X)$. In particular, the rough integral

$$\int \gamma(X)d\mathbf{W} := \int (0, \gamma(X))d\mathbf{Z}$$

is well-defined, as is - with regard to the Feynman–Kac formula (12.12) - the random variable

$$g(X_T) \exp\left(\int_s^T c(X_t)dt + \int_s^T \gamma(X_t)d\mathbf{W}_t\right)(\omega). \tag{12.18}$$

One can see, similar to (11.10), but now also relying on RDE growth estimates as established in Proposition 8.2), with $p = 1/\alpha'$,

$$\left|\int_s^t \gamma(X)d\mathbf{W}\right| \lesssim \|\mathbf{Z}\|_{p\text{-var};[s,t]}$$

whenever $\|\mathbf{Z}\|_{p\text{-var};[s,t]}$ is of order one. An application of the generalised Fernique Theorem 11.7, similar to the proof of Theorem 11.13 but with $\varrho = 1$ in the present context, then shows that the number of consecutive intervals on which $\mathbf{Z}$ accumulates unit p-variation has Gaussian tails; in fact, uniformly in $\varepsilon \in (0, 1]$, if $\mathbf{W}$ is replaced by W^ε with limit $\mathbf{W}$.) This implies that (12.18) is integrable (and uniformly integrable with respect to ε when $\mathbf{W}$ is replaced by W^ε). It follows that

$$u(s, x) := \mathbf{E}^{s,x}\left[g(X_T) \exp\left(\int_s^T c(X_t)dt + \int_s^T \gamma(X_t)d\mathbf{W}_t\right)\right] \tag{12.19}$$

is indeed well-defined and the pointwise limit of u^ε (defined in the same way, with $\mathbf{W}$ replaced by W^ε). By an Arzela–Ascoli argument, the limit is locally uniform. At last, the claimed continuity of the solution map follows from the same arguments, essentially by replacing W^ϵ by $\mathbf{W}^\epsilon$ everywhere in the above argument, and of course using (12.19) with g, $\mathbf{W}$ replaced by g^ε, $\mathbf{W}^\varepsilon$, respectively. $\quad\square$

Remark 12.12. The proof actually shows that our solution $u = u(s, x; \mathbf{W})$ to the linear RDPE (12.10) enjoys a Feynman–Kac type representation, namely (12.19), in terms of the process constructed as solution to the hybrid Itô-rough differential equation (12.15). Assume now W is a Brownian motion, independent of B, and $\mathbf{W}(\omega) = \mathbf{W}^{\text{Strat}} = (W, \mathbb{W}^{\text{Strat}}) \in \mathscr{C}_g^{0,\alpha}$ a.s. It is not difficult to show that $u = u(.,., \mathbf{W}^{\text{Strat}}(\omega))$ coincides with the Feynman–Kac SPDE solution derived by Pardoux [Par79] or Kunita [Kun82], via conditional expectations given $\sigma(\{W_{u,v} : s \leq u \leq v \leq T\})$, and so provides an identification with classical SPDE theory. In conjunction with continuity of the solution map $\mathcal{S} = \mathcal{S}[\mathbf{W}; g]$ one obtains, along the lines of Sections 9.2, SPDE limit theorems of Wong–Zakai type,

Stroock–Varadhan type support statements and Freidlin–Wentzell type small noise large deviations.

Remark 12.13. It is easy to quantify the required regularity of the coefficients. The argument essentially relies on solving (12.17) as bona fide rough differential equation. It is then clear that we need to impose $\mathcal{C}_b^3$-regularity for the vector fields σ and β. The drift vector field b may be taken to be Lipschitz and $c \in \mathcal{C}_b$.

Remark 12.14. We have not given meaning to the actual equation (12.10) which we here reproduce equivalently (cf. Remark 12.10) in the form

$$du = L[u]dt + \Gamma[u]d\mathbf{W}, \qquad u(0, \cdot) = u_0. \tag{12.20}$$

Indeed, in the absence of ellipticity or Hörmander type conditions on L, the solution may not be any more regular than the initial data g so that in general (for $g \in \mathcal{C}_b$, say) the action of the first order differential operator $\Gamma = (\Gamma_1, \ldots, \Gamma_d)$ on u has no pointwise meaning, let alone its rough integral against $\mathbf{W}$. On the other hand, we can (at least formally) test the equation against $\varphi \in \mathcal{D} = \mathcal{C}_c^\infty(\mathbf{R}^n)$ and so arrive the following "analytically weak" formulation: call $u = u(s, x; \mathbf{X})$ a weak solution to (12.20) if for every $\varphi \in \mathcal{D}$ and $0 \le t \le T$ the following integral formula holds:

$$\langle u_t, \varphi \rangle = \langle u_0, \varphi \rangle + \int_0^t \langle u_s, L^*\varphi \rangle ds + \int_0^t \langle u_s, \Gamma^*\varphi \rangle d\mathbf{W}_s. \tag{12.21}$$

In Exercise 12.1 the reader is invited to check that our Feynman–Kac solution is indeed a weak solution in this sense. In particular, the final integral term is a bona fide rough integral of the controlled rough path $(Y, Y') \in \mathscr{D}_W^{2\alpha}$ against $\mathbf{W}$, where

$$Y_t = \langle u_t, \Gamma^*\varphi \rangle, \qquad Y_t' = \langle u_t, \Gamma^*\Gamma^*\varphi \rangle. \tag{12.22}$$

It is seen in [DFS17] that a uniqueness result holds for such weak RPDE solutions holds, provided in the definition a suitable uniformity over the test function φ is required. The strategy is a very similar to what was seen in Section 12.1.2: arguing (for convenience) on the terminal value formulation (12.10), we construct a *regular* forward solution and then employ a forward-backward argument to obtain uniqueness. This is essentially the uniqueness argument employed in Theorem 12.8, with switched roles of forward and backward evolution. Alternatively, in [HH18] the unbounded rough driver framework of [DGHT19b] has been adapted to linear second order RPDEs with L in divergence form.

Remark 12.15. Let $u = u(t, x; \mathbf{X})$ be a weak solution in the sense of (12.21), and W be a Brownian motion with Stratonovich rough path lift $\mathbf{W} = \mathbf{W}^{\text{Strat}}(\omega)$. Then, thanks to Theorem 5.14, it follows that $u(t, x; \omega) := u(t, x; \mathbf{W}^{\text{Strat}}(\omega))$ yields an analytically weak SPDE solution in the sense that for every $\varphi \in \mathcal{D}$ and $0 \le t \le T$ one has, with probability one,

$$\langle u_t, \varphi \rangle = \langle u_0, \varphi \rangle + \int_0^t \langle u_s, L^*\varphi \rangle ds + \int_0^t \langle u_s, \Gamma^*\varphi \rangle \circ dW_s,$$

where the existence of the Stratonovich integral is implied by Corollary 5.2.

12.2.2 Mild solutions to semilinear RPDEs

We now turn to a class of "abstract" rough evolution problems introduced by Gubinelli–Tindel [GT10], although our exposition is taken from [GH19]. Following a familiar picture in PDE theory, we would like to view an RPDE solution as a controlled path with values in a Hilbert space H which solves an RDE of the form

$$du_t = Lu_t dt + F(u_t)d\mathbf{X}_t \quad \text{and} \quad u_0 = \xi \in H \ . \tag{12.23}$$

Here, $\mathbf{X} = (X, \mathbb{X}) \in \mathscr{C}^\gamma([0, T], \mathbf{R}^d), \gamma \in (1/3, 1/2]$, not necessarily geometric. L is a negative definite self-adjoint operator, $F = (F_1, \ldots, F_d)$ are suitable (essentially 0-order) operators. In particular, no transport noise is covered by our setup so that – in contrast to previous sections – there is no restriction here to geometric rough paths.

Remark 12.16. Unlike Section 12.2.1 (Feynman–Kac) and Section 12.2.4 below (maximum principle), the present section is not really restricted to second order equations, even though these constitute the typical examples we have in mind.

To fix ideas, we give an example that will fit into the framework described below.

Example 12.17. Consider the *rough reaction diffusion equation*[2]

$$du_t(x) = \Delta u(x)\, dt + f(u_t(x))\, dt + p(u_t(x))\, d\mathbf{X}_t, \tag{12.24}$$

with $u_t : \mathbf{T}^n \to \mathbf{R}^l$ where $\mathbf{T}^n$ is the n-dimensional torus with Laplace operator Δ, and polynomial nonlinearities f and $p = (p_1, \ldots, p_d)$ on $\mathbf{R}^l$. As as typical in PDE theory, one looks for solutions $u_t \in H^k(\mathbf{T}^n, \mathbf{R}^l) =: H$, where H^k is the L^2-based Sobolev space with k weak derivates in L^2. Of course, Δ is negative definite self-adjoint on H, with dense domain $\text{Dom}(\Delta) = H_1$, where we set (in agreement with a later abstract interpolation space definition) $H_\alpha = H^{k+2\alpha}(\mathbf{T}^n, \mathbf{R}^l)$, and also note that the heat semigroup $(e^{\Delta t})_{t \geq 0}$ acts naturally on this Sobolev scale. The nonlinearity in this example is given by composition with a polynomial. Smoothness of this operation requires H to be an algebra, which, by basic Sobolev theory, requires $k > n/2$. The main theorem below requires each nonlinearity (as operator, here: $u \mapsto p_i \circ u$) to be $\mathcal{C}^3$ in Fréchet sense as map from $H_{-2\gamma} = H^{k-4\gamma}$ into itself. Therefore we have the requirement on k to satisfy $k > n/2 + 4\gamma$. This means that $\gamma = 1/3^+$ is the optimal choice (in a level-2 rough path setting). Of course, this covers the case of Brownian rough paths so that $\mathbf{X}$ above can be replaced by $\mathbf{W}^{\text{Itô}}(\omega)$ or $\mathbf{W}^{\text{Strat}}(\omega)$.

---

[2] As in the case of RDEs with additional drift vector field, Exercise 8.5, the extra nonlinearity $(f \circ u_t)\, dt$ can be absorbed in the $\mathbf{X}$-term, by working with the space-time extensions of $\mathbf{X}$. Less trivially, a direct analysis allows for more general nonlinearities in (12.23) such as to handle 2D Navier–Stokes with rough noise.

We want to give meaning to the rough partial differential equation (12.23). Similar to (12.21), there is a natural – still formal – analytically weak formulation: for every $h \in \mathrm{Dom}(L) \subset H$ and $0 \le t \le T$ the following integral formula holds (angle brackets denote the inner product in H):

$$\langle u_t, h \rangle = \langle \xi, h \rangle + \int_0^t \langle u_s, Lh \rangle ds + \int_0^t \langle F(u_s), h \rangle d\mathbf{X}_s . \qquad (12.25)$$

On the other hand, if $(S_t)_{t \ge 0}$ denotes the associated semigroup $S_t = e^{Lt}$ (which is analytic since L is assumed to be selfadjoint) one expects a mild formulation of the form, for all $0 \le t \le T$

$$u_t = S_t \xi + \int_0^t S_{t-s} F(u_s) d\mathbf{X}_s , \qquad (12.26)$$

where the identity holds between elements in H. The regularity of F will be measured in Fréchet sense, as map from H_α to itself, for a to be specified range of $\alpha \in \mathbf{R}$.[3] Here, for $\alpha \ge 0$, the interpolation space $H_\alpha = \mathrm{Dom}((-L)^\alpha)$ is a Hilbert space when endowed with the norm $\| \cdot \|_{H_\alpha} = \|(-L)^\alpha \cdot \|_H$. Similarly, $H_{-\alpha}$ is defined as the completion of H with respect to the norm $\| \cdot \|_{H_{-\alpha}} = \|(-L)^{-\alpha} \cdot \|_H$. Note that this setting is compatible with that of Exercise 4.16.

The weak formulation requires of course that $s \mapsto \langle F(u_s), h \rangle$ has meaning as a controlled rough path, so that (12.25) is well-defined. In the mild formulation (12.26) on the other hand we recognise the rough convolution integral previously defined in (4.47), provided that $s \mapsto F(u_s)$ is mildly controlled in the sense of (4.46). It can be seen that weak and mild solutions coincide [GH19]. (The proof of this involves a simple variant of the rough Fubini theorem from Exercise 4.11.) In what follows we only consider the mild formulation.

We introduce the following spaces which are a slight strengthening of the spaces $\mathscr{D}_{S,X}^{2\gamma}$ introduced in Exercise 4.17:

$$\mathcal{D}_X^{2\gamma}([0,T], H_\alpha) = \mathscr{D}_{S,X}^{2\gamma}([0,T], H_\alpha) \cap \left(\hat{C}^\gamma([0,T], H_{\alpha+2\gamma}) \times L^\infty([0,T], H_{\alpha+2\gamma}) \right) .$$

The basic ingredients, stability of mildly controlled rough paths under rough convolution and composition with regular functions were already established in Exercises 4.17 and 7.3. Taken together, they show that the image of $(Y, Y') \in \mathcal{D}_X^{2\gamma}([0,T], H)$ under the map

$$\mathscr{M}_T(Y, Y')_t := \left(S_t \xi + \int_0^t S_{t-u} F(Y_u) d\mathbf{X}_u, \, F(Y_t) \right) \qquad (12.27)$$

yields again an element of $\mathcal{D}_X^{2\gamma}([0,T], H)$. We now show that for small enough times this map has a unique fixed point:

---

[3] This rules out taking any derivatives in F. In particular, the previously considered transport noise, involved $D_x u$, is not accommodated in this setting.

Theorem 12.18 (Rough Evolution Equation). *Let $\xi \in H$, $F_1, \ldots, F_d : H \to H$, bounded in $\mathcal{C}^3(H_\beta, H_\beta)$ on bounded sets for every $\beta \geq -2\gamma$, for some $\gamma \in (1/3, 1/2]$, and $\mathbf{X} = (X, \mathbb{X}) \in \mathscr{C}^\gamma(\mathbf{R}_+, \mathbf{R}^d)$. Then there exists $\tau > 0$ and a unique element $(Y, Y') \in \mathcal{D}_X^{2\gamma}([0, \tau), H)$ such that $Y' = F(Y)$ and*

$$Y_t = S_t \xi + \int_0^t S_{t-u} F(Y_u) d\mathbf{X}_u \,, \quad t < \tau. \tag{12.28}$$

Proof. First note $\mathbf{X} = (X, \mathbb{X}) \in \mathscr{C}^\gamma \subset \mathscr{C}^\eta$ for $1/3 < \eta < \gamma \leq 1/2$. Fixing $T < 1$, we will find a solution $(Y, Y') \in \mathcal{D}_X^{2\eta}([0, T], H_{2\eta - 2\gamma})$ as a fixed point of the map $\mathscr{M}_T$ given by (12.27). In the end we will briefly describe how one can make an improvement and show that one actually has $(Y, Y') \in \mathcal{D}_X^{2\gamma}([0, T], H)$. The proof is analogous to Theorem 8.3, the only difference being that we have two different scales of space regularity for which we need to be able to obtain the bound (7.14), as prepared in Exercise 4.17. We will therefore show only invariance of the solution map (12.27), because proving it already contains all the techniques that are not present in the Theorem 8.3.

Note that if (Y, Y') is such that $(Y_0, Y_0') = (\xi, F(\xi))$ then the same is true for $\mathscr{M}_T(Y, Y')$, so we can view $\mathscr{M}_T$ as a map on the complete metric space

$$B_T = \{(Y, Y') \in \mathcal{D}_X^{2\eta}([0, T], H_{2\eta - 2\gamma}) : Y_0 = \xi, \ Y_0' = F(\xi), \|(Y, Y')\|_{X, 2\eta; -2\gamma}^\wedge$$
$$+ \|Y - S.F(\xi)X_{0,\cdot}\|_{\eta; 2\eta - 2\gamma} + \|Y' - S.F(\xi)\|_{\infty; 2\eta - 2\gamma} \leq 1\} \,.$$

(We use the same notational convention as in Exercise 4.17, namely indices after a semicolon indicate in which one of the H_α norms are taken.) Note that by the triangle inequality for $(Y, Y') \in B_T$ we have

$$\|(Y, Y')\|_{\mathcal{D}_X^{2\eta}} \lesssim (1 + \|\xi\| + \|F(\xi)\|)(1 + \|X\|_\gamma) \lesssim 1.$$

Here and below we write $A \lesssim B$ as a shorthand for $A \leq CB$ for a constant C that may depend on $\gamma, \eta, X, \mathbb{X}, F$ and ξ, but is uniform over $T \in (0, 1]$.

It remains to show that for T small enough $\mathscr{M}_T$ leaves B_T invariant and is contracting there, so that the claim follows from the Banach fixed point theorem. We will consider the simpler case when F is $\mathcal{C}_b^3$. For $(Z_t, Z_t') = (F(Y_t), DF(Y_t) \circ Y_t')$ we have by Exercise 7.3

$$\|(Z, Z')\|_{X, 2\eta} \lesssim (1 + \|(Y, Y')\|_{\mathcal{D}_X^{2\eta}})^2 \lesssim (1 + \|\xi\| + \|F(\xi)\|)^2 \lesssim 1 \,,$$

and from Exercise 4.17

$$\|\mathscr{M}_T(Y, Y')\|_{X, 2\eta} = \left\| \left(\int_0^\cdot S_{\cdot - u} Z_u d X_u, Z \right) \right\|_{X, 2\eta}$$
$$\lesssim \|Z\|_{\eta, -2\gamma} + (\|Z_0'\|_{H_{-2\gamma}} + \|(Z, Z')\|_{X, 2\eta; -2\gamma}^\wedge) \varrho_\eta(0, \mathbf{X})$$
$$\lesssim \|Z\|_{\eta, -2\gamma} + (\|Z_0'\|_{H_{-2\gamma}} + \|(Z, Z')\|_{X, 2\eta; -2\gamma}^\wedge) T^{\gamma - \eta}.$$

Since $(Y, Y') \in B_T$, we have the bound $\|Y\|_{\eta, -2\gamma} \leq (\|X\|_\gamma + 1)T^{\gamma - \eta}$. One can also show along the same lines as in Exercise 7.3 that

$$
\begin{aligned}
\|\hat{\delta} Z_{s,t}\|_{H_{-2\gamma}} &\lesssim \|\hat{\delta} Y_{s,t}\|_{H_{-2\gamma}} + \|S_{t-s} Y_s - Y_s\|_{H_{-2\gamma}} + |t - s|^{2\eta} \|F(Y_s)\|_{H_{2\eta - 2\gamma}} \\
&\lesssim \left(T^{\gamma - \eta} |t - s|^\eta + |t - s|^{2\eta} \|Y_s\|_{H_{2\eta - 2\gamma}} + T^\eta |t - s|^\eta \right) \\
&\lesssim \left(T^{\gamma - \eta} + T^{\gamma + \eta} + T^\eta \right) |t - s|^\eta.
\end{aligned}
$$

Therefore since $T < 1$ we conclude that $\|Z\|_{\eta, -2\gamma} \lesssim T^{\gamma - \eta}$.

To estimate $\|\mathscr{M}_T(Y) - S.F(\xi)X_{0,\cdot}\|_{\eta, 2\eta - 2\gamma}$ we use the identity

$$
\hat{\delta}(S.F(\xi)X_{0,\cdot})_{t,s} = S_t F(\xi) X_{s,t}
$$

and since $2\eta < 1$ we can use a better bound from (4.48) to deduce:

$$
\begin{aligned}
\|\hat{\delta}(\mathscr{M}_T(Y) - S.F(\xi)X_{0,\cdot})_{t,s}\|_{H_{2\eta - 2\gamma}} &= \left\| \int_s^t S_{t-u} F(Y_u) dX_u - S_t F(\xi) X_{s,t} \right\|_{H_{2\eta - 2\gamma}} \\
&\leq (\|F(\xi)\|_H + \|Z\|_{\infty; -2\gamma}) \|X\|_\eta |t - s|^\eta + \|Z'\|_{\infty; -2\gamma} \|\mathbb{X}\|_{2\eta} |t - s|^{2\eta} \\
&\quad + C(\|X\|_\eta |R^Z|_{2\eta} + \|\mathbb{X}\|_{2\eta} \|Z'\|_\eta) |t - s|^{3\eta - 2\eta} \\
&\lesssim (\|F(\xi)\|_H + \|Z_0'\|_{H_{-2\gamma}} + \|(Z, Z')\|_{X, 2\eta; -2\gamma}^\wedge) |t - s|^\eta \\
&\lesssim T^{\gamma - \eta} |t - s|^\eta.
\end{aligned}
$$

Finally we estimate the term $\|\mathscr{M}_T(Y)'_t - S_t F(\xi)\|_{H_{2\eta - 2\gamma}}$:

$$
\begin{aligned}
\|\mathscr{M}_T(Y)'_t - S_t F(\xi)\|_{H_{2\eta - 2\gamma}} &= \\
&= \|F(Y_t) - F(S_t \xi) + F(S_t \xi) - F(\xi) + F(\xi) - S_t F(\xi)\|_{H_{2\eta - 2\gamma}} \\
&\lesssim \|Y_t - S_t \xi\|_{H_{2\eta - 2\gamma}} + \|S_t \xi - \xi\|_{H_{2\eta - 2\gamma}} + \|F(\xi) - S_t F(\xi)\|_{H_{2\eta - 2\gamma}} \\
&\lesssim \|Y_t - S_t \xi - S_t F(\xi) X_{t,0}\|_{H_{2\eta - 2\gamma}} + \|F(\xi)\|_H \|X\|_\gamma T^\gamma \\
&\quad + t^{2\gamma - 2\eta} \|\xi\|_H + t^{2\gamma - 2\eta} \|F(\xi)\|_H \\
&\lesssim (\|Y - S.F(\xi)X_{0,\cdot}\|_{\eta, 2\eta - 2\gamma} T^\eta + T^\gamma + T^{2\gamma - 2\eta}) \lesssim T^{\gamma - \eta}.
\end{aligned}
$$

Putting it all together we obtain the bound

$$
\|\mathscr{M}_T(Y) - S.F(\xi)X_{0,\cdot}\|_{\eta; 2\eta - 2\gamma} + \|\mathscr{M}_T(Y)' - S.F(\xi)\|_{\infty; 2\eta - 2\gamma} \\
+ \|\mathscr{M}_T(Y, Y')\|_{X, 2\eta; -2\gamma}^\wedge \lesssim T^{\gamma - \eta}.
$$

If T is small enough we guarantee that the left-hand side of the above expression is smaller than 1, thus proving that B_T is invariant under $\mathscr{M}_T$. In order to show contractivity of $\mathscr{M}_T$, one can use analogous steps to first show

$$
\|\mathscr{M}_T(Y, Y') - \mathscr{M}_T(V, V')\|_{\mathcal{D}_X^{2\eta}} \lesssim \|(Y - V, Y' - V')\|_{\mathcal{D}_X^{2\eta}} T^{\gamma - \eta}.
$$

This guarantees contractivity for small enough T, completing the fixed point argument and thus showing the existence of the unique maximal solution to (12.28).

Let now $(Y, Y') \in \mathcal{D}_X^{2\eta}([0, T], H_{2\eta-2\gamma})$ be the solution constructed above, we sketch an argument showing that in fact it belongs to $\mathcal{D}_X^{2\gamma}([0, T], H)$. We know that

$$Y_t = S_t \xi + S_t F(\xi) X_{0,t} + S_t DF(\xi) F(\xi) + R_{0,t}, \qquad (12.29)$$

$$Y_t - S_{t-s} Y_s = S_{t-s} F(Y_s) X_{t,s} + S_{t-s} DF(Y_s) F(Y_s) \mathbb{X}_{s,t} + R_{s,t}. \quad (12.30)$$

Here $R_{s,t} = \int_s^t S_{t-r} F(Y_r) dX_r - S_{t-s} F(Y_s) X_{s,t} - S_{t-s} DF(Y_s) F(Y_s) \mathbb{X}_{s,t}$. From the estimate on $R_{0,t}$ using (4.48) and since $\xi \in H$, we see that (12.29) implies $Y \in L^\infty([0, T], H)$. Moreover (12.30) implies $Y \in \hat{\mathcal{C}}^\gamma([0, T], H_{-2\gamma})$ which, together with $Y \in L^\infty([0, T], H)$, implies $F(Y) \in \hat{\mathcal{C}}^\gamma([0, T], H^d_{-2\gamma}) \cap L^\infty([0, T], H^d_{2\eta-2\gamma})$. This itself implies that $(Y, F(Y)) \in \mathscr{D}_{S,X}^{2\gamma}([0, T], H_{-2\gamma})$ (using again (12.30)) and $(F(Y), DF(Y) F(Y)) \in \mathscr{D}_{S,X}^{2\gamma}([0, T], H_{-2\gamma})$ which enables us to get an estimate for every $\beta < 3\gamma$:

$$\|R_{s,t}\|_{H_\beta} \lesssim \|F(Y), DF(Y) F(Y)\|_{X,2\gamma;-2\gamma}^{\wedge} |t - s|^{3\gamma-\beta}.$$

Taking $\beta = 2\gamma$ and using (12.30) again we show that $Y \in \hat{\mathcal{C}}^\gamma([0, T], H)$, which completes the proof that $(Y, Y') \in \mathcal{D}_X^{2\gamma}([0, T], H)$. $\quad\square$

12.2.3 Fully nonlinear equations with semilinear rough noise

We now consider nonlinear rough partial differential equations of the form

$$du = F[u]dt + \sum_{i=1}^d H_i[u] \circ dW_t^i(\omega), \qquad u(0, \cdot) = g, \qquad (12.31)$$

with fully nonlinear, possibly degenerate, operator

$$F[u] = F(x, u, Du, D^2 u),$$

and semilinear

$$H_i[u] = H_i(x, u, Du), \qquad i = 1, \dots, d.$$

We essentially rule out nonlinear dependence on Du, hence the terminology "semilinear noise", which makes a (global) flow transformation method work. In a stochastic setting such transformation (at least in the linear case) are attributed to Kunita. As already noted in the context of first order equations, the case $H_i = H_i(x, Du)$ requires a subtle local version of such as transformation and is topic of the pathwise Lions–Souganidis theory of stochastic viscosity theory for fully nonlinear SPDEs; [LS98a, LS98b, LS00b] and [Sou19] for a recent overview.

As in the previous section we aim to replace $\circ dW$ by a "rough" differential $d\mathbf{W}$, for some geometric rough path $\mathbf{W} \in \mathscr{C}_g^{0,\alpha}([0, T], \mathbf{R}^d)$, and show that an RPDE solution arises as the unique limit under approximations $(W^\varepsilon, \mathbb{W}^\varepsilon) \to \mathbf{W}$. Of course, there is little one can say at this level of generality and we have not even clarified in which sense we mean to solve (12.31) when $W \in \mathcal{C}^1$! Let us postpone this discussion and assume momentarily that F and H are sufficiently "nice" so that, for every $W \in \mathcal{C}^1$ and $g \in \mathcal{BC}$, say, there is a classical solution $u = u(t, x)$ for $t > 0$.

With noise of the form $H[u]\dot{W} = \sum_i H_i(x, u, Du)\dot{W}^i$, we shall focus on the following three cases.

a) **Transport noise.** For sufficienly nice vector fields β_i on $\mathbf{R}^n$,

$$H_i[u] = \beta_i(x) \cdot Du \ ;$$

b) **Semilinear noise.** For a sufficienly nice function H_i on $\mathbf{R}^n \times \mathbf{R}$,

$$H_i[u] = H_i(x, u);$$

c) **Linear noise.** With β_i as above and sufficiently nice functions γ_i on $\mathbf{R}^n$

$$H_i[u] = \Gamma_i[u] := \beta_i(x) \cdot Du + \gamma_i(x)u.$$

We now develop the "calculus" for the transformations associated to each of the above cases. All proofs consist of elementary computations and are left to the reader.

Proposition 12.19 (Case a). *Assume that $\psi = \psi^W$ is a $\mathcal{C}^3$ solution flow of diffeomorphisms associated to the ODE $\dot{Y} = -\beta(Y)\dot{W}$, where $W \in \mathcal{C}^1$. (This is the case if $\beta \in \mathcal{C}_b^3$.) Then u is a classical solution to*

$$\partial_t u = F\left(x, u, Du, D^2 u\right) + \langle \beta(x), Du \rangle \dot{W}$$

if and only if $v(t, x) = u(t, \psi_t(x))$ is a classical solution to

$$\partial_t v - F^\psi\left(t, x, v, Dv, D^2 v\right) = 0$$

where F^ψ is determined from

$$F^\psi(t, \psi_t(x), r, p, X) \\ \stackrel{def}{=} F\left(x, r, \langle p, D\psi_t^{-1} \rangle, \langle X, D\psi_t^{-1} \otimes D\psi_t^{-1} \rangle + \langle p, D^2 \psi_t^{-1} \rangle\right) .$$

Proposition 12.20 (Case b). *For any fixed $x \in \mathbf{R}^n$, assume that the one-dimensional ODE*

$$\dot{\varphi} = H(x, \varphi)\dot{W} , \qquad \varphi(0; x) = r ,$$

has a unique solution flow $\varphi = \varphi^W = \varphi(t, r; x)$ which is of class $\mathcal{C}^2$ as a function of both r and x. Then u is a classical solution to

$$\partial_t u = F\left(x, u, Du, D^2 u\right) + H(x, u)\dot{W}$$

if and only if $v(t, x) = \varphi^{-1}(t, u(t, x), x)$, or equivalently $\varphi(t, v(t, x), x) = u(t, x)$, is a solution of

$$\partial_t v - {}^{\varphi}F\big(t, x, r, Dv, D^2v\big) = 0 \,,$$

with

$${}^{\varphi}F(t, x, r, p, X) \stackrel{\text{def}}{=} \frac{1}{\varphi'} F(t, x, \varphi, D\varphi + \varphi'p, \tag{12.32}$$

$$\varphi''p \otimes p + D\varphi' \otimes p + p \otimes D\varphi' + D^2\varphi + \varphi'X\big) \,,$$

where φ' denotes the derivative of $\varphi = \varphi(t, r, x)$ with respect to r.

Remark 12.21. It is worth noting that the "quadratic gradient" term $\varphi''p \otimes p$ disappears in (12.32) whenever $\varphi'' = 0$. This happens when $H(x, u)$ is linear in u, i.e. when

$$H_i[u] = \gamma_i(x)u \,, \qquad i = 1, \ldots, d \,.$$

in which case we have

$$\varphi(t, r, x) = r \exp\left(\int_0^t \gamma(x) dW_s\right) = r \exp\left(\sum_{i=1}^d \gamma_i(x) W_{0,t}^i\right). \tag{12.33}$$

Remark 12.22. Note that all dependence on $\dot{W}$ has disappeared in (12.33), and consequently (12.32). In the SPDE / filtering context this is known as *robustification*: the transformed PDE $(\partial_t - {}^{\varphi}F)v = 0$ can be solved for any $W \in \mathcal{C}([0, T], \mathbf{R}^d)$. This provides a way to solve SPDEs of the form $du = F[u]dt + \sum_{i=1}^d \gamma_i(x)u \circ dW_t$ pathwise, so that u depends continuously on W in uniform topology.

We now turn our attention to case c). The point here is that the "inner" and "outer" transformation seen above, namely

$$v(t, x) = u(t, \psi_t(x)) \,, \qquad v(t, x) = \varphi^{-1}(t, u(t, x), x) \,,$$

respectively, can be combined to handle noise coefficients obtained by adding those from cases a) and b), i.e. noise coefficients of the type $\langle \beta_i(x), Du \rangle + H_i(x, u)$. We content ourselves with the linear case

$$H_i[u] = \langle \beta_i(x), Du \rangle + \gamma_i(x)u \,.$$

Proposition 12.23 (Case c). Let $\psi = \psi^W$ be as in case a) and set $\varphi(t, r, x) = r \exp\left(\int_0^t \gamma(\psi_s(x)) dW_s\right)$. Then u is a (classical) solution to

$$\partial_t u = F\big(x, u, Du, D^2u\big) + \big(\langle \beta(x), Du \rangle + \gamma(x)u\big)\dot{W} \,,$$

if and only if $v(t, x) = u(t, \psi_t(x)) \exp\left(-\int_0^t \gamma(\psi_s(x)) dW_s\right)$ *is a (classical) solution to*

$$\partial_t v - {}^{\varphi}(F^{\psi})\big(t, x, v, Dv, D^2v\big) = 0.$$

Remark 12.24. It is worth noting that the outer transformation $F \to F^{\psi}$ preserves the class of linear operators. That is, if $F[u] = L[u]$ as given in (12.9), then F^{ψ} is again a linear operator. Because of the appearance of quadratic terms in Du, this is not true for the inner transformation $F \to {}^{\varphi}F$ *unless* $\varphi'' = 0$. Fortunately, this happens in the linear case and it follows that the transformation $F \to {}^{\varphi}(F^{\psi})$ used in case c) above does preserve the class of linear operators.

Let us reflect for a moment on what has been achieved. We started with a PDE that involves $\dot{W}$ and in all cases we managed to transform the original problem to a PDE where all dependence on $\dot{W}$ has been isolated in some auxiliary ODEs. In the stochastic context ($\circ dW$ instead of $dW = \dot{W} dt$) this is nothing but the reduction, via stochastic flows, from a *stochastic* PDE to a *random* PDE, to be solved ω-wise. In the same spirit, the rough case is now handled with the aid of flows for RDEs and their stability properties.

Given $\mathbf{W} \in \mathscr{C}_g^{0,\alpha}$, we pick an approximating sequence (W^{ε}), and transform

$$\partial_t u^{\varepsilon} = F[u^{\varepsilon}] + H[u^{\varepsilon}]\dot{W}^{\varepsilon} \tag{12.34}$$

to a PDE of the form

$$\partial_t v^{\varepsilon} = F^{\varepsilon}[v^{\varepsilon}], \tag{12.35}$$

e.g. with $F^{\varepsilon} = F^{\psi}$ and $\psi = \psi^{W^{\varepsilon}}$ in case a) and accordingly in the other cases. Then

$$F^{\varepsilon}[w] = F^{\varepsilon}[t, x, w, Dw, D^2 w]$$

(in abusive notation) and the function F^{ε} which appears on the right-hand side above converges (e.g. locally uniformly) as $\varepsilon \to 0$, due to stability properties of flows associated to RDEs as discussed in Section 8.10.

All one now needs is a (deterministic) PDE framework with a number of good properties, along the following "wish list".

1. All approximate problems, i.e. with $W^{\varepsilon} \in C^1([0, T], \mathbf{R}^d)$

$$\partial_t u^{\varepsilon} = F[u^{\varepsilon}] + \sum_{i=1}^d H_i[u^{\varepsilon}]\dot{W}_t^{\varepsilon,i}, \qquad u^{\varepsilon}(0, \cdot) = g^{\varepsilon},$$

should admit a unique solution, in a suitable class $\mathcal{U}$ of functions on $[0, T] \times \mathbf{R}^n$, for a suitable class of initial conditions in some space $\mathcal{G}$.
2. The change of variable calculus (Propositions 12.19–12.23) should remain valid, so that $u^{\varepsilon} \in \mathcal{U}$ is a solution to (12.34) if and only if its transformation $v^{\varepsilon} \in \mathcal{U}$ is a solution to (12.35).
3. There should be a good stability theory, so that $g^{\varepsilon} \to g^0$ in $\mathcal{G}$ and $F^{\varepsilon} \to F^0$ (in a suitable sense) allows to obtain convergence in $\mathcal{U}$ of solutions v^{ε} to (12.35) with intitial data g^{ε} to the (unique) solution of the limiting problem $\partial_t v^0 = F^0[v^0]$ with initial data g^0.

4. At last, the topology of $\mathcal{U}$ should be weak enough to make sure that $v^\epsilon \to v^0$ implies that the "back-transformed" u^ϵ converges in $\mathcal{U}$, with limit u^0 being v^0 back-transformed.[4]

The final point suggests to *define* a solution to

$$du = F[u]dt + H[u]d\mathbf{W} , \qquad u(0, \cdot) = g , \qquad (12.36)$$

as an element in $\mathcal{U}$ which, under the correct flow transformation associated to $\mathbf{W}$ and H, solves the transformed equation $\partial_t v = F^0[v]$, $v(0, \cdot) = g$. To make this more concrete, consider the transport case a). As before, $\psi = \psi^{\mathbf{W}}$ is the flow associated to the RDE $dY = -\beta(Y)d\mathbf{W}$ and u solves the above RPDE (with $H[u] = \langle \beta(x), Du \rangle$) if, by definition, $v(t, x) := u(t, \psi_t(x))$ solves $\partial_t v = F^\psi[v]$, with $v(0, \cdot) = g$. The same logic applies to cases b) and c).

We then have the following (meta-)theorem, subject to a PDE framework with the above properties.

Theorem 12.25. *Let* $\alpha \in (\frac{1}{3}, \frac{1}{2}]$. *Given a geometric rough path* $\mathbf{W} = (W, \mathbb{W}) \in \mathscr{C}_g^{0,\alpha}([0, T], \mathbf{R}^d)$, *pick* $W^\epsilon \in C^1([0, T], \mathbf{R}^d)$ *so that*

$$(W^\epsilon, \mathbb{W}^\epsilon) := \left(W^\epsilon, \int_0^{\cdot \cdot} W_{0,t}^\epsilon \otimes dW_t^\epsilon \right) \to \mathbf{W}$$

in α-*Hölder rough path metric. Consider unique solutions* $u^\epsilon \in \mathcal{U}$ *to the PDEs*

$$\begin{cases} \partial_t u^\epsilon = F[u^\epsilon] + H[u^\epsilon]\dot{W}^\epsilon \\ u^\epsilon(0, \cdot) = g \in \mathcal{G}. \end{cases} \qquad (12.37)$$

Then there exists $u = u(t, x) \in \mathcal{U}$, *not dependent on the approximating* (W^ϵ) *but only on* $\mathbf{W} \in \mathscr{C}_g^{0,\alpha}([0, T], \mathbf{R}^d)$, *so that*

$$u^\epsilon = \mathcal{S}[W^\epsilon; g] \to u =: \mathcal{S}[\mathbf{W}; g]$$

as $\epsilon \to 0$ *in* $\mathcal{U}$. *This* u *is the unique solution to the RPDE (12.36) in the sense of the above definition. Moreover, the resulting solution map,*

$$\mathcal{S} : \mathscr{C}_g^{0,\alpha}([0, T], \mathbf{R}^d) \times \mathcal{G} \to \mathcal{U}$$

is continuous.

It remains to identify suitable PDE frameworks, depending on the nonlinearity F. When $\partial_t u = F[u]$ is a scalar conservation law, entropy solutions actually provide a suitable framework to handle additional rough noise, at least of (linear) type c), [FG16b]. On the other hand, when $F = F[u]$ is a fully nonlinear second order operator, say of Hamilton–Jacobi–Bellman (HJB) or Isaacs type, the natural framework is viscosity theory [CIL92, FS06] and the problem of handling additional "rough"

---

[4] Given the roughness in t of our transformations, typically α-Hölder, it would not be wise to incorporate temporal C^1-regularity in the definition of the space $\mathcal{U}$.

noise, in the sense of $W \notin C^1$, also with nonlinear $H = H(Du)$, was first raised by Lions–Sougandis [LS98a, LS98b, LS00a, LS00b].

12.2.4 Rough viscosity solutions

Consider a real-valued function $u = u(x)$ with $x \in \mathbf{R}^m$ and assume $u \in C^2$ is a classical supersolution,

$$-G(x, u, Du, D^2 u) \geq 0,$$

where G is continuous and *degenerate elliptic* in the sense that $G(x, u, p, A) \leq G(x, u, p, A + B)$ whenever $B \geq 0$ in the sense of symmetric matrices. The idea is to consider a (smooth) test function φ which touches u from below at some interior point $\bar{x}$. Basic calculus implies that $Du(\bar{x}) = D\varphi(\bar{x})$, $D^2 u(\bar{x}) \geq D^2\varphi(\bar{x})$ and, from degenerate ellipticity,

$$-G(\bar{x}, \varphi, D\varphi, D^2\varphi) \geq 0. \tag{12.38}$$

This motivates the definition of a *viscosity supersolution* (at the point $\bar{x}$) to $-G = 0$ as a (lower semi-)continuous function u with the property that (12.38) holds for any test function which touches u from below at $\bar{x}$. Similarly, *viscosity subsolutions* are (upper semi-)continuous functions defined via test functions touching u from above and by reversing inequality in (12.38); *viscosity solutions* are both super- and subsolutions. Observe that this definition covers (completely degenerate) first order equations as well as parabolic equations, e.g. by considering $\partial_t - F = 0$ on $[0, T] \times \mathbf{R}^n$ where F is degenerate elliptic. Let us mention a few key results of viscosity theory, with special regard to our "wish list".

1. One has existence and uniqueness results in the class of $\mathcal{BC}$ solutions to the initial value problem $(\partial_t - F)u = 0$, $u(0, \cdot) = g \in \mathcal{BUC}(\mathbf{R}^n)$[5], provided $F = F(t, x, u, Du, D^2 u)$ is continuous, degenerate elliptic, there exists $\gamma \in \mathbf{R}$ such that, uniformly in t, x, p, X,

    $$\gamma(s - r) \leq F(t, x, r, p, X) - F(t, x, s, p, X) \text{ whenever } r \leq s, \tag{12.39}$$

    and some technical conditions hold.[6] Without going into technical details, the conditions are met for $F = L$ as in (12.9) and are robust under taking inf and sup (provided the regularity of the coefficients holds uniformly). As a consequence, HJB and Isaacs type nonlinearities, where F takes the form $\inf_a L_a$, $\inf_a \sup_{a'} L_{a,a'}$, are also covered.
2. The change of variables "calculus" of Propositions 12.19–12.23 remains valid for (continuous) viscosity solutions. This can be checked directly from the definition of a viscosity solution.

---

[5] the space of bounded uniformly continuous functions

[6] ...the most important of which is [CIL92, (3.14)]. Additional assumptions on F are necessary, however, in particular due to the unboundedness of the domain $\mathbf{R}^n$, and these are not easily found in the literature; see [DFO14]. One can also obtain existence and uniqueness result in $\mathcal{BUC}$.

3. In fact, the technical conditions mentioned in 1. imply a particularly strong form of uniqueness, known as *comparison*: assume u (resp. v) is a subsolution (resp. supersolution) and $u_0 \leq v_0$; then $u \leq v$ on $[0, T] \times \mathbf{R}^n$. A key feature of viscosity theory is what workers in the field simply call *stability*, a powerful incarnation of which is known as *Barles and Perthame procedure* [FS06, Section VII.3] and relies on comparison for (semicontinuous) sub- and super-solutions. In the form relevant for us, one assumes comparison for $\partial_t - F^0$ and considers viscosity solutions to $(\partial_t - F^\varepsilon)v^\varepsilon = 0$, with $v^\varepsilon(0, \cdot) = g^\varepsilon$, assuming locally uniform boundedness of v^ε and $g^\varepsilon \to g^0$ locally uniformly. Then $v^\varepsilon \to v^0$ locally uniformly where v^0 is the (unique) solution to the limiting problem $(\partial_t - F^0)v^0 = 0$, with $v^0(0, \cdot) = g^0$.

In the context of RPDEs above, again with focus on the transport case a) for the sake of argument, $F^0 = F^\psi$ where $\psi = \psi^{\mathbf{W}}$, where ψ is a flow of $\mathcal{C}^3$-diffeomorphisms (associated to the RDE $dY = -\beta(Y)d\mathbf{W}$ thereby leading to the assumption $\beta \in \mathcal{C}_b^5$). As a structural condition on F, we may simply assume "ψ-invariant comparison" meaning that comparison holds for $\partial_t - F^\psi$, for any $\mathcal{C}^3$-diffeomorphism with bounded derivatives. Checking this condition turns out to be easy. First, when $F = L$ is linear, we have $F^\psi = L^\psi$ also linear, with similar bounds on the coefficients as L due to the stringent assumptions on the derivatives of ψ. From the above discussion, and in particular from what was said in 1., it is then clear that L satisfies ψ-invariant comparison. In fact, stability of the condition in 1. under taking inf and sup, also implies that HJB and Isaacs type nonlinearities satisfy ψ-invariant comparison.

It is now possible to implement the arguments of the previous Theorem 12.25 in the viscosity framework [CFO11], see also [FO11] for applications to splitting methods. We tacitly assume that all approximate problems of the form (12.40) below have a viscosity solution, for all $W^\varepsilon \in \mathcal{C}^1$ and $g \in \mathcal{BUC}$, but see Remark 12.27.

Theorem 12.26. *Let $\alpha \in (\frac{1}{3}, \frac{1}{2}]$. Given a geometric rough path $\mathbf{W} = (W, \mathbb{W}) \in \mathscr{C}_g^{0,\alpha}([0, T], \mathbf{R}^d)$, pick $W^\varepsilon \in \mathcal{C}^1([0, T], \mathbf{R}^d)$ so that $(W^\varepsilon, \mathbb{W}^\varepsilon) \to \mathbf{W}$ in α-Hölder rough path metric. Consider unique $\mathcal{BC}$ viscosity solutions u^ε to*

$$\begin{cases} \partial_t u^\epsilon = F[u^\epsilon] + \langle \beta(x), Du \rangle \dot{W}^\epsilon \\ u^\epsilon(0, \cdot) = g \in \mathcal{BUC}(\mathbf{R}^n) \end{cases} \tag{12.40}$$

where F satisfies ψ-invariant comparison. Then there exists $u = u(t, x) \in \mathcal{BC}$, not dependent on the approximating (W^ε) but only on $\mathbf{W} \in \mathscr{C}_g^{0,\alpha}([0, T], \mathbf{R}^d)$, so that

$$u^\varepsilon = \mathcal{S}[W^\varepsilon; g] \to u =: \mathcal{S}[\mathbf{W}; g]$$

as $\varepsilon \to 0$ in local uniform sense. This u is the unique solution to the RPDE (12.36) with transport noise $H[u] = \langle \beta(x), Du \rangle$ in the sense of the definition given previous to Theorem 12.25. Moreover, we have continuity of the solution map,

$$\mathcal{S} : \mathscr{C}_g^{0,\alpha}([0, T], \mathbf{R}^d) \times \mathcal{BUC}(\mathbf{R}^n) \to \mathcal{BC}([0, T] \times \mathbf{R}^n).$$

Remark 12.27. In the above theorem, existence of RPDE solutions actually relies on existence of approximate solutions u^ε, which one of course expects from standard viscosity theory. Mild structural conditions on F, satisfied by HJB and Isaacs examples, which imply this existence are reviewed in [DFO14]. One can also establish a modulus of continuity for RPDE solutions, so that $u \in \mathcal{BUC}$ after all.

Remark 12.28. Rough partial differential equations as considered here, $du = F[u]dt + \langle \beta(x), Du \rangle d\mathbf{W}$, with $F = \inf_a L_a$ of HJB form, arise in pathwise stochastic control [LS98b, BM07, DFG17], also in conjunction with filtering [AC19].

Unfortunately, in case b), it turns out the structural assumptions one has to impose on F in order to have the necessary comparison for $\partial_t - F^0 = 0$ is rather restrictive, although semilinear situations are certainly covered. Even in this case, due to the appearance of a quadratic nonlinearity in Du, the argument is involved and requires a careful analysis on consecutive small time intervals, rather than $[0, T]$; see [LS00a, DF12]. A nonlinear Feynman–Kac representation, in terms of *rough backward stochastic differential equations* is given in [DF12].

At last, we return to the fully linear case of Section 12.2.3. That is, we consider the (linear noise) case c) with linear $F = L$. With some care [FO14], the double transformation leading to the transformed equation $\partial_t - {}^\varphi(F^\psi) = 0$ can be implemented with the aid of coupled flows of rough differential equations. We can then recover Theorem 12.11, but with somewhat different needs concerning the regularity of the coefficients. (For instance, in the aforementioned theorem we really needed $\sigma, \beta \in \mathcal{C}_b^3$ whereas now, using flow decomposition, we need $\beta \in \mathcal{C}_b^5$ but only $\sigma \in \mathcal{C}_b^1$.

Remark 12.29. By either approach, case c) with linear $F = L$ or Theorem 12.11, we obtain a robust view on classes SPDEs which contain the Zakai equation from filtering theory, provided the initial law admits a $\mathcal{BUC}$-density. Robustness is an important issue in filtering theory, see also Exercise 12.3.

12.3 Stochastic heat equation as a rough path

Nonlinear stochastic partial differential equations driven by very singular noise, say space-time white noise, may suffer from the fact that their nonlinearities are ill-posed. For instance, even in space dimension one, there is no obvious way of giving "weak" meaning to Burgers-like stochastic PDEs of the type

$$\partial_t u^i = \partial_x^2 u^i + f(u) + \sum_{j=1}^n g_j^i(u)\partial_x u^j + \xi^i , \qquad i = 1, \ldots, n , \qquad (12.41)$$

where $\xi = (\xi^i)$ denotes space-time white noise (strictly speaking, n independent copies of scalar space-time white noise). Recall that, at least formally, space-time white noise is a Gaussian generalised stochastic process such that

$$\mathbf{E}\xi^i(t,x)\xi^j(s,y) = \delta_{ij}\delta(t-s)\delta(x-y) \ .$$

As a consequence of the lack of regularity of ξ, it turns out that the solution to the stochastic heat equation (i.e. the case $f = g = 0$ in (12.41) above) is only α-Hölder continuous in the spatial variable x for any $\alpha < 1/2$. In other words, one would not expect any solution u to (12.41) to exhibit spatial regularity better than that of a Brownian motion.

As a consequence, even when aiming for a weak solution theory, it is not clear how to define the integral of a spatial test function φ against the nonlinearity. Indeed, this would require us to make sense of expressions of the type

$$\int \varphi(x)g^i_j(u)\partial_x u^j(t,x)\,dx \ ,$$

for fixed t. When g happens to be a gradient, such an integral can be defined by postulating that the chain rule holds and integrating by parts. For a general g, as arising in applications from path sampling [HSV07], this approach fails. This suggests to seek an understanding of $u(t,\cdot)$ as a spatial rough path. Indeed, this would solve the problem just explained by allowing us to define the nonlinearity in a weak sense as

$$\int \varphi(x)g^i_j(u)\,d\mathbf{u^j}(t,x) \ ,$$

where $\mathbf{u}$ is the rough path associated to u.

In the particular case of (12.41), it is actually sufficient to be able to associate a rough path to the solution ψ to the stochastic heat equation

$$\partial_t\psi = \partial^2_x\psi + \xi \ .$$

Indeed, writing $u = \psi + v$ and proceeding formally for the moment, we then see that v should solve

$$\partial_t v^i = \partial^2_x v^i + f(v+\psi) + \sum_{j=1}^{n} g^i_j(v+\psi)\big(\partial_x\psi^j + \partial_x v^j\big) \ .$$

If we were able to make sense of the term appearing in the right-hand side of this equation, one would expect it to have the same regularity as $\partial_x\psi$ so that, since $\psi(t,\cdot)$ turns out to belong to $\mathcal{C}^\alpha$ for every $\alpha < 1/2$, one would expect $v(t,\cdot)$ to be of regularity $\mathcal{C}^{\alpha+1}$ for every $\alpha < 1/2$. In particular, we would not expect the term involving $\partial_x v^j$ to cause any trouble, so that it only remains to provide a meaning for the term $g^i_j(v+\psi)\partial_x\psi^j$. If we know that $v \in \mathcal{C}^1$ and we have an interpretation of $\psi(t,\cdot)$ as a rough path $\boldsymbol{\psi}$ (in space), then this can be interpreted as the distribution whose action, when tested against a test function φ, is given by

$$\int \varphi(x)g^i_j(\psi+v))\,d\boldsymbol{\psi^j}(t,x) \ .$$

This reasoning can actually be made precise, see the original article [Hai11b]. In this section we limit ourselves to providing the construction of ψ and giving some of its basic properties.

12.3.1 The linear stochastic heat equation

We now study the model problem in this context - the construction of a spatial rough path associated, in essence, to the above SPDE in the case $f = g = 0$. More precisely, we are considering stationary (in time) solution to the stochastic heat equation[7],

$$d\psi_t = -A\psi_t dt + \sigma dW_t, \qquad (12.42)$$

where, for fixed $\lambda > 0$

$$Au = -\partial_x^2 u + \lambda u;$$

and W is a cylindrical Wiener process over $L^2(\mathbf{T})$, the L^2-space over the one-dimensional torus $\mathbf{T} = [0, 2\pi]$, endowed with periodic boundary conditions. Let $(e_k : k \in \mathbf{Z})$ denote the standard Fourier-basis of $L^2(\mathbf{T})$

$$e_k(x) = \begin{cases} \frac{1}{\sqrt{\pi}} \sin(kx) & \text{for } k > 0 \\ \frac{1}{\sqrt{2\pi}} & \text{for } k = 0 \\ \frac{1}{\sqrt{\pi}} \cos(kx) & \text{for } k < 0 \end{cases}$$

which diagonalises the operator A in the sense that

$$Ae_k = \mu_k e_k, \qquad mu_k = k^2 + \lambda, \qquad k \in \mathbf{Z}.$$

Thanks to the fact that we chose $\lambda > 0$, the stochastic heat equation (12.42) has indeed a stationary solution which, by taking Fourier transforms, may be decomposed as $\psi(x, t; \omega) = \sum_k Y_t^k(\omega) e_k(x)$. The components Y_t^k are then a family of independent stationary one-dimensional Ornstein-Uhlenbeck processes given by

$$dY_t^k = -\mu_k Y_t^k dt + \sigma dB_t^k,$$

where $(B^k : k \in \mathbf{Z})$ is a family of i.i.d. standard Brownian motions. An explicit calculation yields

$$\mathbf{E}(Y_s^k Y_t^k) = \frac{\sigma^2}{2\mu_k} \exp(-\mu_k|t - s|),$$

so that in particular, for any fixed time t,

$$\mathbf{E}(Y_t^k)^2 = \frac{\sigma^2}{2\mu_k}.$$

---

[7] With $\lambda = 0$, the 0th mode of ψ behaves like a Brownian motion and ψ cannot be stationary in time, unless one identifies functions that only differ by a constant.

Lemma 12.30. *For each fixed t, the spatial covariance of ψ is given by*

$$\mathbf{E}(\psi(x,t)\psi(y,t)) = K(|x-y|)$$

where K is given by

$$K(u) := \frac{1}{4\pi}\sigma^2 \sum_{k\in\mathbf{Z}} \frac{\cos(ku)}{\mu_k} = \frac{\sigma^2}{4\sqrt{\lambda}\sinh(\sqrt{\lambda}\pi)} \cosh\left(\sqrt{\lambda}(u-\pi)\right).$$

Here, the second equality holds for u restricted to $[0, 2\pi]$. In fact, the cosine series is the periodic continuation of the r.h.s. restricted to $[0, 2\pi]$.

Proof. From the basic identity $\cos(\alpha - \beta) = \cos\alpha\cos\beta + \sin\alpha\sin\beta$,

$$e_{-k}(x)e_{-k}(y) + e_k(x)e_k(y) = \frac{1}{\pi}\cos(k(x-y)), \ k\in\mathbf{Z}.$$

Inserting the respective expansion in $R(x,y) := \mathbf{E}(\psi(x,t)\psi(y,t))$, and using the independence of the $(Y^k : k\in\mathbf{Z})$, gives

$$R(x,y) = \sum_{k\in\mathbf{Z}} e_k(x)e_k(y)\mathbf{E}(Y_t^k)^2 = \frac{1}{2\pi}\mathbf{E}(Y_t^0)^2 + \frac{1}{\pi}\sum_{k=1}^{\infty}\cos(k(x-y))\mathbf{E}(Y_t^k)^2$$

$$= \frac{\sigma^2}{4\pi}\sum_{k\in\mathbf{Z}}\frac{\cos(k(x-y))}{\lambda+k^2},$$

and then $R(x,y) = K(|x-y|)$ where

$$K(x) = \frac{\sigma^2}{4\pi}\sum_{k\in\mathbf{Z}}\frac{\cos(kx)}{\lambda+k^2}.$$

At last, expand the (even) function $\cosh(\sqrt{\lambda}(|\cdot|-\pi))$ in its (cosine) Fourier-series to get the claimed equality. $\square$

Proposition 12.31. *Fix $t \geq 0$. Then $\psi_t(x;\omega) = \psi(t,x;\omega)$, indexed by $x \in [0, 2\pi]$, is a centred Gaussian process with covariance of finite 1-variation. More precisely,*

$$\left\|R_{\psi(t,\cdot)}\right\|_{1;[x,y]^2} \leq 2\pi\|K\|_{C^2;[0,2\pi]}|x-y|,$$

and so (cf. Theorem 10.4), for each fixed $t \geq 0$, the $\mathbf{R}^d$-valued process

$$[0, 2\pi] \ni x \mapsto (\psi_t^1(x),\dots,\psi_t^d(x)),$$

consisting of d i.i.d. copies of ψ_t, lifts canonically to a Gaussian rough path $\boldsymbol{\psi}_t(\cdot) \in \mathscr{C}_g^{0,\alpha}([0, 2\pi], \mathbf{R}^d).$

Proof. This follows immediately from Exercise 10.4. $\square$

Remark 12.32. There are ad-hoc ways to construct a (spatial) rough path lift associated to the stochastic heat-equation, for instance be writing $\psi(t, \cdot)$ as Brownian bridge plus a random smooth function. In this way, however, one ignores the large body of results available for general Gaussian rough paths: for instance, rough path convergence of hyper-viscosity or Galerkin approximation, extensions to fractional stochastic heat equations, concentration of measure can all be deduced from general principles.

We now show that solutions to the stochastic heat equation induces a continuous stochastic evolution in rough path space.

Theorem 12.33. *There exists a continuous modification of the map* $t \mapsto \boldsymbol{\psi}_t$ *with values in* $\mathscr{C}_g^\alpha\big([0, 2\pi], \mathbf{R}^d\big)$.

Proof. Fix s and t. The proof then proceeds in two steps. First, we will verify the assumptions of Corollary 10.6, namely we will show that

$$|\varrho_\alpha(\boldsymbol{\psi}_s, \boldsymbol{\psi}_t)|_{L^q} \leq C \sup_{x,y \in [0,2\pi]} \left[\mathbf{E}\big(|\psi_s(x,y) - \psi_t(x,y)|^2 \big) \right]^\theta,$$

for some constant C that is independent of s and t. In the second step, we will show that (here we may assume $d = 1$), with $\psi_s(x, y) := \psi_s(y) - \psi_s(x)$, one has the bound

$$\sup_{x,y \in [0,2\pi]} \mathbf{E}\left[|\psi_s(x,y) - \psi_t(x,y)|^2 \right] = \mathrm{O}\big(|t - s|^{1/2} \big).$$

The existence of a continuous (and even Hölder) modification is then a consequence of the classical Kolmogorov criterion.

For the first step, we write $X = \big(\psi_s^1(\cdot), \ldots, \psi_s^d(\cdot) \big)$ and $Y = \big(\psi_t^1(\cdot), \ldots, \psi_t^d(\cdot) \big)$. Note that one has independence of $\big(X^i, Y^i \big)$ with $\big(X^j, Y^j \big)$ for $i \neq j$. We have to verify finite 1-variation (in the 2D sense) of the covariance of (X, Y). In view of Proposition 12.31, it remains to establish finite 1-variation of

$$(x, y) \mapsto R_{(X^1, Y^1)}(x, y) = \mathbf{E}\big[\psi_s^1(x) \psi_t^1(y) \big] = \sum_{k \in \mathbf{Z}} e_k(x) e_k(y) \mathbf{E}\big(Y_s^k Y_t^k \big)$$

$$= \frac{\sigma^2}{4\pi} \sum_{k \in \mathbf{Z}} \frac{\cos\big(k(x - y) \big)}{\lambda + k^2} e^{-(\lambda + k^2)|t - s| \cdot} =: R_\tau(x, y).$$

For every $\tau > 0$, exponential decay of the Fourier-modes implies smoothness of R_τ. We claim

$$\|R_\tau\|_{1\text{-var};[u,v]^2} \leq C|v - u| < \infty,$$

uniformly in $\tau \in (0, 1]$ and u, v. To see this, write

$$\|R_\tau\|_{1\text{-var};[u,v]^2} = \int_u^v \int_u^v |\partial_{xy} R_\tau| \, dx \, dy$$

$$\sim \int_u^v \int_u^v \left| \sum k^2 \frac{e^{ik(x-y)}}{\lambda + k^2} e^{-(\lambda + k^2)\tau} \right| dx \, dy$$

$$\sim \int_u^v \int_u^v \left| \sum e^{ik(x-y)} e^{-k^2\tau} \right| dx\, dy$$

$$= \int_u^v \int_u^v p_\tau(x-y) dy\, dx \le |v-u|\,,$$

where we used the trivial estimate $\int_u^v p_\tau(x-y)dy \le \int_0^{2\pi} p_\tau(x-y)dy = 1$. In this expression, p denotes the (positive) transition kernel of the heat semigroup on the torus. The step above, between second and third line, where we effectively set $\lambda = 0$ is harmless. The factor $e^{-\lambda\tau}$ may simply be taken out, and

$$\left| \sum_k \left(1 - \frac{k^2}{\lambda+k^2}\right) e^{ik(x-y)} e^{-k^2\tau} \right| \le \sum_k \left|1 - \frac{k^2}{\lambda+k^2}\right| = \sum_k \frac{\lambda}{\lambda+k^2} < \infty\,.$$

After integrating over $[u,v]^2$, we see that the error made above is actually of order $O(|v-u|^2)$. This is more than enough to conclude that

$$\left\| R_{(X^1,Y^1)} \right\|_{1\text{-var};[u,v]^2} \le C|v-u| < \infty\,,$$

uniformly in $\tau \in (0,1]$ and u,v.

We now turn to the second step of our proof. We claim that $\mathbf{E}|\psi_s^1(x,y) - \psi_t^1(x,y)|^2 = O(|t-s|^{1/2})$, uniformly in $x,y \in [0,2\pi]$. Since

$$\left| \psi_s^1(x,y) - \psi_t^1(x,y) \right| \le \left| \psi_s^1(x) - \psi_t^1(x) \right| + \left| \psi_s^1(y) - \psi_t^1(y) \right|\,,$$

the question reduces to a similar bound on $\mathbf{E}|\psi_s^1(x) - \psi_t^1(x)|^2$, uniform in $x \in [0,2\pi]$. This quantity is equal to

$$\mathbf{E}\left[\psi_s^1(x)\psi_s^1(x)\right] - 2\mathbf{E}\left[\psi_s^1(x)\psi_t^1(x)\right] + \mathbf{E}\left[\psi_t^1(x)\psi_t^1(x)\right]$$

$$= \frac{\sigma^2}{4\pi} \sum_{k\in\mathbf{Z}} \frac{2\left(1 - e^{-(\lambda+k^2)|t-s|}\right)}{\lambda+k^2}\,.$$

$$\le \frac{\sigma^2}{4\pi} \sum_{|k|<N} 2|t-s| + 2\frac{\sigma^2}{4\pi} \sum_{k\ge N} \frac{2\left(1 - e^{-(\lambda+k^2)|t-s|}\right)}{\lambda+k^2}\,,$$

where we used that $1 - e^{-cx} \le cx$ for $c,x > 0$ in the first sum. We then take $N \sim |t-s|^{-1/2}$, so that the first sum is of order $O(|t-s|^{1/2})$. For the second sum, we use the trivial bound $1 - e^{-(\lambda+k^2)|t-s|} \le 1$. It then suffices to note that

$$\sum_{k\ge N} \frac{1}{\lambda+k^2} \le \sum_{k\ge N} \frac{1}{k^2} = O(1/N) = O(|t-s|^{1/2})\,,$$

which completes the proof. $\square$

Remark 12.34. The final estimate in the above proof, namely

$$\mathbf{E}\left|\psi_s^1(x) - \psi_t^1(x)\right|^2 = O\left(|t-s|^{1/2}\right),$$

also implies "almost $\frac{1}{4}$-Hölder" temporal regularity of the stochastic heat equation.

12.4 Exercises

Exercise 12.1 (From [DFS17]) *a) Assume $W \in C^1$. Show that the Feynman–Kac (or equivalently viscosity) solution to (12.11) is an analytically weak solution in the sense of (12.21) with $d\mathbf{W}$ replaced by $\dot{W}\,dt$.*
b) Assume now $\mathbf{W} = (W, \mathbb{W}) \in \mathscr{C}_g^{0,\alpha}$. Show that $(Y, Y') \in \mathscr{D}_W^{2\alpha}$.
c) Show that the Feynman–Kac solution constructed in Theorem 12.11 is an analytically weak solution in the sense of (12.21).

Exercise 12.2 (From [CDFO13]) *A crucial role in the proof of Theorem 12.11 was played by a hybrid Itô-rough differential equation of the form*

$$dX_t = \sigma(X_t)dB + \beta(X_t)d\mathbf{W},\tag{12.43}$$

ultimately solved as (random) rough differential equation, subject to $\sigma, \beta \in C_b^3$. Give an alternative construction to the hybrid equation based on flow decomposition. That is, use the flow associated to the RDE $dY = \beta(Y)d\mathbf{W}$ and transform (12.43) into a bona fide Itô differential equation.

Hint: *When $\mathbf{W}$ is replaced by a C^1 path W^ε this is a straightforward computation. Use the stability of RDE flows, combined with stability results for Itô SDEs to conclude. Specify the regularity requirements on σ, β.*

Exercise 12.3 (Robust filtering, [CDFO13]) *Consider a pair of processes (X, Y) with dynamics*

$$dX_t = V_0(X_t, Y_t)dt + \sum_k Z_k(X_t, Y_t)dW_t^k + \sum_j V_j(X_t, Y_t)dB_t^j,\tag{12.44}$$

$$dY_t = h(X_t, Y_t)dt + dW_t,\tag{12.45}$$

with $X_0 \in L^\infty$ and $Y_0 = 0$. For simplicity, assume coefficients $V_0, V_1, \ldots, V_{d_B}$: $\mathbf{R}^{d_X + d_Y} \to \mathbf{R}^{d_X}$, $Z_1, \ldots, Z_{d_Y}$: $\mathbf{R}^{d_X + d_Y} \to \mathbf{R}^{d_X}$ and $h = (h^1, \ldots, h^{d_Y})$: $\mathbf{R}^{d_X + d_Y} \to \mathbf{R}^{d_Y}$ to be bounded with bounded derivatives of all orders; W and B are independent Brownian motions of the correct dimension. We now interpret X as a signal and Y as noisy and incomplete observation. The filtering problem consists in computing the conditional distribution of the unobserved component X, given the observation Y. Equivalently, one is interested in computing

$$\pi_t(g) = \mathbf{E}[g(X_t, Y_t)|\mathcal{Y}_t],$$

where $\mathcal{Y}_t$ *is the observation filtration and g is a suitably chosen test function. Measure theory tells us that there exists a Borel-measurable map* $\theta_t^g : C([0,t],\mathbf{R}^{d_Y}) \to \mathbf{R}$, *such that a.s.* $\pi_t(g) = \theta_t^g(Y)$ *where we consider* $Y = Y(\omega)$ *as a* $C([0,t],\mathbf{R}^{d_Y})$-*valued random variable. Note that θ_t^g is not uniquely determined (after all, modifications on null sets are always possible). On the other hand, there is obvious interest to have a robust filter, in the sense of having a* continuous *version of θ_t^g, so that close observations lead to nearby conclusions about the signal.*

a) *Give an example showing that, in general, θ_t^g does not admit a continuous version.*

b) *Let $\alpha \in (1/2, 1/3)$. Show that there exists a continuous map on rough path space*

$$\Theta_t^g : \mathscr{C}_g^{0,\alpha}([0,t],\mathbf{R}^{d_Y}) \to \mathbf{R},$$

such that a.s.

$$\pi_t(g) = \Theta_t^g(\mathbf{Y}), \tag{12.46}$$

where $\mathbf{Y}$ is the random geometric rough path obtained from Y by iterated Stratonovich integration.

Hint: *You may use the "Kallianpur–Striebel formula", a standard result in filtering theory which asserts that*

$$\pi_t(g) = \frac{p_t(g)}{p_t(1)}, \quad p_t(g) := \mathbf{E}_0[g(X_t, Y_t)v_t|\mathcal{Y}_t]$$

where

$$\left.\frac{d\mathbf{P}_0}{d\mathbf{P}}\right|_{\mathcal{F}_t} = \exp\left(-\sum_i \int_0^t h^i(X_s, Y_s)dW_s^i - \frac{1}{2}\int_0^t \|h(X_s, Y_s)\|^2 ds\right)$$

and $v = \{v_t, t > 0\}$ is defined as the right-hand side above with $-W$ replaced by Y.

Exercise 12.4 *Show almost sure "$(\frac{1}{4} - \varepsilon)$-Hölder" temporal regularity of $\psi = \psi_t(x; \omega)$, solution to the stochastic heat equation. Show that, for fixed x, $\psi_t(x; \omega)$ is not a semimartingale.*

Exercise 12.5 (Spatial Itô–Stratonovich correction [HM12]) *Writing $\mathbf{T}$ for the interval $[0, 2\pi]$ with periodic boundary, let us say that*

$$u = u(t, x; \omega) : [0, T] \times \mathbf{T} \times \Omega \to \mathbf{R}$$

is a (analytically) weak solution to

$$\partial_t u = \partial_{xx} u - u + \frac{1}{2}\partial_x(u^2) + \xi, \tag{$\star$}$$

if and only if $u = v + \psi$ where ψ is the stationary solution to $\partial_t \psi = \partial_{xx}\psi - \psi + \xi$ and, for all test functions $\varphi \in C^\infty(\mathbf{T})$,

$$\partial_t \langle v, \varphi \rangle = \langle v, \partial_{xx}\varphi \rangle - \langle v, \varphi \rangle - \left\langle \frac{1}{2}u^2, \partial_x\varphi \right\rangle .$$

a) Replace $\frac{1}{2}\partial_x(u^2)$ in $(\star)$ by a (spatially right) finite-difference approximation,

$$\frac{1}{2}\frac{u(.+\varepsilon)^2 - u^2}{\varepsilon};$$

write u^ε for a solution to the resulting equation. Assume $u^\varepsilon \to u$ locally uniformly in probability. Show that u is a solution to $(\star)$.

b) At least formally, $\partial_x\left(\frac{1}{2}u^2\right) = u\partial_x u$ in $(\star)$, which suggests an alternative finite difference approximation, namely,

$$u\frac{(u(.+\varepsilon) - u)}{\varepsilon};$$

Assume $v^\varepsilon = u^\varepsilon - \psi \to v := u - \psi$ and its first (spatial) derivatives converge locally uniformly in probability. Show that u is an analytically weak solution to the perturbed equation

$$\partial_t u = \partial_{xx} u + \frac{1}{2}\partial_x(u^2) + C + \xi$$

with $C \neq 0$. Determine the value of C. **Hint:** *Use Exercise 10.6.*

Solution.   a) By switching to suitable subsequences, we may assume $u^\varepsilon \to u$ locally uniformly with probability one. Write $D_{\varepsilon,l}, D_{\varepsilon,r}$ for a discrete (left, right) finite difference approximation. Note

$$\left\langle D_{\varepsilon,r}\left(\frac{1}{2}u^2\right), \varphi \right\rangle = -\left\langle \frac{1}{2}u^2, D_{\varepsilon,l}\varphi \right\rangle \to -\left\langle \frac{1}{2}u^2, \partial_x\varphi \right\rangle.$$

Given that $v^\varepsilon = u^\varepsilon - \psi \to v := u - \psi$ locally uniform it then suffices to pass to the limit in the (integral formulation) of

$$\partial_t \langle v^\varepsilon, \varphi \rangle = \langle v^\varepsilon, \partial_{xx}\varphi \rangle - \langle v^\varepsilon, \varphi \rangle + \left\langle \frac{1}{2}u^2, D_{\varepsilon,l}\varphi \right\rangle.$$

b) We note

$$D_{\varepsilon,r}\left(\frac{1}{2}u^2\right) = \frac{1}{2}\frac{u(.+\varepsilon)^2 - u^2}{\varepsilon} = \frac{(u(.+\varepsilon) + u)}{2}\frac{(u(.+\varepsilon) - u)}{\varepsilon}$$

$$= u\frac{(u(.+\varepsilon) - u)}{\varepsilon} + \frac{1}{2\varepsilon}(u(.+\varepsilon) - u)^2 .$$

It follows that

$$\partial_t \langle v^\varepsilon, \varphi \rangle = \langle v^\varepsilon, \partial_{xx}\varphi \rangle - \langle v^\varepsilon, \varphi \rangle + \left\langle u^\varepsilon \frac{(u^\varepsilon(.+\varepsilon) - u^\varepsilon)}{\varepsilon}, \varphi \right\rangle .$$

$$= \langle v^\varepsilon, \partial_{xx}\varphi \rangle - \langle v^\varepsilon, \varphi \rangle$$
$$- \left\langle \frac{1}{2}(u^\varepsilon)^2, D_{\varepsilon,l}\varphi \right\rangle - \left\langle \frac{1}{2\varepsilon}(u^\varepsilon(.+\varepsilon) - u^\varepsilon)^2, \varphi \right\rangle .$$

In order to pass to the $\varepsilon \to 0$ limit, we must understand the final "quadratic variation" term. By assumption v^ε are of class $\mathcal{C}^1$, uniformly in ε. Hence

$$[u^\varepsilon(.+\varepsilon) - u^\varepsilon] = \psi(.+\varepsilon) - \psi + v^\varepsilon(.+\varepsilon) - v^\varepsilon$$
$$= \psi(.+\varepsilon) - \psi + O(\varepsilon)$$

and so, with osc $(\psi; \varepsilon)O(1) + O(\varepsilon) = o(1)$ as $\varepsilon \to 0$,

$$\frac{1}{2\varepsilon}(u^\varepsilon(.+\varepsilon) - u^\varepsilon)^2 = \frac{1}{2\varepsilon}(\psi(.+\varepsilon) - \psi)^2 + o(1)$$

we have

$$\left\langle \frac{1}{2\varepsilon}(u^\varepsilon(.+\varepsilon) - u^\varepsilon)^2, \varphi \right\rangle = \left\langle \frac{1}{2\varepsilon}(\psi(.+\varepsilon) - \psi)^2, \varphi \right\rangle + o(1) .$$

From Lemma 12.30 we know that

$$\mathbf{E}[\psi_{x,x+\varepsilon}^2] = 2(K(0) - K(\varepsilon)) = -2K'(0)\varepsilon + o(\varepsilon) = C\varepsilon + o(\varepsilon) .$$

Since $K(u) = \frac{\cosh(u-\pi)}{4\sinh(\pi)}$, we have $C = -2K'(0) = \frac{1}{2}$, and it follows from Exercise 10.6 that

$$\left\langle \frac{1}{2\varepsilon}(\psi(.+\varepsilon) - \psi)^2, \varphi \right\rangle = \frac{1}{2} \int \varphi(x) \frac{\psi_{x,x+\varepsilon}^2}{\varepsilon} dx$$
$$\to \frac{1}{2} \int \varphi(x) C dx = \left\langle \frac{1}{4}, \varphi \right\rangle ,$$

where the convergence takes place in probability. It follows that u is a solution (in the above analytically weak sense) of

$$\partial_t u = \partial_{xx}u - u + \frac{1}{2}\partial_x(u^2) + \frac{1}{4} + \xi .$$

12.5 Comments

Section 12.1: The explicit solution of the rough transport equation in Section 12.1.1 is a (geometric) rough-pathification of the classical method of characteristics and Kunita's (Stratonovich) stochastic version thereof [Kun84], first pointed out in [CF09].

Our intrinsic definition of (regular vs. weak / measure-valued) RPDE solution is essentially taken from Diehl et al. [DFS17] and Bellingeri et al. [BDFT20], which also treats the low regularity case. Bailleul–Gubinelli [BG17] suggest an abstract framework of (unbounded) rough drivers in which $(\Gamma[\cdot]W_{s,t}, \Gamma^2[\cdot]\mathbb{W}_{s,t})$, with Γ as in (12.2), are viewed as (s, t)-indexed familiy of *unbounded operators*

$$\mathbf{A}_{s,t} = (A_{s,t}, \mathbb{A}_{s,t})$$

on a suitable scale of Banach spaces, which satisfy an operator Chen relation and then the (operator) geometricity condition $A_{s,t}^2/2 = \mathbb{A}_{s,t}$. The rough transport equation, say $du_t = \Gamma u_t d\mathbf{W}$ if written as initial value problem, then fits into an abstract rough linear equation of the form

$$du_t = \mathbf{A}(dt)u_t \ .$$

An analytically weak formulation (somewhat similar to our Section 12.1.2, but now formulated via Banach duals) then allows them to obtain existence and uniqueness under $\mathcal{C}_b^3$ assumptions on the vector fields, at the price of a doubling of variables argument related in the spirit to Di Perna–Lions [DL89].

Entropy solutions to scalar conservation laws with rough forcing are studied by Friz–Gess [FG16b]; in [HNS20] Hocquet et al. study a generalized Burgers equation with rough transport noise. A different class of *rough scalar conservation laws*, closely related to rough transport, is given by

$$du + \mathrm{div}_x(A(x, u))d\mathbf{W} = 0 \ , \quad u = u_0 \ , \tag{12.47}$$

where $u : [0, T] \times \mathbf{R}^n \to \mathbf{R}$, with $A = (A_j^i : 1 \le i \le n, 1 \le j \le d)$ sufficiently smooth, matrix valued functions and $\mathbf{W}$ a geometric Hölder rough path over $\mathbf{R}^d$. (The case of linear $A(x, u) = f(x)u$ is precisely the rough continuity equation treated in Section 12.1.2.)

Such equations were studied from a "pathwise" point of view (essentially possible when $A = A(u)$ has no x-dependence or when $d = 1$) in Lions, Perthame and Souganidis [LPS13] and [LPS14], followed by Gess–Souganidis [GS15] who treat the general case (12.47) and then Hofmanová [Hof16]. When $d\mathbf{W} = \dot{W}dt$, this falls into the well established theories of entropy solutions and kinetic solutions. The latter formulation related to rough transport as follows. With

$$\chi(x, \xi, t) := \chi(u(x, t), \xi) := \begin{cases} +1 & \text{if } 0 \le \xi \le u(x, t), \\ -1 & \text{if } u(x, t) \le \xi \le 0, \\ 0 & \text{otherwise,} \end{cases} \tag{12.48}$$

one can rewrite (12.47) in its (formal) kinetic form: for $T > 0$ fixed,

$$d_t\chi + \big(\partial_u A(x, \xi) \cdot D_x\chi - \mathrm{div}_x A(x, \xi)\partial_\xi\chi\big)d\mathbf{W} = (\partial_\xi m)dt \ , \tag{12.49}$$

on $\mathbf{R}^n \times \mathbf{R} \times (0, T]$ with initial data $\chi(\cdot, *, 0) = \chi(u_0(\cdot), *)$ where $\mathrm{div}_x A = (\mathrm{div}_x A_1, \ldots, \mathrm{div}_x A_d)$ and m is a bounded nonnegative measure on $\mathbf{R}^n \times \mathbf{R} \times [0, T]$, known as *defect measure*, which is part of the solution. The definition of rough kinetic solution [GS15] is then given as analytically weak solution of (12.49), with test functions obtained as (spatially) regular solutions to an auxilary rough transport equation, similar in spirit to Section 12.1.2. See also Gess et al. [GPS16] for a semi-discretisation. The idea of test functions with (here: temporal) structure tailor-made to a realisation of the noise (a.k.a. rough path) is central to RPDEs. A well-posedness result for rough kinetic solutions was also obtained by Deya et al. [DGHT19b], in an extended setting of RPDEs with (unbounded) rough drivers, of the form

$$du_t = \mu(dt) + \mathbf{A}(dt)u_t \,,$$

where the abstract assumptions on the drift term μ are seen to accommodate the defect measure. *Rough Hamilton–Jacobi equations* are of the form

$$du + H(Du, x)d\mathbf{W} = 0 \,, \qquad u(0, \cdot) = u_0 \,, \qquad (12.50)$$

on $(0, T] \times \mathbf{R}^n$, with Hamiltonians $H = (H_1, \ldots, H_d)$. When $d\mathbf{W} = \dot{W}dt$, this falls into the well established theory of viscosity solutions, with intrinsic notion of sub (resp. super) solutions via "touching" test functions $\varphi = \varphi(t, x) \in \mathcal{C}^{1,1}$. Short-time regular solutions via the method of "rough" characteristics then supply the correct class of test functions (depending on the noise realisation modelled by $\mathbf{W}$): when inserted in the equation, they at least formally "eliminate" the rough part, this is basically a local change of the unknown. (A global change of coordinates is sometimes possible, notably in the case of transport noise when $H(p, x)$ is linear in p, cf. Section 12.2.3 below.) These ideas form the basis of Lions–Souganidis' stochastic viscosity theory [LS98a, LS98b, LS00b] which predates most works on rough paths, the resulting "pathwise" theory essentially requires $H = H(p)$ with no x-dependence, or $d = 1$; see also [FGLS17] (x-dependent quadratic Hamiltonian) and [GGLS20] (speed of propagation). In spatial dimension $n = 1$, there is a noteworthy connection with rough conservation laws: if v solves the rough HJ equation $dv + A(\partial_x v, x)d\mathbf{W} = 0$, then, at least formally, $u = \partial_x v$ satisfies the rough conservation law $du + \partial_x(A(u, x))d\mathbf{W} = 0$.

Section 12.2: Linear stochastic partial differential equations go back at least to Krylov–Rozovskii [KR77] and play an important problem in filtering theory (Zakai equation). A Feynman–Kac representation appears in Pardoux [Par79] and Kunita [Kun82]. Kunita also has flow decompositions of SPDE solutions. Caruana–Friz [CF09] implement this in the rough path setting in a framework of classical PDE solutions. The construction of hybrid stochastic / rough differential equations which underlies the "rough" Feynman–Kac approach, Theorem 12.11, is taken from [DOR15] (see also [FHL20]). Diehl et al. [DFS17] establish existence and uniqueness, based on an intrinsic definition for (linear) RPDEs, numerical algorithms are given by Bayer et al. [BBR$^+$18]. Hofmanova–Hocquet [HH18] study (linear) RPDEs from a variational perspective and unbounded rough driver perspective, as does Hofmanová et al. [HLN19] for the Navier–Stokes equation perturbed by rough transport noise.

An extension of Lions, Perthame and Souganidis [LPS13, LPS14] to rough, scalar, degenerate parabolic-hyperbolic equation is given in [GS17].

In the context of Crandall–Ishii–Lions viscosity setting, by nature a theory for second order equations with a maximum principle, *stochastic (pathwise) viscosity solutions* for fully non-linear equations were introduced by Lions–Souganidis [LS98a, LS98b, LS00a, LS00b]. Caruana, Friz and Oberhauser [CFO11] introduce *rough viscosity solutions* by a limiting procedure for classes of nonlinear SPDEs with transport noise; an intrinsic definition (via global transformaion) is given e.g. in [DFO14]. An adaption of the original intrinsic definition of (pathwise) viscosity solutions to fully non-linear equations [LS98a] is given in Seeger [See18b]. Extensions to different noise situations are due to Diehl–Friz, [DF12] and then [FO14]. Nonlinear noise, x-dependent and quadratic in Du is considered by Friz, Gassiat, Lions and Souganidis [FGLS17]. Approximation schemes for (pathwise) viscosity solutions of fully nonlinear problems are studied [See18a].

A nonlinear Feynman–Kac representation (with relations to "rough BSDEs") is given in [DF12]. In a filtering context, a (rough path) robustified Kalianpur–Striebel formula (cf. Exercise 12.3) was given by Crisan, Diehl, Friz and Oberhauser [CDFO13], which is also the first source of hybrid differential equations. At last, we refer to Gubinelli–Tindel, Deya et al. and Teichmann [GT10, DGT12, Tei11] for some other rough path approaches to SPDEs. Theorem 12.18 is essentially due to [GH19], but very closely related to the earlier results of [GT10]. Compared to the latter, we restrict ourselves to finite-dimensional drivers, but allow for a more natural class of nonlinearities thanks to a slightly different use of the various interpolation spaces.

Section 12.3: The construction of a spatial rough path associated to the stochastic heat equation is due to Hairer [Hai11b] and allows to deal with otherwise ill-posed SPDEs of stochastic Burgers type, see also Hairer–Weber [HW13] and Friz, Gess, Gulisashvili, Riedel [FGGR16] for various extensions (including multiplicative noise, and fractional Laplacian / non-periodic boundary respectively). This construction is also an ingredient in one construction for solutions to the KPZ equation, see Hairer [Hai13] and Chapter 15. Exercise 12.5, in the spirit of Föllmer – rather than rough path – integration, is taken from Hairer–Maas [HM12]. Similar results are available for rough SPDEs of type (12.41), see Hairer, Maas and Weber [HMW14], but this is beyond the scope of these notes. Bellingeri [Bel20] uses regularity structures to establish an Itô formula for the stochastic heat equation.

Chapter 13
Introduction to regularity structures

We give a short introduction to the main concepts of the general theory of regularity structures. This theory unifies the theory of (controlled) rough paths with the usual theory of Taylor expansions and allows to treat situations where the underlying space is multidimensional.

13.1 Introduction

While a full exposition of the theory of regularity structures is well beyond the scope of this book, we aim to give a concise overview to most of its concepts and to show how the theory of controlled rough paths fits into it. In most cases, we will only state results in a rather informal way and give some ideas as to how the proofs work, focusing on conceptual rather than technical issues. The only exception is the "reconstruction theorem", Theorem 13.12 below, which is one of the linchpins of the whole theory. Since its proof (or rather a slightly simplified version of it) is relatively concise, we provide a fully self-contained version. For precise statements and complete proofs of most of the results exposed here, we refer to the original article [Hai14b]. See also the review articles [Hai15, Hai14a] for shorter expositions that complement the one given here.

It should be clear by now that a controlled rough path $(Y, Y') \in \mathscr{D}_W^{2\alpha}$ bears a strong resemblance to a differentiable function, with the Gubinelli derivative Y' describing the coefficient in front of a "first-order Taylor expansion" of the type

$$Y_t = Y_s + Y_s' W_{s,t} + \mathrm{O}(|t-s|^{2\alpha}) \,. \tag{13.1}$$

Compare this to the fact that a function $f : \mathbf{R} \to \mathbf{R}$ is of class $\mathcal{C}^\gamma$ with $\gamma \in (k, k+1)$ if for every $s \in \mathbf{R}$ there exist coefficients $f_s^{(1)}, \ldots, f_s^{(k)}$ such that

$$f_t = f_s + \sum_{\ell=1}^k f_s^{(\ell)} (t-s)^\ell + \mathrm{O}(|t-s|^\gamma) \,. \tag{13.2}$$

© Springer Nature Switzerland AG 2020
P. K. Friz, M. Hairer, *A Course on Rough Paths*, Universitext,
https://doi.org/10.1007/978-3-030-41556-3_13

Of course, $f_s^{(\ell)}$ is nothing but the ℓth derivative of f at the point s, divided by $\ell!$. In this sense, one should really think of a controlled rough path $(Y, Y') \in \mathscr{D}_W^{2\alpha}$ as a 2α-Hölder continuous function, but with respect to a "model" given by W, rather than the usual Taylor polynomials. This formal analogy between controlled rough paths and Taylor expansions suggests that it might be fruitful to systematically investigate what are the "right" objects that could possibly take the place of Taylor polynomials, while still retaining many of their nice properties.

13.2 Definition of a regularity structure and first examples

The first step in such an endeavour is to set up an algebraic structure reflecting the properties of Taylor expansions. First of all, such a structure should contain a vector space T that will contain the coefficients of our expansion. It is natural to assume that T has a graded structure: $T = \bigoplus_{\alpha \in A} T_\alpha$, for some set A of possible "homogeneities". For example, in the case of the usual Taylor expansion (13.2), it is natural to take for A the set of natural numbers and to have T_ℓ contain the coefficients corresponding to the derivatives of order ℓ. In the case of controlled rough paths however, it is natural to take $A = \{0, \alpha\}$, to have again T_0 contain the value of the function Y at any time s, and to have T_α contain the Gubinelli derivative Y_s'. This reflects the fact that the "monomial" $t \mapsto X_{s,t}$ only vanishes at order α near $t = s$, while the usual monomials $t \mapsto (t - s)^\ell$ vanish at integer order ℓ.

This however isn't the full algebraic structure describing Taylor-like expansions. Indeed, one of the characteristics of Taylor expansions is that an expansion around some point x_0 can be re-expanded around any other point x_1 by writing

$$(x - x_0)^m = \sum_{k + \ell = m} \frac{m!}{k! \ell!} (x_1 - x_0)^k \cdot (x - x_1)^\ell . \tag{13.3}$$

(In the case when $x \in \mathbf{R}^d$, k, ℓ and m denote multi-indices and $k! = k_1! \ldots k_d!$.) Somewhat similarly, in the case of controlled rough paths, we have the (rather trivial) identity

$$W_{s_0, t} = W_{s_0, s_1} \cdot 1 + 1 \cdot W_{s_1, t} . \tag{13.4}$$

What is a natural abstraction of this fact? In terms of the coefficients of a "Taylor expansion", the operation of reexpanding around a different point is ultimately just a linear operation from $\Gamma \colon T \to T$, where the precise value of the map Γ depends on the starting point x_0, the endpoint x_1, and possibly also on the details of the particular "model" that we are considering. In view of the above examples, it is natural to impose furthermore that Γ has the property that if $\tau \in T_\alpha$, then $\Gamma \tau - \tau \in \bigoplus_{\beta < \alpha} T_\beta$. In other words, when reexpanding a homogeneous monomial around a different point, the leading order coefficient remains the same, but lower order monomials may appear.

These heuristic considerations can be summarised in the following definition of an abstract object we call a *regularity structure*:

Definition 13.1. A *regularity structure* $\mathscr{T} = (T, G)$ consists of the following elements:

- A *structure space* given as graded vector space $T = \bigoplus_{\alpha \in A} T_\alpha$ where each T_α is a Banach space, with *index set* $A \subset \mathbf{R}$ bounded from below and locally finite.[1] Elements of T_α are said to have *degree* α and we write $\deg \tau = \alpha$ for $\tau \in T_\alpha$. Given $\tau \in T$, we will write $\|\tau\|_\alpha$ for the norm of its component in T_α.
- A *structure group* G of continuous linear operators acting on T such that, for every $\Gamma \in G$, every $\alpha \in A$, and every $\tau_\alpha \in T_\alpha$, one has

$$\Gamma \tau_\alpha - \tau_\alpha \in T_{<\alpha} \stackrel{\text{def}}{=} \bigoplus_{\beta < \alpha} T_\beta \ . \tag{13.5}$$

A *sector* V of $\mathscr{T}$ is a linear subspace $V = \bigoplus_{\alpha \in A} V_\alpha \subset T$, with closed linear subspaces $V_\alpha \subset T_\alpha$, invariant under G, such that $(V, G|_V)$ is a regularity structure in its own right.

Remark 13.2. In principle, the index set A can be infinite. By analogy with the polynomials, it is then natural to interpret T as the set of all formal series of the form $\sum_{\alpha \in A} \tau_\alpha$, where only finitely many of the τ_α's are non-zero. This also dovetails nicely with the particular form of elements in G. In practice however we will only ever work with finite subsets of A so that the precise topology on T does not matter as long as each of the T_α is finite-dimensional, which is the case in all of the examples we will consider here.

The space T should be thought of as consisting of "abstract" Taylor expansions (or "jets"), where each element of T_α would correspond to a "homogeneous polynomial of degree α" (this will be made in combination with the definition of a model in Definition 13.5 below). To avoid confusion between "abstract" elements of T and "concrete" associated functions (or distributions), we will use colour to denote elements of T, e.g. τ. Typically, T will be generated (as a free vector space) by a set of "basis symbols", so that T consists of all formal (finite) linear combination obtained from regarding these symbols as basis vectors. Given basis symbols / vectors $\tau_1, \tau_2, \ldots$ we indicate this by

$$T = \langle \tau_1, \tau_2, \ldots \rangle. \tag{13.6}$$

Important convention: basis symbols will always by listed in order of increasing homogeneities. That is, $\tau_i \in T_{\alpha_i}$ with $\alpha_1 \leq \alpha_2 \leq \ldots$ in (13.6). We now turn to some first examples of regularity structures.

---

[1] In [Hai14b], T was called *model space*, somewhat in clash with the space of *models*.

13.2.1 The polynomial structure

We start with two simple special cases followed by the general polynomial structure. Fix $\gamma \in (0, 1)$ and consider a real-valued function belonging to the Hölder space of exponent γ, say $f \in C^\gamma$. In other words, $f : \mathbf{R} \to \mathbf{R}$, and $|f_x - f_y| \lesssim |y - x|^\gamma$ uniformly for x, y on compacts. The trivial regularity structure

$$T = T_0 = \langle \mathbf{1} \rangle \cong \mathbf{R} , \qquad G = \{\mathrm{Id}\} ,$$

allows us to interpret the function f as a T-valued map

$$x \mapsto f(x) := f_x \mathbf{1}.$$

Consider next a real-valued function $f : \mathbf{R} \to \mathbf{R}$ of class $C^{2+\gamma}$, with $\gamma \in (0, 1)$. By this we mean that continuous derivatives Df and $D^2 f$ exist, with $D^2 f$ locally γ-Hölder continuous. The minimal regularity structure allowing to capture the fact that $f \in C^{2+\gamma}$ is

$$T = T_0 \oplus T_1 \oplus T_2 = \langle \mathbf{1}, X, X^2 \rangle \cong \mathbf{R}^3 ,$$

with structure group $G = \{\Gamma_h \in \mathcal{L}(T, T) : h \in (\mathbf{R}, +)\}$ where Γ_h is given, with respect to the ordered basis $\mathbf{1}, X, X^2$, by the matrix

$$\Gamma_h \cong \begin{pmatrix} 1 & h & h^2 \\ 0 & 1 & 2h \\ 0 & 0 & 1 \end{pmatrix}.$$

In other words,

$$\Gamma_h \mathbf{1} = \mathbf{1} , \quad \Gamma_h X = X + h\mathbf{1} , \quad \Gamma_h X^2 = (X + h\mathbf{1})^2 ,$$

with the obvious abuse of notation in the last expression.

Note that $\Gamma_g \circ \Gamma_h = \Gamma_{g+h}$, so that G inherits its group structure from $(\mathbf{R}, +)$. Moreover, the triangular form, with ones on the diagonal, expresses exactly the requirement (13.5). This structure allows to represent the function f and its first two derivatives as a truncated Taylor series, namely as the T-valued map

$$x \mapsto f(x) := f_x \mathbf{1} + Df_x X + \frac{1}{2} D^2 f_x X^2.$$

It is now an easy matter to generalise the above considerations to general Hölder maps of several variables, say $f : \mathbf{R}^d \to \mathbf{R}$ in the Hölder space $C^{n+\gamma}$, which is defined by the obvious generalisation of (13.2) to functions on $\mathbf{R}^d$. In this case, we would take T to be the space of polynomials of degree at most n in d commuting indeterminates $X_1, \ldots, X_d$. This motivates the following definition.

Definition 13.3. The *polynomial regularity structure* on $\mathbf{R}^d$ is given by

- $T = \mathbf{R}[X_1, \ldots, X_d]$ is the space of real polynomials in d commuting indeterminates and T_α is given by the homogeneous polynomials of degree $\alpha \in \mathbf{N}$.
- The structure group $G \sim (\mathbf{R}^d, +)$ acts on T via

$$\Gamma_h P(X) = P(X + h\mathbf{1}), \qquad h \in \mathbf{R}^d ,$$

for any polynomial P.

Given an arbitrary multi-index $k = (k_1, \ldots, k_d)$, we write X^k as a shorthand for $X_1^{k_1} \cdots X_d^{k_d}$, and we write $|k| = k_1 + \cdots + k_d$. With this notation, for any $\alpha \in A = \mathbf{N}$,

$$T_\alpha = \langle X^k : |k| = \alpha \rangle. \tag{13.7}$$

Note that $T_{\leq \alpha} = T_0 \oplus T_1 \oplus \cdots \oplus T_\alpha$, i.e. the space of polynomials of degree at most α, any $\alpha \in A = \mathbf{N}$, is a sector of the polynomial regularity structure.

13.2.2 The rough path structure

We start again from simple examples. What structure would be appropriate for Young integration? Fix $\alpha \in (0, 1)$ and consider the problem of integrating a (continuous) path Y against a scalar $W \in \mathcal{C}^\alpha$. In the case of smooth W, the indefinite integral $Z = \int Y \, dW$ exists in Riemann–Stieltjes' sense and one has $\dot{Z} = Y \dot{W}$. In general, $\dot{W}$ only exists as a distribution, more precisely an element of the negative Hölder space $\mathcal{C}^{\alpha-1}$. A regularity structure allowing to describe this situation is given by

$$T = T_{\alpha-1} \oplus T_0 = \langle \dot{W} \rangle \oplus \langle \mathbf{1} \rangle \cong \mathbf{R}^2 , \qquad G = \{\mathrm{Id}\} . \tag{13.8}$$

The potentially ill-defined product $\dot{Z} = Y \dot{W}$ can now be replaced by the perfectly well-defined T-valued map

$$s \mapsto \dot{Z}(s) := Y_s \dot{W} .$$

We shall see later how $\dot{Z}$ gives rise to $\dot{Z}$, the distributional derivative of the indefinite Young integral $\int Y \, dW$, provided of that Y is sufficiently regular, namely $Y \in \mathcal{C}^\beta$ with $\alpha + \beta > 1$.

Let us next consider the "task" of representing a controlled rough path in a suitable regularity structure. More precisely, consider $\alpha \in (1/3, 1/2]$, a path $W \in \mathcal{C}^\alpha$ with values in $\mathbf{R}$, say, and $(Y, Y') \in \mathscr{D}_W^{2\alpha}$ so that

$$Y_t \approx Y_s + Y_s' W_{s,t} . \tag{13.9}$$

The right-hand side above is some sort of Taylor expansion, based on $W \in \mathcal{C}^\alpha$, which describes Y well near the (time) point s. We want to formalise this by attaching to each time s the "jet"

$$Y(s) := Y_s \mathbf{1} + Y_s' W .$$

Performing the substitution $\mathbf{1} \mapsto 1$, $W \mapsto W_{s,\cdot}$ gets us back to the right-hand side of (13.9). This suggests to define the following regularity structure

$$T = T_0 \oplus T_\alpha = \langle \mathbf{1} \rangle \oplus \langle W \rangle \cong \mathbf{R}^2 \,,$$

with structure group $G = \{ \Gamma_h \in \mathcal{L}(T,T) : h \in (\mathbf{R},+) \}$ where Γ_h acts as

$$\Gamma_h \mathbf{1} = \mathbf{1} \,, \qquad \Gamma_h W = W + h\mathbf{1} \,.$$

The regularity structure relevant for rough integration is essentially a combination of the two previous ones. Let $\mathbf{W} = (W, \mathbb{W}) \in \mathscr{C}^\alpha$ and $(Y, Y') \in \mathscr{D}_W^{2\alpha}$ and consider the rough integral $Z := \int Y \, d\mathbf{W}$. Since, for $s \approx t$, we have

$$Z_{s,t} = \int_s^t Y \, d\mathbf{W} \approx Y_s W_{s,t} + Y'_s \mathbb{W}_{s,t} \,,$$

this suggests (rather informally at this stage), that in the vicinity of any fixed time s, the distributional derivative of Z should have an expansion of the type

$$\dot{Z} \approx Y_s \dot{W} + Y'_s \dot{\mathbb{W}}_s \,, \tag{13.10}$$

where $\dot{W} := \partial_t W_t$ and $\dot{\mathbb{W}}_s := \partial_t \mathbb{W}_{s,t}$ are distributional derivatives. This suggests to attach the following "jet" at each point s,

$$\dot{Z}(s) := Y_s \dot{W} + Y'_s \dot{\mathbb{W}} \,. \tag{13.11}$$

The case of multi-component rough paths just needs more basis vectors $\dot{W}^i$, $\dot{\mathbb{W}}^{j,k}$, W^l (with $1 \leq i,j,k,l \leq e$). This suggests the following definition.

Definition 13.4. Let $\alpha \in (1/3, 1/2]$. The *regularity structure for α-Hölder rough paths (over $\mathbf{R}^e$)* is given by $T = T_{\alpha-1} \oplus T_{2\alpha-1} \oplus T_0 \oplus T_\alpha \cong \mathbf{R}^{e+e^2+1+e}$ with

$$T_0 = \langle \mathbf{1} \rangle \,, \qquad\qquad T_\alpha = \langle W^1, \dots, W^e \rangle \,,$$
$$T_{\alpha-1} = \langle \dot{W}^1, \dots, \dot{W}^e \rangle \,, \qquad T_{2\alpha-1} = \langle \dot{\mathbb{W}}^{ij} : 1 \leq i,j \leq e \rangle \,,$$

and structure group $G \sim (\mathbf{R}^e, +)$ acting on T by

$$\begin{aligned} \Gamma_h \mathbf{1} &= \mathbf{1} \,, & \Gamma_h W^i &= W^i + h^i \mathbf{1} \,, \\ \Gamma_h \dot{W}^i &= \dot{W}^i \,, & \Gamma_h \dot{\mathbb{W}}^{ij} &= \dot{\mathbb{W}}^{ij} + h^i \dot{W}^j \,. \end{aligned} \tag{13.12}$$

It will be seen later in Proposition 13.21 that in this framework the function $\dot{Z}$ defined in (13.11) does indeed give rise naturally to $\dot{Z}$, the distributional derivative of the indefinite rough integral $\int Y \, d\mathbf{W}$.

In a Brownian (rough path) context, one has Hölder regularity with exponent $\alpha = 1/2 - \kappa$, for arbitrarily small $\kappa > 0$. The above index set A, relevant for a "regularity structure view" on stochastic integration, then becomes $A = \{ -\frac{1}{2} - \kappa, -2\kappa, 0, \frac{1}{2} - \kappa \}$, which, in abusive but convenient notation, we write as

$$A = \left\{ -\frac{1}{2}^{-}, 0^{-}, 0, \frac{1}{2}^{-} \right\} .$$

Index sets of this form ("half-integers^{-}") will also be typical in later SPDE situations driven by spatial or space-time white noise.

13.3 Definition of a model and first examples

At this stage, a regularity structure is a completely abstract object. It only becomes useful when endowed with a *model*, which is a concrete way of associating to any $\tau \in T$ and $x \in \mathbf{R}^d$, the actual "Taylor polynomial based at x" represented by τ. Furthermore, we want elements $\tau \in T_\alpha$ to represent functions (or possibly distributions!) that "vanish at order α" around the given point x, thereby justifying our terminology of calling α a degree.

Since we would like to allow A to contain negative values and therefore allow elements in T to represent actual distributions, we need a suitable notion of "vanishing at order α". We achieve this by considering the size of our distributions, when tested against test functions that are localised around the given point x_0. Given a test function φ on $\mathbf{R}^d$, we write φ_x^λ as a shorthand for

$$\varphi_x^\lambda(y) = \lambda^{-d}\varphi\big(\lambda^{-1}(y - x)\big) .$$

Given $r \in \mathbf{N}$, we also denote by $\mathcal{B}_r$ the set of all *smooth* test functions $\varphi \colon \mathbf{R}^d \to \mathbf{R}$ such that $\varphi \in \mathcal{C}^r$ with $\|\varphi\|_{\mathcal{C}^r} \leq 1$ that are furthermore supported in the unit ball around the origin; clearly $\mathcal{B}_r \subset \mathcal{D}(\mathbf{R}^d)$, the test function space for $\mathcal{D}'(\mathbf{R}^d)$, the space of distributions on $\mathbf{R}^d$. With these notations, our definition of a model for a given regularity structure $\mathscr{T}$ is as follows.

Definition 13.5. Given a regularity structure $\mathscr{T} = (T, G)$ and an integer $d \geq 1$, a *model* $\mathrm{M} = (\Pi, \Gamma)$ for $\mathscr{T}$ on $\mathbf{R}^d$ consists of maps

$$\begin{aligned} \Pi \colon \mathbf{R}^d &\to \mathcal{L}\big(T, \mathcal{D}'(\mathbf{R}^d)\big) & \Gamma \colon \mathbf{R}^d \times \mathbf{R}^d &\to G \\ x &\mapsto \Pi_x & (x, y) &\mapsto \Gamma_{xy} \end{aligned}$$

such that $\Gamma_{xy}\Gamma_{yz} = \Gamma_{xz}$ and $\Pi_x\Gamma_{xy} = \Pi_y$. Write r for the smallest integer such that $r > |\min A| \geq 0$ and impose that for every compact set $\mathfrak{K} \subset \mathbf{R}^d$ and every $\gamma > 0$, there exists a constant $C = C(\mathfrak{K}, \gamma)$ such that the bounds

$$\big|(\Pi_x\tau)(\varphi_x^\lambda)\big| \leq C\lambda^\alpha\|\tau\|_\alpha , \qquad \|\Gamma_{xy}\tau\|_\beta \leq C|x - y|^{\alpha-\beta}\|\tau\|_\alpha , \qquad (13.13)$$

hold uniformly over $x, y \in \mathfrak{K}$, $\lambda \in (0, 1]$, $\varphi \in \mathcal{B}_r$, $\tau \in T_\alpha$ with $\alpha \leq \gamma$ and $\beta < \alpha$.

We then call Π the *realisation map*, since $\Pi_x\tau$ realises an element $\tau \in T$ as a distribution, and Γ the *reexpansion map*.

One very important remark is that the space $\mathcal{M}$ of all models for a given regularity structure is *not* a linear space. However, it can be viewed as a closed subset (determined by the nonlinear constraints $\Gamma_{xy} \in G$, $\Gamma_{xy}\Gamma_{yz} = \Gamma_{xz}$, and $\Pi_y = \Pi_x\Gamma_{xy}$) of the linear space with seminorms (indexed by the compact set $\mathfrak{K}$ and the upper bound γ) given by the smallest constant C in (13.13). In particular, there is a natural collection of "distances" between models (Π, Γ) and $(\bar{\Pi}, \bar{\Gamma})$ given by the smallest constant C in (13.13), when replacing Π_x by $\Pi_x - \bar{\Pi}_x$ and Γ_{xy} by $\Gamma_{xy} - \bar{\Gamma}_{xy}$. Since this collection is essentially countable (consider for example the sequence of pseudometrics d_n corresponding to the choices $(\mathfrak{K}_n, \gamma_n)$ with $\mathfrak{K}_n$ the centred ball of radius n and $\gamma_n = n$), it determines a metrisable topology (take for example $d = \sum_{n \geq 1} 2^{-n}(d_n \wedge 1)$).

Remark 13.6. The precise choice of r in Definition 13.5 is not very important, as one can see that any other choice $r > |\min A| \geq 0$ leads to the same definition. See Lemma 14.13 for a similar statement in the context of Hölder spaces.

Remark 13.7. The test functions appearing in (13.13) are smooth. It turns out that if these bounds hold for smooth elements of $\mathcal{B}_r$, then $\Pi_x\tau$ can be extended canonically to allow any $\mathcal{C}^r$ test function with compact support.

Remark 13.8. The identity $\Pi_x\Gamma_{xy} = \Pi_y$ reflects the fact that Γ_{xy} is the linear map that takes an expansion around y and turns it into an expansion around x. The first bound in (13.13) states what we mean precisely when we say that $\tau \in T_\alpha$ represents a term that vanishes at order α. (See Exercise 13.2; note that α can be negative, so that this may actually not vanish at all!) The second bound in (13.13) is very natural in view of both (13.3) and (13.4). It states that when expanding a monomial of order α around a new point at distance h from the old one, the coefficient appearing in front of lower-order monomials of order β is of order at most $h^{\alpha-\beta}$.

Remark 13.9. In many cases of interest, it is natural to scale the different directions of $\mathbf{R}^d$ in a different way. This is the case for example when using the theory of regularity structures to build solution theories for parabolic stochastic PDEs, in which case the time direction "counts as" two space directions. This "parabolic scaling" can be formalised by the integer vector $(2, 1, \ldots, 1)$. More generally, one can introduce a scaling $\mathfrak{s}$ of $\mathbf{R}^d$, which is just a collection of d scalars $\mathfrak{s}_i \in [1, \infty)$ and to define φ_x^λ in such a way that the ith direction is scaled by $\lambda^{\mathfrak{s}_i}$. The polynomial structure introduced earlier, in particular (13.7), should be changed accordingly by postulating that the degree of X^k is given by $|k|_\mathfrak{s} = \sum_{i=1}^d \mathfrak{s}_i k_i$. In this case, the Euclidean distance between two points $x, y \in \mathbf{R}^d$ should be replaced everywhere by the corresponding scaled distance $|x - y|_\mathfrak{s} = \sum_i |x_i - y_i|^{1/\mathfrak{s}_i}$. See [Hai14b] for more details.

With these definitions at hand, it is then natural to define an analogue in this context of the space of γ-Hölder continuous functions in the following way.

Definition 13.10. Given a regularity structure $\mathscr{T}$ equipped with a model $\mathrm{M} = (\Pi, \Gamma)$ over $\mathbf{R}^d$, the space $\mathscr{D}_\mathrm{M}^\gamma$ is given by the set of functions $f \colon \mathbf{R}^d \to T_{<\gamma}$ such that, for every compact set $\mathfrak{K}$ and every $\alpha < \gamma$, there exists a constant C with

$$\|f(x) - \Gamma_{xy} f(y)\|_\alpha \leq C|x - y|^{\gamma - \alpha} \qquad (13.14)$$

uniformly over $x, y \in \mathfrak{K}$. Such functions f are called *modelled distributions*. For fixed $\mathfrak{K}$, a seminorm $\|f\|_{M, \gamma; \mathfrak{K}}$ is defined as the smallest constant C in the bound (13.14). The space $\mathscr{D}_M^\gamma$ endowed with this family of seminorms is then a Fréchet space.

It is furthermore convenient to be able to compare two modelled distributions defined over two different models. In this case, a natural way of comparing them is to take as a "metric" the smallest constant C in the bound

$$\|f(x) - \Gamma_{xy} f(y) - \bar{f}(x) + \bar{\Gamma}_{xy} \bar{f}(y)\|_\alpha \leq C|x - y|^{\gamma - \alpha} .$$

Remark 13.11. (Compare with Remark 4.8 in the rough path context.) It is important to note that while the space of models $\mathscr{M}$ is not a linear space, the space $\mathscr{D}_M^\gamma$ is a linear (in fact: Fréchet) space given a model $\mathrm{M} \in \mathscr{M}$. The twist of course is that the space in question depends in a crucial way on the choice of M. The total space then is the disjoint union

$$\mathscr{M} \ltimes \mathscr{D}^\gamma \overset{\text{def}}{=} \bigsqcup_{\mathrm{M} \in \mathscr{M}} \{\mathrm{M}\} \times \mathscr{D}_M^\gamma,$$

with base space $\mathscr{M}$ and "fibres" $\mathscr{D}_M^\gamma$.

The most fundamental result in the theory of regularity structures then states that given $f \in \mathscr{D}^\gamma$ with $\gamma > 0$, there exists a *unique* distribution $\mathcal{R} f$ on $\mathbf{R}^d$ such that, for every $x \in \mathbf{R}^d$, $\mathcal{R} f$ "looks like $\Pi_x f(x)$ near x". More precisely, one has

Theorem 13.12 (Reconstruction). *Let* $\mathrm{M} = (\Pi, \Gamma)$ *be a model for a regularity structure* $\mathscr{T}$ *on* $\mathbf{R}^d$. *Assume* $f \in \mathscr{D}_M^\gamma$ *with* $\gamma > 0$. *Then, there exists a unique linear map*

$$\mathcal{R} = \mathcal{R}_M \colon \mathscr{D}_M^\gamma \to \mathcal{D}'(\mathbf{R}^d)$$

such that

$$\left|(\mathcal{R} f - \Pi_x f(x))(\varphi_x^\lambda)\right| \lesssim \lambda^\gamma , \qquad (13.15)$$

uniformly over $\varphi \in \mathcal{B}_r$ *and* λ *as before, and locally uniformly in* x. *For* $\gamma < 0$, *everything remains valid but uniqueness of* $\mathcal{R}$.

Remark 13.13. With a look to Remark 13.11, and $\mathrm{M} = (\Pi, \Gamma) \in \mathscr{M}$, one should really view $\mathcal{R} = \mathcal{R}_M f$ as a map from $\mathscr{M} \ltimes \mathscr{D}^\gamma$ into $\mathcal{D}'$. Since the space $\mathscr{M} \ltimes \mathscr{D}^\gamma$ is *not* a linear space, this shows that the map $\mathcal{R}$ isn't actually linear, despite appearances. However, the map $(\Pi, \Gamma, f) \mapsto \mathcal{R} f$ turns out to be locally Lipschitz continuous provided that the distance between (Π, Γ, f) and $(\bar{\Pi}, \bar{\Gamma}, \bar{f})$ is given by the smallest constant C such that

$$\|f(x) - \bar{f}(x) - \Gamma_{xy} f(y) + \bar{\Gamma}_{xy} \bar{f}(y)\|_\alpha \leq C|x - y|^{\gamma - \alpha} ,$$
$$\left|(\Pi_x \tau - \bar{\Pi}_x \tau)(\varphi_x^\lambda)\right| \leq C \lambda^\alpha \|\tau\| ,$$
$$\|\Gamma_{xy} \tau - \bar{\Gamma}_{xy} \tau\|_\beta \leq C|x - y|^{\alpha - \beta} \|\tau\| .$$

Here, in order to obtain bounds on $(\mathcal{R}f - \bar{\mathcal{R}}\bar{f})(\psi)$ for some smooth compactly supported test function ψ, the above bounds should hold uniformly for x and y in a neighbourhood of the support of ψ. The proof that this stronger continuity property also holds is actually crucial when showing that sequences of solutions to mollified equations all converge to the same limiting object. However, its proof is somewhat more involved which is why we chose not to give it here but refer instead to [Hai14b, Thm 3.10].

Remark 13.14. There are obvious analogies between the construction of the reconstruction operator $\mathcal{R}$ and that of the "rough integral" in Section 4. As a matter of fact, there exists a slightly more abstract formulation of the reconstruction theorem which can be interpreted as a multidimensional analogue to the sewing lemma, Lemma 4.2, see [Hai14b, Prop. 3.25].

Remark 13.15. The reconstruction theorem with $\gamma < 0$ allows one to recover the Lyons–Victoir extension theorem previously obtained in Exercise 2.14, see also Exercise 13.6. Note that the reconstruction theorem does *not* hold for $\gamma = 0$ (even if we forego uniqueness of $\mathcal{R}$), for the same reason that the Lyons–Victoir extension theorem fails for $\alpha = \frac{1}{2}$ (and more generally when $1/\alpha \in \mathbf{N}$).

In the particular case where $\Pi_x \tau$ happens to be a continuous function for every $\tau \in T$ (and every $x \in \mathbf{R}^d$), we will see in Remark 13.27 that $\mathcal{R}f$ is also a continuous function and $\mathcal{R}$ is given by the somewhat trivial explicit formula

$$(\mathcal{R}f)(x) = \big(\Pi_x f(x)\big)(x) .$$

We postpone the proof of the reconstruction theorem to Section 13.4 and turn instead to our previous list of regularity structures, now adding the relevant models and indicating the interest of the reconstruction map.

13.3.1 The polynomial model

Recall the polynomial regularity structure in d variables defined in Section 13.2.1. In this context, the polynomial model P is given by

$$\big(\Pi_x X^k\big) = (y \mapsto (y - x)^k) , \qquad \Gamma_{xy} = \Gamma_h \big|_{h=x-y} .$$

We leave it as an exercise to the reader to verify that this does indeed satisfy the bounds and relations of Definition 13.5.

In the sense of the following proposition, modelled distributions in the context of the polynomial model are nothing but classical Hölder functions.

Proposition 13.16. *Let $\beta = n + \gamma$ with $n \in \mathbf{N}$ and $\gamma \in (0, 1)$. If f belongs to the Hölder space $\mathcal{C}^\beta$, then $f \in \mathcal{D}_P^\beta$ with*

$$f(x) = f(x)\mathbf{1} + \sum_{1 \le |k| \le n} \frac{f^{(k)}(x)}{k!} X^k \ .$$

Conversely, if $\hat{f} \in \mathscr{D}_P^\beta$ then $f := \langle \hat{f}, \mathbf{1} \rangle$ is in $\mathcal{C}^\beta$ and necessarily $\hat{f} = f$. $\square$

This proposition is essentially a consequence of the (well-known) fact that $f \in \mathcal{C}^\beta$ if and only if for every $x \in \mathbf{R}^d$, there exists a polynomial $P_x = P_x(y)$ of degree n, such that, locally uniformly in x, y, one has $|f(y) - P_x(y)| \lesssim |y - x|^\beta$. Necessarily then, such a function f is n times continuously differentiable, and P_x is its Taylor polynomial of degree n. This characterisation and the above proposition remain valid for integer values of β with the caveat that in this context $\mathcal{C}^\beta$ means $\beta - 1$ times continuously differentiable with the highest order derivatives locally Lipschitz continuous.

It will be convenient for the sequel to introduce a suitable notion of "negative" Hölder spaces. In fact, the definition of a model (see also Exercise 13.2) suggests that a very natural space of distributions is obtained in the following way. Given $\alpha > 0$, we denote by $\mathcal{C}^{-\alpha}$ the space of all distributions η such that, with r the smallest integer such that $r > \alpha$,

$$\left| \eta(\varphi_x^\lambda) \right| \lesssim \lambda^{-\alpha} \ ,$$

uniformly over all $\varphi \in \mathcal{B}_r$ and $\lambda \in (0, 1]$, and locally uniformly in x. Given any compact set $\mathfrak{K}$, the best possible constant such that the above bound holds uniformly over $x \in \mathfrak{K}$ yields a seminorm. The collection of these seminorms endows $\mathcal{C}^{-\alpha}$ with a Fréchet space structure.

Remark 13.17. In terms of the scale of classical Besov spaces, the space $\mathcal{C}^{-\alpha}$ is a local version of $\mathcal{B}_{\infty,\infty}^{-\alpha}$. It is in some sense the largest space of distributions that is invariant under the scaling $\varphi(\cdot) \mapsto \lambda^{-\alpha}\varphi(\lambda^{-1}\cdot)$, see for example [BP08].

Let us now give a very simple application of the reconstruction theorem. It is a classical result in the "folklore" of harmonic analysis (see for example [BCD11, Thm 2.52] for a very similar statement) that the product extends naturally to $\mathcal{C}^\beta \times \mathcal{C}^{-\alpha}$ into $\mathcal{D}'(\mathbf{R}^d)$ if and only if $\beta > \alpha$, which can also be seen as higher-dimensional version of the Young integral, cf. Exercise 13.1. We illustrate how to use the reconstruction theorem in order to obtain a straightforward proof of the "if" part of this result:

Theorem 13.18. *For $\beta > \alpha > 0$, there is a continuous bilinear map*

$$B \colon \mathcal{C}^\beta \times \mathcal{C}^{-\alpha} \to \mathcal{D}'(\mathbf{R}^d)$$

such that $B(f, g) = fg$ for any two continuous functions f and g.

Proof. Assume from now on that $g = \xi \in \mathcal{C}^{-\alpha}$ for some $\alpha > 0$ and that $f \in \mathcal{C}^\beta$ for some $\beta > \alpha$. We then build a regularity structure $\mathscr{T}$ in the following way. For the index set A, we take $A = \mathbf{N} \cup (\mathbf{N} - \alpha)$ and for T, we set $T = V \oplus W$, where each one of the spaces V and W is a copy of the polynomial regularity structure (in

d commuting variables). We also choose Γ as in the polynomial case above, acting simultaneously and identically on each of the two instances.

As before, we denote by X^k the canonical basis vectors in V. We also use the suggestive notation "ΞX^k" for the corresponding basis vector in W, but we postulate that $\Xi X^k \in T_{|k|-\alpha}$ rather than $\Xi X^k \in T_{|k|}$. Given any distribution $\xi \in \mathcal{C}^{-\alpha}$, we then define a model (Π^ξ, Γ), where Γ is as in the canonical model, while Π^ξ acts as

$$\left(\Pi_x^\xi X^k\right)(y) = (y-x)^k\,, \qquad \left(\Pi_x^\xi \Xi X^k\right)(y) = (y-x)^k\xi(y)\,,$$

with the obvious abuse of notation in the second expression. It is then straightforward to verify that $\Pi_y = \Pi_x \circ \Gamma_{xy}$ and that the relevant analytical bounds are satisfied, so that this is indeed a model.

Denote now by $\mathcal{R}^\xi$ the reconstruction map associated to the model (Π^ξ, Γ) and, for $f \in \mathcal{C}^\beta$, denote by f the element in $\mathcal{D}^\beta$ given by the local Taylor expansion of f of order β at each point. Note that even though the space $\mathcal{D}^\beta$ does in principle depend on the choice of model, in our situation $f \in \mathcal{D}^\beta$ for any choice of ξ. It follows immediately from the definitions that the map $x \mapsto \Xi f(x)$ belongs to $\mathcal{D}^{\beta-\alpha}$ so that, provided that $\beta > \alpha$, one can apply the reconstruction operator to it. This suggests that the multiplication operator we are looking for can be defined as

$$B(f,\xi) = \mathcal{R}^\xi\left(\Xi f\right)\,.$$

By Theorem 13.12, this is a jointly continuous map from $\mathcal{C}^\beta \times \mathcal{C}^{-\alpha}$ into $\mathcal{D}'(\mathbf{R}^d)$, provided that $\beta > \alpha$. If ξ happens to be a smooth function, then it follows immediately from the remark after Theorem 13.12 that $B(f,\xi) = f(x)\xi(x)$, so that B is indeed the requested continuous extension of the usual product.   □

Remark 13.19. In the context of this theorem, one can actually show that $B(f,g) \in \mathcal{C}^{-\alpha}$. More generally, denoting by $-\alpha$ the smallest degree arising in a given regularity structure $\mathscr{T}$, i.e. $\alpha = -\min A$, it is possible to show that the reconstruction operator $\mathcal{R}$ takes values in $\mathcal{C}^{-\alpha}$.

The reader may notice that one can also work with a finite-dimensional regularity structure, based on index set $\tilde{N} \cup (\tilde{N} - \alpha)$, with $\tilde{N} = \{0, 1, \ldots, n\}$ and $\beta = n + \gamma$. In particular, if $n = 0$, the regularity structure used here is exacty the one already encountered in (13.8).

13.3.2 The rough path model

Let us see now how some of the results of Section 4 can be reinterpreted in the light of this theory. Fix $\alpha \in (1/3, 1/2]$ and let $\mathscr{T}$ be the rough path regularity structure put forward in Definition 13.4. Recall that this means that $T_0 = \langle \mathbf{1} \rangle$, T_α and $T_{\alpha-1}$ are copies of $\mathbf{R}^e$ with respective basis vectors W^j and $\dot{W}^j$, and $T_{2\alpha-1}$ is a copy of $\mathbf{R}^{e \times e}$ with basis vectors $\dot{\mathbb{W}}^{ij}$. The structure group G is isomorphic to $\mathbf{R}^e$ and, for $h \in \mathbf{R}^e$, acts on T via

$$\Gamma_h 1 = 1 \,, \quad \Gamma_h \dot{W}^i = \dot{W}^i \,, \quad \Gamma_h W^i = W^i + h^i 1 \,, \quad \Gamma_h \dot{\mathbb{W}}^{ij} = \dot{\mathbb{W}}^{ij} + h^i \dot{W}^j \,.$$

$$(13.16)$$

Let now $\mathbf{W} = (W, \mathbb{W})$ be an α-Hölder continuous rough path over $\mathbf{R}^e$. It turns out that this defines a model for $\mathscr{T}$ in the following way:

Lemma 13.20. *Given an α-Hölder continuous rough path $\mathbf{W}$, one can define a model $M = M_{\mathbf{W}}$ for $\mathscr{T}$ on $\mathbf{R}$ by setting $\Gamma_{t,s} = \Gamma_{W_{s,t}}$ and*

$$\left(\Pi_s 1 \right)(t) = 1 \,, \qquad\qquad \left(\Pi_s W^j \right)(t) = W^j_{s,t}$$

$$\left(\Pi_s \dot{W}^j \right)(\psi) = \int \psi(t)\, dW^j_t \,, \qquad \left(\Pi_s \dot{\mathbb{W}}^{ij} \right)(\psi) = \int \psi(t)\, d\mathbb{W}^{ij}_{s,t} \,.$$

Here, both integrals are perfectly well-defined Riemann integrals, with the differential in the second case taken with respect to the variable t. Given a controlled rough path $(Y, Y') \in \mathscr{D}^{2\alpha}_W$, this then defines an element $Y \in \mathscr{D}^{2\alpha}_M$ by

$$Y(s) = Y(s)\, 1 + Y'_i(s)\, W^i \,,$$

with summation over i implied.

Proof. We first check that the algebraic properties of Definition 13.5 are satisfied. It is clear that $\Gamma_{s,u} \Gamma_{u,t} = \Gamma_{s,t}$ and that $\Pi_s \Gamma_{s,u} \tau = \Pi_u \tau$ for $\tau \in \{1, W^j, \dot{W}^j\}$. Regarding $\dot{\mathbb{W}}^{ij}$, we differentiate Chen's relations (2.1) which yields the identity

$$d\mathbb{W}^{i,j}_{s,t} = d\mathbb{W}^{i,j}_{u,t} + W^i_{s,u}\, dW^j_t \,.$$

The last missing algebraic relation then follows at once. The required analytic bounds follow immediately (exercise!) from the definition of the rough path space $\mathscr{C}^\alpha$.

Regarding the function Y defined in the statement, we have

$$\| Y(s) - \Gamma_{s,u} Y(u) \|_0 = |Y(s) - Y(u) + Y'_i(u) W^i_{s,u}| \,,$$
$$\| Y(s) - \Gamma_{s,u} Y(u) \|_\alpha = |Y'(s) - Y'(u)| \,,$$

so that the condition (13.14) with $\gamma = 2\alpha$ does indeed coincide with the definition of a controlled rough path. □

Theorems 4.4 and 4.10 can then be recovered as a particular case of the reconstruction theorem in the following way.

Proposition 13.21. *In the same context as above, let $\alpha \in (\frac{1}{3}, \frac{1}{2}]$, and consider the modelled distribution $Y \in \mathscr{D}^{2\alpha}_{M_{\mathbf{W}}}$ built as above from a controlled rough path $(Y, Y') \in \mathscr{D}^{2\alpha}_W$. Then, the map $Y\dot{W}^j$ given by*

$$\left(Y \dot{W}^j \right)(s) := Y(s)\, \dot{W}^j + Y'_i(s)\, \dot{\mathbb{W}}^{ij}$$

belongs to $\mathscr{D}^{3\alpha-1}$. Furthermore, there exists a function Z, unique up to addition of constants, such that

$$\left(\mathcal{R}Y\dot{W}^j\right)(\psi) = \int \psi(t)\, dZ(t)\,,$$

and such that $Z_{s,t} = Y(s)\, W_{s,t}^j + Y_i'(s)\, \mathbb{W}_{s,t}^{i,j} + \mathrm{O}(|t-s|^{3\alpha}).$

Proof. The fact that $Y\dot{W}^j \in \mathscr{D}^{3\alpha-1}$ is an immediate consequence of the definitions. Since $\alpha > \frac{1}{3}$ by assumption, we can apply the reconstruction theorem to it, from which it follows that there exists a unique distribution η such that, if ψ is a smooth compactly supported test function, one has

$$\eta(\psi_s^\lambda) = \int \psi_s^\lambda(t)Y(s)\, dW_t^j + \int \psi_s^\lambda(t)Y_i'(s)\, d\mathbb{W}_{s,t}^{i,j} + \mathrm{O}(\lambda^{3\alpha-1})\,.$$

By a simple approximation argument, see Exercise 13.10, one can take for ψ the indicator function of the interval $[0,1]$, so that

$$\eta(\mathbf{1}_{[s,t]}) = Y(s)\, W_{s,t}^j + Y_i'(s)\, \mathbb{W}_{s,t}^{i,j} + \mathrm{O}(|t-s|^{3\alpha})\,.$$

Here, the reason why one obtains an exponent 3α rather than $3\alpha - 1$ is that it is really $|t-s|^{-1}\mathbf{1}_{[s,t]}$ that scales like an approximate δ-distribution as $t \to s$. $\quad\square$

Remark 13.22. Using the formula (13.26), it is straightforward to verify that if W happens to be a smooth function and $\mathbb{W}$ is defined from W via (2.2), but this time viewing it as a definition for the right-hand side, with the left-hand side given by a usual Riemann integral, then the function Z constructed in Proposition 13.21 coincides with the usual Riemann integral of Y against W^j.

Remark 13.23. The theory of (controlled) rough paths of lower regularity already hinted at in Section 2.4 can be recovered from the reconstruction operator and a suitable choice of regularity structure (essentially two copies of the truncated tensor algebra) in virtually the same way.

13.4 Proof of the reconstruction theorem

The proof of the reconstruction theorem originally given in [Hai14b] relied on wavelet analysis, in particular on the existence of compactly supported wavelets of arbitrary regularity [Dau88]. More recently, Otto and Weber [OSSW18] and then Moinat and Weber [MW18] obtained a version of the reconstruction theorem that bypasses this theory and is completely self-contained. The version of the proof given here is inspired by their work and has the advantage of being purely local: although we state the result for models and modelled distributions that are assumed to be defined on all of $\mathbf{R}^d$, the proof generalises immediately to arbitrary domains. The proof given here also generalises immediately to non-Euclidean scalings, even in situations where the ratios between scaling exponents are irrational.

A crucial ingredient is the following remark. Fix $\alpha > 0$ and let $\varrho : \mathbf{R}^d \to \mathbf{R}$ be even, smooth, compactly supported in the ball of radius 1, such that

$$\int x^k \varrho(x)\, dx = \delta_{k,0}\,, \qquad 0 < |k| \le \alpha\,, \tag{13.17}$$

where k denotes a d-dimensional multi-index and δ denotes Kronecker's delta. Note that such a function necessarily exists, since otherwise one would be able to find a polynomial P of degree at most α such that $\int P(x)\varphi(x)\, dx = 0$ for *every* smooth and compactly supported φ, which is clearly absurd. (See also Exercise 13.8 for a constructive proof.)

Given such a function ϱ, we define $\varrho^{(n)}(x) = 2^{nd}\varrho(2^n x)$, as well as

$$\varrho^{(n,m)} = \varrho^{(n)} * \varrho^{(n+1)} * \cdots * \varrho^{(m)}\,, \tag{13.18}$$

where $*$ denotes convolution. We also set $\varphi^{(n)} = \lim_{m\to\infty} \varrho^{(n,m)}$, so that in particular $\varphi^{(n)} = \varrho^{(n)} * \varphi^{(n+1)}$ and we write $\varrho_x^{(n)}(y) = \varrho^{(n)}(y-x)$ and similarly for $\varphi_x^{(n)}$; see Exercise 13.7 to see that the limit $\varphi^{(n)}$ exists and belongs to $\mathcal{C}_c^\infty$. We then have the following preliminary lemma.

Lemma 13.24. *Let $\alpha > 0$, let ϱ be as above and let $\xi_n \colon \mathbf{R}^d \to \mathbf{R}$ be a sequence of functions such that for every compact $\mathfrak{K}$ there exists $C_{\mathfrak{K}}$ such that $\sup_{x\in\mathfrak{K}} |\xi_n(x)| \le C_{\mathfrak{K}} 2^{\alpha n}$, and such that furthermore $\xi_n = \varrho^{(n)} * \xi_{n+1}$. Then, the sequence ξ_n is Cauchy in $\mathcal{C}^{-\beta}$ for every $\beta > \alpha$ and its limit ξ satisfies $\xi_n = \varphi^{(n)} * \xi$.*

If furthermore, for some $x \in \mathbf{R}^d$ and $\gamma > -\alpha$ one has the bound $|\xi_n(y)| \le 2^{\alpha n}\big(|x-y|^{\gamma+\alpha} + 2^{-(\gamma+\alpha)n}\big)$, uniformly over $n \ge 0$ and $|y-x| \le 1$, then $|\xi(\psi_x^\lambda)| \lesssim \lambda^\gamma$ for $\lambda \le 1$.

Proof. Let $\lambda \in (0,1]$ and let ψ_λ be a test function that is supported in the ball of radius λ and such that $|D^k\psi| \le \lambda^{-d-|k|}$ for all $|k| \le \alpha + 1$. In order to show that ξ_n is Cauchy in $\mathcal{C}^{-\beta}$ it then suffices to exhibit a bound of the type

$$|\psi_\lambda * (\xi_n - \xi_{n+1})| \lesssim \lambda^{-\beta} 2^{(\alpha-\beta)n}\,, \tag{13.19}$$

locally uniformly in x, for a proportionality constant independent of ψ_λ. Since there exists $\bar{C} > 0$ such that $\int |\psi_\lambda(x)|\, dx \le \bar{C}$, uniformly over λ and ψ_λ, it follows from the assumption $|\xi_n(x)| \le C 2^{\alpha n}$ that the left-hand side of (13.19) is bounded by $(1 + 2^\alpha) C \bar{C} 2^{\alpha n}$, so that the bound (13.19) holds whenever $\lambda \le 2^{-n}$.

To deal with the converse case $2^{-n} \le \lambda$, we rewrite the left-hand side of (13.19) as $|(\psi_\lambda * \varrho^{(n)} - \psi_\lambda) * \xi_{n+1}|$ and we note that, by Taylor's remainder theorem,

$$\left|\psi_\lambda(y) - T_x^{(\alpha)}(y)\right| \overset{\text{def}}{=} \left|\psi_\lambda(y) - \sum_{|k|\le\alpha} \frac{D^k\psi_\lambda(x)}{k!}(y-x)^k\right| \lesssim \lambda^{-N-d}|y-x|^N\,, \tag{13.20}$$

where $N = \lceil \alpha \rceil$. Since, by (13.17), one has $\varrho^{(n)} * T_x^{(\alpha)} = T_x^{(\alpha)}$ and since $T_x^{(\alpha)}(x) = \psi_\lambda(x)$, one has

$$\big(\psi_\lambda * \varrho^{(n)} - \psi_\lambda\big)(x) = \big(\varrho^{(n)} * (\psi_\lambda - T_x^{(\alpha)})\big)(x)\,,$$

which is bounded by $\lambda^{-N-d}2^{-nN}$ as an immediate consequence of (13.20). Since furthermore the support of this function has diameter at most 2λ, it follows that its integral is at most $\lambda^{-N}2^{-nN}$ so that, combining this with the a priori bound $|\xi_{n+1}| \lesssim 2^{\alpha n}$, we conclude that

$$|\psi_\lambda * (\xi_n - \xi_{n+1})| \lesssim \lambda^{-N}2^{(\alpha-N)n} \ .$$

Since $N \geq \alpha$, the bound (13.19) then follows for $2^{-n} \leq \lambda$ as required.

Since we have just shown that the sequence ξ_n is Cauchy, it has a limit $\xi \in \mathcal{C}^{-\beta}$. Given a test function ψ, we have

$$\xi_n(\psi) = \xi_{n+1}(\varrho^{(n)} * \psi) = \xi_m(\varrho^{(n,m)} * \psi) = \xi(\varphi^{(n)} * \psi) \ ,$$

showing that $\xi_n = \varphi^{(n)} * \xi$ as required. (Here we use the fact that the convergence $\varrho^{(n,m)} \to \varphi^{(n)}$ takes place in $\mathcal{C}^r$ for $r = r_\beta$ by Exercise 13.7.)

The proof of the second claim follows the same lines. We write

$$\xi(\psi_x^\lambda) = \xi_n(\psi_x^\lambda) + \sum_{k \geq n}(\xi_{k+1} - \xi_k)(\psi_x^\lambda) \ ,$$

where n is chosen in such a way that $\lambda \in [2^{-(n+1)}, 2^{-n}]$. As a consequence of this choice and of our assumption on ξ_n, one has the bound

$$|\xi_n(\psi_x^\lambda)| \lesssim \lambda^{-d} \int_{B_x(\lambda)} 2^{\alpha n}\big(|x - y|^{\gamma+\alpha} + 2^{-(\gamma+\alpha)n}\big)\, dy$$
$$\lesssim \lambda^{\gamma+\alpha}2^{\alpha n} + 2^{-\gamma n} \lesssim \lambda^\gamma \ .$$

To bound $(\xi_{k+1} - \xi_k)(\psi_x^\lambda)$ we proceed as above so that

$$\big|(\xi_{k+1} - \xi_k)(\psi_x^\lambda)\big| \lesssim \lambda^{-N-d}2^{-nN} \int_{B_x(2\lambda)} |\xi_{n+1}(y)|\, dy$$
$$\lesssim \lambda^{\gamma+\alpha-N}2^{(\alpha-N)n} + \lambda^{-N}2^{-(\gamma+N)n} \ .$$

Since $N > \alpha$ and $N > -\gamma$, this is summable and its sum is again of order λ^γ, thus concluding the proof. □

Remark 13.25. Note the strong similarity of this setting with that of multiresolution analysis [Mey92]: the image of the convolution operator with $\varphi^{(n)}$ plays the role of V_n and convolution with $\varrho^{(n)}$ plays the role of the projection $V_{n+1} \to V_n$.

Let us now restate the reconstruction theorem for the reader's convenience. (We only consider the case $\gamma > 0$ here.)

Theorem 13.26. *Let $\mathscr{T}$ be a regularity structure as above and let (Π, Γ) a model for $\mathscr{T}$ on $\mathbf{R}^d$. Then, for $\gamma > 0$, there exists a unique linear map $\mathcal{R}: \mathscr{D}^\gamma \to \mathcal{D}'(\mathbf{R}^d)$ such that*

$$\big|(\mathcal{R}f - \Pi_x f(x))(\psi_x^\lambda)\big| \lesssim \lambda^\gamma \ ,$$

uniformly over $\psi \in \mathcal{B}_r$ and $\lambda \in (0,1]$, and locally uniformly in x. The statement still holds for $\gamma < 0$, except that uniqueness fails.

Proof. We first define operators $\mathcal{R}^{(m,m)}$ by

$$\left(\mathcal{R}^{(m,m)} f\right)(y) = \left(\varphi^{(m)} * \Pi_y f(y)\right)(y) = \left(\Pi_y f(y)\right)\left(\varphi_y^{(m)}\right). \qquad (13.21)$$

The idea then is to obtain $\mathcal{R}$ as the limit of $\mathcal{R}^{(m,m)}$ as $m \to \infty$. This however turns out not to be that easy to obtain directly. Instead, we try to make use of Lemma 13.24 and define, for $m > n$,

$$\mathcal{R}^{(n,m)} f = \varrho^{(n,m-1)} * \mathcal{R}^{(m,m)} f,$$

so that, as a consequence of the identity $\Pi_z = \Pi_y \Gamma_{yz}$,

$$\left(\mathcal{R}^{(n,m)} f - \mathcal{R}^{(n,m+1)} f\right)(x) = \int \varrho_x^{(n,m-1)}(y)$$

$$\int \varrho_z^{(m)}(y)\left(\Pi_y\left(f(y) - \Gamma_{yz} f(z)\right)\right)\left(\varphi_z^{(m+1)}\right) dz\, dy.$$

At this stage we note that, as a consequence of the analytical bounds (13.13) imposed in the definition of a model, the quantity $\left(\Pi_y \tau\right)\left(\varphi_z^{(m+1)}\right)$ is bounded by $C2^{-\alpha m}\|\tau\|_\alpha$, uniformly over $|y - z| \lesssim 2^{-m}$ and $\tau \in T_\alpha$. On the other hand, the definition of the spaces $\mathscr{D}^\gamma$ guarantees that the component of $f(y) - \Gamma_{yz} f(z)$ in T_α is bounded by $2^{(\alpha-\gamma)m}$, again uniformly over $|y - z| \lesssim 2^{-m}$. Since $\int |\varrho_x^{(n,m-1)}(y)|\, dy \lesssim 1$, uniformly over m and n, we conclude that

$$\left\|\left(\mathcal{R}^{(n,m)} - \mathcal{R}^{(n,m+1)}\right) f\right\|_{L^\infty} \lesssim 2^{-\gamma m}, \qquad (13.22)$$

uniformly over $n \geq 0$ and $m \geq n$. Furthermore, it is straightforward to check that

$$\left\|\mathcal{R}^{(n,n)} f\right\|_{L^\infty} \lesssim 2^{-\underline{\alpha} n}, \qquad (13.23)$$

where $\underline{\alpha}$ denotes the smallest degree in the ambient regularity structure. It follows that $\mathcal{R}^{(n)} f = \lim_{m \to \infty} \mathcal{R}^{(n,m)} f$ is well-defined and also satisfies the bound (13.23). Since the identity

$$\mathcal{R}^{(n,m)} f = \varrho^{(n)} * \mathcal{R}^{(n+1,m)} f$$

holds for every $m \geq n + 1$, it follows that $\mathcal{R}^{(n)} f = \varrho^{(n)} * \mathcal{R}^{(n+1)} f$, so that $\mathcal{R} f = \lim_{n \to \infty} \mathcal{R}^{(n)} f$ exists in $\mathcal{C}^\alpha$ for every $\alpha < \underline{\alpha}$ by Lemma 13.24.

It remains to show that one has the bound

$$\left|\left(\mathcal{R} f - \Pi_x f(x)\right)\left(\psi_x^\lambda\right)\right| \lesssim \lambda^\gamma. \qquad (13.24)$$

For this, we note first that if we define $f_x \in \mathscr{D}^\gamma$ by $f_x(y) = \Gamma_{yx} f(x)$, then one has $\mathcal{R}^{(n)} f_x = \varphi^{(n)} * \Pi_x f(x)$, so that (13.24) can be written as

$$\left|\mathcal{R}\left(f - f_x\right)\left(\psi_x^\lambda\right)\right| \lesssim \lambda^\gamma. \qquad (13.25)$$

Since $\|(f - f_x)(y)\|_\alpha \lesssim |y - x|^{\gamma-\alpha}$, it follows from the definition (13.14) of $\mathscr{D}^\gamma$ that

$$\left|\left(\mathcal{R}^{(n,n)}(f - f_x)\right)(y)\right| = \left|\left(\Pi_y(f - f_x)\right)(\varphi_y^{(n)})\right| \lesssim \sum_{\alpha<\gamma} 2^{-\alpha n}|y - x|^{\gamma-\alpha}$$

$$\lesssim 2^{-\underline{\alpha} n}\left(|y - x|^{\gamma-\underline{\alpha}} + 2^{(\underline{\alpha}-\gamma)n}\right) .$$

By (13.22) the same bound also holds for $\mathcal{R}^{(n)}$, so that the claim follows from the second part of Lemma 13.24.

The case $\gamma < 0$ works in a similar way, but this time we explicitly define

$$\mathcal{R}f = \mathcal{R}^{(0,0)}f + \sum_n (\varrho^{(n)} - \delta) * \mathcal{R}^{(n,n)}f ,$$

where δ denotes the Dirac delta-distribution. We leave it as an exercise for the reader to verify that this sum does indeed converge in $\mathcal{C}^\alpha$ for every $\alpha < \underline{\alpha}$ and that the limit satisfies the required bound. $\square$

Remark 13.27. In the particular case where $\Pi_x\tau$ happens to be a continuous function for every $\tau \in T$ (and every $x \in \mathbf{R}^d$), $\mathcal{R}f$ is also a continuous function and one has the identity

$$\left(\mathcal{R}f\right)(x) = \left(\Pi_x f(x)\right)(x) . \tag{13.26}$$

We leave it as an exercise to show that this is the case, taking (13.21) as a starting point.

13.5 Exercises

Exercise 13.1   a) *Relate Theorem 13.18, in case $d = 1$, with the Young integral.*
   b) *Draw inspiration from Weierstrass's construction of a continuous nowhere differentiable function to construct examples demonstrating the "only if" part of Theorem 13.18.*

Exercise 13.2 (Hölder spaces) *For $k \in \mathbf{N}$ and $\alpha \in (0, 1)$, it is customary to define $\mathcal{C}^{k+\alpha}$ as the space of k times continuously differentiable functions $f : \mathbf{R}^d \to \mathbf{R}$ such that their derivatives of order k are α-Hölder continuous. Show that this agrees with the obvious extension to $\mathbf{R}^d$ of the definition given earlier in (13.2).*

Exercise 13.3 *Show that in general, the function Z from Proposition 13.21 coincides, up to an additive constant, with the rough integral $\int_0^t Y(s)\, dX_s^j$, in the sense of Remark 4.12.*

♯ **Exercise 13.4** *Let $\bar\gamma \geq \gamma > 0$ and let $f \in \mathcal{C}(\mathbf{R}^d, T_{<\bar\gamma})$ such the "modelled distribution" bound (13.14) holds for every $\alpha < \gamma$.*

$$\|f\|_{\mathscr{D}^\gamma} < \infty .$$

Show that the projection of f on $T_{<\gamma}$ belongs to $\mathscr{D}^\gamma$.

Exercise 13.5 *Let (Π, Γ) be a model for the "rough path" regularity structure given in Definition 13.4 with the additional property that $\Pi_s \dot{W}^i$ is the distributional derivative of $\Pi_s W^i$ for every s. Show that it is then necessarily of the form $\mathbf{M_W}$ for some α-Hölder rough path $\mathbf{W}$ as in Lemma 13.20.*

Exercise 13.6 *Using the regularity structure defined in Section 13.3.2, give a proof of the Lyons–Victoir extension theorem using the case $\gamma < 0$ of the reconstruction theorem.* **Hint:** *A useful fact is that, for any symbol τ of degree α and any model (Π, Γ), the function $y \mapsto f_x^\tau(y) = \Gamma_{yx}\tau - \tau$ belongs to $\mathscr{D}^\alpha$.*

* **Exercise 13.7** *Show that the limit $\varphi^{(n)} = \lim_{m\to\infty} \varrho^{(n,m)}$ with $\varrho^{(n,m)}$ as in (13.18) exists and belongs to C_c^∞, with the limit being taken in C^r for any $r > 0$. Show furthermore that, despite the fact that one necessarily has $\int |\varrho(x)|\, dx > 1$ (why?), there exists a constant C such that $\int |\varrho^{(n,m)}(x)|\, dx < C$, uniformly over $n, m \in \mathbf{N}$.* **Hint:** *Work in Fourier space to show existence and smoothness of the limit and in direct space to show that it has compact support.*

Exercise 13.8 *Show that it is possible to find a smooth compactly supported function ϱ such that (13.17) holds.* **Hint:** *Note first that for any ψ integrating to 1 one can find a differential operator $\mathcal{L}$ of order α with constant coefficients and without constant term such that $\int \psi(x)P(x)\, dx = ((\mathrm{Id} - \mathcal{L})P)(0)$ for all polynomials P of degree α. Show that then $\varrho = \sum_{k \le \alpha} (\mathcal{L}^*)^k \psi$ does the trick, where $\mathcal{L}^*$ denotes the formal adjoint of $\mathcal{L}$.*

Exercise 13.9 *Show that the construction of Section 2.4 determines a regularity structure with $T = T^{(p)}(\mathbf{R}^d)$, structure group $G^{(p)}(\mathbf{R}^d)$, and such that $\deg e_w = \alpha|w|$. Show also that every rough path $\mathbf{X}$ determines a model for this regularity structure and that the definition of a controlled path given in Definition 4.18 coincides with the definition of the space $\mathscr{D}^{p\alpha}$ for the model associated to the rough path $\mathbf{X}$.*

Exercise 13.10 *Show that one can indeed take $\varphi = \mathbf{1}_{[0,1]}$ in the last step of the proof of Proposition 13.21.* **Hint:** *show first that one can write*

$$\mathbf{1}_{[0,1]} = \sum_{n \ge 0} (\varphi_n + \psi_n),$$

where φ_n is supported on $[0, 2^{-n}]$, ψ_n is supported on $[1 - 2^{-n}, 1]$, all of these functions are smooth, and $\|D^k \varphi_n\|_\infty + \|D^k \psi_n\|_\infty \le C2^{kn}$ for some $C > 0$, uniformly over $n \ge 0$ and $k \in [0, r]$.

Exercise 13.11 *Given a fixed regularity structure and model, given $\gamma > 0$, $\tau \in T_\gamma$ and $x \in \mathbf{R}^d$, define a function $f_{x,\tau}\colon \mathbf{R}^d \to T_{<\gamma}$ by*

$$f_{x,\tau}(y) = \Gamma_{yx}\tau - \tau\,.$$

Show that $f_{x,\tau} \in \mathscr{D}^\gamma$ and that one has $\mathcal{R}f_{x,\tau} = \Pi_x\tau$. Use this to give another proof of Lyons' extension theorem (Exercise 4.6).

13.6 Comments

All basic definitions (regularity structure, model, modelled distribution, ...) are taken from [Hai14b]. An alternative theory to the theory of regularity structures was introduced more or less simultaneously in Gubinelli–Imkeller–Perkowski [GIP15]. Instead of the reconstruction theorem, that theory builds on properties of Bony's paraproduct [Bon81, BMN10, BCD11] and it introduces a notion of "paracontrolled distribution" which replaces the notion of "modelled distribution". This theory is also able to deal with stochastic PDEs like the KPZ equation or the dynamical Φ^4_3 equation, see Catellier–Chouk [CC18b], but its scope is not as wide as that of the theory of regularity structures. For example, as it stands it does not appear to be able to deal with classical one-dimensional parabolic SPDEs driven by space-time white noise with a diffusion coefficient depending on the solution or the type of equation arising as natural evolutions on the space of loops with values in a manifold [Hai16, BGHZ19]. This is however evolving rapidly as a number of recent results show that paracontrolled calculus can alternatively be used as the foundation for the analytical aspects of the theory of regularity structures. We refer to [BB19, BH18, MP18, BH19, BM19] for more details.

One advantage of the paraproduct-based theory is that one generally deals with globally defined objects rather than the "jets" used in the theory of regularity structures. It also uses some already well-studied objects, so that it can rely on a substantial body of existing literature. On the flip side, it usually achieves a less clean break between the analytical and the algebraic aspects of a given problem. Furthermore, while the probabilistic aspects of the theory are expected to be equivalent to some extent, it is not completely clear how an analogue of the results [CH16] would even be formulated in the paracontrolled setting, although the results mentioned above may provide a hint. A third approach, closer in spirit to Wilson's renormalisation group ideas, was developed by Kupiainen [Kup16] who used it to give an alternative construction of the solutions to the dynamical Φ^4_3 equation.

The regularity structure view on rough paths, Sections 13.2.2 and 13.3.2, is further explored in [BCFP19]; see also [Hai14b, Sec. 4.4]. As already mentioned, the original proof of the reconstruction theorem given in [Hai14b] (also reproduced in the first edition of this book) relies on wavelet analysis, in particular on the existence of compactly supported wavelets of arbitrary regularity [Dau88]. The new proof in Section 13.4 was inspired by [OSSW18, MW18] and has the advantage of being entirely self-contained. One additional advantage is that the current proof immediately generalises to scalings $\mathfrak{s}$ that are not necessarily rational. (Rationality of $\mathfrak{s}$ was required in the original articles in order to be able to build a suitable wavelet basis by tensorisation of one-dimensional wavelet bases.)

One advantage of the proof using wavelets is that it implies that a model is uniquely determined by the actions of Π_x and Γ_{xy} on countably many translates and scalings of a finite number of functions and for a countable number of values of x, y. It also makes it very easy to prove a Kolmogorov-type criterion for models, see [Hai14b, Prop. 3.32 & Thm. 10.7].

Chapter 14
Operations on modelled distributions

The original motivation for the development of the theory of regularity structures was to provide robust solution theories for singular stochastic PDEs like the KPZ equation or the dynamical Φ_3^4 model. The idea is to reformulate them as fixed point problems in some space $\mathscr{D}^\gamma$ (or rather a slightly modified version that takes into account possible singular behaviour near time 0) based on a suitable random model in a regularity structure purpose-built for the problem at hand. In order to achieve this this chapter provides a systematic way of formulating the standard operations arising in the construction of the corresponding fixed point problem (differentiation, multiplication, composition by a regular function, convolution with the heat kernel) as operations on the spaces $\mathscr{D}^\gamma$.

14.1 Differentiation

Being a local operation, differentiating a modelled distribution is straightforward, provided that the model one works with is sufficiently rich. Denote by $\mathcal{L}$ some (formal) differential operator with constant coefficients that is homogeneous of degree m, i.e. it is of the form

$$\mathcal{L} = \sum_{|k|=m} a_k D^k \,,$$

where k is a d-dimensional multi-index, $a_k \in \mathbf{R}$, and D^k denotes the kth mixed derivative in the distributional sense.

Given a regularity structure (T, G), it is convenient to define "abstract" differentiation only on suitable substructures. The appropriate notion of sector was already introduced in Definition 13.1. We have

Definition 14.1. Consider a sector $V \subset T$. A linear operator $\partial \colon V \to T$ is said to *realise* $\mathcal{L}$ (of degree m) for the model (Π, Γ) if

© Springer Nature Switzerland AG 2020
P. K. Friz, M. Hairer, *A Course on Rough Paths*, Universitext,
https://doi.org/10.1007/978-3-030-41556-3_14

- one has $\partial \tau \in T_{\alpha-m}$ for every $\tau \in V_\alpha$,
- one has $\Gamma \partial \tau = \partial \Gamma \tau$ for every $\tau \in V$ and every $\Gamma \in G$.
- one has $\Pi_x \partial \tau = \mathcal{L} \Pi_x \tau$ for every $\tau \in V$ and every $x \in \mathbf{R}^d$.

Writing $\mathscr{D}^\gamma(V)$ for those elements in $\mathscr{D}^\gamma$ taking values in the sector V, it then turns out that one has the following fact:

Proposition 14.2. *Assume that ∂ realises $\mathcal{L}$ for the model (Π, Γ) and let $f \in \mathscr{D}^\gamma(V)$ for some $\gamma > m$. Then, $\partial f \in \mathscr{D}^{\gamma-m}$ and the identity $\mathcal{R} \partial f = \mathcal{L} \mathcal{R} f$ holds.*

Proof. The fact that $\partial f \in \mathscr{D}^{\gamma-m}$ is an immediate consequence of the definitions, so we only need to show that $\mathcal{R} \partial f = \mathcal{L} \mathcal{R} f$.

By the "uniqueness" part of the reconstruction theorem, this on the other hand follows immediately if we can show that, for every fixed test function ψ and every $x \in \mathbf{R}^d$, one has

$$\left(\Pi_x \partial f(x) - \mathcal{L} \mathcal{R} f \right)(\psi_x^\lambda) \lesssim \lambda^\delta \,,$$

for some $\delta > 0$. Here, we defined ψ_x^λ as before. By the assumption on the model Π, we have the identity

$$\left(\Pi_x \partial f(x) - \mathcal{L} \mathcal{R} f \right)(\psi_x^\lambda) = \left(\mathcal{L} \Pi_x f(x) - \mathcal{L} \mathcal{R} f \right)(\psi_x^\lambda) = - \left(\Pi_x f(x) - \mathcal{R} f \right)(\mathcal{L}^* \psi_x^\lambda) \,,$$

where $\mathcal{L}^*$ is the formal adjoint of $\mathcal{L}$. Since, as a consequence of the homogeneity of $\mathcal{L}$, one has the identity $\mathcal{L}^* \psi_x^\lambda = \lambda^{-m} \left(\mathcal{L}^* \psi \right)_x^\lambda$, it then follows immediately from the reconstruction theorem that the right-hand side of this expression is of order $\lambda^{\gamma-m}$, as required. $\square$

14.2 Products and composition by regular functions

One of the main purposes of the theory presented here is to give a robust way to multiply distributions (or functions with distributions) that goes beyond the barrier illustrated by Theorem 13.18. Provided that our functions / distributions are represented as elements in $\mathscr{D}^\gamma$ for some model and regularity structure, we can multiply their "Taylor expansions" pointwise, provided that we give ourselves a table of multiplication on T.

It is natural to consider products with the following properties.

Definition 14.3. Given a regularity structure (T, G) and two sectors $V, \bar{V} \subset T$, a *product* on $(V, \bar{V})$ is a bilinear map $\star \colon V \times \bar{V} \to T$ such that, for any $\tau \in V_\alpha$ and $\bar{\tau} \in \bar{V}_\beta$, one has $\tau \star \bar{\tau} \in T_{\alpha+\beta}$ and such that, for any element $\Gamma \in G$, one has $\Gamma(\tau \star \bar{\tau}) = \Gamma \tau \star \Gamma \bar{\tau}$.

Remark 14.4. The condition that degrees add up under multiplication is very natural, bearing in mind the case of the polynomial regularity structure. The second condition is also very natural since it merely states that if one reexpands the product of two "polynomials" around a different point, one should obtain the same result as if one reexpands each factor first and then multiplies them together.

Given such a product, we can ask ourselves when the pointwise product of an element $\mathscr{D}^{\gamma_1}$ with an element in $\mathscr{D}^{\gamma_2}$ again belongs to some $\mathscr{D}^{\gamma}$. In order to answer this question, we introduce the notation $\mathscr{D}_\alpha^\gamma$ to denote those elements $f \in \mathscr{D}^\gamma$ such that furthermore

$$f(x) \in T_{\geq\alpha} = \bigoplus_{\beta \geq \alpha} T_\beta ,$$

for every x. With this notation at hand, it is not hard to show:

Theorem 14.5. Let $f_1 \in \mathscr{D}_{\alpha_1}^{\gamma_1}(V)$, $f_2 \in \mathscr{D}_{\alpha_2}^{\gamma_2}(\bar{V})$, and let $\star$ be a product on $(V, \bar{V})$. Then, the function f given by $f(x) = f_1(x) \star f_2(x)$ belongs to $\mathscr{D}_\alpha^\gamma$ with

$$\alpha = \alpha_1 + \alpha_2 , \qquad \gamma = (\gamma_1 + \alpha_2) \wedge (\gamma_2 + \alpha_1) . \tag{14.1}$$

Proof. It is clear that $f(x) \in T_{\geq\alpha}$, so it remains to show that it belongs to $\mathscr{D}^\gamma$. Furthermore, since we are only interested in showing that $f_1 \star f_2 \in \mathscr{D}^\gamma$, we discard all of the components in T_β for $\beta \geq \gamma$.

By the properties of the product $\star$, it remains to obtain a bound of the type

$$\|\Gamma_{xy} f_1(y) \star \Gamma_{xy} f_2(y) - f_1(x) \star f_2(x)\|_\beta \lesssim |x - y|^{\gamma - \beta} .$$

By adding and subtracting suitable terms, we obtain

$$\begin{aligned}
\|\Gamma_{xy} f(y) - f(x)\|_\beta \leq &\; \|(\Gamma_{xy} f_1(y) - f_1(x)) \star (\Gamma_{xy} f_2(y) - f_2(x))\|_\beta \\
&+ \|(\Gamma_{xy} f_1(y) - f_1(x)) \star f_2(x)\|_\beta \\
&+ \|f_1(x) \star (\Gamma_{xy} f_2(y) - f_2(x))\|_\beta .
\end{aligned} \tag{14.2}$$

It follows from the properties of the product $\star$ that the first term in (14.2) is bounded by a constant times

$$\sum_{\beta_1 + \beta_2 = \beta} \|\Gamma_{xy} f_1(y) - f_1(x)\|_{\beta_1} \|\Gamma_{xy} f_2(y) - f_2(x)\|_{\beta_2}$$

$$\lesssim \sum_{\beta_1 + \beta_2 = \beta} \|x - y\|^{\gamma_1 - \beta_1} \|x - y\|^{\gamma_2 - \beta_2} \lesssim \|x - y\|^{\gamma_1 + \gamma_2 - \beta} .$$

Since $\gamma_1 + \gamma_2 \geq \gamma$, this bound is as required. The second term is bounded by a constant times

$$\sum_{\beta_1 + \beta_2 = \beta} \|\Gamma_{xy} f_1(y) - f_1(x)\|_{\beta_1} \|f_2(x)\|_{\beta_2} \lesssim \sum_{\beta_1 + \beta_2 = \beta} \|x - y\|^{\gamma_1 - \beta_1} \mathbf{1}_{\beta_2 \geq \alpha_2}$$

$$\lesssim \|x - y\|^{\gamma_1 + \alpha_2 - \beta} ,$$

where the second inequality uses the identity $\beta_1 + \beta_2 = \beta$. Since $\gamma_1 + \alpha_2 \geq \gamma$, this bound is again of the required type. The last term is bounded similarly by reversing the roles played by f_1 and f_2. $\quad\square$

Remark 14.6. Strictly speaking, it is the projection of $f(x) = f_1(x) \star f_2(x)$ to $T_{<\gamma}$ that belongs to $\mathscr{D}_\alpha^\gamma$, see Exercise 13.4.

Remark 14.7. It is clear that the formula (14.1) for γ is optimal in general as can be seen from the following two "reality checks". First, consider the case of the polynomial model and take $f_i \in C^{\gamma_i}$. In this case, the (abstract) truncated Taylor series f_i for f_i belong to $\mathscr{D}_0^{\gamma_i}$. It is clear that in this case, the product cannot be expected to have better regularity than $\gamma_1 \wedge \gamma_2$ in general, which is indeed what (14.1) states. The second reality check comes from (the proof of) Theorem 13.18. In this case, with $\beta > \alpha \geq 0$, one has $f \in \mathscr{D}_0^\beta$, while the constant function $x \mapsto \Xi$ belongs to $\mathscr{D}_{-\alpha}^\infty$ so that, according to (14.1), one expects their product to belong to $\mathscr{D}_{-\alpha}^{\beta-\alpha}$, which is indeed the case.

It turns out that if we have a product on a regularity structure, then in many cases this also naturally yields a notion of composition with regular functions. Of course, one could in general not expect to be able to compose a regular function with a distribution of negative order. As a matter of fact, we will only define the composition of regular functions with elements in some $\mathscr{D}^\gamma$ for which it is guaranteed that the reconstruction operator yields a continuous function. One might think at this case that this would yield a triviality, since we know of course how to compose arbitrary continuous function. The subtlety is that we would like to design our composition operator in such a way that the result is again an element of $\mathscr{D}^\gamma$.

For this purpose, we say that a given sector $V \subset T$ is *function-like* if $\alpha < 0 \implies V_\alpha = 0$ and if V_0 is one-dimensional. (Denote the unit vector of V_0 by **1**.) We will furthermore always assume that our models are *normal* in the sense that $(\Pi_x \mathbf{1})(y) = 1$. In this case, it turns out that if $f \in \mathscr{D}^\gamma(V)$ for a function-like sector V, then $\mathcal{R}f$ is a continuous function and one has the identity $(\mathcal{R}f)(x) = \langle \mathbf{1}, f(x) \rangle$, where we denote by $\langle \mathbf{1}, \cdot \rangle$ the element in the dual of V which picks out the prefactor of **1**.

Assume now that we are given a regularity structure with a function-like sector V and a product $\star \colon V \times V \to V$. For any smooth function $G \colon \mathbf{R} \to \mathbf{R}$ and any $f \in \mathscr{D}^\gamma(V)$ with $\gamma > 0$, we can then *define* $G \circ f$ (also denoted $G(f)$) to be the V-valued function given by

$$
(G \circ f)(x) = \sum_{k \geq 0} \frac{G^{(k)}(\bar{f}(x))}{k!} \mathcal{Q}_{<\gamma} \tilde{f}(x)^{\star k} ,
$$

where we have set

$$
\bar{f}(x) = \langle \mathbf{1}, f(x) \rangle , \qquad \tilde{f}(x) = f(x) - \bar{f}(x) \mathbf{1} ,
$$

and weher $\mathcal{Q}_{<\gamma} \colon T \to T_{<\gamma}$ is the natural projection. Here, $G^{(k)}$ denotes the kth derivative of G and $\tau^{\star k}$ denotes the k-fold product $\tau \star \cdots \star \tau$. We also used the usual conventions $G^{(0)} = G$ and $\tau^{\star 0} = \mathbf{1}$.

Note that as long as G is C^∞, this expression is well-defined. Indeed, by assumption, there exists some $\alpha_0 > 0$ such that $\tilde{f}(x) \in T_{\geq \alpha_0}$. By the properties of the product, this implies that one has $\tilde{f}(x)^{\star k} \in T_{\geq k\alpha_0}$. As a consequence, when considering the component of $G \circ f$ in T_β for $\beta < \gamma$, the only terms that give a contribution are those with $k < \gamma/\alpha_0$. Since we cannot possibly hope in general that $G \circ f \in \mathscr{D}^{\gamma'}$ for some $\gamma' > \gamma$, this is all we really need.

It turns out that if G is sufficiently regular, then the map $f \mapsto G \circ f$ enjoys similarly nice continuity properties to what we are used to from classical Hölder spaces. The following result is the analogue in this context to Lemma 7.3:

Proposition 14.8. *In the same setting as above, provided that G is of class C^k with $k > \gamma/\alpha_0$, the map $f \mapsto G \circ f$ is continuous from $\mathscr{D}^\gamma(V)$ into itself. If $k > \gamma/\alpha_0 + 1$, then it is locally Lipschitz continuous.*

The proof of the first statement can be found in [Hai14b], while the second statement was shown in [HP15]. It is a somewhat lengthy, but ultimately rather straightforward calculation.

14.3 Classical Schauder estimates

One of the reasons why the theory of regularity structures is very successful at providing detailed descriptions of the small-scale features of solutions to semilinear (S)PDEs is that it comes with very sharp Schauder estimates. A full proof of the Schauder estimates for regularity structures is beyond the scope of this book, but we want to convey the flavour of the proof. The aim of this section is therefore to give a self-contained proof of the classical Schauder estimates which state that for any (compactly supported) kernel K that is approximately homogeneous of degree $\beta - d$, the convolution map $\zeta \mapsto K * \zeta$ is continuous from $\mathcal{C}^\alpha$ to $\mathcal{C}^{\alpha+\beta}$, provided that $\alpha + \beta$ is not a positive integer. We first make precise our assumptions on the kernel K.

Definition 14.9. Given $\beta > 0$, a kernel $K : \mathbf{R}^d \backslash \{0\} \to \mathbf{R}$, smooth except for a singularity at the origin, is said to be β-*regularising* if it is supported in the unit ball around the origin and, for every $k \in \mathbf{N}^d$, there exists a constant C such that $|D^k K(x)| \leq C|x|^{\beta-d-|k|}$.

Immediate examples are (smooth truncations of) the Newton potential in dimension $d \geq 3$, proportional to $1/|x|^{d-2}$ and hence 2-regularising, the fractional Volterra kernel $(x^{H-1/2} 1_{x>0})$ with $d = 1$ and $\beta = H + 1/2$. The heat kernel on space-time $\mathbf{R}^{n+1}$, proportional to $(t, x) \mapsto t^{-n/2} \exp(-\frac{|x|^2}{4t}) 1_{t>0}$, also fits in this setting (and is 2-regularising), provided one works with "parabolic" scaling (cf. Remark 13.9).

As in Section 13.3, and for any $r \in \mathbf{N}$, we work with $\mathcal{B}_r \subset \mathcal{D}$, the set of smooth test functions with C^r-norm bounded by 1 and supported in the unit ball. It will be convenient for the purpose of this section to write $\mathcal{B}_{r,x}^\lambda$ for the set of all test functions of the form φ_x^λ with $\varphi \in \mathcal{B}_r$. Such $\psi \in \mathcal{B}_{r,x}^\lambda$ are characterised by having support in the ball of radius λ centred at x and derivatives bounds $|D^k \psi| \leq \lambda^{-d-|k|}$ for $|k| \leq r$. We also note that, for any real $s \in [0, r]$, the estimate $\|\psi\|_{\mathcal{C}^s} \lesssim \lambda^{-d-s}$ holds true.

Lemma 14.10. *Given a β-regularising kernel K and $r \geq 0$, one can write $K = \sum_{n \geq -1} K_n$ in such a way that $2^{\beta n} K_n \in C\mathcal{B}_{r,0}^{2^{-n}}$ for some $C > 0$.*

Proof. As is common in the construction of Paley–Littlewood blocks, we work with a dyadic partitions of unity, based on a smooth "cutoff" function" $\varphi : \mathbf{R}_+ \to [0,1]$, supported in $[2^{-1}, 2^1]$, such that $\sum_{n \geq 0} \varphi_n \equiv 1$ on $(0,1]$, where $\varphi_n := \varphi(2^n \cdot)$ is supported in $[2^{-n-1}, 2^{-n+1}]$. Since K is supported in $\{x : |x| \leq 1\}$, the stated decomposition clearly holds with (smooth) $K_n(x) := \varphi_{n+1}(|x|)K(x)$, supported in the ball of radius 2^{-n} centred at the origin. To see that $2^{\beta n} K_n \in C\mathcal{B}_{r,0}^{2^{-n}}$, for given $r \geq 0$, it remains to see that $|D^j K_n| \lesssim (2^{-n})^{\beta - d - |j|}$ for $|j| \leq r$. This estimate holds, with K_n replaced by K, by the defining property of a β-regularising kernel, restricted to $x \asymp 2^{-n}$. On the other hand, $|D^i \varphi_n| = |(2^n)^{|i|} D^i \varphi| \lesssim (2^n)^{|i|}$, and we conclude with Leibnitz' product rule. $\quad\square$

The following simple proposition is the first crucial ingredient in our approach. Loosely speaking, it states that the convolution of two test functions localised at two distinct scales is localised at the sum (or equivalently maximum) of the two scales and that one gains in amplitude if the tighter of the two test functions annihilates polynomials of a certain degree.

Proposition 14.11. *There exists $C > 0$ such that, for all $\varphi \in \mathcal{B}_{r,x}^\lambda$ and $\psi \in \mathcal{B}_{r,y}^\mu$, one has $\psi * \varphi \in C\mathcal{B}_{r,x+y}^{\lambda+\mu}$. If furthermore $\lambda \leq \mu$ and $\int P(z)\varphi(z)\,dz = 0$ for every polynomial P with $\deg P < \gamma \leq r$, some $\gamma \in \mathbf{R}_+$, then $\psi * \varphi \in C(\lambda/\mu)^\gamma \mathcal{B}_{\lfloor r - \gamma \rfloor, x+y}^{2\mu}$.*

Proof. Clearly, $\psi * \varphi$ is supported in the ball of radius $\lambda + \mu$ centred at $x + y$. For the first claim, by swapping the roles of φ and ψ if necessary, we may assume $\lambda \leq \mu$. To see that the convolution yields an element in $\mathcal{B}_{r,x+y}^{\lambda+\mu}$, in view of the characterisation of such spaces, it suffices to estimate, for $|k| \leq r$, $D^k(\psi * \varphi) = (D^k \psi) * \varphi$ using $|(D^k \psi)| \lesssim \mu^{-d-|k|} \asymp (\lambda + \mu)^{-d-|k|}$ and $\int |\varphi(z)|\,dz \leq C$ (independent of λ). Regarding the second claim, we write

$$
\begin{aligned}
D^k(\psi * \varphi)(\cdot) &= \int \psi^{(k)}(\cdot - z)\,\varphi(z)\,dz \\
&= \int \left(\psi^{(k)}(\cdot - z) - P_\cdot^{\gamma;(k)}(\cdot - z) \right) \varphi(z)\,dz \, ,
\end{aligned}
$$

for $0 \leq |k| \leq r - \gamma$, where $P_\cdot^{\gamma;(k)}$ denotes the Taylor expansion (at the dotted base-point) of $\psi^{(k)} \equiv D^k \psi$ of integer degree $\gamma - \{\gamma\} < \gamma$ (annihilated by φ). It remains to be seen that, for all such k,

$$
|D^k(\varphi * \psi)(\cdot)| \lesssim (\lambda/\mu)^\gamma \mu^{-d-|k|} \, .
$$

To this end, using that $\gamma + |k| \leq r$, one has the estimate

$$
|\psi^{(k)}(\cdot - z) - P_\cdot^{k,\gamma}(\cdot - z)| \lesssim \|\psi\|_{C^{\gamma + |k|}} |z|^\gamma \lesssim \mu^{-d-\gamma-|k|} |z|^\gamma \, .
$$

We only need to consider z in the support of φ, and in fact can assume without loss of generality that $x = 0$ (otherwise subtract another annihilated Taylor polynomial...), so that $\int |z|^\gamma |\varphi(z)| \, dz \le \lambda^\gamma \int |\varphi(z)| \, dz \lesssim \lambda^\gamma$. The desired estimate now follows. □

Our second crucial ingredient is a characterisation of Hölder spaces that is well adapted to our approach. For this, we define the following scale of spaces of distributions.

Definition 14.12. For $\alpha \in \mathbf{R}$, write $r = r_o(\alpha)$ for the smallest non-negative integer such that $r + \alpha > 0$. We then define $\mathcal{Z}^\alpha$ as the space of distributions on $\mathbf{R}^d$ such that for every compact set $\mathfrak{K} \subset \mathbf{R}^d$ there exists a constant C such that the bound

$$|\zeta(\varphi)| \le C\lambda^\alpha ,$$

holds uniformly $\lambda \in (0, 1]$, $x \in \mathfrak{K}$ and all $\varphi \in \mathcal{B}^\lambda_{r,x}$ such that $\int \varphi(z) P(z) \, dz = 0$ for all polynomials P with $\deg P \le \alpha$. For any compact set $\mathfrak{K}$, the best possible constant such that the above bound holds uniformly over $x \in \mathfrak{K}$ yields a seminorm. The collection of these seminorms endows $\mathcal{Z}^\alpha$ with a Fréchet space structure.

The precise choice of r in Definition 14.12 is not very important, as one could have taken any other choice $r \ge r_o(\alpha)$. More precisely, one has the following result.

Lemma 14.13. *For $r \ge r_o(\alpha)$, write $\mathcal{Z}^\alpha_r$ for $\mathcal{Z}^\alpha$ as defined above, but with $r_o(\alpha)$ replaced by r. Then $\mathcal{Z}^\alpha_r = \mathcal{Z}^\alpha$.*

Proof. We fix a partition of unity $\{\chi_y\}_{y \in \Lambda}$ for $\mathbf{R}^d$ such that all the χ_y are translates of χ_0 by $y \in \mathbf{R}^d$ and $\Lambda \subset \mathbf{R}^d$ is a lattice. In particular, we make sure that $\chi_y \in \mathcal{B}^\lambda_{r,y}$. Given any $\lambda > 0$, we write $\chi_{y,\lambda}(x) = \chi_{y/\lambda}(x/\lambda)$ and we set $\Lambda_\lambda = \Lambda/\lambda$. We also fix a function $\psi \in \mathcal{C}^\infty$ with support in the centred unit ball and such that

$$\int_{\mathbf{R}^d} x^k \psi(x) \, dx = \delta_{k,0} , \qquad \forall k : |k| \le r . \tag{14.3}$$

(Such functions exist by Exercise 13.8.) We then write $\tilde{\psi}(x) = 2^d \psi(2x) - \psi(x)$ and note that by (14.3) one has $\int_{\mathbf{R}^d} x^k \tilde{\psi}(x) \, dx = 0$ for $|k| \le r$.

Let now $\alpha < 0$ and take $\zeta \in \mathcal{Z}^\alpha_r$, we want to show that $\zeta \in \mathcal{Z}^\alpha$. Given $\varphi \in \mathcal{B}^\lambda_{r_o,x}$ and setting $\lambda_n = 2^{-n}\lambda$, we write

$$\varphi = \varphi * \psi^\lambda + \sum_{n \ge 0} \sum_{y \in \Lambda_{\lambda_n}} \varphi_{n,y} , \qquad \varphi_{n,y} = \left(\varphi * \tilde{\psi}^{\lambda_n}\right) \cdot \chi_{y,\lambda_n} . \tag{14.4}$$

As a simple consequence of the Taylor remainder theorem, one has the bound

$$\left\| \varphi * D^k \tilde{\psi}^{\lambda_n} \right\|_\infty \lesssim \lambda^{-d} 2^{-r_o n} \lambda_n^{-|k|} = 2^{-(d+r_o)n} \lambda_n^{-d-|k|} ,$$

so that there exists a constant C independent of φ such that $\varphi_{n,y} \in C 2^{-(d+r_o)n} \mathcal{B}^{\lambda_n}_{r,y}$, which in particular implies that

$$|\zeta(\varphi_{n,y})| \lesssim \lambda^\alpha 2^{-(d+r_o+\alpha)n} \ . \tag{14.5}$$

Since the number of terms in Λ_{λ_n} such that $\varphi_{n,y}$ is non-zero is of order 2nd, we conclude that

$$|\zeta(\varphi)| \lesssim \lambda^\alpha + \sum_{n \geq 0} \lambda^\alpha 2^{-(r_o+\alpha)n} \lesssim \lambda^\alpha \ ,$$

where we used the fact that $r_o + \alpha > 0$ by definition.

Note that the assumption $\alpha < 0$ was used in order to obtain the bound (14.5) since there is no reason for $\varphi_{n,y}$ to annihilate polynomials even if φ does. The case $\alpha > 0$ is easier, noting that the definition of $\mathcal{Z}_r^\alpha$ implies that $\zeta * \tilde{\psi}^{\lambda_n}$ is a continuous function bounded by $\mathcal{O}(\lambda_n^\alpha)$. We then use the fact that

$$\zeta(\varphi) = \zeta(\varphi * \psi^\lambda) + \sum_{n \geq 0} \langle \zeta * \tilde{\psi}^{\lambda_n}, \varphi \rangle \ ,$$

with $\langle \cdot, \cdot \rangle$ denoting the L^2 scalar product, combined with the fact that φ integrates to $\mathcal{O}(1)$, to conclude that $|\zeta(\varphi)| \lesssim \lambda^\alpha(1 + \sum_{n \geq 0} 2^{-\alpha n}) \lesssim \lambda^\alpha$ as required.

The case $\alpha = 0$ is a bit more delicate and we leave it as Exercise 14.3. $\quad \square$

Remark 14.14. Validity of the stated bounds implies that distributions in $\mathcal{Z}^\alpha \subset \mathcal{D}'$ can be extended canonically to test functions in $\mathcal{C}_c^r$ (elements in $\mathcal{C}^r$ with compact support). In this sense, $\mathcal{Z}^\alpha$ is contained in the topological dual of $\mathcal{C}_c^r$. (The situation is similar in the definition of models, cf. Remark 13.7.)

For $\alpha < 0$, the polynomial-annihilation condition is void and there is no additional condition on φ besides $\varphi \in \mathcal{B}_{r,x}^\lambda$. In this case $\mathcal{Z}^\alpha$ is precisely the negative Hölder space $\mathcal{C}^\alpha$ introduced in Section 13.3.1. The following proposition shows that to some extent this is also true in case of positive Hölder spaces, as previously encountered in Section 13.3.1.

Proposition 14.15. *For $\alpha \notin \mathbf{N}$, one has $\mathcal{Z}^\alpha = \mathcal{C}^\alpha$.*

Proof. There is nothing to prove for $\alpha < 0$, so let $\alpha > 0$. We first show that $\mathcal{C}^\alpha \subset \mathcal{Z}^\alpha$, this inclusion also being valid for integer values of α. In fact, it suffices to note that, given $f \in \mathcal{C}^\alpha$ and $\varphi \in \mathcal{B}_{r,x}^\lambda$ as in Definition 14.12, one has

$$\int f(y)\varphi(y)\, dy = \int \left(f(y) - P_x^\alpha(y-x)\right)\varphi(y)\, dy \lesssim \lambda^\alpha \ ,$$

where the identity follows from the fact that φ annihilates P_x^α, the Taylor expansion at order α of f, based at x, and the bound is as in the proof of Proposition 14.11.

For the converse inclusion, we first consider the case $\alpha \in (0,1)$ and let $\zeta \in \mathcal{Z}^\alpha$. Let $\varrho \colon \mathbf{R}^d \to \mathbf{R}$ be a smooth function that is compactly supported in the unit ball around the origin and such that $\int \varrho(z)\, dz = 1$. Note first that, for any $x \in \mathbf{R}^d$ and $\lambda \in (0,1]$, it follows from the definition of $\mathcal{Z}^\alpha$ that one has the bound

$$|\zeta(\varrho_x^{2^{-n}\lambda}) - \zeta(\varrho_x^{2^{-n-1}\lambda})| = |\zeta(\varrho_x^{2^{-n}\lambda} - \varrho_x^{2^{-n-1}\lambda})| \leq C\lambda^\alpha 2^{-\alpha n} \ .$$

It follows that $f(x) = \lim_{n \to \infty} \zeta(\varrho_x^{2^{-n}\lambda})$ is well-defined and that

$$|f(x) - \zeta(\varrho_x^\lambda)| \lesssim \lambda^\alpha .$$

As a consequence, one has

$$|f(x) - f(y)| \lesssim \lambda^\alpha + |\zeta(\varrho_x^\lambda - \varrho_y^\lambda)| .$$

Choosing $\lambda = |x - y|$, it follows that $f \in \mathcal{C}^\alpha$. The fact that $f = \zeta$ in the sense that $\zeta(\varphi) = \int f(z)\,\varphi(z)\,dz$ follows immediately from the fact that

$$\zeta(\varphi) = \lim_{\lambda \to 0} \zeta(\varphi * \varrho^\lambda) = \lim_{\lambda \to 0} \int \zeta(\varrho_x^\lambda)\,\varphi(x)\,dx .$$

The claim for general non-integer α can then be seen from the fact that $\zeta \in \mathcal{Z}^\alpha$ implies $D^k \zeta \in \mathcal{Z}^{\alpha - |k|}$ (interpreted as distributional derivatives) for every multi-index k. Details are left to the reader. □

Remark 14.16. For $n \in \mathbf{N}$, the spaces $\mathcal{Z}^n$ are usually called *Hölder–Zygmund spaces* in the literature (thus our choice of symbol $\mathcal{Z}$). They are distinct from the usual Hölder spaces since one can check that $x \mapsto x^n \log x$ belongs to $\mathcal{Z}^n$, but not to $\mathcal{C}^n$.

With all of these preliminaries in place, we can give a very simple proof of Schauder's theorem. (See for example [Sim97] for an alternative proof of a very similar statement.)

Theorem 14.17. *For any β-regularising kernel K, the map $\zeta \mapsto K * \zeta$ is continuous from $\mathcal{Z}^\alpha$ to $\mathcal{Z}^{\alpha + \beta}$ for every $\alpha \in \mathbf{R}$.*

Proof. Let $\zeta \in \mathcal{Z}^\alpha$ and let $\varphi \in \mathcal{B}_{r,x}^\lambda$ where we will (and can by Lemma 14.13) work with suitable $r \geq r_o(\alpha + \beta)$, chosen below, such that $\int \varphi(z)P(z)\,dz = 0$ for every P with $\deg P \leq \alpha + \beta$. Lemma 14.10 yields a decomposition $(K_n : n \geq -1)$ for $\check{K}(x) = K(-x)$, so that

$$(K * \zeta)(\varphi) = \zeta(\check{K} * \varphi). = \sum_n \zeta(K_n * \varphi) = \sum_n 2^{-\beta n} \zeta(2^{\beta n} K_n * \varphi) , \quad (14.6)$$

with $2^{\beta n} K_n \in C\mathcal{B}_{r,0}^{2^{-n}}$ for some $C > 0$. It then follows from Proposition 14.11 (applied with $\mu = 2^{-n}$, noting that $K_n * \varphi$ also annihilates polynomials of degree up to $\alpha + \beta$) and the definition of $\mathcal{Z}^\alpha$ that

$$|\zeta(2^{\beta n} K_n * \varphi)| \lesssim \begin{cases} \lambda^\alpha & \text{if } 2^{-n} \leq \lambda, \\ (2^n \lambda)^\gamma 2^{-\alpha n} & \text{otherwise,} \end{cases}$$

provided $\lfloor r - \gamma \rfloor \geq r_o(\alpha + \beta)$. We will also need $\gamma > \alpha + \beta$, so that for instance $r := 2(|\alpha| + \beta) + 2$ is a safe choice. Inserting this bound into (14.6), and using $\beta > 0, \gamma > \alpha + \beta$ to estimate the geometric sums, one has the bounds

$$\sum_{\substack{n \geq 0 \\ 2^{-n} \leq \lambda}} 2^{-\beta n} \lambda^\alpha \lesssim \lambda^{\alpha+\beta}, \qquad \sum_{\substack{n \geq 0 \\ 2^{-n} \geq \lambda}} 2^{(\gamma-\alpha-\beta)n} \lambda^\gamma \lesssim \lambda^{\alpha+\beta},$$

it follows that $|(K * \zeta)(\varphi)| \lesssim \lambda^{\alpha+\beta}$, whence the claim follows. $\qquad \square$

Remark 14.18. The proof is (much) simpler in the "negative" case, with Hölder exponents $\alpha < \alpha + \beta < 0$. In essence, this is due to the absence of polynomial vanishing conditions. More specifically, one can take $r = r_o(\alpha + \beta)$ in the above proof, and then $\gamma = 0$ later on, so that only the easy (first) part of Proposition 14.15 is used. A reduction of the general to the negative case, in dimension $d = 1$, is discussed in Exercise 14.2.

Remark 14.19. One can verify that the proof never made explicit use of the Euclidean scaling and can be adapted mutatis mutandis to the case of arbitrary scalings as mentioned in Remark 13.9, provided that the notion of "β-regularising kernel" is adjusted accordingly (replace the exponent $\beta - d - |k|$ by $\beta - |\mathfrak{s}| - |k|_\mathfrak{s}$).

14.4 Multilevel Schauder estimates and admissible models

As we saw in the previous section, the classical Schauder estimates state that if $K \colon \mathbf{R}^d \to \mathbf{R}$ is a kernel that is smooth everywhere, except for a singularity at the origin that is approximately homogeneous of degree $\beta - d$ for some fixed $\beta > 0$ (i.e. it is β-regularising in the sense of Definition 14.9), then the operator $f \mapsto K * f$ maps $\mathcal{C}^\alpha$ into $\mathcal{C}^{\alpha+\beta}$ for every $\alpha \in \mathbf{R}$, except for those values for which $\alpha + \beta \in \mathbf{N}$.

It turns out that similar Schauder estimates hold in the context of general regularity structures in the sense that it is in general possible to build an operator $\mathcal{K} \colon \mathscr{D}^\gamma \to \mathscr{D}^{\gamma+\beta}$ with the property that $\mathcal{R}\mathcal{K}f = K * \mathcal{R}f$. We call such a statement a "multi-level Schauder estimate" since it is a form of Schauder estimate for all the components of f in T_α for all $\alpha < \gamma$. Of course, such a statement can only be expected to hold if our regularity structure contains not only the objects necessary to describe $\mathcal{R}f$ up to order γ, but also those required to describe $K * \mathcal{R}f$ up to order $\gamma + \beta$. What are these objects? At this stage, it might be useful to reflect on the effect of the convolution of a singular function (or distribution) with K.

Let us assume for a moment that a given real-valued function f is smooth everywhere, except at some point x_0. It is then straightforward to convince ourselves that $K * f$ is also smooth everywhere, except at x_0. Indeed, for any $\delta > 0$, we can write $K = K_\delta + K_\delta^c$, where K_δ is supported in a ball of radius δ around 0 and K_δ^c is a smooth function. Similarly, we can decompose f as $f = f_\delta + f_\delta^c$, where f_δ is supported in a δ-ball around x_0 and f_δ^c is smooth. Since the convolution of a smooth function with an arbitrary distribution is smooth, it follows that the only non-smooth component of $K * f$ is given by $K_\delta * f_\delta$, which is supported in a ball of radius 2δ around x_0. Since δ was arbitrary, the statement follows. By linearity, this strongly suggests that the local structure of the singularities of $K * f$ can be described completely by only using knowledge on the local structure of the singularities of f.

It also suggests that the "singular part" of the operator $\mathcal{K}$ should be local, with the non-local parts of $\mathcal{K}$ only contributing to the "regular part".

This discussion suggests that we need the following ingredients to build an operator $\mathcal{K}$ with the desired properties:

- The polynomial structure should be part of our regularity structure in order to be able to describe the "regular parts".
- We should be given an "abstract integration operator" $\mathcal{I}$ (of order β) on T which describes how the "singular parts" of $\mathcal{R}f$ transform under convolution by K.
- We should restrict ourselves to models which are "compatible" with the action of $\mathcal{I}$ in the sense that the behaviour of $\Pi_x \mathcal{I}\tau$ should relate in a suitable way to the behaviour of $K * \Pi_x \tau$ near x.

One way to implement these ingredients is to assume first that our regularity structure contains abstract polynomials in the following sense.

Assumption 14.20 *There exists a sector $\bar{T} \subset T$ isomorphic to the polynomial regularity structure. In other words, $\bar{T}_\alpha \neq 0$ if and only if $\alpha \in \mathbf{N}$, and one can find basis vectors X^k of $T_{|k|}$ such that every element $\Gamma \in G$ acts on $\bar{T}$ by $\Gamma X^k = (X + h\mathbf{1})^k$ for some $h \in \mathbf{R}^d$.*

Furthermore, we assume that there exists an abstract integration operator $\mathcal{I}$, of fixed order $\beta > 0$, with the following properties.

Assumption 14.21 *There exists a linear map $\mathcal{I}: V \to T$ for some sector $V \subset T$ such that $\mathcal{I}V_\alpha \subset T_{\alpha+\beta}$ and, for every $\Gamma \in G$ and $\tau \in T$,*

$$\Gamma \mathcal{I}\tau - \mathcal{I}\Gamma\tau \in \bar{T} . \tag{14.7}$$

Remark 14.22. We do not want to assume $\Gamma \mathcal{I} = \mathcal{I}\Gamma$. This is already seen in case of the rough path structure given by Definition 13.4. The map $\mathcal{I}: W^i \mapsto W^i$, $1 \leq i \leq e$, constitutes an abstract integration operator (defined on the sector $T_{\alpha-1}$). Since a generic $\Gamma_h \in G$ maps W^i to $W^i + h^i\mathbf{1}$, we see that $\Gamma \mathcal{I} - \mathcal{I}\Gamma \neq 0$ (for $h \neq 0$) and takes values in $T_0 = \langle\mathbf{1}\rangle$.

Finally, we want to restrict our attention to models that are compatible with this structure for a given kernel K in the following sense.

Definition 14.23. Given a β-regularising kernel K and a regularity structure $\mathscr{T}$ satisfying Assumptions 14.20 and 14.21, we say that a model (Π, Γ) is *admissible* if the identities

$$\left(\Pi_x X^k\right)(y) = (y - x)^k , \qquad \Pi_x \mathcal{I}\tau = K * \Pi_x \tau - \Pi_x \mathcal{J}_x \tau , \tag{14.8}$$

hold for every $\tau \in V$. Here, $\mathcal{J}_x : V \to \bar{T}$ is the linear map given on homogeneous elements by

$$\mathcal{J}_x \tau = \sum_{|k| < \deg \tau + \beta} \frac{X^k}{k!} \int D^k K(x - y) \left(\Pi_x \tau\right)(dy) . \tag{14.9}$$

Remark 14.24. In some cases, it will be convenient to introduce a whole family $\mathcal{I}_k$ of integration operators of order $\beta - |k|$. The notion of admissibility is then defined similarly, with $\mathcal{I}$ replaced by $\mathcal{I}_k$ and K replaced by $D^k K$, to the extent that these symbols are included in the structure space.

Remark 14.25. If ξ is smooth and we furthermore impose that Π_x is multiplicative (which is not enforced in general!), this yields a recursion to define the *canonical model* associated to ξ provided one manages to construct Γ_{xy} at the same time. The correct recursion to do this is

$$\Gamma_{xy}(\mathcal{I} + \mathcal{J}_y)\tau = (\mathcal{I} + \mathcal{J}_x)\Gamma_{xy}\tau \,, \tag{14.10}$$

which is clearly consistent with the constraint (14.7) and which one can show guarantees that $\Pi_x \Gamma_{xy} \mathcal{I}\tau = \Pi_y \mathcal{I}\tau$. See also Exercise 14.6.

Remark 14.26. Recall that if P is a polynomial and K is a compactly supported function, then $K * P$ is again a polynomial of the same degree as P. Since, for $\Pi_x \tau$ smooth enough, the term $\Pi_x \mathcal{J}_x \tau$ appearing in (14.8) is nothing but the Taylor expansion of $K * \Pi_x \tau$ around x, it follows that one has $\Pi_x \mathcal{I} X^k = 0$ for any multi-index k and any admissible model, which would suggest that one could have imposed the identity $\mathcal{I} X^k = 0$ already at the algebraic level. This would however create inconsistencies later on when incorporating renormalisation, unless we assume that $\int K(x)P(x)\,dx = 0$ for every polynomial P of degree N, for some sufficiently large value of N. Here, we chose to simply add instead $\mathcal{I} X^k$ as separate symbols to our regularity structure and to then set $\mathcal{I} X^k = \mathcal{I} X^k$.

Remark 14.27. While $K * \xi$ is well-defined for any distribution ξ, it is not so clear *a priori* whether the operator $\mathcal{J}_x$ given in (14.9) is also well-defined. It turns out that the axioms of a model do ensure that this is the case. The correct way of interpreting (14.9) is by

$$\mathcal{J}_x \tau = \sum_{|k| < \deg \tau + \beta} \sum_{n \geq 0} \frac{X^k}{k!}\left(\Pi_x \tau\right)\left(D^k K_n(x - \bullet)\right) \,,$$

with K_n as in Lemma 14.10. The scaling properties of the K_n ensure that the function $2^{(\beta - |k|)n} D^k K_n(x - \bullet)$ is a test function that is localised around x at scale 2^{-n}. As a consequence, one has

$$\left|\left(\Pi_x \tau\right)\left(D^k K_n(x - \bullet)\right)\right| \lesssim 2^{(|k| - \beta - \deg \tau)n} \,,$$

so that this expression is indeed summable as long as $|k| < \deg \tau + \beta$.

Remark 14.28. As a matter of fact, it turns out that the above definition of an admissible model dovetails very nicely with our axioms defining a general model. Indeed, starting from *any* regularity structure $\mathscr{T}$, *any* model (Π, Γ) for $\mathscr{T}$, and a β-regularising kernel K, it is usually possible to build a larger regularity structure $\hat{\mathscr{T}}$ containing $\mathscr{T}$ (in the "obvious" sense that $T \subset \hat{T}$ and the action of $\hat{G}$ on T is compatible with that of G) and endowed with an abstract integration map $\mathcal{I}$, as well as an admissible model $(\hat{\Pi}, \hat{\Gamma})$ on $\hat{\mathscr{T}}$ which reduces to (Π, Γ) when restricted to T. See [Hai14b] for more details.

The only exception to this rule arises when the original structure T contains some homogeneous element τ which does not represent a polynomial and which is such that $\deg \tau + \beta \in \mathbf{N}$. Since the bounds appearing both in the definition of a model and in that of a β-regularising kernel are only upper bounds, it is in practice easy to exclude such a situation by slightly tweaking the definition of either the exponent β or of the original regularity structure $\mathscr{T}$.

With all of these definitions in place, we can finally build the operator $\mathcal{K} \colon \mathscr{D}^{\gamma} \to \mathscr{D}^{\gamma+\beta}$ announced at the beginning of this section. Recalling the definition of $\mathcal{J}$ from (14.9), we set

$$(\mathcal{K}f)(x) = \mathcal{I}f(x) + \mathcal{J}_x f(x) + (\mathcal{N}f)(x) , \tag{14.11}$$

where the operator $\mathcal{N}$ is given by

$$(\mathcal{N}f)(x) = \sum_{|k|<\gamma+\beta} \frac{X^k}{k!} \int D^k K(x-y) \left(\mathcal{R}f - \Pi_x f(x)\right)(dy) . \tag{14.12}$$

Note first that thanks to the reconstruction theorem, it is possible to verify that the right-hand side of (14.12) does indeed make sense for every $f \in \mathscr{D}^{\gamma}$ in virtually the same way as in Remark 14.27. One has:

Theorem 14.29. *Let K be a β-regularising kernel, let $\mathscr{T} = (T, G)$ be a regularity structure satisfying Assumptions 14.20 and 14.21, and let (Π, Γ) be an admissible model for $\mathscr{T}$. Then, for every $f \in \mathscr{D}^{\gamma}$ with $\gamma \in (0, N - \beta)$ and $\gamma + \beta \notin \mathbf{N}$, the function $\mathcal{K}f$ defined in (14.11) belongs to $\mathscr{D}^{\gamma+\beta}$ and satisfies $\mathcal{R}\mathcal{K}f = K * \mathcal{R}f$.*

Proof. The complete proof of this result can be found in [Hai14b] and will not be given here. Since it is rather straightforward, we will however give a proof of Schauder's estimate in the classical case (i.e. that of the polynomial regularity structure) in Section 14.3 below.

Let us simply show that one has indeed $\mathcal{R}\mathcal{K}f = K * \mathcal{R}f$ in the particular case when our model consists of continuous functions so that Remark 13.27 applies. In this case, one has

$$(\mathcal{R}\mathcal{K}f)(x) = \left(\Pi_x(\mathcal{I}f(x) + \mathcal{J}_x f(x)))\right)(x) + \left(\Pi_x(\mathcal{N}f)(x))\right)(x) .$$

As a consequence of (14.8), the first term appearing in the right-hand side of this expression is given by

$$\left(\Pi_x(\mathcal{I}f(x) + \mathcal{J}_x f(x)))\right)(x) = \left(K * \Pi_x f(x)\right)(x) .$$

On the other hand, the only term contributing to the second term is the one with $k = 0$ (which is always present since $\gamma > 0$ by assumption) which then yields

$$\big(\Pi_x\big(\mathcal{N}f\big)(x)\big)(x) = \int K(x-y)\,\big(\mathcal{R}f - \Pi_x f(x)\big)(dy)\,.$$

Adding both of these terms, we see that the expression $\big(K * \Pi_x f(x)\big)(x)$ cancels, leaving us with the desired result. □

We are now in principle in possession of all of the ingredients required to formulate fixed point problems for a large number of semilinear stochastic PDEs: multiplication, composition by regular functions, differentiation, and integration against the Green's function of the linearised equation. Before we show how this can be leveraged in practice in order to build a robust solution theory for the KPZ equation, we briefly explore some of main concepts in setting of (very) rough paths.

14.5 Rough volatility and robust Itô integration revisited

Recent applications from mathematical finance, where $\sigma(t,\omega) = \sigma(\widehat{W}_t)$ models *rough stochastic volatility*, involve (standard) Itô integrals of the form

$$\int_0^T \sigma(\widehat{W}_t)d(W_t, \bar{W}_t) \equiv \int_0^T f(\widehat{W}_t)dW_t + \int_0^T \bar{f}(\widehat{W}_t)d\bar{W}_t\,, \qquad (14.13)$$

where $\sigma = (f, \bar{f}) : \mathbf{R} \to \mathbf{R}^2$ is a sufficiently smooth map, $(W, \bar{W})$ is a 2-dimensional standard Brownian motion, and $\widehat{W}_t$ given by

$$\int K^H(t-s)\,dW_s\,, \qquad (14.14)$$

with Riemann–Liouville kernel $K^H(x) = x^{H-1/2}1_{x>0}$. Since $K^H \in L^2_{\mathrm{loc}}(\mathbf{R})$ but not in $L^2(\mathbf{R})$, we replace it in the sequel by a compactly supported K, smooth away from zero and equal to K^H in some neighbourhood of zero. We then require W to be a two-sided Brownian motion, so that $\xi := \dot{W}$ defines Gaussian white noise on $\mathbf{R}$, and

$$\widehat{W} = K * \xi\,. \qquad (14.15)$$

Alternatively, as done in [BFG$^+$19], see also [BFG20], one can restrict integration in (14.14) to $[0, t]$ with the benefit of exactly recovering Brownian motion $\widehat{W} = W$ for $H = 1/2$ in which case the integral (14.13) fits squarely into rough integration theory (namely Theorem 4.4, applied with the Itô Brownian rough path from Proposition 3.4). However, for $H \in (0, 1/2)$ rough integration must fail. Indeed, K is $(1/2 + H)$-regularising so that it follows from Schauder's Theorem 14.17 that $\widehat{W}$ and then $\sigma(\widehat{W})$ have generically H^--Hölder regularity and hence cannot be expected to be controlled by $W \in \mathcal{C}^{1/2^-}$. We can make (minor) progress by noting that $(\widehat{W}, \bar{W})$ is a 2-dimensional Gaussian process with *independent* components. At least for $H > 1/3$, the results of Section 10.3 for Gaussian rough paths apply essentially directly to the final integral $\int \bar{f}(\widehat{W})d\bar{W}$ above and Exercise 14.8 allows to deal with arbitrary $H > 0$.

The remainder of this section will focus on the other, seemingly harmless, one-dimensional Itô integral, with $\widehat{W}$ as given in (14.15),

$$\int_0^T f(\widehat{W})dW \ . \tag{14.16}$$

We are interested in a robust form of this Itô stochastic integral. In case of $\widehat{W} = W$ we can in fact express (14.16) via Itô's formula, which immediately gives a version of this integral which is continuous in W, even in uniform topology. Certainly, this trick fails when $\widehat{W} \neq W$.

In this section we set up a regularity structure that provides a full solution to this problem. Needless to say, this structure is much simpler than what is needed for the KPZ equation in the next chapter. Yet, it showcases a number of features omnipresent for singular SPDEs, but without some of the added complexity coming from PDE theory.

Recall that the Hölder exponent of $\widehat{W}$ is $H - \kappa$ for any $\kappa > 0$. As a result, we have $|\widehat{W}_{s,t}^m| \lesssim |t - s|^{m(H-\kappa)}$ and the building blocks for a robust representation of (14.16) are

$$\mathbb{W}_{s,t}^m = \int_s^t (\widehat{W}_{s,r})^m \, dW_r \ , \tag{14.17}$$

with $m = 0, 1, 2, \ldots, M$ where M is the smallest integer such that $(M + 1)H + 1/2 > 1$, which reflects the analytic redundancy of $\mathbb{W}^{M+1}$ in the sense of

$$|\mathbb{W}^{M+1}(s,t)| \lesssim |t - s|^{(M+1)(H-\kappa)+1/2} = o(t - s) \ ,$$

for small enough $\kappa > 0$. For definiteness, let us focus on the case

$$H > \frac{1}{8}, \quad M = 3 \ .$$

We first define symbols (these will be the basis vectors of our regularity structure) to represent $(\widehat{W}_{s,t})^m, 0 \leq m \leq 3$. If $\Xi \equiv \circ$ is the symbol for white noise $\xi \equiv \dot{W}$, we can write the required symbols indifferently as

$$\{\mathbf{1}, \mathcal{I}(\Xi), \mathcal{I}(\Xi)^2, \mathcal{I}(\Xi)^3\} \ \equiv \ \{\mathbf{1}, \mathord{\uparrow}, \mathord{\vee}, \mathord{\vee\!\!\!\vee}\}.$$

The map $\mathcal{I} : \Xi \mapsto \mathcal{I}(\Xi)$ represents convolution with K and is graphically represented by a downfacing plain line; multiplication (which we postulate to be commutative and associative) is depicted by joining trees at their roots. For instance, $\mathord{\uparrow} \star \mathord{\vee} = \mathord{\vee\!\!\!\vee}$ (we will omit $\star$ in the sequel). Similarly, the symbols denoting $(\widehat{W}_{s,t})^m \dot{W}_t$, defined as the generalised derivative $\partial \mathbb{W}_{s,\cdot}^m$, are given in the same pictorial representation as $\{\circ, \mathord{\updownarrow}, \mathord{\vee}, \mathord{\vee\!\!\!\vee}\}$ (with for example $\mathord{\vee} = \mathcal{I}(\Xi)^2 \Xi$). We then define the structure space of our regularity structure as the free vector space generated by these symbols, namely

$$T = \langle \circ, \mathfrak{l}, \Upsilon, \Psi, \mathbf{1}, \mathbf{l}, \mathbf{V}, \mathbf{\Psi} \rangle . \tag{14.18}$$

The partial product defined on T (for example $\circ \Upsilon = \Upsilon$) does not extend to all of T.[1] It is natural to postulate that Ξ has degree $\deg \Xi = -\frac{1}{2}^-$ (the presence of the exponent '$-$' reflects the fact that in order for the bound (13.13) to be satisfied when $\Pi_t \Xi$ is given by white noise, we need to make sure that $\deg \Xi$ is strictly smaller than $-\frac{1}{2}$, but by how much exactly is irrelevant as long as it is a small enough quantity), that $\mathcal{I}$ increases degree by $H + \frac{1}{2}$, and that the degree is additive under multiplication. Since it is natural to take $\deg \mathbf{1} = 0$ to retain consistency with the polynomial regularity structure, this uniquely determines the degree of each of the basis vectors of T, for instance

$$\deg \Psi = \deg \circ + 3 \deg \mathfrak{l} = (3H - \tfrac{1}{2})^- .$$

To understand the structure group, we shift from a base point s to a new base point t. Basic additivity properties of the integral in (14.17) show that

$$\mathrm{W}^3_{s,\cdot} = \mathrm{W}^3_{t,\cdot} + 3\mathrm{W}^2_{t,\cdot}\widehat{W}_{s,t} + 3\mathrm{W}^1_{t,\cdot}\widehat{W}^2_{s,t} + \mathrm{W}^0_{t,\cdot}\widehat{W}^3_{s,t} + \mathrm{W}^3_{s,t} .$$

Considering the (generalised) derivative in the free variable, we have

$$\partial \mathrm{W}^3_{s,\cdot} = \partial \mathrm{W}^3_{t,\cdot} + 3(\partial \mathrm{W}^2_{t,\cdot})\widehat{W}_{s,t} + 3(\partial \mathrm{W}^1_{t,\cdot})\widehat{W}^2_{s,t} + (\partial \mathrm{W}^0_{t,\cdot})\widehat{W}^3_{s,t} . \tag{14.19}$$

This suggests to "break up" the symbol Ψ (for $\partial \mathrm{W}^3_{*,\cdot}$) in the form

$$\Delta^+(\Psi) := \Psi \otimes \mathbf{1} + 3\Upsilon \otimes \mathbf{l} + 3\mathfrak{l} \otimes \mathbf{V} + \circ \otimes \mathbf{\Psi} \in T \otimes T^+ ,$$

where the introduction of a new space T^+ is justified by the fact that elements in T^+ represent functions of two variables (s and t here), while elements of T represent functions of one variable (the base point s resp. t) that are distributions in the remaining free variable. In particular, it is rather natural that T^+ (unlike T) contains no symbols of negative degree and that elements of T^+ can be multiplied freely. In other words, it is natural in this context to define T^+ as the free commutative algebra generated by the single element $\mathbf{l} \stackrel{\mathrm{def}}{=} \mathcal{J}(\circ)$. The difference between T^+ and T is emphasised in our notation by drawing basis vectors of T^+ in black.

The action of the linear map $\Delta^+ : T \to T \otimes T^+$ has the appealing graphical interpretation of *cutting off positive branches*: for instance, the summand $3\Upsilon \otimes \mathbf{l} = \Upsilon \otimes 3\mathbf{l}$ in $\Delta^+(\Psi)$ is explained as follows: there are three ways to "cut off" a "lollipop" $\mathfrak{l}$ from Ψ, which are then painted black and put as $3\mathbf{l} \in T^+$ to the right-hand side; the remaining "pruned" tree $\Upsilon \in T$ goes to the left. Similarly, there are three ways to cut off two lollipops from Ψ, which then appear as $3\mathbf{V} \in T^+$ on the right-hand side, while the pruned remainder $\mathfrak{l} \in T$ appears on the left.

---

[1] For instance, we do *not* want our regularity structure to contain a symbol Ξ^2 denoting the square of white-noise. We also have no need for trees with ≥ 4 branches so that products like $\mathbf{l}\Psi, \mathbf{V}\mathbf{V}$ etc. remain deliberately undefined within T.

A concise recursive algebraic description of Δ^+ starts with

$$\Delta^+ \mathbf{1} = \mathbf{1} \otimes \mathbf{1} \,, \quad \Delta^+ \Xi = \Xi \otimes \mathbf{1} \,,$$

followed by an extension to all of T by imposing the identities[2]

$$\Delta^+ (\tau \bar{\tau}) = \Delta^+ \tau \cdot \Delta^+ \bar{\tau} \,,$$
$$\Delta^+ \mathcal{I}(\tau) = (\mathcal{I} \otimes \mathrm{Id}) \Delta^+ \tau + \mathbf{1} \otimes \mathcal{J}(\tau) \,.$$

Here, $\mathcal{J}(\tau)$ is the element in T^+ obtained from a (then painted black) symbol τ. In our pictorial representation $\mathcal{J}$ is visualised by a (black) downfacing line. The tree associated to $\mathcal{J}(\tau)$ has exactly one line emerging from the root (such trees are called *planted*). In the present example, $\tau = \circ$ is the only symbol in T, as given in (14.18), with image under $\mathcal{I}$ in T, so that the second relation above can only produce $\mathord{\text{\textyogh}} = \mathcal{J}(\circ) \in T^+$; whereas the first relation leads to powers thereof (in T^+).

Let now G_+ denote the set of characters on T^+, i.e. all linear maps $g : T^+ \to \mathbf{R}$ with the property that $g(\sigma \bar{\sigma}) = g(\sigma) g(\bar{\sigma})$ for any two elements σ and $\bar{\sigma}$ in T^+. There is not much choice here, since $c = g(\mathord{\text{\textyogh}}) \in \mathbf{R}$ fully determines any such map. In order to get back to (14.19), we introduce $\Gamma_g : T \to T$ by

$$\Gamma_g \tau = (\mathrm{Id} \otimes g) \Delta^+ \tau \,, \tag{14.20}$$

so that, for instance, $\Gamma_g(\mathord{\text{\textyogh}\kern-0.1em\text{\textyogh}}) = \mathord{\text{\textyogh}\kern-0.1em\text{\textyogh}} + 3c\mathord{\text{\textyogh}} + 3c^2\mathord{\text{\textyogh}} + c^3 \circ \in T$, and with $c = g(\mathord{\text{\textyogh}}) = \widehat{W}_{s,t}$ this precisely captures (14.19) as an abstract shift map $\Gamma_{st} = \Gamma_{g_{s,t}}$ with $g_{s,t}(\mathord{\text{\textyogh}}) = \widehat{W}_{s,t}$. In principle, (14.20) makes sense for every $g \in (T^+)^*$, but it turns out that the set of those maps Γ_g with $g \in G_+$ forms a group, which is precisely our structure group:

$$G := \{\Gamma_g : g \in G_+\}. \tag{14.21}$$

Written in matrix form, with respect to the ordered basis of T consisting of 4 negative and 4 non-negative symbols, each Γ_g is block-diagonal with two (4×4)-blocks of the form

$$\begin{pmatrix} 1 & c & c^2 & c^3 \\ 0 & 1 & 2c & 3c^2 \\ 0 & 0 & 1 & 3c \\ 0 & 0 & 0 & 1 \end{pmatrix} =: N_c$$

One can check that $N_c N_{\bar{c}} = N_{c+\bar{c}}$ with $c, \bar{c} \in \mathbf{R}$ so that, as a group, G is isomorphic to $(\mathbf{R}, +)$. This completes the construction of the regularity structure (T, G). We leave it to the reader to identify pairs of sectors on which (the usually omitted) $\star$ defines a product in the sense of Section 14.2 and to show that $\mathcal{I}$ is indeed an abstract integration operator[3] in the sense of Definition 14.21.

---

[2] The multiplicative property is understood for all symbols $\tau, \bar{\tau} \in T$ which can be multiplied in T.

[3] In the present setting there is no need to include higher order abstract polynomials $X, X^2, \ldots$ as part of T.

As already hinted at, the natural *Itô model* $\mathrm{M}^{\mathrm{Itô}} := (\Pi, \Gamma)$ in this context is defined by setting

$$\Pi_s 1 = 1, \quad \Pi_s \Xi = \dot{W}, \quad \Pi_s(\mathcal{I}(\Xi)^m) = \widehat{W}_{s,\cdot}^m, \quad \Pi_s(\Xi \mathcal{I}(\Xi)^m) = \partial \mathbb{W},$$

as well as $\Gamma_{st} = \Gamma_{g_{s,t}}$ with $g_{s,t}(\mathfrak{l}) = \widehat{W}_{s,t}$. We leave it to the reader to check that $\mathrm{M}^{\mathrm{Itô}}$ satisfies the required bounds (13.13) and therefore really defines a random model for the regularity structure (T, G). We also note that the model is *admissible* in the sense of Definition 14.23: in essence, this is seen from the identity

$$\Pi_s \mathcal{I} \Xi = K * \Pi_s \Xi - \Pi_s \mathcal{J}(s) \Xi = K * \dot{W} - (K * \dot{W})(s) = \widehat{W}_{s,\cdot}. \qquad (14.22)$$

where we used that only $k = 0$ figures in the sum of (14.9), so that

$$\mathcal{J}_s \Xi = 1 \int K(s - t) \left(\Pi_t \Xi\right)(dt) = (K * \dot{W})(s) \, 1 \, .$$

On the other hand, we can replace white noise $\dot{W} = \dot{W}(\omega)$ by a mollification $\dot{W}^\varepsilon := \delta^\varepsilon * \dot{W}$ with $\delta^\varepsilon(t) = \varepsilon^{-1} \varrho(\varepsilon^{-1} t)$, for some $\varrho \in \mathcal{C}_c^\infty$ with $\int \varrho = 1$, or indeed any smooth function ξ, and define the associated *canonical model* $\mathscr{L}(\xi) = (\Pi, \Gamma)$ by prescribing

$$\Pi_s \Xi = \xi, \quad \Pi_s(\mathcal{I}(\Xi)^m) = (K * \xi)_{s,\cdot}^m, \quad \Pi_s(\Xi \mathcal{I}(\Xi)^m) = \xi(\cdot)(K * \xi)_{s,\cdot}^m \, ,$$

as well as $g_{s,t}(\mathfrak{l}) = (K * \xi)_{s,t}$. We again leave it to the reader to check that $\mathscr{L}(\xi)$ is indeed an admissible model for our regularity structure.

It is interesting to consider the canonical model $\mathscr{L}(\dot{W}^\varepsilon)$ as $\varepsilon \to 0$. Formally, one would expect convergence to a "Stratonovich model", but this does not exist because of an *infinite Itô–Stratonovich correction*. To wit, assume the approximate bracket

$$[W, \widehat{W}]^\pi := \sum_{[s,t] \in \pi} W_{s,t} \widehat{W}_{s,t}$$

converges, say in L^1, upon refinement $|\pi| \to 0$. Then the mean would have to convergence, which is contradicted by the computation, using Itô isometry,

$$\mathbf{E} W_{s,t} \widehat{W}_{s,t} = \int_s^t K(t - r) dr = \int_0^{t-s} K(r) dr$$

$$\sim \int_0^{t-s} K^H(r) dr = c_H (t - s)^{H + \frac{1}{2}} \, ,$$

and the standing assumption that $H < 1/2$. As a consequence, the canonical model $\mathscr{L}(\dot{W}^\varepsilon)$ will not converge as $\varepsilon \to 0$, although the previous discussion suggests to "cure" this by subtracting a diverging term, namely to consider[4]

---

[4] This is an instance of *Wick renormalisation* where one replaces the product of two scalar Gaussian random variables X, Y by $X \diamond Y := XY - \mathbf{E}[XY]$.

$$\int \widehat{W}^\varepsilon \, dW^\varepsilon - \mathbf{E}\Big(\int \widehat{W}^\varepsilon \, dW^\varepsilon \Big) \,, \tag{14.23}$$

with integration understood over $[s,t]$ with re-centred integrand $\widehat{W}_{s,\cdot}$. However, such Wick renormalisation at the level of generalised increments may destroy the algebraic Chen relations. (Indeed, they only hold when the expectation is proportional to $[s,t]$, which has no reason to be the case in general.)

In fact, our admissible model $(\varPi, \varGamma)$ here can be described in terms of a single "base-point free" realisation map $\boldsymbol{\varPi} : T \to \mathcal{D}'$ which enjoys somewhat more natural relations, such as

$$\boldsymbol{\varPi}\mathcal{I}\varXi = K * \boldsymbol{\varPi}\varXi = K * \dot{W} = K * \xi$$

instead of (14.22) in the Itô-model case, and similarly for $\boldsymbol{\varPi}^\varepsilon$ with $\dot{W}$ replaced by $\dot{W}^\varepsilon = \xi^\varepsilon$. The full specification reads[5]

$$\boldsymbol{\varPi}^\varepsilon \mathbf{1} = 1, \qquad\qquad\qquad \boldsymbol{\varPi}^\varepsilon \varXi = \xi^\varepsilon,$$
$$\boldsymbol{\varPi}^\varepsilon (\mathcal{I}(\varXi)^m) = (K * \xi^\varepsilon)^m, \qquad \boldsymbol{\varPi}^\varepsilon (\varXi\mathcal{I}(\varXi)^m) = \xi^\varepsilon (K * \xi^\varepsilon)^m \,. \tag{14.24}$$

Remark 14.30. Define a character f_t on T^+ by specifying (in the Itô model[6])

$$f_t(\mathbf{1}) = f_t(\mathcal{J}(\varXi)) := \int K(t-s)\,(\varPi_t\varXi)(s) = (K * \xi)_t \,, \tag{14.25}$$

and also a linear map $F_t \colon T \to T$ by $F_t \tau = (\mathrm{Id} \otimes f_t)\Delta^+\tau$. One checks without difficulty that F_t is an invertible map, $\varGamma_{ts} = F_t^{-1} \circ F_s$ and

$$\boldsymbol{\varPi} = \varPi_s F_s^{-1} = \varPi_t F_t^{-1} \implies \varPi_s = \varPi_t F_t^{-1} \circ F_s = \boldsymbol{\varPi} \circ F_s \,.$$

At the level of the canonical model $\boldsymbol{\varPi}^\varepsilon$, switching to $\varPi_t^\varepsilon = \boldsymbol{\varPi}^\varepsilon F_t$, this construction merely replaces $K * \xi^\varepsilon$ with the "base-pointed" expression $(K * \xi^\varepsilon)_{t,\cdot}$ and tracks the induced changes to the higher levels.

The Wick renormalisation in (14.23) points us to the (divergent) quantity[7]

$$\mathcal{D} \stackrel{\mathrm{def}}{=} \mathbf{E}(\boldsymbol{\varPi}^\varepsilon(\overset{\circ}{\mathcal{V}})) = \mathbf{E}[(K * \delta^\varepsilon * \xi)(t)(\bar{\delta}^\varepsilon * \xi)(t)]$$
$$= \int_{\mathbf{R}} (K * \delta^\varepsilon)(t-s)\bar{\delta}^\varepsilon(t-s)\,ds = (K * \bar{\delta}^\varepsilon)(0) \,.$$

where we recall $\delta^\varepsilon = \varepsilon^{-1}\varrho(\varepsilon^{-1}\cdot)$; and similarly for $\bar{\delta}^\varepsilon$ with $\bar{\varrho} = \varrho(-(\cdot)) * \varrho$. Since $K(x) = x^{H-1/2}\mathbf{1}_{x>0}$ in a neighbourhood of zero, there is no loss of generality in assuming that this includes the support of $\bar{\varrho}$. For $\varepsilon \in (0,1]$, it follows that[8]

---

[5] One defines $\boldsymbol{\varPi}(\varXi\mathcal{I}(\varXi)^m)$ as the distributional derivative of an Itô integral.

[6] ...and similarly in the canonical one, with $(K * \xi)_t$ replaced by $(K * \xi^\varepsilon)_t$...

[7] Thanks to stationarity, this quantity is independent of t. In particular, one could immediately take $t = 0$.

[8] In the case of $H = 1/2$, so that $K^H \equiv 1$, noting that $\varrho(\cdot)$, and hence $\bar{\varrho} = \varrho(-(\cdot)) * \varrho$, has unit mass, the constant equals $1/2$, which is the same $1/2$ appearing in the Itô–Stratonovich correction.

$$\mathfrak{d} = (K * \bar{\delta}^\varepsilon)(0) = \int_0^\infty K^H(s) \frac{1}{\varepsilon} \bar{\varrho}\Big(\frac{s}{\varepsilon}\Big)\,ds = \varepsilon^{H-1/2} \int_0^\infty K^H(s)\, \bar{\varrho}(s)\,ds\ .$$

We can now replace the informal (14.23) by defining a "renormalised" (admissible) model

$$\boldsymbol{\Pi}^{\varepsilon;\text{ren}}(\Xi\mathcal{I}(\Xi)) := \boldsymbol{\Pi}^\varepsilon(\Xi\mathcal{I}(\Xi) + c_1^\varepsilon \mathbf{1})\ ,$$

with diverging constant

$$c_1^\varepsilon = -\mathbf{E}(\boldsymbol{\Pi}^\varepsilon(\overset{\circ}{\mathfrak{d}})) = -\mathfrak{d}\ .$$

In essence, we can leave it to the algebra to handle the correct shifting to different base points (in other words: to recover $(\Pi^{\varepsilon;\text{ren}}, \Gamma^{\varepsilon;\text{ren}})$ from knowledge of $\boldsymbol{\Pi}^{\varepsilon;\text{ren}}$) in the same spirit as Chen's relation allows to work out increments $\mathbf{X}_{s,t}$ of a given rough path $t \mapsto \mathbf{X}_t$.) On the analytic side, we note that the right-hand side still has controlled blow up of order deg $\Xi\mathcal{I}(\Xi) = (-1/2 + H)^- < 0$. This further suggests that the renormalisation procedure can be described by suitable (linear) maps, say $M : T \to T$, which are (only) allowed to produces additional terms (of higher degrees) as, for instance, $M_{c_1} : \Xi\mathcal{I}(\Xi) \mapsto \Xi\mathcal{I}(\Xi) + c_1\mathbf{1}$ in our present example.

At this stage we could proceed "by hand" and try to work out the correct fixes for all $\Pi_s^\varepsilon(\Xi\mathcal{I}(\Xi)^m)$, $m = 1, 2, 3$, but care is necessary since "curing" level $m = 1$, as done above, will spill over to the higher levels. This is already seen in the instructive case when $m = 0$, i.e. for $\Pi_s(\Xi) = \dot{W}$. Indeed, if one "renormalises" $\dot{W} \implies \dot{W} + c_0$, then writing $V(t) := t$, this leads to[9]

$$\mathbb{W}_{s,t}^m = \int_s^t (\widehat{W}_{s,r})^m\, dW_r \mapsto \int_s^t (\widehat{W}_{s,r} + c_0 \widehat{V}_{s,r})^m\,(dW_r + c_0 dV_r)\ .$$

and hence affects all higher levels ($m = 1, 2, \ldots$). While $\dot{V} = 1$ naturally has $\mathbf{1}$ as associated symbol, $\widehat{V}$ leads to a new symbol, indifferently written as $\mathcal{I}\mathbf{1} \equiv \mathcal{I}()$ or $|$, in agreement with out earlier convention to represent action of $\mathcal{I}$ as single downfacing line.

$$\Xi(\mathcal{I}\Xi)^m \mapsto (\Xi + c_0\mathbf{1})(\mathcal{I}\Xi + c_0\mathcal{I}\mathbf{1})^m\ .$$

Provided we manage to define all these "fixes" (for $m = 0, 1, 2, 3$) consistently, we can expect a family of linear maps $M = M_c$ indexed by $c = (c_0, c_1, c_2, c_3) \in \mathbf{R}^4$ which furthermore constitutes a group in the sense that of (the matrix identity) $M_c M_{\bar{c}} = M_{c+\bar{c}}$ with $c, \bar{c} \in \mathbf{R}^4$. This is the *renormalisation group*, here isomorphic to $(\mathbf{R}^4, +)$. There was a cheat here, in that our initial collection of symbols (with linear span T) was not rich enough to define M_c as linear map from T into itself. In this sense T was incomplete, and one should work on a space $\tilde{T} \supset T$ which contains required symbols such as $|$ or $\vee$. (The notion of *complete rule* put forward in [BHZ19] formalises this.) However, in the present example this was really a consequence of the (analytically unnecessary!) level-0 renormalisation. In fact, $c_0 = 0$ is the only possible choice that respects the symmetry of the noise, in the sense that $\dot{W}$ and $-\dot{W}$

---

[9] This is nothing but a variation of the concept of *translation of rough paths*.

have identical law. This reduces the renormalisation group to $(\mathbf{R}^3, +)$ and reflects a general principle: symmetries help to reduce the dimension of the renormalisation group. See [BGHZ19] for an example where this principle takes centre stage in a striking manner.

In general one proceeds as follows. Define T^- as the free commutative algebra generated by all negative symbols in T; that is,

$$T^- := \mathrm{Alg}(\{\circ, \mathfrak{l}, \mathbb{V}, \mathbb{V}\}) . \tag{14.26}$$

(Similarly to before, we colour basis elements of T^- differently to distinguish them from those of T and/or T^+.) Elements in T^- are naturally represented as linear combination of (unordered) *forests*; for instance

$$-\tfrac{1}{2}\mathbf{1} - 3\mathfrak{l} + \circ\circ + \tfrac{4}{3}\circ\mathbb{V}\,\mathbb{V} \in T^- ,$$

where $\mathbf{1}$ denotes the empty forest. As before, it is useful to introduce a linear map $\Delta^- : T \to T^- \otimes T$ which iterates over all possible ways of extracting possibly empty collections of subtrees of negative degree, putting them as a forest on the left-hand side, and leaving the remaining tree (where all "extracted" subtrees have now been contracted to a point) on the T-valued right-hand side. For instance,

$$\Delta^-(\mathbb{V}) = \mathbf{1} \otimes \mathbb{V} + \ldots + 3\mathfrak{l} \otimes \mathbb{V} + \ldots + 3\circ\circ \otimes \mathbb{V} + \ldots + \mathbb{V} \otimes \mathbf{1} .$$

The resulting renormalisation maps $M : T \to T$ are then parametrised by characters on T^-, similar to the construction of the structure group. Consider for instance the case of a character $g = g^\varepsilon$ defined by $g(\mathfrak{l}) = c_1^\varepsilon$, $g(\mathbb{V}) = c_3^\varepsilon$, and set to vanish on the remaining two generators $\circ$ and $\mathbb{V}$. Then, the map M_g given by

$$M_g = (g \otimes \mathrm{Id})\Delta^-$$

acts as the identity on all symbols of T other than

$$M_g\mathfrak{l} = \mathfrak{l} + c_1^\varepsilon\mathbf{1}, \quad M_g\mathbb{V} = \mathbb{V} + 2c_1^\varepsilon\mathfrak{l}, \quad M_g\mathbb{V} = \mathbb{V} + 3c_1^\varepsilon\mathbb{V} + c_3^\varepsilon\mathbf{1} . \tag{14.27}$$

The resulting renormalised model $\boldsymbol{\Pi}^{\varepsilon;\mathrm{ren}} \equiv \boldsymbol{\Pi}^\varepsilon M_{g^\varepsilon}$ realises, for instance, the symbol $\mathbb{V}$ as

$$\boldsymbol{\Pi}^{\varepsilon;\mathrm{ren}}\mathbb{V} = \boldsymbol{\Pi}^\varepsilon M_g\mathbb{V} = \xi^\varepsilon (K * \xi^\varepsilon)^3 + 3c_1^\varepsilon(K * \xi^\varepsilon)^2 + c_3^\varepsilon .$$

It is a non-trivial but nevertheless fairly general fact that it is possible to choose the character g^ε in such a way that the model $\boldsymbol{\Pi}^{\varepsilon;\mathrm{ren}}$ converges to a limiting model. This is the case if we choose g^ε as the *BPHZ character* (see [BHZ19, Thm 6.18]) associated to $\boldsymbol{\Pi}^\varepsilon$. This is defined in general as the *unique* character g^ε of T^- such that the renormalised model $\boldsymbol{\Pi}^{\varepsilon;\mathrm{ren}}$ satisfies $\mathbf{E}\boldsymbol{\Pi}^{\varepsilon;\mathrm{ren}}\tau = 0$ for every symbol τ of strictly negative degree. With our earlier choice

$$c_1^\varepsilon = -\mathbf{E}(\boldsymbol{\Pi}^\varepsilon\mathfrak{l})(0) = -\mathcal{D}$$

it is immediate from (14.27) that one has indeed $\mathbf{E}(\boldsymbol{\Pi}^\varepsilon M_{g^\varepsilon}\,\mathfrak{I}) = 0$. Furthermore, since first and third moments of centred Gaussians vanish, we also have $\mathbf{E}(\boldsymbol{\Pi}^\varepsilon M_{g^\varepsilon}\,\mathfrak{V}) = \mathbf{E}(\boldsymbol{\Pi}^\varepsilon M_{g^\varepsilon}\,\mathfrak{V}) = 0$ as a consequence of the fact that we set $g(\circ) = g(\mathfrak{V}) = 0$. Finally, it follows from Wick's formula that

$$\mathbf{E}\boldsymbol{\Pi}^\varepsilon M_{g^\varepsilon}\,\mathfrak{V} = \mathbf{E}[\xi^\varepsilon(K*\xi^\varepsilon)^3] + 3c_1^\varepsilon\mathbf{E}(K*\xi^\varepsilon)^2 + c_3^\varepsilon$$
$$= 3\Big(\mathbf{E}[\xi^\varepsilon(K*\xi^\varepsilon)] + c_1^\varepsilon\Big)\mathbf{E}(K*\xi^\varepsilon)^2 + c_3^\varepsilon$$
$$= 3\big(\mathfrak{I} + c_1^\varepsilon\big)\,\mathfrak{V} + c_3^\varepsilon = c_3^\varepsilon\,,$$

so that $\boldsymbol{\Pi}^\varepsilon M_{g^\varepsilon}\,\mathfrak{V}$ has vanishing mean if and only if we also choose $c_3^\varepsilon = 0$.

We have made it plausible that

$$\mathbf{M}^{\varepsilon;\mathrm{ren}} := \big(\Pi^{\varepsilon;\mathrm{ren}}, \Gamma^{\varepsilon;\mathrm{ren}}\big) \leftrightarrow \boldsymbol{\Pi}^{\varepsilon;\mathrm{ren}},$$

indeed gives rise to an (admissible) model, with all analytic bounds and algebraic constraints intact, and such that in the sense of model convergence,

$$\mathbf{M}^{\varepsilon;\mathrm{ren}} \to \mathbf{M}^{\mathrm{BPHZ}} = \mathbf{M}^{\mathrm{It\hat{o}}}\,. \tag{14.28}$$

The main result of [CH16] is that the convergence $\mathbf{M}^{\varepsilon;\mathrm{ren}} \to \mathbf{M}^{\mathrm{BPHZ}}$ remains true in vastly greater generality and that the limiting model is independent of the specific choice of $\mathbf{M}^\varepsilon$ for a large class of stationary approximations ξ^ε to the noise ξ.

At last, we leave it to the reader to adapt the material of Section 13.3.2 to define the modelled distribution that allows to reconstruct the Itô integral $\int_0^t f(\widehat{W}_s)dW_s$ and further deduce from (14.28) the following (renormalised) *Wong–Zakai result*,

$$\int_0^t f(\widehat{W}_s^\varepsilon)dW_s^\varepsilon - c_1^\varepsilon \int_0^t f'(\widehat{W}_s^\varepsilon)ds \to \int_0^t f(\widehat{W}_s)dW_s \tag{14.29}$$

where we recall that $c_1^\varepsilon = \varepsilon^{H-1/2}\int_0^\infty K^H(s)\,\bar{\varrho}(s)ds$. Noting that $\bar{\varrho} = \varrho(-(\cdot)) * \varrho$ is even and has unit mass, we see that $c_1^\varepsilon = \frac{1}{2}$ when $H = 1/2$. We can then pass to the limit for each term on the right-hand side of (14.29) separately. This allows us to recover the identity

$$\int_0^t f(W_s) \circ dW_s - \frac{1}{2}\int_0^t f'(W_s)ds = \int_0^t f(W_s)dW_s\,,$$

in agreement with the usual Itô–Stratonovich correction familiar from stochastic calculus.

14.6 Exercises

Exercise 14.1 *a) Construct an example of a regularity structure with trivial group G, as well as a model and modelled distributions f_i such that both $\mathcal{R}f_1$ and $\mathcal{R}f_2$ are continuous functions but the identity*

$$\mathcal{R}(f_1 \star f_2)(x) = (\mathcal{R}f_1)(x)\,(\mathcal{R}f_2)(x)$$

fails.
b) Transfer Exercise 2.10 to the present context.

Solution. (We only address the first part.) Consider for instance the regularity structure given by $A = (-2\kappa, -\kappa, 0)$ for fixed $\kappa > 0$ with each T_α being a copy of $\mathbf{R}$ given by $T_{-n\kappa} = \langle \Xi^n \rangle$. We furthermore take for G the trivial group. This regularity structure comes with an obvious product by setting $\Xi^m \star \Xi^n = \Xi^{m+n}$ provided that $m + n \le 2$.

Then, we could for example take as a model for $\mathscr{T} = (T, G)$:

$$\left(\Pi_x \Xi^0\right)(y) = 1\,, \quad \left(\Pi_x \Xi\right)(y) = 0\,, \quad \left(\Pi_x \Xi^2\right)(y) = c\,, \tag{14.30}$$

where c is an arbitrary constant. Let furthermore

$$f_1(x) = f_1(x)\Xi^0 + \tilde{f}_1(x)\Xi\,, \qquad f_2(x) = f_2(x)\Xi^0 + \tilde{f}_2(x)\Xi\,.$$

Since our group G is trivial, one has $f_i \in \mathcal{D}^\gamma$ provided that each of the f_i belongs to $\mathcal{C}^\gamma$ and each of the $\tilde{f}_i$ belongs to $\mathcal{C}^{\gamma+\kappa}$. (And one has $\gamma + \kappa < 1$.) One furthermore has the identity $\left(\mathcal{R}f_i\right)(x) = f_i(x)$.

However, the pointwise product is given by

$$\left(f_1 \star f_2\right)(x) = f_1(x)f_2(x)\Xi^0 + \left(\tilde{f}_1(x)f_2(x) + \tilde{f}_2(x)f_1(x)\right)\Xi + \tilde{f}_1(x)\tilde{f}_2(x)\Xi^2\,,$$

which by Theorem 14.5 belongs to $\mathcal{D}^{\gamma-\kappa}$. Provided that $\gamma > \kappa$, one can then apply the reconstruction operator to this product and one obtains

$$\mathcal{R}\left(f_1 \star f_2\right)(x) = f_1(x)f_2(x) + c\tilde{f}_1(x)\tilde{f}_2(x)\,,$$

which is obviously quite different from the pointwise product $(\mathcal{R}f_1)(x) \cdot (\mathcal{R}f_2)(x)$.

How should this be interpreted? For $n > 0$, we could have defined a model $\Pi^{(n)}$ by

$$\left(\Pi_x^{(n)}\Xi^0\right)(y) = 1,\quad \left(\Pi_x^{(n)}\Xi\right)(y) = \sqrt{2c}\,\sin(ny),\quad \left(\Pi_x^{(n)}\Xi^2\right)(y) = 2c\sin^2(ny).$$

Denoting by $\mathcal{R}^{(n)}$ the corresponding reconstruction operator, we have the identity

$$\left(\mathcal{R}^{(n)}f_i\right)(x) = f_i(x) + \sqrt{2c}\tilde{f}_i(x)\sin(nx)\,,$$

as well as $\mathcal{R}^{(n)}(f_1 \star f_2) = \mathcal{R}^{(n)} f_1 \cdot \mathcal{R}^{(n)} f_2$. As a model, the model $\Pi^{(n)}$ actually converges to the limiting model Π defined in (14.30). As a consequence of the continuity of the reconstruction operator, this implies that

$$\mathcal{R}^{(n)} f_1 \cdot \mathcal{R}^{(n)} f_2 = \mathcal{R}^{(n)}(f_1 \star f_2) \to \mathcal{R}(f_1 \star f_2) \neq \mathcal{R} f_1 \cdot \mathcal{R} f_2 \,,$$

which is of course also easy to see "by hand". This shows that in some cases, the "non-canonical" models as in (14.30) can be interpreted as limits of "canonical" models for which the usual rules of calculus hold. Even this is however not always the case (think of the Itô Brownian rough path).

Exercise 14.2 *Consider* $\mathcal{Z}^\alpha = \mathcal{Z}^\alpha(\mathbf{R}^d)$.

a) *Show that distributional derivatives satisfy $D^k \mathcal{Z}^\alpha \subset \mathcal{Z}^{\alpha-|k|}$ for any multi-index k. Show that for $d=1$ equality holds. That is, any $g \in \mathcal{Z}^{\alpha-k}$, with $k \in \mathbf{N}$, is the kth distributional derivative of some $f \in \mathcal{Z}^\alpha$.*

b) *The proof of Schauder's theorem in Section 14.3 was more involved in the "positive" case, when $0 \leq \alpha + \beta \in [n-1, n)$, some $n \in \mathbf{N}$. Give an easier proof in the case $d=1$ by reducing the positive to the negative case.*

** **Exercise 14.3** *Provide a proof of the case $\alpha = 0$ in Lemma 14.13.*

Solution. As in Lemma 14.13, we aim to bound $|\zeta(\varphi)|$ for $\varphi \in \mathcal{B}^\lambda_{r_o, x}$ and $\zeta \in \mathcal{Z}^\alpha_r$ for some $r \geq r_o$. One strategy is to consider a compactly supported wavelet basis of regularity r and to separately bound the terms in the wavelet expansion of φ.

If we wish to rely purely on elementary arguments, one strategy goes as follows.

a) Show first that $\zeta \in \mathcal{Z}^\alpha_r$ if and only if $\zeta \chi \in \mathcal{Z}^\alpha_r$ for every smooth compactly supported function χ. This allows us to reduce ourselves to the case when ζ itself is compactly supported and we assume this from now on.

b) Show that if $\zeta \in \mathcal{Z}^0_r$ is supported in a ball of radius 1 and if ψ is such that $\int \psi(x)\,dx = 0$ and such that $|D^k \psi(x)| \leq (1+|x|)^{-\beta-|k|}$ for $|k| \leq r$ and some large enough exponent k, then $|\zeta(\psi^\lambda_x)| \lesssim 1$, uniformly over such ψ and over $x \in \mathbf{R}^d$ and $\lambda \in (0,1]$.

c) Choose a function ψ with the property that its Fourier transform is smooth, identically 1 in the ball of radius 1, and identically 0 outside of the ball of radius 2 and define $\tilde\psi$ as in the proof of Lemma 14.13. Write

$$\varphi = \varphi * \psi^\lambda + \sum_{n \geq 0} \varphi * \tilde\psi^{\lambda_n}$$

as in the proof of Lemma 14.13.

d) Choose χ such that its Fourier transform is smooth, identically equal to 1 on the annulus of radii in $[1,4]$ and vanishes outside the annulus of radii in $[1/2, 5]$. Note that this implies that $\tilde\psi^{\lambda_n} = \tilde\psi^{\lambda_n} * \chi^{\lambda_n}$ and conclude that

$$\zeta(\varphi * \tilde\psi^{\lambda_n}) = \langle \zeta * \tilde\psi^{\lambda_n}, \varphi * \chi^{\lambda_n} \rangle \,.$$

e) Use the fact that $\varphi \in \mathcal{C}^1$ and χ integrates to 0 to conclude that $|\varphi * \chi^{\lambda_n}| \lesssim 2^{-n}\lambda^{-d}$ and therefore that $|\zeta(\varphi * \tilde{\psi}^{\lambda_n})| \lesssim 2^{-n}$, which is summable as required.

* **Exercise 14.4** *Show that, for g smooth enough, one has $K * (g\eta) - g(K * \eta) \in \mathcal{C}^{\alpha+\beta+1}$ for every β-regularising kernel K and $\eta \in \mathcal{C}^\alpha$ with $\alpha < 0$. How smooth is smooth enough? Compare the following two strategies.*

Strategy 1: *Go through the proof of the Schauder estimate in Section 14.3 and estimate the difference $\langle K_n * (g\eta) - g(K_n * \eta), \psi_\lambda \rangle$.*

Strategy 2: *Consider the regularity structure T spanned by the Taylor polynomials and an additional symbol Ξ of degree α, with the structure group acting trivially on Ξ. We extend this by adding an integration operator of order β and all products with Taylor polynomials. We also consider on it the natural model mapping Ξ to η. Writing $g \in \mathcal{D}^\gamma$ for the Taylor lift of g as in Proposition 13.16, verify that $g\Xi \in \mathcal{D}^{\gamma+\alpha}$. The multilevel Schauder estimate then shows that, provided that $\gamma + \alpha > 0$, one has $\mathcal{K}(g\Xi) \in \mathcal{D}^{\gamma+\alpha+\beta}$ and $g\mathcal{K}(\Xi) \in \mathcal{D}^{\gamma+\min\{0,\alpha+\beta\}}$, so in particular*

$$F \stackrel{\text{def}}{=} \mathcal{K}(g\Xi) - g\mathcal{K}(\Xi) \in \mathcal{D}^{1+\alpha+\beta},$$

provided that $\gamma > \max\{1, -\alpha, 1 + \alpha + \beta\}$. Furthermore, the explicit expression for $\mathcal{K}$ shows that

$$\mathcal{K}(g\Xi) = g\mathcal{I}(\Xi) + g'\mathcal{I}(X\Xi) + (\dots), \qquad g\mathcal{K}(\Xi) = g\mathcal{I}(\Xi) + (\dots),$$

where $(\dots)$ denotes terms that either belong to the polynomial part of the regularity structure or are of degree strictly greater than $\alpha + \beta + 1$ (which is the degree of $\mathcal{I}(X\Xi)$). In particular, the truncation of F at level $\alpha + \beta + 1$ belongs to $\mathcal{D}_{\mathrm{P}}^{\alpha+\beta+1}$, and we conclude by the second part of Proposition 13.16.

Exercise 14.5 *Consider space-time $\mathbf{R}^d$ with one temporal and $(d-1)$ spatial dimensions, under the parabolic scaling $(2, 1, \dots, 1)$, as introduced in Remark 13.9. Denote by $\mathcal{G}$ the heat kernel (i.e. the Green's function of the operator $\partial_t - \partial_x^2$). Show that one has the decomposition*

$$\mathcal{G} = K + \hat{K},$$

where the kernel K satisfies all of the assumptions of Section 14.4 (with $\beta = 2$) and the remainder $\hat{K}$ is smooth and bounded.

Exercise 14.6 (From [Bru18]) *In the context of Remark 14.25, establish the recursion*

$$\Gamma_{xy}\mathcal{I}\tau = \mathcal{I}(\Gamma_{xy}\tau) - \Gamma_{xy}\mathcal{J}_{xy}\tau, \qquad (14.31)$$

with

$$\mathcal{J}_{xy}\tau := \sum_{|k|<\deg \tau+\beta} \frac{X^k}{k!} \Pi_x(\mathcal{I}_k(\Gamma_{xy}\tau))(y).$$

Exercise 14.7 *Show that if one defines $\Gamma_{xy}\mathcal{I}\tau$ in such a way that (14.10) holds, then it guarantees that $\Pi_x\Gamma_{xy}\mathcal{I}\tau = \Pi_y\mathcal{I}\tau$.*

Exercise 14.8 *Adapt the material in Section 14.5 and construct a suitable regularity structure and model so that the two-dimensional Itô integral (14.13) is obtained as reconstruction of a suitable modelled distribution.*

14.7 Comments

The material on differentiation, products and admissible models follows essentially [Hai14b], although the conditions on the kernel K – previously assumed to annihilate certain polynomials – are now more flexible. In particular, we do not enforce the identity $\mathcal{I}(X^k) = 0$ and instead allow for the possibility of simply including symbols $\mathcal{I}(X^k)$ as basis vectors of our regularity structure. It is the case that any admissible model will necessarily satisfy $\Pi_x \mathcal{I}(X^k) = 0$, but in general $\Gamma_{xy} \mathcal{I}(X^k) \neq 0$. The material of Section 14.5 is essentially taken from [BFG+19], with a viewpoint similar to [BCFP19].

Chapter 15
Application to the KPZ equation

We show how the theory of regularity structures can be used to build a robust solution theory for the KPZ equation. We also give a very short survey of the original approach to the same problem using controlled rough paths and we discuss how the two approaches are linked.

15.1 Formulation of the main result

Let us now briefly explain how the theory of regularity structures can be used to make sense of solutions to very singular semilinear stochastic PDEs. We will keep the discussion in this chapter at a very informal level without attempting to make mathematically precise statements. The interested reader may find more details in [Hai13, Hai14b].

For definiteness, we focus on the case of the KPZ equation [KPZ86], which is formally given by

$$\partial_t h = \partial_x^2 h + (\partial_x h)^2 + \xi - C , \tag{15.1}$$

where ξ denotes space-time white noise, the spatial variable takes values in the one-dimensional torus $\mathbb{T}$, i.e. in the interval $[0, 2\pi]$ endowed with periodic boundary conditions, and C is a fixed constant. The problem with such an equation is that even the solution to the linear part of the equation, namely

$$\partial_t \Psi = \partial_x^2 \Psi + \xi ,$$

is not differentiable as a function of the spatial variable. As a matter of fact, as already noted in Section 12.3, for any fixed time t, Ψ has the regularity of Brownian motion as a function of the spatial variable x. As a consequence, the only way of possibly giving meaning to (15.1) is to "renormalise" the equation by subtracting from its right-hand side an "infinite constant", which counteracts the divergence of the term $(\partial_x h)^2$.

© Springer Nature Switzerland AG 2020
P. K. Friz, M. Hairer, *A Course on Rough Paths*, Universitext,
https://doi.org/10.1007/978-3-030-41556-3_15

This has usually been interpreted in the following way. Assuming for a moment that ξ is a smooth function, a simple consequence of the change of variables formula shows that if we define $h = \log Z$, then Z satisfies the PDE

$$\partial_t Z = \partial_x^2 Z + Z \xi \, .$$

The only ill-posed product appearing in this equation now is the product of the solution Z with white noise ξ. As long as Z takes values in L^2, this product can be given a meaning as a classical Itô integral, so that the equation for Z can be interpreted as the Itô equation

$$dZ = \partial_x^2 Z \, dt + Z \, dW \, , \tag{15.2}$$

were W is an L^2-cylindrical Wiener process. It is well known [DPZ92] that this equation has a unique (mild) solution and we can then go backwards and *define* the solution to the KPZ equation as $h = \log Z$. The expert reader will have noticed that this argument appears to be flawed: since (15.2) is interpreted as an Itô equation, we should really use Itô's formula to find out what equation h satisfies. If one does this a bit more carefully, one notices that the Itô correction term appearing in this way is indeed an infinite constant! This is the case in the following sense. If W_ε is a Wiener process with spatial covariance given by $x \mapsto \varepsilon^{-1} \varrho(\varepsilon^{-1} x)$ for some smooth compactly supported function ϱ integrating to 1 and Z_ε solves (15.2) with W replaced by W_ε, then $h_\varepsilon = \log Z_\varepsilon$ solves

$$dh = \partial_x^2 h \, dt + (\partial_x h)^2 \, dt + dW_\varepsilon - \varepsilon^{-1} C_\varrho \, dt \, , \tag{15.3}$$

for some constant C_ϱ depending on ϱ. Since Z_ε converges to a strictly positive limit Z, this shows that the sequence of functions h_ε solving (15.3) converges to a limit h. This limit is called the Hopf–Cole solution to the KPZ equation [Hop50, Col51, BG97, Qua11].

This notion of solution is of course not very satisfactory since it relies on a nonlinear transformation and provides no direct interpretation of the term $(\partial_x h)^2$ appearing in the right-hand side of (15.1). Furthermore, many natural growth models lead to equations that structurally "look like" (15.1), rather than (15.2). Since perturbations are usually rather badly behaved under exponentiation and since there is no really good approximation theory for (15.2) either (for example it had been an open problem for some time whether space-time regularisations of the noise lead to the same notion of solution), one would like to have a robust solution theory for (15.1) directly.

Such a robust solution theory is precisely what the theory of regularity structures provides. More precisely, it provides spaces $\mathcal{M}$ (a suitable space of "admissible models") and $\mathscr{D}^\gamma$, maps $\mathcal{S}_a$ (an abstract "solution map"), $\mathcal{R}$ (the reconstruction operator) and $\mathscr{L}$ (a "canonical lift map"), as well as a finite-dimensional group $\mathfrak{R}$ acting both on $\mathbf{R}$ and $\mathcal{M}$ such that the following diagram commutes:

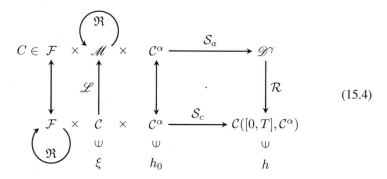

Here, $\mathcal{S}_c$ denotes the classical solution map $\mathcal{S}_c(C, \xi, h_0)$ which provides the solution (up to some fixed final time T) to the equation

$$\partial_t h = \partial_x^2 h + (\partial_x h)^2 + \xi - C\,, \quad h(0, x) = h_0(x)\,, \tag{15.5}$$

for regular instances of the noise ξ. The space $\mathcal{F}$ of "formal right-hand sides" is in this case just a copy of $\mathbf{R}$ which holds the value of the constant C appearing in (15.5). The diagram commutes in the sense that if $M \in \mathfrak{R}$, then

$$\mathcal{S}_c(M(C), \xi, h_0) = \mathcal{R}\mathcal{S}_a(C, M(\mathcal{L}(\xi)), h_0)\,,$$

where we identify M with its respective actions on $\mathbf{R}$ and $\mathcal{M}$. A full justification of these considerations for a very large class of systems of SPDEs is beyond the scope of this text. The construction of $\mathfrak{R}$ in full generality and its action on the space of admissible models was obtained in [BHZ19]. Its adjoint action on a suitable space of equations $\mathcal{F}$ as well as the commutativity of the above diagram were then obtained in [BCCH17]. Important additional features of this picture are the following:

- If ξ_ε denotes a "natural" regularisation of space-time white noise, then there exists a sequence M_ε of elements in $\mathfrak{R}$ such that $M_\varepsilon \mathcal{L}(\xi_\varepsilon)$ converges to a limiting random element $(\Pi, \Gamma) \in \mathcal{M}$. This element can also be characterised directly without resorting to specific approximation procedures and $\mathcal{R}\mathcal{S}_a(0, (\Pi, \Gamma), h_0)$ coincides almost surely with the Hopf–Cole solution to the KPZ equation. The fact that an analogous statement "always" holds for subcritical equations was shown in the work [CH16].
- The maps $\mathcal{S}_a$ and $\mathcal{R}$ are both continuous, unlike the map $\mathcal{S}_c$ which is discontinuous in its second argument for any topology for which ξ_ε converges to ξ.
- As an abstract group, the "renormalisation group" $\mathfrak{R}$ is simply equal to $(\mathbf{R}^3, +)$. However, it is possible to extend the picture to deal with much larger classes of approximations, which has the effect of increasing both $\mathfrak{R}$ and the space $\mathcal{F}$ of possible right-hand sides. See for example [HQ18] for a proof of convergence to KPZ for a much larger class of interface growth models.

Remark 15.1. An important condition for the convergence result in [CH16] to hold is that T does *not* contain any symbol τ with $\deg \tau \le -\frac{d}{2}$ and such that τ contains more than one noise as a subsymbol. This in particular explains why fractional Brownian motion B^H with Hurst parameter H can only be lifted to a rough path when $H > \frac{1}{4}$ even though SDEs driven by fractional Brownian motion are "subcritical" for every $H > 0$. Indeed, for $H = \frac{1}{4}$, the natural degree of the symbol $\mathbb{W}$ of Section 13.2.2 (which would be represented by $\mathord{\text{\rotatebox{90}{?}}}$ in the graphical notation used earlier and contains two instances of the noise) would be $(2H - 1)^- = -\frac{1}{2}^- < -\frac{d}{2}$.

An example of statement that can be proved from these considerations (see [Hai13, Hai14b, HQ18]) is the following.

Theorem 15.2. *Consider the sequence of equations*

$$\partial_t h_\varepsilon = \partial_x^2 h_\varepsilon + (\partial_x h_\varepsilon)^2 + \xi_\varepsilon - C_\varepsilon , \tag{15.6}$$

where $\xi_\varepsilon = \delta_\varepsilon * \xi$ with $\delta_\varepsilon(t,x) = \varepsilon^{-3}\varrho(\varepsilon^{-2}t, \varepsilon^{-1}x)$, for some smooth compactly supported function ϱ with $\int \varrho = 1$, and ξ denotes space-time white noise. Then, there exists a (diverging) choice of constants C_ε such that the sequence h_ε converges in probability to a limiting process h.

Furthermore, one can ensure that the limiting process h does not depend on the choice of mollifier ϱ and that it coincides with the Hopf–Cole solution to the KPZ equation.

Remark 15.3. It is important to note that although the limiting process is independent of the choice of mollifier ϱ, the constant C_ε *does* very much depend on this choice, as we already alluded to earlier.

Remark 15.4. Regarding the initial condition, one can take $h_0 \in \mathcal{C}^\beta$ for any fixed $\beta > 0$. Unfortunately, this result does not cover the case of "infinite wedge" initial conditions, see for example [Cor12].

The aim of this section is to sketch how the theory of regularity structures can be used to obtain this kind of convergence results and how (15.4) is constructed. First of all, we note that while our solution h will be a Hölder continuous space-time function (or rather an element of $\mathscr{D}^\gamma$ for some regularity structure with a model over $\mathbf{R}^2$), the "time" direction has a different scaling behaviour from the three "space" directions. As a consequence, it turns out to be effective to slightly change our definition of "localised test functions" by setting

$$\varphi_{(s,x)}^\lambda(t,y) = \lambda^{-3}\varphi\big(\lambda^{-2}(t-s), \lambda^{-1}(y-x)\big) .$$

Accordingly, the "effective dimension" of our space-time is actually 3, rather than 2. The theory presented in Chapter 13 extends *mutatis mutandis* to this setting. (Note however that when considering the degree of a regular monomial, powers of the time variable should now be counted double.) Note also that with this way of measuring regularity, space-time white noise belongs to $\mathcal{C}^{-\alpha}$ for every $\alpha > \frac{3}{2}$. This is because of the bound

$$\big(\mathbf{E}\langle\xi, \varphi_x^\lambda\rangle^2\big)^{1/2} = \|\varphi_x^\lambda\|_{L^2} \approx \lambda^{-\frac{3}{2}} ,$$

combined with an argument somewhat similar to the proof of Kolmogorov's continuity lemma.

15.2 Construction of the associated regularity structure

Our first step is to build a regularity structure that is sufficiently large to allow to reformulate (15.1) as a fixed point in $\mathscr{D}^\gamma$ for some $\gamma > 0$. Denoting by $\mathcal{G}$ the heat kernel (i.e. the Green's function of the operator $\partial_t - \partial_x^2$), we can rewrite the solution to (15.1) with initial condition h_0 as

$$ h = \mathcal{G} * \left((\partial_x h)^2 + \xi \right) + \mathcal{G} h_0 , \tag{15.7} $$

where $*$ denotes space-time convolution and where we denote by $\mathcal{G} h_0$ the harmonic extension of h_0. (That is the solution to the heat equation with initial condition h_0.)

Remark 15.5. We view (15.7) as an equation on the whole space by considering its periodic extension.

In order to have a chance of fitting this into the framework described above, we first decompose the heat kernel $\mathcal{G}$ as in Exercise 14.5 as

$$ \mathcal{G} = K + \hat{K} , $$

where the kernel K satisfies all of the assumptions of Section 14.4 (with $\beta = 2$) and the remainder $\hat{K}$ is smooth. If we consider any regularity structure containing the usual Taylor polynomials and equipped with an admissible model, is straightforward to associate to $\hat{K}$ an operator $\hat{\mathcal{K}} \colon \mathscr{D}^\gamma \to \mathscr{D}^\infty$ via

$$ (\hat{\mathcal{K}} f)(z) = \sum_k \frac{X^k}{k!} (D^k \hat{K} * \mathcal{R} f)(z) , $$

where z denotes a space-time point and k runs over all possible 2-dimensional multiindices. Similarly, the harmonic extension of h_0 can be lifted to an element in $\mathscr{D}^\infty$ which we denote again by $\mathcal{G} h_0$ by considering its Taylor expansion around every space-time point. At this stage, we note that we actually cheated a little: while $\mathcal{G} h_0$ is smooth in $\{(t, x) : t > 0, x \in \mathbb{T}\}$ and vanishes when $t < 0$, it is of course singular on the time-0 hyperplane $\{(0, x) : x \in \mathbb{T}\}$. This problem can be cured by introducing weighted versions of the spaces $\mathscr{D}^\gamma$ allowing for singularities on a given hyperplane. A precise definition of these *singular model* spaces and their behaviour under multiplication and the action of the integral operator $\mathcal{K}$ can be found in [Hai14b]; but see Exercise 4.12 for the (singular, controlled) rough path analogue. For the purpose of the informal discussion given here, we will simply ignore this problem.

This suggests that the "abstract" formulation of (15.1) should be given by

$$H = \mathcal{K}\big((\partial H)^2 + \Xi\big) + \hat{\mathcal{K}}\big((\partial H)^2 + \Xi\big) + \mathcal{G}h_0\,, \qquad (15.8)$$

where it still remains to be seen how to define an "abstract differentiation operator" ∂ realising the spatial derivative ∂_x as in Section 14.1. In view of (14.11), this equation is of the type

$$H = \mathcal{I}\big((\partial H)^2 + \Xi\big) + (\ldots)\,, \qquad (15.9)$$

where the terms $(\ldots)$ consist of functions that take values in the subspace $\bar{T}$ of T spanned by regular Taylor polynomials in the time variable X_0 and the space variable X_1. (As previously, X denotes the collection of both.) In order to build a regularity structure in which (15.9) can be formulated, it is then natural to start with the structure $\bar{T}$ given by these abstract polynomials (again with the parabolic scaling which causes the abstract "time" variable to have degree 2 rather than 1), and to then add a symbol Ξ to it which we postulate to have degree $-\frac{3}{2}^-$, where we denote by α^- an exponent strictly smaller than, but arbitrarily close to, the value α. As a consequence of our definitions, it will also turn out that the symbol ∂ is always immediately followed by the symbol $\mathcal{I}$, so that it makes sense to introduce the shorthand $\mathcal{I}' = \partial \mathcal{I}$. This is also suggestive of the fact that $\mathcal{I}'$ can itself be considered an abstract integration map, associated to the kernel $K' = \partial_x K$. Comparing this to Remark 14.24, we see that we could alternatively view $\mathcal{I}'$ as the operator $\mathcal{I}_{(0,1)}$.

Remark 15.6. In order to avoid a proliferation of inconsequential terms, we impose from the start the identity $\mathcal{I}'(1) = 0$ in T (we can do this by Remark 15.6). We could also set $\mathcal{I}(1) = 0$ by choosing K appropriately, but this is irrelevant anyway in view of Remark 15.8 below.

We then simply add to T all of the formal expressions that an application of the right-hand side of (15.9) can generate for the description of H, ∂H, and $(\partial H)^2$. The degree of a given expression is furthermore completely determined by the rules $\deg \mathcal{I}\tau = \deg \tau + 2$, $\deg \partial \tau = \deg \tau - 1$ and $\deg \tau \bar{\tau} = \deg \tau + \deg \bar{\tau}$. For example, it follows from (15.9) that the symbol $\mathcal{I}(\Xi)$ is required for the description of H, so that $\mathcal{I}'(\Xi)$ is required for the description of ∂H. This then implies that $\mathcal{I}'(\Xi)^2$ is required for the description of the right-hand side of (15.9), which in turn implies that $\mathcal{I}(\mathcal{I}'(\Xi)^2)$ is also required for the description of H, etc. This "Picard iteration" yields the (formal) expansion, writing z for a generic space-time point,[1]

$$\begin{aligned} H(z) = h(z)\,\mathbf{1} + \mathcal{I}(\Xi) + \mathcal{I}(\mathcal{I}'(\Xi)^2) + h'(z)\,X_1 \\ + 2\mathcal{I}(\mathcal{I}'(\Xi)\mathcal{I}'(\mathcal{I}'(\Xi)^2)) + 2h'(z)\mathcal{I}(\mathcal{I}'(\Xi)) + \ldots \end{aligned}$$

where h and h' are to be considered as independent functions (similar to a controlled rough path). In particular, h may not be differentiable at all.

Remark 15.7. Here we made a distinction between $\mathcal{I}(\Xi)$, interpreted as the linear map $\mathcal{I}$ applied to the symbol Ξ, and the symbol $\mathcal{I}(\Xi)$. Since the map $\mathcal{I}$ is then

---

[1] Note that h' is treated as an independent function (similar to the Gubinelli derivative of a controlled path); we do not even expect h to be differentiable!

defined by $\mathcal{I}(\Xi) := \mathcal{I}(\Xi)$, this distinction is somewhat moot and will be blurred in the sequel. Similarly, the abstract (spatial) differentiation operator ∂ acts on suitable symbols as $\partial(\mathcal{I}(\ldots)) := \mathcal{I}'(\ldots)$, plus of course $\partial(X_0^{k_0} X_1^{k_1}) := k_1 X_0^{k_0} X_1^{k_1-1}$, for every multi-index (k_0, k_1).

More formally, denote by $\mathcal{U}$ the collection of those formal expressions that are required to describe H. This is then defined as the smallest collection containing X^k for all multiindices $k \geq 0$, $\mathcal{I}(\Xi)$, and such that

$$\tau_1, \tau_2 \in \mathcal{U} \implies \mathcal{I}(\partial\tau_1 \partial\tau_2) \in \mathcal{U} .$$

We then set

$$\mathcal{W} = \mathcal{U} \cup \{\Xi\} \cup \{\partial\tau_1 \partial\tau_2 : \tau_i \in \mathcal{U}\} , \tag{15.10}$$

and define T as the set of all linear combinations of elements in a finite subset $\mathcal{W}_0 \subset \mathcal{W}$, sufficiently large to allow close the fixed pointed problem (15.8). Remark that this defines (implicitly!) a multiplication between some (but not all) of the symbols, notably $\partial\tau_1 \star \partial\tau_2 := \partial\tau_1 \partial\tau_2$ so that we can safely omit $\star$ in the sequel. Naturally, T_α consists of those linear combinations that only involve elements in $\mathcal{W}_0$ of degree α. (Already $\mathcal{W}$ contains only finitely many elements of degree less than α, which reflects subcriticality of the problem.)

In order to simplify expressions later, we use again a shorthand graphical notation for elements of $\mathcal{W}$ as we already did in Section 14.5. Similarly to before, Ξ is represented a small circle, while the integration map $\mathcal{I}$ is represented by a downfacing wavy line and $\mathcal{I}' = \partial\mathcal{I}$ is represented by a downfacing plain line. For example, we write

$$\mathcal{I}'(\Xi)^2 = \mathord{\text{\textsci}} \star \mathord{\text{\textsci}} = \mathord{\mathsf{V}} , \qquad (\mathcal{I}'(\mathcal{I}'(\Xi)^2))^2 = \mathord{\mathsf{Y}} \star \mathord{\mathsf{Y}} = \mathord{\mathsf{W}} , \qquad \mathcal{I}(\mathcal{I}'(\Xi)^2) = \mathord{\mathsf{Y}} .$$

Symbols containing factors of X have no particular graphical representation, so we will for example write $X_i \mathcal{I}'(\Xi)^2 = X_i \mathord{\mathsf{V}}$. With this notation,

$$H = h\,\mathbf{1} + \mathord{\text{\textsci}} + \mathord{\mathsf{Y}} + h'\,X_1 + 2\mathord{\mathsf{V}} + 2h'\mathord{\mathsf{C}} + \ldots$$

described with symbols in $\mathcal{U} = \{\mathbf{1}, \mathord{\text{\textsci}}, \mathord{\mathsf{Y}}, X_1, \mathord{\mathsf{V}}, \mathord{\mathsf{C}}, \ldots\}$, here spelled out up to degree $\frac{3}{2}$ (which will turn out to be "enough", cf. Remark 15.8 below). For the "right-hand side" of the equation we need to include Ξ and, spelling out symbols up to degree 0 which is the minimum required to be able to apply the reconstruction operator to it,

$$\{\partial\tau_1 \partial\tau_2 : \tau_i \in \mathcal{U}\} = \{\mathord{\mathsf{V}}, \mathord{\mathsf{V}}, \mathord{\text{\textsci}}, \mathord{\mathsf{W}}, \mathord{\mathsf{W}}, \mathord{\mathsf{Y}}, \mathord{\mathsf{C}}, \mathbf{1}, \ldots\} .$$

As it turns out, provided that we also include the noise itself, the 14 symbols encountered so far already generate a sufficiently large structure space, given by

$$T = T_{\text{KPZ}} \stackrel{\text{def}}{=} \langle \mathcal{W}_0 \rangle = \langle \Xi, \mathord{\mathsf{V}}, \mathord{\mathsf{V}}, \mathord{\text{\textsci}}, \mathord{\mathsf{W}}, \mathord{\mathsf{W}}, \mathord{\mathsf{Y}}, \mathord{\mathsf{C}}, \mathbf{1}, \mathord{\text{\textsci}}, \mathord{\mathsf{Y}}, X_1, \mathord{\mathsf{V}}, \mathord{\mathsf{C}} \rangle . \tag{15.11}$$

Here we ordered symbols by increasing order of degree. In fact, if τ is a tree with l circles, m plain lines and k wavy lines, then $\deg \tau = n \times \frac{3}{2}^- + m + 2k$. Note that deg $X_1 = 1$ for the abstract space variable, whereas due to parabolic scaling the abstract time variable has deg $X_0 = 2$ and does not show up here.

Note that at this stage, we have not defined a regularity structure yet, as we have not described a structure group G acting on T. However, similarly to what was done in (14.24), it is already natural to consider "representations" of the existing structure, which are linear maps $\boldsymbol{\Pi}$ from T into some suitable space of functions / distributions respecting a form of admissibility condition. For the sake of the present discussion, we assume that all objects are smooth. Given a (smooth) realisation of a "driving noise" ξ, we can then define its canonical lift by setting

$$\big(\boldsymbol{\Pi}\Xi\big)(x) = \xi(x)\,, \qquad \big(\boldsymbol{\Pi}X^k\big)(x) = x^k\,, \tag{15.12}$$

and then recursively by

$$\boldsymbol{\Pi}\tau\bar{\tau} = \boldsymbol{\Pi}\tau \cdot \boldsymbol{\Pi}\bar{\tau}\,, \qquad \boldsymbol{\Pi}\mathcal{I}\tau = K * \boldsymbol{\Pi}\tau\,. \tag{15.13}$$

In general, we say that a linear map $\boldsymbol{\Pi}\colon T \to \mathcal{C}(\mathbf{R}^d)$ is *admissible* if one has the relations

$$\boldsymbol{\Pi}\mathcal{I}\tau = K * \boldsymbol{\Pi}\tau\,, \qquad \boldsymbol{\Pi}\mathbf{1} = 1\,, \qquad \boldsymbol{\Pi}X^k\tau = (\cdot)^k\boldsymbol{\Pi}\tau\,. \tag{15.14}$$

(And similarly with $\mathcal{I}$ replaced by $\mathcal{I}'$ and K replaced by $\partial_x K$ in the case of KPZ...)

Such a map $\boldsymbol{\Pi}$ is clearly not a model since it is a single linear map rather than a family of such maps and the admissibility condition (14.8) is replaced by the more "natural" identity $\boldsymbol{\Pi}\mathcal{I}\tau = K * \boldsymbol{\Pi}\tau$. We will see in the next section how to construct the structure group G and how to use this construction to assign in a unique way a model to the linear map $\boldsymbol{\Pi}$.

Remark 15.8 (Where to truncate?). The (14-dimensional) space T_{KPZ} is indeed sufficient to treat the KPZ equation. Indeed, once in possession of an admissible model, thanks to Theorem 14.5, the fixed point problem (15.8) can be solved in $\mathscr{D}^\gamma$ as soon as γ is a little bit greater than $3/2$. This is why we only need to keep track of terms describing the abstract KPZ solution up to degree $3/2$. Regarding the terms required to describe the right-hand side of the fixed point problem, we need to go up to degree 0, which guarantees that the reconstruction operator (and therefore also the integration operator $\mathcal{K}$) is well-defined. This is similar to $T = T_{<1/2}$, as in Definition 13.4, being sufficient to treat rough / stochastic integration (and then SDEs) in a Brownian rough path / model context. Indeed, in that context (Proposition 13.21) consider $Y \in \mathscr{D}_0^{2\alpha}$ (now for α to be determined!) and abstract Brownian noise $\dot{W} \in \mathscr{D}_{-1/2^-}^\infty$. Then $f(Y)$, composition with a nice function f, is also in $\mathscr{D}_0^{2\alpha}$ and the product is in $\mathscr{D}^{2\alpha-1/2^-}$. We needed this exponent to be positive to have a well-defined rough integration which in turn allows to formulate a fixed point problem, so that we need $2\alpha \geq 1/2$. By definition of $\mathscr{D}^{2\alpha}$, this means that we need Y to take values in $T_{<1/2}$ which is of course what we did by working in $\langle \dot{W}, \mathbb{W}, \mathbf{1}, W \rangle$, ignoring all symbols of higher degree.

15.3 The structure group and positive renormalisation

Recall that the purpose of the group G is to provide a class of linear maps $\Gamma : T \to T$ arising as possible candidates for the action of "reexpanding" a "Taylor series" around a different point. In our case, in view of (14.8) and Definition 14.3, the coefficients of these reexpansions will naturally be some polynomials in x and in the expressions appearing in (14.9). This suggests that we should define a space T^+ whose basis vectors consist of formal expressions of the type

$$ X^k \prod_{i=1}^{N} \mathcal{J}_{\ell_i}(\tau_i) , \qquad (15.15) $$

where N is an arbitrary but finite number, the τ_i are canonical basis elements in $\mathcal{W}$ defined in (15.10), and the ℓ_i are d-dimensional multiindices satisfying $|\ell_i| <$ deg $\tau_i + 2$. (The last bound is a reflection of the restriction of the summands in (14.9) with $\beta = 2$.) The space T^+, which also contains the empty product $\mathbf{1}$, is endowed with a natural commutative product, written as $\cdot$ or (usually) omitted. $(T^+, \cdot, \mathbf{1})$ is nothing but the free commutative algebra over the symbols $\{X_i, \mathcal{J}_\ell(\tau)\}$ with $i \in \{1, \ldots, d\}$ and $\tau \in \mathcal{W}$ with deg $\mathcal{J}_\ell(\tau) := \deg \tau + 2 - |\ell| > 0$.)

Remark 15.9. While the canonical basis of T^+ is related to that of T, it should be viewed as a completely disjoint space. We emphasise this by using the notation $\mathcal{J}_\ell$ rather than $\mathcal{I}_\ell$.

The space T^+ also has a natural graded structure $T^+ = \bigoplus T^+_\alpha$ similarly to before by setting

$$ \deg \mathcal{J}_\ell(\tau) = \deg \tau + 2 - |\ell| , \qquad \deg X^k = |k| , $$

and by postulating that the degree of a product is the sum of the degrees of its factors. Unlike in the case of T however, elements of T^+ all have strictly positive degree, except for the empty product $\mathbf{1}$ which we postulate to have degree 0.

Still inspired by (14.8), as well as by the multiplicativity constraint given by Definition 14.3, we consider the following construction. We define a linear map, sometimes called *coaction*, $\Delta^+ : T \to T \otimes T^+$ in the following way. For the basic elements Ξ, $\mathbf{1}$ and X_i ($i \in \{0, 1\}$), we set

$$ \Delta^+ \mathbf{1} = \mathbf{1} \otimes \mathbf{1} , \quad \Delta^+ \Xi = \Xi \otimes \mathbf{1} , \quad \Delta^+ X_i = X_i \otimes \mathbf{1} + \mathbf{1} \otimes X_i . $$

We then extend this recursively to all of T by imposing the following identities

$$ \Delta^+(\tau \bar{\tau}) = \Delta^+ \tau \cdot \Delta^+ \bar{\tau} , $$

$$ \Delta^+ \mathcal{I}(\tau) = (\mathcal{I} \otimes \mathrm{Id}) \Delta^+ \tau + \sum_\ell \frac{X^\ell}{\ell!} \otimes \mathcal{J}_\ell(\tau) , $$

$$ \Delta^+ \mathcal{I}'(\tau) = (\mathcal{I}' \otimes \mathrm{Id}) \Delta^+ \tau + \sum_{\ell,m} \frac{X^\ell}{\ell!} \otimes \mathcal{J}_{\ell+(0,1)}(\tau) . $$

Here, we extend $\tau \mapsto \mathcal{J}_k(\tau)$ to a linear map $\mathcal{J}_k \colon T \to T^+$ by setting $\mathcal{J}_k(\tau) = 0$ for those basis vectors $\tau \in \mathcal{W}$ for which $\deg \tau \le |k| - 2$. This in particular shows that the sums appearing in the above expressions are actually finite.

Let now G_+ denote the set of characters on T^+, i.e. all linear maps $g \colon T^+ \to \mathbf{R}$ with the property that $g(\sigma\bar{\sigma}) = g(\sigma)g(\bar{\sigma})$ for any two elements σ and $\bar{\sigma}$ in T^+. Then, to any such map, we can associate a linear map $\Gamma_g \colon T \to T$ by

$$\Gamma_g \tau = (\mathrm{Id} \otimes g)\Delta^+ \tau . \tag{15.16}$$

In principle, this definition makes sense for every $g \in (T^+)^*$. However, as already seen in (14.21) it turns out that the set of such maps with $g \in G_+$ forms a group, which we take as our structure group G by setting again

$$G \overset{\mathrm{def}}{=} \{\Gamma_g : g \in G_+\} . \tag{15.17}$$

Remark 15.10. A less explicit way to define G is to simply take it as the set of all linear maps that are 'allowed' in the sense that they are upper triangular with the identity on the diagonal as imposed by (13.5), commute with derivatives as in Definition 14.1, are multiplicative with respect to the product as in Definition 14.3, and satisfy (14.7). See for example [Hai16].

Example 15.11 (KPZ structure group). Running through this procedure, and restricting to $T = T_{\mathrm{KPZ}}$ reveals G as a 7-dimensional (non-commutative) matrix group, canonically realised as a subgroup of the invertible maps $T \to T$, themselves representable as 16×16-matrix. Full details are left for Exercise 15.1.

Example 15.12 (KPZ). Recall $T = \langle \Xi, \vee, \vee\hspace{-1pt}\circ, \uparrow, \vee\hspace{-1pt}\vee, \vee\hspace{-1pt}\circ, Y, \circ\hspace{-1pt}\vee, \mathbf{1}, \ldots \rangle$ in the case of KPZ. Then T^+ is linearly spanned by the symbol $\mathbf{1}$ and polynomials in the commuting symbols as (partially!) listed in

$$\{\mathcal{J}'(\vee\hspace{-1pt}\circ), \mathcal{J}'(\uparrow), \ldots, \mathcal{J}(\Xi), \mathcal{J}(\vee), X_1, \mathcal{J}(\vee\hspace{-1pt}\circ), \mathcal{J}(\uparrow), \ldots\}$$

with (non-negative) degrees $\{\frac{1}{2}^{\equiv}, \frac{1}{2}^{-}, \ldots, 1^{=}, 1, \frac{3}{2}^{\equiv}, \frac{3}{2}^{-}, \ldots\}$ and shorthands $\mathcal{J} = \mathcal{J}_{(0,0)}, \mathcal{J}' = \mathcal{J}_{(0,1)}$. We note that all symbols here can be represented by *elementary* trees,[2] where $\mathcal{J}(\tau)$ (resp. $\mathcal{J}'(\tau)$) is represented by attaching a *single* downfacing wavy (resp. plain) line to the root of τ. For instance

$$3 \cdot \mathbf{1} - \mathcal{J}(\Xi) + 2 \cdot \mathcal{J}'(\uparrow) \cdot \mathcal{J}'(\vee\hspace{-1pt}\vee\hspace{-1pt}\vee) \in T^+$$

but the symbol $\mathcal{J}'(\Xi)$ (which would be of negative homogeneity) is not an element of T^+.

Before we show that G does indeed form a group (actually a subgroup of the invertible maps from T to T), we show how to use it to turn an admissible linear

---

[2] With some goodwill this even includes X-factors, which then appear as polynomial decorations of the trees.

map $\boldsymbol{\Pi}: T \to \mathcal{C}^\infty(\mathbf{R}^d)$ (in the sense of (15.14)) into a model (Π, Γ). Consider the recursion

$$f_x(\mathcal{J}_\ell(\tau)) = - \sum_{|k+\ell| < |\tau|+2} \frac{(-x)^k}{k!} \int D^{\ell+k} K(x-y) \left(\Pi_x \tau\right)(dy) ,$$

$$\Pi_x \tau = (\boldsymbol{\Pi} \otimes f_x)\Delta^+ \tau , \tag{15.18}$$

where we furthermore impose that the f_x are characters, namely that they extend to all of $(T^+)^*$ in a multiplicative fashion, $f_x(\sigma\bar{\sigma}) = f_x(\sigma) f_x(\bar{\sigma})$. We leave it as a simple exercise to verify that these two identities are sufficient to define the f_x and the Π_x uniquely.

Remark 15.13. The correspondence $\boldsymbol{\Pi} \Leftrightarrow (\Pi, \Gamma)$ can also be inverted and the two notions of admissibility are consistent, so that these are two completely equivalent ways of looking at admissible models for our regularity structure. Indeed, it suffices to set $\boldsymbol{\Pi}\tau = \mathcal{R}H_\tau$, where the elements $H_\tau \in \mathcal{D}^\infty$ (i.e. one can make sure that $H_\tau \in \mathcal{D}^\gamma$ for any fixed γ) are given by $H_{X^k}(x) = (X+x)^k$, $H_\Xi(x) = \Xi$, and then recursively by

$$H_{\mathcal{I}(\tau)} = \mathcal{K}H_\tau , \qquad H_{\tau\bar{\tau}} = H_\tau \cdot H_{\bar{\tau}} .$$

In particular, this correspondence does not at all rely on the fact that the model was built by lifting a smooth function. Note that this is strongly reminiscent of the construction given in Exercise 13.11. See also Exercise 15.3.

If we now define elements $F_x \in G$ by

$$F_x \tau \stackrel{\text{def}}{=} \Gamma_{f_x} = (\text{Id} \otimes f_x)\Delta^+ \tau , \tag{15.19}$$

and then set (an expression for F_x^{-1} is given below)

$$\Gamma_{xy} = F_x^{-1} F_y , \tag{15.20}$$

it follows immediately from (15.18) that the Π_x and the maps Γ_{xy} do indeed satisfy the desired algebraic relation $\Pi_x \Gamma_{xy} = \Pi_y$. We also note that the coefficients of the linear maps Γ_{xy} are expressed as polynomials of the numbers $f_x(\mathcal{J}_{\ell_i}(\tau_i))$ and $f_y(\mathcal{J}_{\ell_i}(\tau_i))$ for suitable expressions τ_i and multiindices ℓ_i. Note that the linear maps $F_x : T \to T$ perform a kind of "recentering" of $\boldsymbol{\Pi}$ around x in the sense that (15.18) guarantees that, at least when $\boldsymbol{\Pi}$ is sufficiently smooth, $\Pi_x \mathcal{I}(\tau)$ vanishes at the order determined by the degree of τ. As a matter of fact, one could even have taken this as the defining property of the maps F_x (together with the fact that they are of the form (15.19) for some multiplicative functional f_x). We will see in Section 15.5 below that the renormalisation procedure required to give a meaning to singular SPDEs like the KPZ equation can equally be interpreted as a type of recentering procedure, but this time in "probability space". This also explains the terminology "positive renormalisation" which is sometimes encountered for the maps F_x.

We now argue that G as defined above actually forms a group, so that in particular the maps F_x are invertible. To this end, define a linear map $\Delta^+ : T^+ \to T^+ \otimes T^+$, very similarly to the previously defined map $\Delta^+ : T \to T \otimes T^+$, by

$$\Delta^+ \mathbf{1} = \mathbf{1} \otimes \mathbf{1} , \qquad \Delta^+ X = X \otimes \mathbf{1} + \mathbf{1} \otimes X ,$$

extended recursively to all of T^+ by imposing the identities, for all multiindices k,

$$\Delta^+ (\sigma \bar{\sigma}) = (\Delta^+ \sigma)(\Delta^+ \bar{\sigma}) ,$$
$$\Delta^+ \mathcal{J}_k(\tau) = (\mathcal{J}_k \otimes \mathrm{Id}) \Delta^+ \tau + \sum_{\ell \in \mathbf{N}^2} \frac{X^\ell}{\ell!} \otimes \mathcal{J}_{\ell + k}(\tau) . \tag{15.21}$$

It can be verified that Δ^+ is *coassociatve* in the sense

$$(\Delta^+ \otimes \mathrm{Id}) \Delta^+ = (\mathrm{Id} \otimes \Delta^+) \Delta^+ . \tag{15.22}$$

This and the multiplicative property make Δ^+ a *coproduct* and T^+ a (connected, graded) *coalgebra*. From general principles there exists a unique linear map $\mathcal{A}^+ : T^+ \to T^+$, called *antipode*, so that $(T^+, \cdot, \Delta^+, \mathcal{A}^+)$ is a *Hopf algebra*. Moreover, our notational overload is justified by the fact that (15.22) also holds when both sides of the identity are interpreted as linear maps $T \to T \otimes T^+ \otimes T^+$.

We then define a product $\circ$ on the space of linear functionals $f \colon T^+ \to \mathbf{R}$ by

$$(f \circ g)(\sigma) = (f \otimes g) \Delta^+ \sigma , \tag{15.23}$$

noting that coassociativity of Δ^+ implies associativity of $\circ$. Restricted to multiplicative elements, i.e. to G_+, the definition of the antipode implies that G_+ is indeed a group with $f^{-1} = f \mathcal{A}^+$, that is $f^{-1} \circ f = f \circ f^{-1} = e$, where $e \colon T^+ \to \mathbf{R}$ maps every basis vector of the form (15.15) to zero, except for $e(\mathbf{1}) = 1$. This is a general construction for Hopf algebras and G_+ is known as the *character group* of T^+. The product $\circ$ in this context is usually called the *convolution product*. Indeed, the first identity in (15.21), valid by definition for every coproduct in a Hopf algebra, ensures that if f and g belong to G_+, then $f \circ g \in G_+$. (Spelled out, this says if $f, g \in (T^+)^*$ are both multiplicative in the sense that $f(\sigma \bar{\sigma}) = f(\sigma) f(\bar{\sigma})$, then $f \circ g$ is again multiplicative.)

Since, by definition, $\Gamma_f = (\mathrm{Id} \otimes f) \Delta$ we can rewrite (15.19) as $F_x = \Gamma_{f_x}$, and the intertwining identity (15.22) entails that

$$\Gamma_{f \circ g} = \Gamma_f \Gamma_g .$$

Also, the element e is neutral in the sense that Γ_e is the identity operator, and as a consequence $\Gamma_{f^{-1}} = \Gamma_f^{-1}$ whenever $f \in G_+$. In particular then,

$$F_x^{-1} = \Gamma_{f_x^{-1}} = \Gamma_{f_x \mathcal{A}^+}$$

and we can fully spell out (15.20) as

$$\Gamma_{xy} = \Gamma_{f_x \mathcal{A}^+ \circ f_y} = (\mathrm{Id} \otimes \gamma_{x,y}) \Delta^+ , \quad \gamma_{x,y} \overset{\text{def}}{=} f_x \mathcal{A}^+ \circ f_y = (f_x \mathcal{A}^+ \otimes f_y) \Delta^+ .$$

The fact that Δ^+ preserves degree (as can be seen by induction from its definition) and that elements of T^+ all have strictly positive degree, except for $\mathbf{1}$ leads to the conclusion that, for every $\Gamma \in G$ and every $\tau \in T$, $\Gamma\tau$ is indeed of the form (13.5). The multiplicativity property of Δ^+ furthermore guarantees that the constraint mentioned in Definition 14.3 does hold. This justifies our definition of structure group G associated to T as the set of all multiplicative linear functionals on T^+, acting on T via (15.16), as given in (15.17), for G has group structure induced from G_+.

Returning to the relation between Π_x and $\boldsymbol{\Pi}$, we showed actually more, namely that the knowledge of $\boldsymbol{\Pi}$ and the knowledge of (Π, Γ) are equivalent. Indeed, on the one hand one has $\boldsymbol{\Pi} = \Pi_x F_x^{-1}$ and the map F_x can be recovered from Π_x by (15.18) and (15.19). On the other hand however, one also has of course $\Pi_x = \boldsymbol{\Pi} F_x$ and, if we equip T with an adequate recursive structure, then we have already seen that the coefficients f_x are uniquely determined by $\boldsymbol{\Pi}$.

Furthermore, the correspondence $(\Pi, \Gamma) \leftrightarrow \boldsymbol{\Pi}$ outlined above works for *any* admissible model and does not at all rely on the fact that it was built by lifting a continuous function. In particular, it does *not* rely on the fact that Π_x and $\boldsymbol{\Pi}$ are multiplicative. In the general case, the first identity in (15.13) may then of course fail to be true, even if $\boldsymbol{\Pi}\tau$ happens to be a continuous function for every $\tau \in T$. The only reason why our definition of an admissible model does not simply consist of the single map $\boldsymbol{\Pi}$ is that there seems to be no simple way of describing the topology given by Definition 13.5 in terms of $\boldsymbol{\Pi}$.

15.4 Reconstruction for canonical lifts

Recall that, given any sufficiently regular function ξ (say a continuous space-time function), there is a canonical way of lifting ξ to an admissible model $\mathscr{L}\xi = (\Pi, \Gamma)$ for T by imposing (15.12) and (15.13), and then turning $\boldsymbol{\Pi}$ into a model as described in the previous paragraph. With such a model $\mathscr{L}\xi$ at hand, it follows from (15.13) and (13.26) that the associated reconstruction operator satisfies the properties

$$\mathcal{R}\mathcal{K}f = K * \mathcal{R}f , \quad \mathcal{R}(fg) = \mathcal{R}f \cdot \mathcal{R}g ,$$

as long as all the functions to which $\mathcal{R}$ is applied belong to $\mathscr{D}^\gamma$ for some $\gamma > 0$. As a consequence, applying the reconstruction operator $\mathcal{R}$ to both sides of (15.8), we see that if H solves (15.8) then, provided that the model $(\Pi, \Gamma) = \mathscr{L}\xi$ was built as above starting from any *continuous* realisation ξ of the driving noise, the function $h = \mathcal{R}H$ solves the equation (15.1).

At this stage, the situation is as follows. For any *continuous* realisation ξ of the driving noise, we have factorised the solution map $(h_0, \xi) \mapsto h$ associated to (15.1) into maps

$$(h_0, \xi) \mapsto (h_0, \mathscr{L}\xi) \mapsto H \mapsto h = \mathcal{R}H ,$$

where the middle arrow corresponds to the solution to (15.8) in some weighted $\mathscr{D}^\gamma$-space. The advantage of such a factorisation is that the last two arrows yield *continuous* maps, even in topologies sufficiently weak to be able to describe driving noise having the lack of regularity of space-time white noise. The only arrow that isn't continuous in such a weak topology is the first one. At this stage, it should be believable that a similar construction can be performed for a very large class of semilinear stochastic PDEs, provided that certain scaling properties are satisfied. This is indeed the case and large parts of this programme have been carried out in [Hai14b].

Given this construction, one is lead naturally to the following question: given a sequence ξ_ε of "natural" regularisations of space-time white noise, for example as in (15.6), do the lifts $\mathscr{L}\xi_\varepsilon$ converge in probably in a suitable space of admissible models? Unfortunately, unlike in the theory of rough paths where this is very often the case (see Section 10), the answer to this question in the context of SPDEs is often an emphatic **no**. Indeed, if it were the case for the KPZ equation, then one could have been able to choose the constant C_ε to be independent of ε in (15.6), which is certainly not the case.

15.5 Renormalisation of the KPZ equation

One way of circumventing the fact that $\mathscr{L}\xi_\varepsilon$ does not converge to a limiting model as $\varepsilon \to 0$ is to consider instead a sequence of *renormalised* models. The main idea is to exploit the fact that our definition of an admissible model does *not* impose the multiplicative identity

$$\boldsymbol{\Pi}\tau\bar{\tau} = \boldsymbol{\Pi}\tau \cdot \boldsymbol{\Pi}\bar{\tau}\,,$$

used in (15.13) for the canonical lift, even in situations where ξ itself happens to be a continuous function. One question that then imposes itself is: what are the natural ways of "deforming" the usual product which still lead to lifts to an admissible model? It turns out that the regularity structure whose construction was sketched above comes equipped with a natural *finite-dimensional* group of continuous transformations $\mathfrak{R}$ on its space of admissible models (henceforth called the "renormalisation group"), which essentially amounts to the space of all natural deformations of the product. It then turns out that even though the canonical lift $\mathscr{L}\xi_\varepsilon$ does not converge, it is possible to find a sequence M_ε of elements in $\mathfrak{R}$ such that the sequence $M_\varepsilon\mathscr{L}\xi_\varepsilon$ converges to a limiting model $(\hat{\boldsymbol{\Pi}}, \hat{\boldsymbol{\Gamma}})$. Unfortunately, the elements M_ε do *not* preserve the image of $\mathscr{L}$ in the space of admissible models. As a consequence, when solving the fixed point map (15.8) with respect to the model $M_\varepsilon\mathscr{L}\xi_\varepsilon$ and inserting the solution into the reconstruction operator, it is not clear *a priori* that the resulting function (or distribution) can again be interpreted as the solution to some modified PDE. It turns out that in the present setting this is again the case and the modified equation is precisely given by (15.6), where C_ε is some linear combination of the constants appearing in the description of M_ε.

There are now three questions that remain to be answered:

1. How does one construct the renormalisation group $\mathfrak{R}$?
2. How does one derive the new equation obtained when renormalising a model?
3. What is the right choice of M_ε ensuring that the renormalised models converge?

As already pointed out at the start of this chapter, these questions have now been answered in full generality in the series of articles [Hai14b, BHZ19, CH16, BCCH17]. The aim of this section is to illlustrate how the machinery developed there applies to the particular case of the KPZ equation and go give a feeling for how the main steps of the construction generalise to other settings.

15.5.1 The renormalisation group

How does all this help with the identification of a natural class of deformations for the usual product? Throughout this section, we will only consider models constructed from a single map Π by the recursive procedure given in (15.18), combined with (15.20). At this point, we crucially note that if $\Pi : T \to \mathcal{C}^\infty(\mathbf{R}^d)$ is an arbitrary admissible linear map (in the sense that $\Pi \mathcal{I}\tau = K * \Pi\tau$ as before), then there is no reason in general for (15.18) and (15.20) to define a model. The reason is that while these definitions do guarantee that $\Pi_x \mathcal{I}\tau$ satisfies the first bound in (13.13), there is no reason in general for $(\Pi_x \tau)(y)$ to vanish at the right order as $y \to x$ for an arbitrary symbol τ that is not obtained by applying the integration map to some other symbol. It is however the case that these bounds hold whenever Π is obtained as the canonical lift of a smooth function, as can easily be seen from the multiplicativity property of the canonical lift.

This suggests to define a space $\mathscr{M}_\infty$ consisting of those admissible maps $\Pi : T \to \mathcal{C}^\infty(\mathbf{R}^d)$ which do generate a model by the above procedure. By Remark 15.13, there is a canonical bijection between $\mathscr{M}_\infty$ and the set of all smooth admissible models, so we henceforth also call an element $\Pi \in \mathscr{M}_\infty$ simply a model (or an admissible model). Note that even though the space of linear maps $T \to \mathcal{C}^\infty(\mathbf{R}^d)$ is linear, the space $\mathscr{M}_\infty$ is far from being a linear space.

At this stage, we would like to introduce probability into the game. For this, note first that we have a natural action S of the group of translations $(\mathbf{R}^d, +)$ onto T by setting $S_h X^k = (X + h)^k$, $S_h \Xi = \Xi$, and then recursively by

$$S_h \mathcal{I}\tau = \mathcal{I}S_h\tau , \qquad S_h \tau\bar\tau = S_h\tau S_h\bar\tau .$$

We then note that if ξ happens to be a stationary stochastic process and $\Pi = \mathscr{L}\xi$ is its canonical lift as a random model, then Π is a stationary stochastic process in the generalised sense that

$$(\Pi\tau)(\cdot + h) \stackrel{\text{law}}{=} (\Pi S_h\tau)(\cdot) .$$

In order to define the renormalisation group $\mathfrak{R}$, it is then natural to consider only transformations of the space of admissible models that preserve this property. Since we are not in general allowed to multiply components of $\boldsymbol{\Pi}$ and we do not want to "pull arbitrary functions out of a hat", the only remaining operation is to form linear combinations. It is therefore natural to look for linear maps $M: T \to T$ which furthermore preserve $\mathscr{M}_\infty$ in the sense that if, given $\boldsymbol{\Pi} \in \mathscr{M}_\infty$, we define $\boldsymbol{\Pi}^M$ by

$$\boldsymbol{\Pi}^M \tau = \boldsymbol{\Pi} M \tau , \tag{15.24}$$

one would like to have again $\boldsymbol{\Pi}^M \in \mathscr{M}_\infty$. It is clear that in order to guarantee this, M needs to commute with the integration operators $\mathcal{I}$ and $\mathcal{I}'$, but this alone is by no means sufficient.

It turns out that the construction of a natural family of operators with the required properties goes in a way that is strongly reminiscent of the construction of the structure group, but with many aspects of the construction "reversed". A natural starting point of the construction is given by the set $\mathcal{W}_- \subset \mathcal{W}$ consisting of the canonical basis vectors of *strictly negative* degree of our regularity structure T which furthermore have the property that they can be built from products and integrations applied to Ξ, i.e. do not involve any X^k for $k > 0$. We then define T^- similarly to T^+ as the free unital algebra generated by $\mathcal{W}_-$, i.e.[3]

$$T^- \stackrel{\text{def}}{=} \mathrm{Alg}\left(\left\{ \circ, \mathcal{V}, \mathcal{V}, \mathfrak{l}, \mathcal{W}, \mathcal{V}, \Upsilon, \langle \circ \rangle \right\}\right) ,$$

the algebra given by all polynomials with real coefficients and indeterminates in $\mathcal{W}_-$; the unit is denoted by $\mathbf{1}$ (or, equivalently, as the empty forest $\emptyset$). The reason why $\mathcal{W}_-$ is expected to play a major role is that, by combing Exercise 13.11 with admissibility and multiplicativity of the action of Γ, $\boldsymbol{\Pi}\tau$ for $\deg \tau > 0$ is uniquely determined by the knowledge of $\boldsymbol{\Pi}\tau$ for all symbols τ with $\deg \tau \leq 0$.

By analogy with the BPHZ renormalisation procedure in quantum field theory [BP57, Hep69, Zim69], it is natural to look for renormalisation maps that consist in "contracting subtrees of negative degree". In order to formalise such an operation, we take more seriously the interpretation of the canonical basis elements of T as "trees". More precisely, we consider labelled trees $\tau = (V, E, \varrho, \mathfrak{n}, \mathfrak{e})$, where V is a finite vertex set, $E \subset V \times V$ is an edge set, $\varrho \in V$ is a root, $\mathfrak{n}: V \to \mathbf{N}^d$ is a "polynomial label" and $\mathfrak{e}: E \to \{\Xi, \mathcal{I}, \mathcal{I}'\}$ is an "edge label". As usual, we identify labelled trees if they can be related by a tree isomorphism preserving the root and labels. The way this correspondence works is as follows. The symbol X^k is represented as the (unique) tree with a sole vertex $V = \{\varrho\}$ and polynomial label $\mathfrak{n}(\varrho) = k$. The symbol Ξ is represented by the tree with two vertices $V = \{\varrho, \bullet\}$, one (oriented) edge $E = \{e\} = \{(\bullet, \varrho)\}$, and labels $\mathfrak{n} = 0$, $\mathfrak{e}(e) = \Xi$. Integration is then performed by adding an edge of the corresponding type to the root, i.e. we have for example

---

[3] As in the case of rough volatility, cf. 14.26, we colour basis elements of T^- differently to distinguish them from those of T and / or T^+. Elements in T^- are naturally represented as (unordered) *forests*.

$$\mathcal{I}(V, E, \varrho, \mathfrak{n}, \mathfrak{e}) = (V \sqcup \{\bar{\varrho}\}, E \sqcup \{(\varrho, \bar{\varrho})\}, \bar{\varrho}, \mathcal{I}\mathfrak{n}, \mathcal{I}\mathfrak{e}) \,,$$

where $\mathcal{I}\mathfrak{n}(\bar{\varrho}) = 0$ and otherwise agrees with $\mathfrak{n}$, while $\mathcal{I}\mathfrak{e}((\varrho, \bar{\varrho})) = \mathcal{I}$ and again otherwise agrees with $\mathfrak{e}$. Multiplication is obtained by joining roots:

$$(V, E, \varrho, \mathfrak{n}, \mathfrak{e}) \cdot (\bar{V}, \bar{E}, \bar{\varrho}, \bar{\mathfrak{n}}, \bar{\mathfrak{e}}) = ((V \sqcup \bar{V})/\{\varrho, \bar{\varrho}\}, E \sqcup \bar{E}, \{\varrho, \bar{\varrho}\}, \mathfrak{n} \sqcup \bar{\mathfrak{n}}, \mathfrak{e} \sqcup \bar{\mathfrak{e}}) \,,$$

where $(\mathfrak{n} \sqcup \bar{\mathfrak{n}})(\{\varrho, \bar{\varrho}\}) = \mathfrak{n}(\varrho) + \bar{\mathfrak{n}}(\bar{\varrho})$.

Remark 15.14. This is nothing but a formalisation of the graphical notation already used earlier. The notation used in (15.11) for example suggests that one could equivalently have viewed the noise as part of a "vertex label" and this is the viewpoint taken for example in [BCCH17]. It appears however that viewing noises as edges, as for example in [BHZ19], usually yields a more consistent formalism. This is especially the case in situations where one would like to "attach" additional information to noises as done in [CCHS20, Sec. 5].

In a similar way, elements of T^- can be interpreted as elements $A = (V, E, \varrho, \mathfrak{e})$ as above, except that there is no "polynomial label" $\mathfrak{n}$ and (V, E) is allowed to be a forest, with ϱ denoting the set of its roots, one per connected component. In particular, the empty forest $V = \emptyset$ is allowed, which wasn't the case for T.

Given $A = (\bar{V}, \bar{E}, \bar{\varrho}, \bar{\mathfrak{e}}) \in T^-$ and $\tau = (V, E, \varrho, \mathfrak{n}, \mathfrak{e}) \in T$, we say that $A \subset \tau$ if one has an injective map $\iota \colon \bar{V} \sqcup \bar{E} \to V \sqcup E$ preserving connectivity and edge labels. Note that the injectivity of ι implies in particular that the different connected components of A are vertex-disjoint in τ. In such a situation, we then write $\mathcal{R}_A \tau$ for the tree obtained by contracting the connected components of A in τ, i.e. the vertex set of $\mathcal{R}_A \tau$ consists of $V/\sim$ where $v \sim \bar{v}$ if v and $\bar{v}$ are equal or belong to the image of the same connected component of A, while the edge set of $\mathcal{R}_A \tau$ equals $E \setminus \iota \bar{E}$.
We then define an operator $\Delta^- \colon T \to T^- \otimes T$ by

$$\Delta^- \tau = \sum_{A \subset \tau} \mathcal{Q}_- A \otimes \mathcal{R}_A \tau \,, \tag{15.25}$$

where $\mathcal{Q}_- A = A$ if every connected component of A has negative degree and $\mathcal{Q}_- A = 0$ otherwise. Note again the graphical interpretation of extracting possibly empty collections of subtrees of negative degree.

Example 15.15. For the regularity structure associated to the KPZ equation, we have for example[4]

$$\Delta^- \overset{\text{\tiny ❧}}{\mathbf{Y}} = \overset{\text{\tiny ❧}}{\mathbf{Y}} \otimes \mathbf{1} + \mathbf{1} \otimes \overset{\text{\tiny ❧}}{\mathbf{Y}} + 2 \overset{\text{\tiny ❧}}{\mathbf{Y}} \otimes \overset{\text{\tiny ❧}}{\mathbf{Y}} + \overset{\text{\tiny ❧}}{\mathbf{Y}} \otimes \mathbf{Y}$$
$$+ \mathbf{Y} \otimes \overset{\text{\tiny ❧}}{\mathbf{Y}} + 2 \mathbf{\uparrow} \otimes \overset{\text{\tiny ❧}}{\mathbf{Y}} + \mathbf{\uparrow} \otimes \overset{\text{\tiny ❧}}{\mathbf{Y}} + \mathbf{\uparrow} \otimes \overset{\text{\tiny ❧}}{\mathbf{Y}} \tag{15.26}$$
$$+ 2 \mathbf{\uparrow\uparrow} \otimes \overset{\text{\tiny ❧}}{\mathbf{Y}} + 2 \mathbf{\uparrow\uparrow} \otimes \overset{\text{\tiny ❧}}{\mathbf{Y}} + 2 \mathbf{\uparrow\uparrow\uparrow} \otimes \mathbf{\langle} \,,$$

where we used red symbols to denote elements of T^- just as in Section 14.5. In most situations it is natural to only consider characters of T^- that vanish on planted trees,

---

[4] Mind that $\overset{\text{\tiny ❧}}{\mathbf{Y}} \equiv \mathbf{Y} \subset \overset{\text{\tiny ❧}}{\mathbf{Y}}$ in three distinct ways which explains the terms $2 \overset{\text{\tiny ❧}}{\mathbf{Y}} \otimes \overset{\text{\tiny ❧}}{\mathbf{Y}} + \overset{\text{\tiny ❧}}{\mathbf{Y}} \otimes \mathbf{Y}$.

i.e. trees with only one edge incident to the root,[5] in which case this simplifies to

$$\Delta^- \mathord{\Large\text{⋎}} = \mathord{\Large\text{⋎}} \otimes 1 + 1 \otimes \mathord{\Large\text{⋎}} + 2 \mathord{\text{⋎}} \otimes \mathord{\text{⋎}} + \mathord{\text{⋎}} \otimes \mathord{Y} .$$

Note also that there is for example no term $\mathord{\text{⋎}} \otimes \mathord{\text{ı}}$ appearing in (15.26); indeed $\mathord{\text{⋎}}$ fails to have negative degree, hence is not an element of T^- and killed by $\mathcal{Q}_-$.

Remark 15.16. Since $\mathcal{I}'(1) = 0$ by Remark 15.6, there is no term such as $\mathord{V} \otimes \mathord{\text{⋎}}$ appearing in the right-hand side of (15.26).

Remark 15.17. While the present construction is sufficient for KPZ, in full generality, one should also allow polynomial decorations for elements in T^- in which case the expression for Δ^- involves additional combinatorial factors, similarly to the definition of Δ^+.

Our motivation for the definition of Δ^- is as follows. Assigning a number to each $\tau \in W_-$ is equivalent to choosing an algebra morphism $g \colon T^- \to \mathbf{R}$. If we ignore for a moment the labels $\mathfrak{n}$ and $\mathfrak{e}$, an operation of the type $M_g \colon T \to T$ with

$$M_g \tau = (g \otimes \mathrm{Id})\Delta^- \tau , \tag{15.27}$$

then corresponds to iterating over all ways of contracting subtrees of negative degree contained in τ and replacing them by the corresponding constant assigned to it by g. This corresponds to replacing a kernel of possibly several variables by a multiple of a Dirac delta function forcing all arguments to collapse.

Similarly to before, one can also define an operator $\Delta^- \colon T^- \to T^- \otimes T^-$ by setting

$$\Delta^- B = \sum_{A \subset B} \mathcal{Q}_- A \otimes \mathcal{Q}_- \mathcal{R}_A B ,$$

where the notions of inclusion $A \subset B$ and the contraction $\mathcal{R}_A B$ are defined in complete analogy to above.

This yields an algebraic structure very similar to the one given by T and T^+. We will however not describe it in any more detail here, but refer instead to [BHZ19] for additional details. In particular, T^-, with forest product and coproduct Δ^-, admits an antipode $\mathcal{A}^-$ turning it into a commutative Hopf algebra. Its characters then form a group with product analogous to (15.23) and inverse given by $g \mapsto g\mathcal{A}^-$, acting on T by (15.27).

Definition 15.18. The *renormalisation group* $\mathfrak{R}$ for our regularity structure T is defined as the character group of T^-.

Remark 15.19. The original definition of the "renormalisation group" given in [Hai14b] (and in the first edition of this book) is slightly more general. In the situation of the regularity structure built for a two-component KPZ equation, i.e.

---

[5] In essence, extracting negative trees will help to renormalise otherwise ill-posed products. A single edge incident to the root corresponds to convolution with a (compactly supported) kernel, which is always well-posed.

exactly the same as discussed here, except that there are two "noises" Ξ_1 and Ξ_2 and every occurrence of Ξ can be replaced by either of them, the old definition would for example include the map M that swaps the two noises in a consistent way. (Consistency is in the sense that $M\mathcal{I}'(\Xi_2)\mathcal{I}'(\mathcal{I}'(\Xi_1)^2) = \mathcal{I}'(\Xi_1)\mathcal{I}'(\mathcal{I}'(\Xi_2)^2)$ for example.) This is not an operation that is described by a character of T^-. The advantage of the present definition is that it is much more explicit. Furthermore, it follows from the analytical results of [CH16] that it is sufficiently large to serve the purpose of renormalising divergent models.

Example 15.20. Continuing the above example, we have

$$\Delta^- \overset{\circ\circ}{\underset{\circ\circ}{\mathsf{Y}}} = \overset{\circ\circ}{\underset{\circ\circ}{\mathsf{Y}}} \otimes 1 + 1 \otimes \overset{\circ\circ}{\underset{\circ\circ}{\mathsf{Y}}} + 2 \overset{\circ}{\mathsf{V}} \otimes \overset{\circ}{\mathsf{V}} + \overset{\circ}{\mathsf{V}} \otimes \mathsf{Y} + \mathsf{Y} \otimes \overset{\circ}{\mathsf{V}} .$$

Note that we have not considered the simplification of removing planted trees. Instead, the analogues of the remaining terms appearing in (15.26) are killed by the projection $\mathcal{Q}_-$. We also note that this expression is symmetric in the two factors T^- which is the case for all the symbols appearing in the analysis of the KPZ equation. This implies that the KPZ renormalisation group $\mathfrak{R}$ is abelian. (In general though, the presence of "overlapping divergencies" can cause $\mathfrak{R}$ to be non-abelian.)

One of the main results of [BHZ19] is a generalisation of the following statement, which shows that the action of the renormalisation group plays nice with our notion of admissible model.

Theorem 15.21. *Let $g \in \mathfrak{R}$ and define $M_g = (g \otimes \mathrm{Id})\Delta^-$ as in (15.27). Then, for any $\boldsymbol{\Pi} \in \mathscr{M}_\infty$, one has $\boldsymbol{\Pi}^g \overset{\text{def}}{=} \boldsymbol{\Pi} M_g \in \mathscr{M}_\infty$. Furthermore, one has*

$$\Pi_x^g = \Pi_x M_g , \qquad \Gamma_{xy}^g = M_g^{-1}\Gamma_{xy}M_g . \qquad (15.28)$$

Proof. We sketch the proof. Recall that Δ^- has been defined (with notational overload) as map from $T \to T^- \otimes T$ and $T^- \to T^- \otimes T^-$; we now also define $\Delta^- : T^+ \to T^- \otimes T^+$ as multiplicative linear map, determined by

$$\Delta^- X_i = 1 \otimes X_i, \qquad \Delta^- \mathcal{J}_\ell(\tau) = (\mathrm{Id} \otimes \mathcal{J}_\ell(\cdot))\Delta^- \tau .$$

In the special case of KPZ one can check by hand that, thanks in particular to the fact that $\mathcal{I}'(\mathbf{1}) = 0$ by Remark 15.6 (which correctly suggests that we should also impose $\mathcal{J}'(\mathbf{1}) = 0$),

(i) On T one has the *cointeraction formula*

$$M_{13}(\Delta^- \otimes \Delta^-)\Delta^+ = (\mathrm{Id} \otimes \Delta^+)\Delta^- , \qquad (15.29)$$

where $M_{13} : T^- \otimes T \otimes T^- \otimes T^+ \to T^- \otimes T \otimes T^+$ is the map that multiplies the first and third factor (in T^-), and the same holds also on T^+.

(ii) The actions of $\mathfrak{R}$ onto T and T^+ given by M_g do not decrease the degree. (For the relevant set of characters g, this is seen explicitly in Exercise 15.2.)

Recall the correspondence $\boldsymbol{\Pi} \Leftrightarrow (\Pi, \Gamma)$ given in Remark 15.13. With the special properties (i)-(ii) it is straightforward to verify that, for $g \in \mathfrak{R}$ arbitrary, $\boldsymbol{\Pi}^g = \boldsymbol{\Pi} M_g$ defines a model $\boldsymbol{\Pi}^g \Leftrightarrow (\Pi^g, \Gamma^g)$ with

$$\Pi_x^g = \Pi_x M_g = (g \otimes \Pi_x)\Delta^-, \quad \Gamma_{xy}^g = (\text{Id} \otimes \gamma_{x,y}^g)\Delta^+, \quad \gamma_{xy}^g \overset{\text{def}}{=} (g \otimes \gamma_{xy})\Delta^-.$$

(The second identity in (15.28) then follows from the formula for γ_{xy}^g, combined with the cointeraction formula.) To show all this, first write $f_x = f_x^{\boldsymbol{\Pi}}$ for f_x obtained from $\boldsymbol{\Pi}$ as in (15.18). One shows recursively that

$$f_x^{\boldsymbol{\Pi}^g} = f_x^{\boldsymbol{\Pi}} M_g.$$

One then uses (i), on T, to show that the required identity for Π_x^g holds. Finally, one uses (i), on T^+ to show that if one views $M_g = (g \otimes \text{Id})\Delta^-$ as acting on T^+, then its action distributes over the product in the character group defined in (15.23) in the sense that $(M_g f) \circ (M_g \bar{f}) = M_g(f \circ \bar{f})$, which then implies the required identity for γ_{xy}^g. The fact that the action of M_g does not decrease degrees guarantees that (Π^g, Γ^g) is again a model (since (Π, Γ) is). $\square$

Remark 15.22. In general (i.e. in the case of similar regularity structures set up for different examples of subcritical semilinear SPDEs), the cointeraction property (15.29) may fail. It turns out however that it can still be rescued by working in a suitably extended regularity structure, see [Hai16, BHZ19].

One important feature of this theorem is that the last statement provides quantitative bounds on the map $\boldsymbol{\Pi} \mapsto \boldsymbol{\Pi}^g$ which show that it can be extended to a continuous action of $\mathfrak{R}$ onto the space $\mathscr{M}$ of all admissible models. A crucial property of $\mathfrak{R}$ is that it is sufficiently large to allow us to "recenter" models in a natural way.

Definition 15.23. Let ξ be a (smooth) stationary stochastic process and let $\boldsymbol{\Pi}$ be its canonical lift. Then, there exists a unique character $g^{\text{BPHZ}} \in \mathfrak{R}$ such that $\boldsymbol{\Pi}^{\text{BPHZ}} = \boldsymbol{\Pi} M_{g^{\text{BPHZ}}}$ satisfies $\mathbf{E}(\boldsymbol{\Pi}^{\text{BPHZ}} \tau)(0) = 0$ for every canonical basis vector $\tau \in T$ with $\deg \tau < 0$. We call $\boldsymbol{\Pi}^{\text{BPHZ}}$ the *BPHZ lift* of ξ.

Remark 15.24. This is named after Bogoliubow, Parasiuk, Hepp and Zimmermann [BP57, Hep69, Zim69] who introduced an analogous renormalisation procedure in the context of perturbative quantum field theory in the sixties.

Remark 15.25. Note also that while the BPHZ lift of a noise ξ is "canonical", it does depend on the choice of kernel K for our notion of admissibility. In particular, different truncations of the heat kernel will in general lead to different values for the BPHZ renormalisation constants.

A beautiful property of the BPHZ lift is that it is much more stable than the canonical lift. Indeed, it was shown in [CH16] that one can introduce a natural measure of the "size" $N(\xi)$ of a stationary noise ξ which is such that for any sequence ξ_n such that $\sup_n N(\xi_n) < \infty$ and $\xi_n \to \xi$ in probability as random distributions, the corresponding BPHZ lifts $\boldsymbol{\Pi}_n^{\text{BPHZ}}$ converge to a limiting model $\boldsymbol{\Pi}^{\text{BPHZ}}$. This limiting model is furthermore independent of the choice of approximating sequence.

15.5.2 The renormalised equations

As introduced, the renormalisation group $\mathfrak{R}$ for KPZ is a Lie group of dimension 8, equal to the number of symbols ($\circ$, $\mathcal{V}$, $\mathcal{V}$, $\mathfrak{l}$, $\mathcal{V}$, $\mathcal{V}$, $\mathcal{Y}$, $\mathcal{C}$) used to generate T^-. As already hinted in Example 15.15 above, we will not need to renormalise planted trees, nor the noise symbol itself, nor symbols with three leaves (cubic in Gaussian noise, hence of zero mean, so that the BPHZ condition is trivially satisfied). We thus define a character g on T^- by specifying

$$g(\mathcal{C}) = C_0, \qquad g(\mathcal{V}) = C_1, \qquad g(\mathcal{V}) = C_2, \qquad g(\mathcal{V}) = C_3 , \qquad (15.30)$$

and set to vanish on the remaining symbols which require no renormalisation. The resulting renormalisation maps $M : T \to T$ is then given by $M := (g \otimes \mathrm{Id})\Delta^-$. (It turns out that we only need a three-parameter subgroup of $\mathfrak{R}$ to renormalise the equation, but in order to explain the procedure we prefer to work with the larger 4-dimensional subgroup of $\mathfrak{R}$.) It is now rather straightforward to show the following:

Proposition 15.26. *Let* $M := (g \otimes \mathrm{Id})\Delta^-$ *with* g *as specified in (15.30) and let* $(\Pi^M, \Gamma^M) = M\mathscr{L}\xi$, *where* $\mathscr{L}\xi$ *is the canonical lift of some smooth function* ξ. *Let furthermore* H *be the solution to (15.8) with respect to the model* (Π^M, Γ^M). *Then, writing* $\mathcal{R}^M$ *for the reconstruction operator associated to this renormalised model, the function* $h(t, x) = (\mathcal{R}^M H)(t, x)$ *solves the equation*

$$\partial_t h = \partial_x^2 h + (\partial_x h)^2 - 4C_0\,\partial_x h + \xi - (C_1 + C_2 + 4C_3) .$$

Proof. By Theorem 14.5, it turns out that (15.8) can be solved in $\mathscr{D}^\gamma$ as soon as γ is a little bit greater than $3/2$. Therefore, we only need to keep track of its solution H up to terms of degree $3/2$. By repeatedly applying the identity (15.9), we see that the solution $H \in \mathscr{D}^\gamma$ for γ close enough to $3/2$ is necessarily of the form

$$H = h\mathbf{1} + \mathfrak{l} + \mathcal{V} + h' X_1 + 2\mathcal{V} + 2h'\mathcal{C} ,$$

for some real-valued functions h and h'. (Note that h' is treated as an independent function here, we certainly do not suggest that the function h is differentiable! Our notation is only by analogy with the classical Taylor expansion.) As an immediate consequence, ∂H is given by

$$\partial H = \mathfrak{l} + \mathcal{V} + h'\mathbf{1} + 2\mathcal{V} + 2h'\mathcal{C} , \qquad (15.31)$$

as an element of $\mathscr{D}^\gamma$ for γ sufficiently close to $1/2$. Similarly, the right-hand side of the equation is given up to order 0 by

$$(\partial H)^2 + \Xi = \Xi + \mathcal{V} + 2\mathcal{V} + 2h'\,\mathfrak{l} + \mathcal{V} + 4\mathcal{V} + 2h'\mathcal{Y} + 4h'\mathcal{C} + (h')^2\,\mathbf{1} . \quad (15.32)$$

It follows from the definition of M that one then has the identity

$$M\partial H = \partial H - 4C_0\mathcal{C} ,$$

so that, as an element of $\mathscr{D}^\gamma$ with very small (but positive) γ, one has the identity

$$(M\partial H)^2 = (\partial H)^2 - 8C_0 \mathord{\scalebox{0.8}{$\vee$}} .$$

As a consequence, after neglecting all terms of strictly positive order, one has the identity (writing c instead of $c\mathbf{1}$ for real constants c)

$$M\big((\partial H)^2 + \Xi\big) = (\partial H)^2 + \Xi - C_0\big(4\mathord{\scalebox{0.8}{$\uparrow$}} + 4\mathord{\scalebox{0.8}{$\vee$}} + 8\mathord{\scalebox{0.8}{$\vee$}} + 4h'\,\mathbf{1}\big) - C_1 - C_2 - 4C_3$$
$$= (M\partial H)^2 + \Xi - 4C_0\,M\partial H - (C_1 + C_2 + 4C_3) .$$

Combining this with the fact that M and ∂ commute, the claim now follows at once. $\square$

Remark 15.27. It turns out that, thanks to the symmetry $x \mapsto -x$ enjoyed by our problem, the corresponding model can be renormalised by a map M as above, but with $C_0 = 0$. The reason why we considered the general case here is twofold. First, it shows that it is possible to obtain renormalised equations that differ from the original equation in a more complicated way than just by the addition of a large constant. Second, if one tries to approximate the KPZ equation by a microscopic model which is not symmetric under space inversion, then the constant C_0 plays a non-trivial role, see for example [HS17].

15.5.3 Convergence of the renormalised models

It remains to argue why one expects to be able to find constants C_i^ε such that the sequence of renormalised models $M^\varepsilon \mathscr{L}\xi_\varepsilon$ with $M^\varepsilon = \exp(\sum_{i=1}^3 C_i^\varepsilon L_i)$ converges to a limiting model. Instead of considering the actual sequence of models, we only consider the sequence of stationary processes $\hat{\Pi}^\varepsilon \tau := \Pi^\varepsilon M^\varepsilon \tau$, where Π^ε is associated to $(\Pi^\varepsilon, \Gamma^\varepsilon) = \mathscr{L}\xi_\varepsilon$ as in Section 15.5.1.

Remark 15.28. It is important to note that we do *not* attempt here to give a full proof that the renormalised model converges to a limit in the correct topology for the space of admissible models. We only aim to argue that it is *plausible* that $\hat{\Pi}^\varepsilon$ converges to a limit in *some* topology. A full proof of convergence (but in a slightly different setting) can be found in [Hai13], see also [Hai14b, Section 10] and [CH16] for most general statements.

Since there are general arguments available to deal with all the expressions τ of positive degree as well as expressions of the type $\mathcal{I}'(\tau)$ and Ξ itself, we restrict ourselves to those that remain. Inspecting (15.11), we see that they are given by

$$\mathord{\scalebox{0.8}{$\vee$}}, \quad \mathord{\scalebox{0.8}{$\vee$}}, \quad \mathord{\scalebox{0.8}{$\vee$}}, \quad \mathord{\scalebox{0.8}{$\vee$}}, \quad \mathord{\scalebox{0.8}{$\vee$}} .$$

For this part, some elementary notions from the theory of Wiener chaos expansions are required, but we'll try to hide this as much as possible. At a formal level, one has the identity

$$\boldsymbol{\Pi}^{\varepsilon}\hat{?} = K' * \xi_{\varepsilon} = K'_{\varepsilon} * \xi \,,$$

where the kernel K'_{ε} is given by $K'_{\varepsilon} = K' * \delta_{\varepsilon}$. This shows that, at least formally, one has

$$\left(\boldsymbol{\Pi}^{\varepsilon}\hat{\vee}\right)(z) = \left(K' * \xi_{\varepsilon}\right)(z)^2 = \iint K'_{\varepsilon}(z - z_1)K'_{\varepsilon}(z - z_2)\,\xi(z_1)\xi(z_2)\,dz_1\,dz_2 \,.$$

Similar but more complicated expressions can be found for any formal expression τ. This naturally leads to the study of random variables of the type

$$I_k(f) = \int \cdots \int f(z_1, \ldots, z_k)\,\xi(z_1) \cdots \xi(z_k)\,dz_1 \cdots dz_k \,. \tag{15.33}$$

Ideally, one would hope to have an Itô isometry of the type $\mathbf{E}I_k(f)I_k(g) = \langle f^{\mathrm{sym}}, g^{\mathrm{sym}} \rangle$, where $\langle \cdot, \cdot \rangle$ denotes the L^2-scalar product and f^{sym} denotes the symmetrisation of f. This is unfortunately *not* the case. Instead, one should replace the products in (15.33) by *Wick products*, which are formally generated by all possible *contractions* of the type

$$\xi(z_i)\xi(z_j) \mapsto \xi(z_i) \diamond \xi(z_j) + \delta(z_i - z_j) \,.$$

If we then set

$$\hat{I}_k(f) = \int \cdots \int f(z_1, \ldots, z_k)\,\xi(z_1) \diamond \cdots \diamond \xi(z_k)\,dz_1 \cdots dz_k \,,$$

One has indeed

$$\mathbf{E}\hat{I}_k(f)\hat{I}_k(g) = \langle f^{\mathrm{sym}}, g^{\mathrm{sym}} \rangle \,.$$

Furthermore, one has equivalence of moments in the sense that, for every $k > 0$ and $p > 0$ there exists a constant $C_{k,p}$ such that

$$\mathbf{E}|\hat{I}_k(f)|^p \leq C_{k,p}\|f^{\mathrm{sym}}\|^p \,.$$

Finally, one has $\mathbf{E}\hat{I}_k(f)\hat{I}_\ell(g) = 0$ if $k \neq \ell$. Random variables of the form $\hat{I}_k(f)$ for some $k \geq 0$ and some square integrable function f are said to belong to the *kth homogeneous Wiener chaos*.

Returning to our problem, we first argue that it should be possible to choose M^{ε} in such a way that $\hat{\boldsymbol{\Pi}}^{\varepsilon}\hat{\vee}$ converges to a limit as $\varepsilon \to 0$. The above considerations suggest that one should rewrite $\boldsymbol{\Pi}^{\varepsilon}\hat{\vee}$ as

$$\left(\boldsymbol{\Pi}^{\varepsilon}\hat{\vee}\right)(z) = \left(K' * \xi_{\varepsilon}\right)(z)^2 \tag{15.34}$$

$$= \iint K'_{\varepsilon}(z - z_1)K'_{\varepsilon}(z - z_2)\,\xi(z_1) \diamond \xi(z_2)\,dz_1\,dz_2 + C^{(1)}_{\varepsilon} \,,$$

where the constant $C^{(1)}_{\varepsilon}$ is given by the contraction

$$C_\varepsilon^{(1)} = \overline{V} \overset{\text{def}}{=} \int \left(K_\varepsilon'(z) \right)^2 dz \ .$$

Note now that K_ε' is an ε-approximation of the kernel K' which has the same singular behaviour as the derivative of the heat kernel. In terms of the parabolic distance, the singularity of the derivative of the heat kernel scales like $K(z) \sim |z|^{-2}$ for $z \to 0$. (Recall that we consider the parabolic distance $|(t, x)| = \sqrt{|t| + |x|}$, so that this is consistent with the fact that the derivative of the heat kernel is bounded by t^{-1}.) This suggests that one has $\left(K_\varepsilon'(z) \right)^2 \sim |z|^{-4}$ for $|z| \gg \varepsilon$. Since parabolic space-time has scaling dimension 3 (time counts double!), this is a non-integrable singularity. As a matter of fact, there is a whole power of z missing to make it borderline integrable, which suggests that one has

$$C_\varepsilon^{(1)} \sim \frac{1}{\varepsilon} \ .$$

This already shows that one should not expect $\boldsymbol{\Pi}^\varepsilon \mathcal{V}$ to converge to a limit as $\varepsilon \to 0$. However, it turns out that the first term in (15.34) converges to a distribution-valued stationary space-time process, so that one would like to somehow get rid of this diverging constant $C_\varepsilon^{(1)}$. This is exactly where the renormalisation map M^ε (in particular the factor $\exp(-C_1 L_1)$) enters into play. Following the above definitions, we see that one has

$$\left(\hat{\boldsymbol{\Pi}}^\varepsilon \mathcal{V} \right)(z) = \left(\boldsymbol{\Pi}^\varepsilon M \mathcal{V} \right)(z) = \left(\boldsymbol{\Pi}^\varepsilon \mathcal{V} \right)(z) - C_1 \ .$$

This suggests that if we make the choice $C_1 = C_\varepsilon^{(1)}$, then $\hat{\boldsymbol{\Pi}}^\varepsilon \mathcal{V}$ does indeed converge to a non-trivial limit as $\varepsilon \to 0$. This limit is a distribution given, at least formally, by

$$\left(\boldsymbol{\Pi}^\varepsilon \mathcal{V} \right)(\psi) = \iint \psi(z) K'(z - z_1) K'(z - z_2) \, dz \, \xi(z_1) \diamond \xi(z_2) \, dz_1 \, dz_2 \ .$$

Using again the scaling properties of the kernel K', it is not too difficult to show that this yields indeed a random variable belonging to the second homogeneous Wiener chaos for every choice of smooth test function ψ.

The case $\tau = \langle\!\!\!^\circ_\circ$ is treated in a somewhat similar way. This time one has

$$\left(\boldsymbol{\Pi}^\varepsilon \langle\!\!\!^\circ_\circ \right)(z) = \left(K' * \xi_\varepsilon \right)(z) \left(K' * K' * \xi_\varepsilon \right)(z)$$

$$= \iint K_\varepsilon'(z - z_1)(K * K_\varepsilon')(z - z_2) \, \xi(z_1) \diamond \xi(z_2) \, dz_1 \, dz_2 + C_\varepsilon^{(0)} \ ,$$

where the constant $C_\varepsilon^{(0)}$ is given by the contraction

$$C_\varepsilon^{(0)} = \Diamond \overset{\text{def}}{=} \int K_\varepsilon'(z)(K' * K_\varepsilon')(z) \, dz \ .$$

This time however K_ε' is an odd function (in the spatial variable) and $K' * K_\varepsilon'$ is an even function, so that $C_\varepsilon^{(0)}$ vanishes for every $\varepsilon > 0$. This is why we can set $C_0 = 0$ and no renormalisation is required for $\langle\!\!\!^\circ_\circ$.

Turning to our list of terms of negative degree, it remains to consider $\vee$, ψ, and $\vee$. It turns out that the latter two are the more difficult ones, so we only discuss these. Let us first argue why we expect to be able to choose the constant C_2 in such a way that $\hat{\boldsymbol{\Pi}}^{\varepsilon}\psi$ converges to a limit. In this case, the "bad" term comes from the part of $\left(\boldsymbol{\Pi}^{\varepsilon}\psi\right)(z)$ belonging to the homogeneous chaos of order 0. This is simply a constant, which is given by

$$C_{\varepsilon}^{(2)} = 2 \vee \stackrel{\text{def}}{=} 2\int K'(z)K'(\bar{z})Q_{\varepsilon}^2(z-\bar{z})\,dz\,d\bar{z}\,, \qquad (15.35)$$

where the kernel Q_{ε} is given by

$$Q_{\varepsilon}(z) = \int K'_{\varepsilon}(\bar{z})K'_{\varepsilon}(\bar{z}-z)\,d\bar{z}\,.$$

Remark 15.29. The factor 2 comes from the fact that the contraction (15.35) appears twice, since it is equal to the contraction $\vee$. In principle, one would think that the contraction $\vee$ also contributes to $C_{\varepsilon}^{(2)}$. This term however vanishes due to the fact that the integral of K'_{ε} vanishes.

Since K'_{ε} is an ε-mollification of a kernel with a singularity of order -2 and the scaling dimension of the underlying space is 3, we see that Q_{ε} behaves like an ε-mollification of a kernel with a singularity of order $-2-2+3 = -1$ at the origin. As a consequence, the singularity of the integrand in (15.35) is of order -6, which gives rise to a logarithmic divergence as $\varepsilon \to 0$. This suggests that one should choose $C_2 = C_{\varepsilon}^{(2)}$ in order to cancel out this diverging term and obtain a non-trivial limit for $\hat{\boldsymbol{\Pi}}^{\varepsilon}\psi$ as $\varepsilon \to 0$. This is indeed the case.

We finally turn to the case $\tau = \vee$. In this case, there are "bad" terms appearing in the Wiener chaos decomposition of $\boldsymbol{\Pi}^{\varepsilon}\vee$ both in the second and the zeroth Wiener chaos. This time, the constant appearing in the zeroth Wiener chaos is given by

$$C_{\varepsilon}^{(3)} = 2 \vee \stackrel{\text{def}}{=} 2\int K'(z)K'(\bar{z})Q_{\varepsilon}(\bar{z})Q_{\varepsilon}(z+\bar{z})\,dz\,d\bar{z}\,,$$

which diverges logarithmically for exactly the same reason as $C_{\varepsilon}^{(2)}$. Setting $C_2 = C_{\varepsilon}^{(2)}$, this diverging constant can again be cancelled out. The combinatorial factor 2 arises in essentially the same way as for $\vee$ and the contribution of the term where the two top nodes are contracted vanishes for the same reason as previously.

It remains to consider the contribution of $\boldsymbol{\Pi}^{\varepsilon}\vee$ to the second Wiener chaos. This contribution consists of three terms, which correspond to the contractions

It turns out that the first one of these terms does not give raise to any singularity. The last two terms can be treated in essentially the same way, so we focus on the last one, which we denote by η^ε. For fixed ε, the distribution (actually smooth function) η^ε is given by

$$\eta^\varepsilon(\psi) = \int \psi(z_0) K'(z_0 - z_1) Q_\varepsilon(z_0 - z_1) K'(z_2 - z_1)$$
$$\times K'_\varepsilon(z_3 - z_2) K'_\varepsilon(z_4 - z_2)\, \xi(z_3) \diamond \xi(z_4)\, dz \ .$$

The problem with this is that as $\varepsilon \to 0$, the product $\hat{Q}_\varepsilon := K'Q_\varepsilon$ converges to a kernel $\hat{Q} = K'Q$, which has a non-integrable singularity at the origin. In particular, it is not clear *a priori* whether the action of integrating a test function against $\hat{Q}_\varepsilon$ converges to a limiting distribution as $\varepsilon \to 0$. Our saving grace here is that since Q_ε is even and K' is odd, the kernel $\hat{Q}_\varepsilon$ integrates to 0 for every fixed ε.

This is akin to the problem of making sense of the "Cauchy principal value" distribution, which formally corresponds to the integration against $1/x$. For the sake of the argument, let us consider a function $W : \mathbf{R} \to \mathbf{R}$ which is compactly supported and smooth everywhere except at the origin, where it diverges like $|W(x)| \sim 1/|x|$. It is then natural to associate to W a "renormalised" distribution $\mathscr{R}W$ given by

$$(\mathscr{R}W)(\varphi) = \int W(x)\big(\varphi(x) - \varphi(0)\big)\, dx \ .$$

Note that $\mathscr{R}W$ has the property that if $\varphi(0) = 0$, then it simply corresponds to integration against W, which is the standard way of associating a distribution to a function. Furthermore, the above expression is always well-defined, since φ is smooth and therefore the factor $(\varphi(x) - \varphi(0))$ cancels out the singularity of W at the origin. It is also straightforward to verify that if W_ε is a sequence of smooth approximations to W (say one has $W_\varepsilon(x) = W(x)$ for $|x| > \varepsilon$ and $|W_\varepsilon| \lesssim 1/\varepsilon$ otherwise) which has the property that each W_ε integrates to 0, then $W^\varepsilon \to \mathscr{R}W$ in a distributional sense.

In the same way, one can show that $\hat{Q}_\varepsilon$ converges as $\varepsilon \to 0$ to a limiting distribution $\mathscr{R}\hat{Q}$. As a consequence, one can show that η^ε converges to a limiting (random) distribution η given by

$$\eta(\psi) = \int \psi(z_0)\, \mathscr{R}\hat{Q}(z_0 - z_1) K'(z_2 - z_1) K'(z_3 - z_2) K'(z_4 - z_2)\, \xi(z_3) \diamond \xi(z_4)\, dz \ .$$

It should be clear from this whole discussion that while the precise values of the constants C_i depend on the details of the mollifier δ_ε, the limiting (random) model $(\hat{\Pi}, \hat{\Gamma})$ obtained in this way is independent of it. Combining this with the continuity of the solution to the fixed point map (15.8) and of the reconstruction operator $\mathcal{R}$ with respect to the underlying model, we see that the statement of Theorem 15.2 follows almost immediately.

15.6 The KPZ equation and rough paths

In the particular case of the KPZ equation, it turns out that is possible to give a robust solution theory by only using "classical" controlled rough path theory, as exposed in the earlier part of this book. This is actually how it was originally treated in [Hai13]. To see how this can be the case, we make the following crucial remarks:

1. First, looking at the expression (15.31) for ∂H, we see that most symbols come with constant coefficients. The only non-constant coefficients that appear are in front of the term $\mathbf{1}$, which is some kind of renormalised value for ∂H, and in front of the term $\Lsh$. This suggests that the problem of finding a solution h to the KPZ equation (or equivalently a solution h' to the corresponding Burgers' equation) can be simplified considerably by considering instead the function v given by

$$v = \partial_x h - \boldsymbol{\Pi}\big(\Uparrow + \Psi + 2\,\Psi\big)\,, \tag{15.36}$$

   where $\boldsymbol{\Pi}$ is the operator given by (15.12–15.14).
2. The only symbol τ appearing in ∂H such that $\deg \tau + \deg \Lsh < 0$ is the symbol $\Uparrow$. Furthermore, one has

$$\Delta \mathbf{1} = \mathbf{1} \otimes \mathbf{1}\,, \qquad \Delta \Lsh = \Lsh \otimes \mathbf{1} + \mathbf{1} \otimes \mathcal{J}'(\Uparrow)\,,$$
$$\Delta \Uparrow = \Uparrow \otimes \mathbf{1}\,, \qquad \Delta \Psi = \Psi \otimes \mathbf{1} + \Uparrow \otimes \mathcal{J}'(\Uparrow)\,.$$

   It then follows from this and the definition (15.16) of the structure group G that the space $\langle \Uparrow, \Psi, \mathbf{1}, \Lsh \rangle \subset T$ is invariant under the action of G. Furthermore, its action on this subspace is completely described by one real number corresponding to $\mathcal{J}'(\Uparrow)$. Finally, viewing this subspace as a regularity structure in its own right, we see that it is nothing but the regularity structure of Section 13.3.2, provided that we make the identifications $\Uparrow \sim \dot{W}$, $\Lsh \sim W$, and $\Psi \sim \mathbb{W}$.
3. One has the identities

$$\Delta \Psi = \Psi \otimes \mathbf{1} + \Uparrow \otimes \mathcal{J}'(\Psi)\,, \qquad \Delta \Psi = \Psi \otimes \mathbf{1} + \Uparrow \otimes \mathcal{J}'(\Psi)\,,$$

   so that the pair of symbols $\{\Psi, \Psi\}$ could also have played the role of $\{W, \dot{W}\}$ in the previous remark.

Let now ξ be a *smooth* function and let h be given by the solution to the unrenormalised KPZ equation (15.1). Defining $\boldsymbol{\Pi}$ by $\boldsymbol{\Pi}\Xi = \xi$ and then recursively as in (15.13), and defining v by (15.36), we then obtain for v the equation

$$\partial_t v = \partial_x^2 v + \partial_x\big(v\,\boldsymbol{\Pi}\Uparrow + 4\,\boldsymbol{\Pi}\,\Psi\big) + R\,, \tag{15.37}$$

where the "remainder" R belongs to $\mathcal{C}^\alpha$ for every $\alpha < -1$. Similarly to before, it also turns out that if we replace $\boldsymbol{\Pi}$ bi $\hat{\boldsymbol{\Pi}} = \boldsymbol{\Pi}^M$ defined as in (15.24) (with $C_0 = 0$) and h as the solution to the renormalised KPZ equation (15.6) with $C_\varepsilon = C_1 + C_2 + 4C_3$, then v also satisfies (15.37), but with $\boldsymbol{\Pi}$ replaced by the renormalised model $\hat{\boldsymbol{\Pi}}$.

We are now in the following situation. As a consequence of (15.31) we can *guess* that for any fixed time t, the solution v should be controlled by the function $\hat{\Pi}\,\mathord{\scriptstyle\mathrm{<}}$, which we can interpret as one component (say W^1) of some rough path $(W, \mathbb{W})$. Note that here the spatial variable plays the role of time! The time variables merely plays the role of a parameter, so we really have a family of rough paths indexed by time. Furthermore, $\hat{\Pi}\,\mathord{\mathrm{l}}$ can be interpreted as the distributional derivative of another component (say W^0) of the rough path W. Finally, the function $\hat{\Pi}\,\mathord{\scriptstyle\mathrm{V}}$ can be interpreted as a third component W^2 of W.

As a consequence of the second and third remarks above, the two distributions $\hat{\Pi}\,\mathord{\scriptstyle\mathrm{V}}$ and $\hat{\Pi}\,\mathord{\scriptstyle\mathrm{V}}$ can then be interpreted as the distributional derivatives of the "iterated integrals" $\mathbb{W}^{1,0}$ and $\mathbb{W}^{2,1}$. It follows automatically from these algebraic relations combined with the analytic bounds (13.13) that $\mathbb{W}^{1,0}$ and $\mathbb{W}^{2,1}$ then satisfy the required estimates (2.3). Our model does not provide any values for $\mathbb{W}^{1,2}$, but these turn out not to be required. Assuming that v is indeed controlled by $X_1 = \hat{\Pi}\,\mathord{\scriptstyle\mathrm{<}}$, it is then possible to give meaning to the term $v\,\Pi\,\mathord{\mathrm{l}}$ appearing in (15.37) by using "classical" rough integration.

As a consequence, we then see that the right-hand side of (15.37) is of the form $\partial_x^2 Y$, for some function Y controlled by W^0. One of the main technical results of [Hai13] guarantees that if Z solves

$$\partial_t Z = \partial_x^2 Z + \partial_x^2 Y \;,$$

and Y is controlled by W^0, then Z is necessarily controlled by $W^1 = \hat{\Pi}\,\mathord{\scriptstyle\mathrm{<}}$. This "closes the loop" and allows to set up a fixed point equation for v that is stable as a function of the underlying model $\hat{\Pi}$ and therefore also allows to deal with the limiting case of the KPZ equation driven by space-time white noise.

15.7 Exercises

Exercise 15.1 (KPZ Structure Group) *Consider the 16-dimensional KPZ regularity structure with $T = T_{\mathrm{KPZ}}$ given by*

$$T = \big\langle\, \Xi, \mathord{\scriptstyle\mathrm{V}}, \mathord{\scriptstyle\mathrm{V}}, \mathord{\mathrm{l}}, \mathord{\scriptstyle\mathrm{VV}}, \mathord{\scriptstyle\mathrm{V}}, \mathord{\mathrm{Y}}, \mathord{\scriptstyle\mathrm{<}}, \mathbf{1}, \mathord{\scriptstyle\mathrm{V}}, \mathord{\scriptstyle\mathrm{<}}, \mathord{\mathrm{l}}, \mathord{\mathrm{Y}}, X_1, \mathord{\scriptstyle\mathrm{V}}, \mathord{\scriptstyle\mathrm{<}} \,\big\rangle \,.$$

Show that the structure group G is a 7-dimensional (non-commutative) Lie group, an element $\Gamma \in G \subset \mathcal{L}(T, T)$ of which has the upper triangular matrix representation

$$
\begin{pmatrix}
1 & & & & & & & & & & & & & \\
& 1 & & & & & & & & & & & & \\
& & 1 & & & & & & & & & & & \\
& & & 1 & c_1 & c_2 & & & & & & & & \\
& & & & 1 & & & & & & & & & \\
& & & & & 1 & & & & & & & & \\
& & & & & & 1 & & & & & & & \\
& & & & & & & 1 & c_1\ c_2\ c_3\ c_4\ c_5 & c_6\ c_7 & & & & \\
& & & & & & & & 1 & & & & & \\
& & & & & & & & & 1 & & & & \\
& & & & & & & & & & 1 & & & \\
& & & & & & & & & & & 1 & & \\
& & & & & & & & & & & & 1 & c_1\ c_2 \\
& & & & & & & & & & & & & 1 & \\
& & & & & & & & & & & & & & 1
\end{pmatrix}
$$

where empty entries mean zeros. Note that the upper-triangular form reflects the fact that $\Gamma - \mathrm{Id}$ is only allowed to produce lower order terms. (Remark: It is immediate from this representation that $\langle \uparrow, \curlyvee, 1, \curlyvee \rangle$ and $\langle \uparrow, \curlyvee, 1, \curlyvee \rangle$ are indeed sectors, with "rough path" index set $\{-\frac{1}{2}^{-}, 0^{-}, 0, \frac{1}{2}^{-}\}$, and action of the structure group exactly as in the rough path case (13.12) (with "h" replaced by c_1 and c_2, respectively.)

Solution. We first derive the coaction on all the symbols, and here prefer to write Δ for the coaction and keep Δ^{+} for the coproduct on T^{+}. By definition of the coaction, $\Delta(\Xi) = \Xi \otimes \mathbf{1}$ and

$$
\Delta(\uparrow) = \mathcal{I}'(\Xi) \otimes \mathbf{1} + \sum_{k \in \mathbf{N}^2} \frac{X^k}{k!} \otimes \mathcal{J}'_k(\Xi) = \uparrow \otimes \mathbf{1} \,,
$$

since $\deg \mathcal{J}'_k(\Xi) = \deg \mathcal{J}_{k+(0,1)}(\Xi) = \deg \Xi + 1 - |k| < 0$ so that $\mathcal{J}'_k(\Xi) = 0$. Similarly, write Δ instead of Δ^{+} for better readability,

$$
\begin{aligned}
\Delta(\curlyvee) &= \Delta(\uparrow)\Delta(\uparrow) = (\uparrow \star \uparrow) \otimes \mathbf{1} = \curlyvee \otimes \mathbf{1}, \\
\Delta(\curlyvee) &= \Delta \mathcal{I}'(\curlyvee) = \ldots = \curlyvee \otimes \mathbf{1}, \\
\Delta(\curlyvee) &= \Delta(\uparrow)\Delta(\curlyvee) = \ldots = \curlyvee \otimes \mathbf{1}, \\
\Delta(\curlyvee) &= \Delta(\curlyvee)\Delta(\curlyvee) = \curlyvee \otimes \mathbf{1}, \\
\Delta(\curlyvee) &= \curlyvee \otimes \mathbf{1} + \mathbf{1} \otimes \mathcal{J}'(\curlyvee), \\
\Delta(\curlyvee) &= \Delta(\uparrow)\Delta(\curlyvee) = \curlyvee \otimes \mathbf{1} + \uparrow \otimes \mathcal{J}'(\curlyvee).
\end{aligned}
$$

Note the interpretation of *cutting off positive branches*: $\deg \mathcal{J}'(\text{⋎}) = 1 + 3(-\frac{3}{2}^-) + 4 = \frac{1}{2}^- > 0$, and also $\deg \mathcal{J}'(\text{↑}) = \frac{1}{2}^-$ as seen in

$$\Delta(\text{⟨}) = \text{⟨} \otimes \mathbf{1} + \mathbf{1} \otimes \mathcal{J}'(\text{↑}),$$
$$\Delta(\text{⟨ₒ}) = \Delta(\text{⟨})\Delta(\text{↑}) = \text{⟨ₒ} \otimes \mathbf{1} + \text{↑} \otimes \mathcal{J}'(\text{↑}).$$

To deal with $\text{↑} = \mathcal{I}(\Xi)$, note $\deg \mathcal{J}(\Xi) > 0, \deg \mathcal{J}'(\Xi) < 0$ so that the latter term does not figure (same reasoning for $\text{⋎} = \mathcal{I}(\mathbb{V})$), and obtain

$$\Delta(\text{↑}) = \text{↑} \otimes \mathbf{1} + \mathbf{1} \otimes \mathcal{J}(\Xi),$$
$$\Delta(\text{⋎}) = \text{⋎} \otimes \mathbf{1} + \mathbf{1} \otimes \mathcal{J}(\mathbb{V}).$$

By definition, $\Delta X_1 = X_1 \otimes \mathbf{1} + \mathbf{1} \otimes X_1$. Next consider ⋎ and ⟨. In view of $|\mathcal{J}(\text{⋎})|$ and $|\mathcal{J}'(\text{⋎})| > 0$ we have (same reasoning for ⟨),

$$\Delta\left(\text{⋎}\right) = \text{⋎} \otimes \mathbf{1} + \mathbf{1} \otimes \mathcal{J}(\text{⋎}) + X_1 \otimes \mathcal{J}'(\text{⋎}),$$
$$\Delta\left(\text{⟨}\right) = \text{⟨} \otimes \mathbf{1} + \mathbf{1} \otimes \mathcal{J}(\text{↑}) + X_1 \otimes \mathcal{J}'(\text{↑}),$$

Inspecting the above reveals that we need $\mathbf{1}$ and then the following 7 "positive" symbols (also viewable as trees) in T^+,

$$\mathcal{J}'(\text{⋎}), \mathcal{J}'(\text{↑}), \mathcal{J}(\Xi), \mathcal{J}(\mathbb{V}), X_1, \mathcal{J}(\text{⋎}), \mathcal{J}(\text{↑}), \qquad (15.38)$$

of resp. homogeneities $\frac{1}{2}^-, \frac{1}{2}^-, \frac{1}{2}^-, 1^-, 1, \frac{3}{2}^-, \frac{3}{2}^-$. On the other hand, T^+ was introduced abstractly as free commutative algebra generated by all of the above symbols (with unit element 1). Even upon truncation, say $T^+ = T^+_{<3/2}$ with abusive notation, this leaves us with $10 + 4 + 1 = 15$ generating symbols,

$$\mathcal{J}'(\text{⋎}), \mathcal{J}'(\text{↑}), \dots, \mathcal{J}'(\text{↑}); \mathcal{J}(\Xi), \dots, \mathcal{J}(\text{↑}); X_1 \qquad (15.39)$$

(of which only 7 are needed). Of course, T^+ also contains (free) products such as $\mathcal{J}'(\text{⋎})\mathcal{J}'(\text{↑}), X_1\mathcal{J}'(\text{⋎}), \mathcal{J}'(\text{↑})\mathcal{J}(\mathbb{V},)$ (all of degree $< 3/2$), however by working in T these did not appear as "right-hand side"-image of Δ above.

Consider now a *character* of the algebra T^+; that is, an element $g \in (T^+)^*$, so that $g(1) = 1$ and $g(\sigma\bar{\sigma}) = g(\sigma)g(\bar{\sigma})$. (Actually, in view of the truncation we impose this only for $\sigma, \bar{\sigma}$ with $\deg(\sigma\bar{\sigma}) = \deg\sigma + \deg\bar{\sigma} < 3/2$.) Such g is obviously determined by its value on each of the 15 basis symbols listed in (15.39). Now T^+ can be given a Hopf structure, with coproduct Δ^+ and antipode, so that the set of characters forms the group G^+, with product given by

$$(f \circ g)(\sigma) = (f \otimes g)\Delta^+\sigma = \sum_{(\sigma)} \langle f, \sigma'\rangle\langle g, \sigma''\rangle;$$

inverses are given in terms of the antipode. One thus sees that G^+ is a 15-dimensional (Lie) group. However, only a 7-dimensioal subgroup is needed, for we only care about the 7 values arising from (15.38), which we call

$$c_1 = g\big(\mathcal{J}'(\Yup)\big), \ldots, c_7 = g(\mathcal{J}(\Xi)).$$

It then follows from $\Gamma_g := (\mathrm{Id} \otimes g)\Delta$ that $\Gamma_g : T \to T$ acts as identity on all symbols other than

$$\Gamma_g\left(\Yup\right) = \Yup + g(\mathcal{J}'(\Yup))\mathbf{1} \equiv \Yup + c_1\mathbf{1},$$
$$\Gamma_g\left(\Yup\right) = \Yup + g(\mathcal{J}'(\Yup))\,\mathsf{I} = \Yup + c_1\,\mathsf{I},$$
$$\Gamma_g(\Yup) = \Yup + g(\mathcal{J}'(\mathsf{I}))\mathbf{1} = \Yup + c_2\mathbf{1},$$
$$\Gamma_g(\Yup) = \Yup + c_2\,\mathsf{I},$$
$$\Gamma_g(\mathsf{I}) = \mathsf{I} + g(\mathcal{J}(\Xi))\mathbf{1} = \mathsf{I} + c_3\mathbf{1},$$
$$\Gamma_g(\Yup) = \Yup + g(\mathcal{J}(\Yup))\mathbf{1} = \Yup + c_4\mathbf{1},$$
$$\Gamma_g(X_1) = X_1 + c_5\mathbf{1},$$
$$\Gamma_g(\Yup) = \Yup + c_6\mathbf{1} + c_1 X_1,$$
$$\Gamma_g(\Yup) = \Yup + c_7\mathbf{1} + c_2 X_1.$$

The matrix representation of Γ_g is then immediate.

Exercise 15.2 (KPZ Renormalisation Group) *Consider again the 16-dimensional KPZ regularity structure with structure space $T = T_{\mathrm{KPZ}}$. The renormalisation group was given as subgroup $\mathfrak{R} \subset \mathcal{L}(T,T)$, given by $M_g\tau = (g \otimes \mathrm{Id})\Delta^- \tau$, where g ranges over the characters of T^-. Consider more specifically $M = M_g$ with g as specified in (15.30), i.e. $g(\Yup) = C_0, g(\mathbb{V}) = C_1, g(\Yup) = C_2, g(\Yup) = C_3$ and set to vanish on the remaining symbols.*

Show that this gives a subgroup of $\mathfrak{R}$ which is a 4-dimensional (commutative) Lie group, an element $M \in \mathfrak{R} \subset \mathcal{L}(T,T)$ of which has lower triangular matrix representation

$$
\begin{pmatrix}
1 \\
 & 1 \\
 & 2C_0 & 1 \\
 & & & 1 \\
 & & & & 1 \\
 & & & C_0 & 1 \\
 & & & 2C_0 & & 1 \\
 & C_1 & & C_2 & C_3 & C_0 & 1 \\
 & & & & & & 1 \\
 & & & & & & 2C_0 & 1 \\
 & & & & & & & & 1 \\
 & & & & & & & & & 1 \\
 & & & & & & & & & & 1 \\
 & & & & & & & & & & & 1 \\
 & & & & & & & & & & & 2C_0 & 1
\end{pmatrix}
$$

Exercise 15.3 *Show that the two procedures for recovering Π from the knowledge of (Π, Γ) outlined in Remark 15.13 and on page 301 are equivalent.*

15.8 Comments

The original proof [Hai13] of well-posedness of the KPZ equation without using the Cole–Hopf transform did not use regularity structures but instead viewed the solution at any fixed time as a spatial rough path controlled by the solution to the linearised equation, in the spirit of Section 12.3. An alternative approach using paracontrolled distributions as developed in [GIP15] was used in [GP17] to obtain a number of additional properties of the solutions, including a clean variational formulation.

Given that the KPZ equation is expected to enjoy a form of "universality", a very natural question is that of showing that "most" classes of interface fluctuation models converge to in in the weakly asymmetric regime. The first result in this direction was obtained by Bertini–Giacomin [BG97], but this relied crucially on a microscopic version of the Hopf–Cole transform to show that the transformed particle system converges to the multiplicative stochastic heat equation. A first more general result was obtained by Jara–Conçalves [GJ14] who showed that the large scale fluctuations of a large number of particle systems solve the KPZ equation in a relatively weak sense. It has been an open problem for quite some time now whether such a weak notion of solution characterises solutions to the KPZ uniquely. Major progress in this direction was obtained by Gubinelli–Perkowski [GP18] who showed that this is indeed the case at stationarity under an additional structural assumption on the generator of the particle system that can be verified for a number of systems of interest.

On the other hand, a large class of interface fluctuation models that fall outside of this approach is given by solutions to an equation of the type

$$\partial_t h_\varepsilon = \partial_x^2 h_\varepsilon + \sqrt{\varepsilon} F(\partial_x h_\varepsilon) + \eta(t, x) , \tag{15.40}$$

where η is a (smooth) space-time random field with sufficiently good mixing properties, $F : \mathbf{R} \to \mathbf{R}$ is an even function growing at infinity, and $\varepsilon > 0$ is a parameter controlling the asymmetry of the problem. Under rather weak assumptions on η and F one then expects to be able to find constants C_ε such that $\varepsilon^{-1/2} h_\varepsilon(\varepsilon^{-2} t, \varepsilon^{-1} x) - C_\varepsilon t$ converges to solutions to the KPZ equation. This was shown to be indeed the case in various special cases of increasing generality in [HS17, HQ18, HX19, FG19]. (The last reference treats a different class of models but its proofs could be adapted to the setting of (15.40).)

There is a natural generalisation of the KPZ equation going in a completely different direction. Indeed, given a Riemannian manifold $(\mathcal{M}, g)$ (where g denotes the metric tensor), we can ask ourselves what the natural "stochastic heat equation with values in $\mathcal{M}$" looks like. A moment's thought suggests that it should be given, in local coordinates, by an equation of the form

$$\partial_t u^\alpha = \partial_x^2 u^\alpha + \Gamma_{\beta\gamma}^\alpha(u) \, \partial_x u^\beta \, \partial_x u^\gamma + \sigma_i^\alpha(u) \, \xi_i , \tag{15.41}$$

where the ξ_i are i.i.d. space-time white noises, $\Gamma_{\beta\gamma}^\alpha$ are the Christoffel symbols for $\mathcal{M}$, the σ_i are any finite collection of vector fields such that

$$\sigma_i^\alpha \sigma_i^\beta = g , \tag{15.42}$$

and summation over repeated indices is implied. By combining the results of [CH16, BHZ19, BCCH17], it is not difficult to see that there are natural notions of solution to (15.41), but these are of course only well-defined modulo an element of the renormalisation group $\mathfrak{R}$. It turns out that in this case, even after taking into account simplifications due to the symmetry $x \leftrightarrow -x$ and the fact that the noises are i.i.d. Gaussian, the relevant subgroup of $\mathfrak{R}$ is generically (namely for large enough dimension of $\mathcal{M}$) of dimension 54.

This is a good example illustrating the role played by symmetries. In this particular case, there are two additional symmetries one would like to exploit. On the one hand, one would like to enforce equivariance under the group of diffeomorphisms of $\mathcal{M}$. In other words, solutions to (15.41) should be independent of the specific coordinate system used to write (15.41). This is akin to the property of solutions to regular SDEs written in *Stratonovich form* (or indeed those of RDEs driven by a geometric rough path). On the other hand, the derivation of (15.41) implicitly makes use of Itô's isometry to guarantee that, at least in law, its solutions do not depend on the specific choice of the vector fields satisfying (15.42). This in turn is akin to the property of solutions to SDEs written in *Itô form*. It turns out – and this is the main result of [BGHZ19] – that in this context it is possible to find solution theories that do satisfy both properties *simultaneously*! In fact there still exists a two-parameter family of them, but if we restrict ourselves to (15.1) (i.e. with Γ and σ related to the same metric g), then it reduces to a one-parameter family and the corresponding correction term (analogous to the Itô-Stratonovich correction term allowing to switch between solution theories for SDEs) is given by a multiple of the gradient of the scalar curvature of $\mathcal{M}$. This sheds new light on observations that had previously been made in a closely related context both in the physics [Che72, Um74] and in the mathematics [Dar84, IM85, AD99] literatures.

References

[AC17]   A. ANANOVA and R. CONT. Pathwise integration with respect to paths of finite quadratic variation. *J. Math. Pures Appl. (9)* **107**, no. 6, (2017), 737–757. `doi: 10.1016/j.matpur.2016.10.004`.

[AC19]   A. L. ALLAN and S. N. COHEN. Pathwise stochastic control with applications to robust filtering. *arXiv e-prints* (2019), 1–42. Ann. Appl. Probab., to appear. `arXiv:1902.05434`.

[AD99]   L. ANDERSSON and B. K. DRIVER. Finite-dimensional approximations to Wiener measure and path integral formulas on manifolds. *J. Funct. Anal.* **165**, no. 2, (1999), 430–498. `doi:10.1006/jfan.1999.3413`.

[AFS19]   C. AMÉNDOLA, P. FRIZ, and B. STURMFELS. Varieties of signature tensors. *Forum Math. Sigma* **7**, (2019), e10, 54. `doi:10.1017/fms.2019.3`.

[Aid07]   S. AIDA. Semi-classical limit of the bottom of spectrum of a Schrödinger operator on a path space over a compact Riemannian manifold. *J. Funct. Anal.* **251**, no. 1, (2007), 59–121. `doi:10.1016/j.jfa.2007.06.009`.

[Aid15]   S. AIDA. Reflected rough differential equations. *Stochastic Processes Appl.* **125**, no. 9, (2015), 3570–3595. `doi:10.1016/j.spa.2015.03.008`.

[Alm66]   F. J. ALMGREN, JR. *Plateau's problem: An invitation to varifold geometry.* W. A. Benjamin, Inc., New York-Amsterdam, 1966, xii+74.

[App09]   D. APPLEBAUM. *Lévy Processes and Stochastic Calculus.* Cambridge Studies in Advanced Mathematics. Cambridge University Press, 2 ed., 2009. `doi:10.1017/CBO9780511809781`.

[AR91]   S. ALBEVERIO and M. RÖCKNER. Stochastic differential equations in infinite dimensions: solutions via Dirichlet forms. *Probab. Theory Related Fields* **89**, no. 3, (1991), 347–386. `doi:10.1007/BF01198791`.

[BA88]   G. BEN AROUS. Methods de Laplace et de la phase stationnaire sur l'espace de Wiener. *Stochastics* **25**, no. 3, (1988), 125–153. `doi:10.1080/17442508808833536`.

[BA89]   G. BEN AROUS. Flots et séries de Taylor stochastiques. *Probab. Theory Related Fields* **81**, no. 1, (1989), 29–77. `doi:10.1007/BF00343737`.

[Bai14]   I. BAILLEUL. Flows driven by Banach space-valued rough paths. In *Séminaire de Probabilités XLVI*, vol. 2123 of *Lecture Notes in Math.*, 195–205. Springer, Cham, 2014. `doi:10.1007/978-3-319-11970-0_7`.

[Bai15a] I. BAILLEUL. Flows driven by rough paths. *Revista Matemática Iberoamericana* **31**, no. 3, (2015), 901–934. `doi:10.4171/rmi/858`.

[Bai15b] I. BAILLEUL. Regularity of the Itô-Lyons map. *Confluentes Math.* **7**, no. 1, (2015), 3–11. `doi:10.5802/cml.15`.

[Bai19]   I. BAILLEUL. Rough integrators on Banach manifolds. *Bull. Sci. Math.* **151**, (2019), 51–65. `doi:10.1016/j.bulsci.2018.12.001`.

© Springer Nature Switzerland AG 2020
P. K. Friz, M. Hairer, *A Course on Rough Paths*, Universitext,
https://doi.org/10.1007/978-3-030-41556-3

[Bal00]   E. J. BALDER. Lectures on Young measure theory and its applications in economics. *Rend. Istit. Mat. Univ. Trieste* **31**, no. suppl. 1, (2000), 1–69. Workshop on Measure Theory and Real Analysis (Italian) (Grado, 1997).

[Bau04]   F. BAUDOIN. *An introduction to the geometry of stochastic flows.* Imperial College Press, London, 2004, x+140. doi:10.1142/9781860947261.

[BB19]   I. BAILLEUL and F. BERNICOT. High order paracontrolled calculus. *Forum Math. Sigma* **7**, (2019), e44, 94. doi:10.1017/fms.2019.44.

[BBR+18]   C. BAYER, D. BELOMESTNY, M. REDMANN, S. RIEDEL, and J. SCHOENMAKERS. Solving linear parabolic rough partial differential equations. *arXiv e-prints* (2018), 1–36. arXiv:1803.09488.

[BC17]   I. BAILLEUL and R. CATELLIER. Rough flows and homogenization in stochastic turbulence. *J. Differential Equations* **263**, no. 8, (2017), 4894–4928. doi:10.1016/j.jde.2017.06.006.

[BC19]   H. BOEDIHARDJO and I. CHEVYREV. An isomorphism between branched and geometric rough paths. *Ann. Inst. Henri Poincaré Probab. Stat.* **55**, no. 2, (2019), 1131–1148. doi:10.1214/18-aihp912.

[BCCH17]   Y. BRUNED, A. CHANDRA, I. CHEVYREV, and M. HAIRER. Renormalising SPDEs in regularity structures. *arXiv e-prints* (2017), 1–85. J. Eur. Math. Soc., to appear. arXiv:1711.10239.

[BCD11]   H. BAHOURI, J.-Y. CHEMIN, and R. DANCHIN. *Fourier analysis and nonlinear partial differential equations*, vol. 343 of *Grundlehren der Mathematischen Wissenschaften*. Springer, Heidelberg, 2011, xvi+523. doi:10.1007/978-3-642-16830-7.

[BCD19]   I. BAILLEUL, R. CATELLIER, and F. DELARUE. Propagation of chaos for mean field rough differential equations. *arXiv e-prints* (2019), 1–61. arXiv:1907.00578.

[BCD20]   I. BAILLEUL, R. CATELLIER, and F. DELARUE. Solving mean field rough differential equations. *Electron. J. Probab.* **25**, (2020), 51 pp. doi:10.1214/19-EJP409.

[BCEF20]   Y. BRUNED, C. CURRY, and K. EBRAHIMI-FARD. Quasi-shuffle algebras and renormalisation of rough differential equations. *Bull. Lond. Math. Soc.* **52**, no. 1, (2020), 43–63. doi:10.1112/blms.12305.

[BCF18]   Y. BRUNED, I. CHEVYREV, and P. K. FRIZ. Examples of renormalized SDEs. In *Stochastic partial differential equations and related fields, in Honor of Michael Röckner, Bielefeld 2016*, vol. 229 of *Springer Proc. Math. Stat.*, 303–317. Springer, Cham, 2018. doi:10.1007/978-3-319-74929-7_19.

[BCFP19]   Y. BRUNED, I. CHEVYREV, P. K. FRIZ, and R. PREISS. A rough path perspective on renormalization. *J. Funct. Anal.* **277**, no. 11, (2019), 108283, 60. doi:10.1016/j.jfa.2019.108283.

[BD15]   I. BAILLEUL and J. DIEHL. The inverse problem for rough controlled differential equations. *SIAM J. Control Optim.* **53**, no. 5, (2015), 2762–2780. doi:10.1137/140995982.

[BDFT20]   C. BELLINGERI, A. DJURDJEVAC, P. K. FRIZ, and N. TAPIA. Transport and continuity equations with (very) rough noise. *arXiv e-prints* (2020), 1–20. arXiv:2002.10432.

[Bel20]   C. BELLINGERI. An Itô type formula for the additive stochastic heat equation. *Electron. J. Probab.* **25**, (2020), 52 pp. doi:10.1214/19-EJP404.

[BF13]   C. BAYER and P. K. FRIZ. Cubature on Wiener space: pathwise convergence. *Appl. Math. Optim.* **67**, no. 2, (2013), 261–278. doi:10.1007/s00245-012-9187-8.

[BFG+19]   C. BAYER, P. K. FRIZ, P. GASSIAT, J. MARTIN, and B. STEMPER. A regularity structure for rough volatility. *Math. Financ.* (2019), 1–51. doi:10.1111/mafi.12233.

[BFG20]   C. BELLINGERI, P. K. FRIZ, and M. GERENCSÉR. Singular paths spaces and applications. *arXiv e-prints* (2020), 1–15. arXiv:2003.03352.

[BFH09]   E. BREUILLARD, P. FRIZ, and M. HUESMANN. From random walks to rough paths. *Proc. Amer. Math. Soc.* **137**, no. 10, (2009), 3487–3496. doi:10.1090/S0002-9939-09-09930-4.

[BFRS16]   C. BAYER, P. K. FRIZ, S. RIEDEL, and J. SCHOENMAKERS. From rough path estimates to multilevel Monte Carlo. *SIAM J. Numer. Anal.* **54**, no. 3, (2016), 1449–1483. doi:10.1137/140995209.

[BG97]   L. BERTINI and G. GIACOMIN. Stochastic Burgers and KPZ equations from particle systems. *Comm. Math. Phys.* **183**, no. 3, (1997), 571–607. doi:10.1007/s002200050044.

[BG17]   I. BAILLEUL and M. GUBINELLI. Unbounded rough drivers. *Ann. Fac. Sci. Toulouse Math. (6)* **26**, no. 4, (2017), 795–830. doi:10.5802/afst.1553.

[BGHZ19]   Y. BRUNED, F. GABRIEL, M. HAIRER, and L. ZAMBOTTI. Geometric stochastic heat equations. *arXiv e-prints* (2019), 1–83. arXiv:1902.02884.

[BGLY14]   Y. BOUTAIB, L. G. GYURKÓ, T. LYONS, and D. YANG. Dimension-free Euler estimates of rough differential equations. *Rev. Roumaine Math. Pures Appl.* **59**, no. 1, (2014), 25–53.

[BGLY15]   H. BOEDIHARDJO, X. GENG, T. LYONS, and D. YANG. Note on the signatures of rough paths in a Banach space. *arXiv e-prints* (2015), 1–14. arXiv:1510.04172.

[BGLY16]   H. BOEDIHARDJO, X. GENG, T. LYONS, and D. YANG. The signature of a rough path: Uniqueness. *Advances in Mathematics* **293**, (2016), 720–737. doi:10.1016/j.aim.2016.02.011.

[BH91]   N. BOULEAU and F. HIRSCH. *Dirichlet forms and analysis on Wiener space*, vol. 14 of *de Gruyter Studies in Mathematics*. Walter de Gruyter & Co., Berlin, 1991, x+325. doi:10.1515/9783110858389.

[BH07]   F. BAUDOIN and M. HAIRER. A version of Hörmander's theorem for the fractional Brownian motion. *Probab. Theory Related Fields* **139**, no. 3-4, (2007), 373–395. doi:10.1007/s00440-006-0035-0.

[BH18]   I. BAILLEUL and M. HOSHINO. Paracontrolled calculus and regularity structures. *arXiv e-prints* (2018), 1–32. arXiv:1812.07919.

[BH19]   I. BAILLEUL and M. HOSHINO. Regularity structures and paracontrolled calculus. *arXiv e-prints* (2019), 1–29. arXiv:1912.08438.

[BHZ19]   Y. BRUNED, M. HAIRER, and L. ZAMBOTTI. Algebraic renormalisation of regularity structures. *Invent. Math.* **215**, no. 3, (2019), 1039–1156. arXiv:1610.08468. doi:10.1007/s00222-018-0841-x.

[Bis81a]   J.-M. BISMUT. Martingales, the Malliavin calculus and Hörmander's theorem. In *Stochastic integrals (Proc. Sympos., Univ. Durham, Durham, 1980)*, vol. 851 of *Lecture Notes in Math.*, 85–109. Springer, Berlin, 1981.

[Bis81b]   J.-M. BISMUT. Martingales, the Malliavin calculus and hypoellipticity under general Hörmander's conditions. *Z. Wahrsch. Verw. Gebiete* **56**, no. 4, (1981), 469–505. doi:10.1007/bf00531428.

[BL19]   A. BRAULT and A. LEJAY. The non-linear sewing lemma I: weak formulation. *Electron. J. Probab.* **24**, (2019), Paper No. 59, 24. doi:10.1214/19-EJP313.

[BLY15]   H. BOEDIHARDJO, T. LYONS, and D. YANG. Uniform factorial decay estimates for controlled differential equations. *Electron. Commun. Probab.* **20**, (2015), no. 94, 11. doi:10.1214/ECP.v20-4124.

[BM07]   R. BUCKDAHN and J. MA. Pathwise stochastic control problems and stochastic HJB equations. *SIAM Journal on Control and Optimization* **45**, no. 6, (2007), 2224–2256. doi:10.1137/S036301290444335X.

[BM19]   I. BAILLEUL and A. MOUZARD. Paracontrolled calculus for quasilinear singular PDEs. *arXiv e-prints* (2019), 1–32. arXiv:1912.09073.

[BMN10]   Á. BÉNYI, D. MALDONADO, and V. NAIBO. What is . . . a paraproduct? *Notices Amer. Math. Soc.* **57**, no. 7, (2010), 858–860.

[BMSS95]   V. BALLY, A. MILLET, and M. SANZ-SOLE. Approximation and support theorem in Hölder norm for parabolic stochastic partial differential equations. *Ann. Probab.* **23**, no. 1, (1995), 178–222. doi:10.1214/aop/1176988383.

[BNOT16]   F. BAUDOIN, E. NUALART, C. OUYANG, and S. TINDEL. On probability laws of solutions to differential systems driven by a fractional brownian motion. *Ann. Probab.* **44**, no. 4, (2016), 2554–2590. doi:10.1214/15-AOP1028.

[BNQ14]   H. BOEDIHARDJO, H. NI, and Z. QIAN. Uniqueness of signature for simple curves. *J. Funct. Anal.* **267**, no. 6, (2014), 1778–1806. doi:10.1016/j.jfa.2014.06.006.

[Boe18]   H. BOEDIHARDJO. Decay rate of iterated integrals of branched rough paths. *Annales de l'Institut Henri Poincaré C, Analyse non linéaire* **35**, no. 4, (2018), 945 – 969. doi:10.1016/j.anihpc.2017.09.002.

[Bog98]   V. I. BOGACHEV. *Gaussian measures*, vol. 62 of *Mathematical Surveys and Monographs*. American Mathematical Society, Providence, RI, 1998, xii+433. doi:10.1090/surv/062.

[Bon81]   J.-M. BONY. Calcul symbolique et propagation des singularités pour les équations aux dérivées partielles non linéaires. *Ann. Sci. École Norm. Sup. (4)* **14**, no. 2, (1981), 209–246. doi:10.24033/asens.1404.

[Bor75]   C. BORELL. The Brunn-Minkowski inequality in Gauss space. *Invent. Math.* **30**, no. 2, (1975), 207–216. doi:10.1007/bf01425510.

[BP57]   N. N. BOGOLIUBOW and O. S. PARASIUK. Über die Multiplikation der Kausal-funktionen in der Quantentheorie der Felder. *Acta Math.* **97**, (1957), 227–266. doi:10.1007/BF02392399.

[BP08]   J. BOURGAIN and N. PAVLOVIĆ. Ill-posedness of the Navier-Stokes equations in a critical space in 3D. *J. Funct. Anal.* **255**, no. 9, (2008), 2233–2247. doi:10.1016/j.jfa.2008.07.008.

[BR19]   I. BAILLEUL and S. RIEDEL. Rough flows. *J. Math. Soc. Japan* **71**, no. 3, (2019), 915–978. doi:10.2969/jmsj/80108010.

[BRS17]   I. BAILLEUL, S. RIEDEL, and M. SCHEUTZOW. Random dynamical systems, rough paths and rough flows. *J. Differential Equations* **262**, no. 12, (2017), 5792–5823. doi:10.1016/j.jde.2017.02.014.

[Bru18]   Y. BRUNED. Recursive formulae in regularity structures. *Stoch. Partial Differ. Equ. Anal. Comput.* **6**, no. 4, (2018), 525–564. doi:10.1007/s40072-018-0115-z.

[But72]   J. C. BUTCHER. An algebraic theory of integration methods. *Math. Comp.* **26**, (1972), 79–106. doi:10.1090/S0025-5718-1972-0305608-0.

[Car10]   M. CARUANA. Peano's theorem for rough differential equations in infinite-dimensional Banach spaces. *Proc. Lond. Math. Soc. (3)* **100**, no. 1, (2010), 177–215. doi:10.1112/plms/pdp028.

[CC18a]   G. CANNIZZARO and K. CHOUK. Multidimensional SDEs with singular drift and universal construction of the polymer measure with white noise potential. *Ann. Probab.* **46**, no. 3, (2018), 1710–1763. doi:10.1214/17-AOP1213.

[CC18b]   R. CATELLIER and K. CHOUK. Paracontrolled distributions and the 3-dimensional stochastic quantization equation. *Ann. Probab.* **46**, no. 5, (2018), 2621–2679. doi:10.1214/17-aop1235.

[CCHS20]   A. CHANDRA, I. CHEVYREV, M. HAIRER, and H. SHEN. Langevin dynamic for the 2D Yang-Mills measure, 2020. In preparation.

[CDFM18]   M. COGHI, J.-D. DEUSCHEL, P. FRIZ, and M. MAURELLI. Pathwise McKean–Vlasov theory with additive noise. *arXiv e-prints* (2018), 1–41. Ann. Appl. Probab., to appear. arXiv:1812.11773.

[CDFO13]   D. CRISAN, J. DIEHL, P. K. FRIZ, and H. OBERHAUSER. Robust filtering: correlated noise and multidimensional observation. *Ann. Appl. Probab.* **23**, no. 5, (2013), 2139–2160. doi:10.1214/12-AAP896.

[CDL15]   T. CASS, B. K. DRIVER, and C. LITTERER. Constrained rough paths. *Proc. Lond. Math. Soc. (3)* **111**, no. 6, (2015), 1471–1518. doi:10.1112/plms/pdv060.

[CDLL16]   T. CASS, B. K. DRIVER, N. LIM, and C. LITTERER. On the integration of weakly geometric rough paths. *J. Math. Soc. Japan* **68**, no. 4, (2016), 1505–1524. doi:10.2969/jmsj/06841505.

[CDM01]   M. CAPITAINE and C. DONATI-MARTIN. The Lévy area process for the free Brownian motion. *J. Funct. Anal.* **179**, no. 1, (2001), 153–169. doi:10.1006/jfan.2000.3679.

[CF09]   M. CARUANA and P. FRIZ. Partial differential equations driven by rough paths. *J. Differential Equations* **247**, no. 1, (2009), 140–173. doi:10.1016/j.jde.2009.01.026.

[CF10]   T. CASS and P. FRIZ. Densities for rough differential equations under Hörmander's condition. *Ann. of Math. (2)* **171**, no. 3, (2010), 2115–2141. doi:10.4007/annals.2010.171.2115.

[CF18]   K. CHOUK and P. K. FRIZ. Support theorem for a singular SPDE: the case of gPAM. *Ann. Inst. Henri Poincaré Probab. Stat.* **54**, no. 1, (2018), 202–219. doi:10.1214/16-AIHP800.

[CF19]   I. CHEVYREV and P. K. FRIZ. Canonical rdes and general semimartingales as rough paths. *Ann. Probab.* **47**, no. 1, (2019), 420–463. doi:10.1214/18-AOP1264.

[CFG17]   G. CANNIZZARO, P. K. FRIZ, and P. GASSIAT. Malliavin calculus for regularity structures: the case of gPAM. *J. Funct. Anal.* **272**, no. 1, (2017), 363–419. doi:10.1016/j.jfa.2016.09.024.

[CFK⁺19a] I. CHEVYREV, P. K. FRIZ, A. KOREPANOV, I. MELBOURNE, and H. ZHANG. Deterministic homogenization for discrete-time fast-slow systems under optimal moment assumptions. *arXiv e-prints* (2019), 1–24. arXiv:1903.10418.

[CFK⁺19b] I. CHEVYREV, P. K. FRIZ, A. KOREPANOV, I. MELBOURNE, and H. ZHANG. Multiscale systems, homogenization, and rough paths. In *Probability and analysis in interacting physical systems, In Honor of S.R.S. Varadhan, Berlin, August, 2016*, vol. 283 of *Springer Proc. Math. Stat.*, 17–48. Springer, Cham, 2019. doi:10.1007/978-3-030-15338-0.

[CFKM19]   I. CHEVYREV, P. K. FRIZ, A. KOREPANOV, and I. MELBOURNE. Superdiffusive limits for deterministic fast-slow dynamical systems. *arXiv e-prints* (2019), 1–35. arXiv:1907.04825.

[CFO11]   M. CARUANA, P. K. FRIZ, and H. OBERHAUSER. A (rough) pathwise approach to a class of non-linear stochastic partial differential equations. *Ann. Inst. H. Poincaré Anal. Non Linéaire* **28**, no. 1, (2011), 27–46. doi:10.1016/j.anihpc.2010.11.002.

[CFV07]   L. COUTIN, P. FRIZ, and N. VICTOIR. Good rough path sequences and applications to anticipating stochastic calculus. *Ann. Probab.* **35**, no. 3, (2007), 1172–1193. doi:10.1214/009117906000000827.

[CFV09]   T. CASS, P. FRIZ, and N. VICTOIR. Non-degeneracy of Wiener functionals arising from rough differential equations. *Trans. Amer. Math. Soc.* **361**, no. 6, (2009), 3359–3371. doi:10.1090/S0002-9947-09-04677-7.

[CH16]   A. CHANDRA and M. HAIRER. An analytic BPHZ theorem for regularity structures. *arXiv e-prints* (2016), 1–129. arXiv:1612.08138.

[Che54]   K.-T. CHEN. Iterated integrals and exponential homomorphisms. *Proc. London Math. Soc. (3)* **4**, (1954), 502–512. doi:10.1112/plms/s3-4.1.502.

[Che57]   K.-T. CHEN. Integration of paths, geometric invariants and a generalized Baker-Hausdorff formula. *Ann. of Math. (2)* **65**, no. 1, (1957), 163–178. doi:10.2307/1969671.

[Che58]   K.-T. CHEN. Integration of paths—a faithful representation of paths by non-commutative formal power series. *Trans. Amer. Math. Soc.* **89**, (1958), 395–407. doi:10.2307/1993193.

[Che71]   K.-T. CHEN. Algebras of iterated path integrals and fundamental groups. *Trans. Amer. Math. Soc.* **156**, (1971), 359–379. doi:10.2307/1995617.

[Che72]   K. CHENG. Quantization of a general dynamical system by Feynman's path integration formulation. *J. Math. Phys.* **13**, no. 11, (1972), 1723–1726. doi:10.1063/1.1665897.

[Che18]   I. CHEVYREV. Random walks and Lévy processes as rough paths. *Probab. Theory Related Fields* **170**, no. 3-4, (2018), 891–932. doi:10.1007/s00440-017-0781-1.

[CHLT15]   T. CASS, M. HAIRER, C. LITTERER, and S. TINDEL. Smoothness of the density for solutions to Gaussian rough differential equations. *Ann. Probab.* **43**, no. 1, (2015), 188–239. doi:10.1214/13-AOP896.

[Cho39]   W.-L. CHOW. Über Systeme von linearen partiellen Differentialgleichungen erster Ordnung. *Math. Ann.* **117**, (1939), 98–105. doi:10.1007/BF01450011.

[CIL92]   M. G. CRANDALL, H. ISHII, and P.-L. LIONS. User's guide to viscosity solutions of second order partial differential equations. *Bull. Amer. Math. Soc. (N.S.)* **27**, no. 1, (1992), 1–67. doi:10.1090/s0273-0979-1992-00266-5.

[CK00]   A. CONNES and D. KREIMER. Renormalization in quantum field theory and the Riemann-Hilbert problem. I. The Hopf algebra structure of graphs and the main theorem. *Comm. Math. Phys.* **210**, no. 1, (2000), 249–273. doi:10.1007/s002200050779.

[CK16]   I. CHEVYREV and A. KORMILITZIN. A primer on the signature method in machine learning. *arXiv e-prints* (2016), 1–45. arXiv:1603.03788.

[CL05]   L. COUTIN and A. LEJAY. Semi-martingales and rough paths theory. *Electron. J. Probab.* **10**, (2005), no. 23, 761–785. doi:10.1214/EJP.v10-162.

[CL14]   L. COUTIN and A. LEJAY. Perturbed linear rough differential equations. *Ann. Math. Blaise Pascal* **21**, no. 1, (2014), 103–150. doi:10.5802/ambp.338.

[CL15]   T. CASS and T. LYONS. Evolving communities with individual preferences. *Proc. Lond. Math. Soc. (3)* **110**, no. 1, (2015), 83–107. doi:10.1112/plms/pdu040.

[CL16]   I. CHEVYREV and T. LYONS. Characteristic functions of measures on geometric rough paths. *Ann. Probab.* **44**, no. 6, (2016), 4049–4082. doi:10.1214/15-AOP1068.

[CL18]   L. COUTIN and A. LEJAY. Sensitivity of rough differential equations: an approach through the omega lemma. *J. Differential Equations* **264**, no. 6, (2018), 3899–3917. doi:10.1016/j.jde.2017.11.031.

[Cla66]   M. CLARK. *The representation of non-linear stochastic systems with applications to filtering.* Ph.D. thesis, Imperial College, 1966.

[CLL12]   T. CASS, C. LITTERER, and T. LYONS. Rough paths on manifolds. In *New trends in stochastic analysis and related topics*, vol. 12 of *Interdiscip. Math. Sci.*, 33–88. World Sci. Publ., Hackensack, NJ, 2012. doi:10.1142/9789814360920_0002.

[CLL13]   T. CASS, C. LITTERER, and T. LYONS. Integrability and tail estimates for Gaussian rough differential equations. *Ann. Probab.* **41**, no. 4, (2013), 3026–3050. doi:10.1214/12-AOP821.

[CN19]   M. COGHI and T. NILSSEN. Rough nonlocal diffusions. *arXiv e-prints* (2019), 1–54. arXiv:1905.07270.

[CO17]   T. CASS and M. OGRODNIK. Tail estimates for Markovian rough paths. *Ann. Probab.* **45**, no. 4, (2017), 2477–2504. doi:10.1214/16-AOP1117.

[CO18]   I. CHEVYREV and M. OGRODNIK. A support and density theorem for Markovian rough paths. *Electron. J. Probab.* **23**, (2018), Paper No. 56, 16. doi:10.1214/18-ejp184.

[Col51]   J. D. COLE. On a quasi-linear parabolic equation occurring in aerodynamics. *Quart. Appl. Math.* **9**, (1951), 225–236. doi:10.1090/qam/42889.

[Com19]   G. COMI. *Semi-Linear Heat Equation and Singular Volterra Equation.* Ph.D. thesis, Università degli studi di Milano Bicocca, Università degli studi di Pavia, 2019.

[Cor12]   I. CORWIN. The Kardar-Parisi-Zhang equation and universality class. *Random Matrices Theory Appl.* **1**, no. 1, (2012), 1130001, 76. doi:10.1142/S2010326311300014.

[CP19]   R. CONT and N. PERKOWSKI. Pathwise integration and change of variable formulas for continuous paths with arbitrary regularity. *Trans. Am. Math. Soc., Ser. B* **6**, (2019), 161–186. doi:10.1090/btran/34.

[CQ02]   L. COUTIN and Z. QIAN. Stochastic analysis, rough path analysis and fractional Brownian motions. *Probab. Theory Related Fields* **122**, no. 1, (2002), 108–140. doi:10.1007/s004400100158.

[CW16]   T. CASS and M. P. WEIDNER. Tree algebras over topological vector spaces in rough path theory. *arXiv e-prints* (2016), 1–25. arXiv:1604.07352.

[Dar84]   R. W. R. DARLING. On the convergence of Gangolli processes to Brownian motion on a manifold. *Stochastics* **12**, no. 3-4, (1984), 277–301. doi:10.1080/17442508408833305.

[Dau88]   I. DAUBECHIES. Orthonormal bases of compactly supported wavelets. *Comm. Pure Appl. Math.* **41**, no. 7, (1988), 909–996. doi:10.1002/cpa.3160410705.

[Dav08]   A. M. DAVIE. Differential equations driven by rough paths: an approach via discrete approximation. *Appl. Math. Res. Express. AMRX* **2008**, no. 2, (2008), 1–40. doi:10.1093/amrx/abm009.

[DD16]   F. DELARUE and R. DIEL. Rough paths and 1d SDE with a time dependent distributional drift: application to polymers. *Probab. Theory Related Fields* **165**, no. 1-2, (2016), 1–63. doi:10.1007/s00440-015-0626-8.

[Der10]   S. DEREICH. Rough paths analysis of general Banach space-valued Wiener processes. *J. Funct. Anal.* **258**, no. 9, (2010), 2910–2936. doi:10.1016/j.jfa.2010.01.018.

[DF12]   J. DIEHL and P. FRIZ. Backward stochastic differential equations with rough drivers. *Ann. Probab.* **40**, no. 4, (2012), 1715–1758. doi:10.1214/11-AOP660.

[DFG17]   J. DIEHL, P. K. FRIZ, and P. GASSIAT. Stochastic control with rough paths. *Applied Mathematics & Optimization* **75**, no. 2, (2017), 285–315. doi:10.1007/s00245-016-9333-9.

[DFM16]   J. DIEHL, P. FRIZ, and H. MAI. Pathwise stability of likelihood estimators for diffusions via rough paths. *Ann. Appl. Probab.* **26**, no. 4, (2016), 2169–2192. doi:10.1214/15-AAP1143.

[DFMS18]   J.-D. DEUSCHEL, P. K. FRIZ, M. MAURELLI, and M. SLOWIK. The enhanced Sanov theorem and propagation of chaos. *Stochastic Process. Appl.* **128**, no. 7, (2018), 2228–2269. doi:10.1016/j.spa.2017.09.010.

[DFO14]   J. DIEHL, P. K. FRIZ, and H. OBERHAUSER. Regularity theory for rough partial differential equations and parabolic comparison revisited. In *Stochastic Analysis and Applications, In Honour of Terry Lyons*, vol. 100 of *Springer Proc. Math. Stat.*, 203–238. Springer, Cham, 2014. doi:10.1007/978-3-319-11292-3_8.

[DFS17]   J. DIEHL, P. FRIZ, and W. STANNAT. Stochastic partial differential equations: a rough paths view on weak solutions via Feynman-Kac. *Ann. Fac. Sci. Toulouse Math. (6)* **26**, no. 4, (2017), 911–947. doi:10.5802/afst.1556.

[DGHT19a] A. DEYA, M. GUBINELLI, M. HOFMANOVÁ, and S. TINDEL. One-dimensional reflected rough differential equations. *Stochastic Process. Appl.* **129**, no. 9, (2019), 3261–3281. doi:10.1016/j.spa.2018.09.007.

[DGHT19b] A. DEYA, M. GUBINELLI, M. HOFMANOVÁ, and S. TINDEL. A priori estimates for rough PDEs with application to rough conservation laws. *J. Funct. Anal.* **276**, no. 12, (2019), 3577–3645. doi:10.1016/j.jfa.2019.03.008.

[DGT12]   A. DEYA, M. GUBINELLI, and S. TINDEL. Non-linear rough heat equations. *Probab. Theory Related Fields* **153**, no. 1-2, (2012), 97–147. doi:10.1007/s00440-011-0341-z.

[DL89]   R. J. DIPERNA and P.-L. LIONS. Ordinary differential equations, transport theory and Sobolev spaces. *Invent. Math.* **98**, no. 3, (1989), 511–547. doi:10.1007/BF01393835.

[DMT12]   Y. DO, C. MUSCALU, and C. THIELE. Variational estimates for paraproducts. *Rev. Mat. Iberoam.* **28**, no. 3, (2012), 857–878. doi:10.4171/RMI/694.

[DNT12]   A. DEYA, A. NEUENKIRCH, and S. TINDEL. A Milstein-type scheme without Lévy area terms for SDEs driven by fractional Brownian motion. *Ann. Inst. Henri Poincaré Probab. Stat.* **48**, no. 2, (2012), 518–550. doi:10.1214/10-AIHP392.

[DOP19]   J.-D. DEUSCHEL, T. ORENSHTEIN, and N. PERKOWSKI. Additive functionals as rough paths. *arXiv e-prints* (2019), 1–30. arXiv:1912.09819.

[DOR15]   J. DIEHL, H. OBERHAUSER, and S. RIEDEL. A Lévy area between Brownian motion and rough paths with applications to robust nonlinear filtering and rough partial differential equations. *Stochastic Process. Appl.* **125**, no. 1, (2015), 161–181. doi:10.1016/j.spa.2014.08.005.

[Dos77]   H. Doss. Liens entre équations différentielles stochastiques et ordinaires. *Ann. Inst. H. Poincaré Sect. B (N.S.)* **13**, no. 2, (1977), 99–125.

[DPD03]   G. DA PRATO and A. DEBUSSCHE. Strong solutions to the stochastic quantization equations. *Ann. Probab.* **31**, no. 4, (2003), 1900–1916. doi:10.1214/aop/1068646370.

[DPZ92]   G. DA PRATO and J. ZABCZYK. *Stochastic equations in infinite dimensions*, vol. 44 of *Encyclopedia of Mathematics and its Applications*. Cambridge University Press, Cambridge, 1992, xviii+454. doi:10.1017/CBO9780511666223.

[DT09]   A. DEYA and S. TINDEL. Rough Volterra equations. I. The algebraic integration setting. *Stoch. Dyn.* **9**, no. 3, (2009), 437–477. doi:10.1142/S0219493709002737.

[Faw04]   T. FAWCETT. *Non-commutative harmonic analysis*. Ph.D. thesis, University of Oxford, 2004.

[FdLP06]   D. FEYEL and A. DE LA PRADELLE. Curvilinear integrals along enriched paths. *Electron. J. Probab.* **11**, (2006), no. 34, 860–892. doi:10.1214/EJP.v11-356.

[FDM08]   D. FEYEL, A. DE LA PRADELLE, and G. MOKOBODZKI. A non-commutative sewing lemma. *Electron. Commun. Probab.* **13**, (2008), 24–34. doi:10.1214/ECP.v13-1345.

[FG16a]   P. FRIZ and P. GASSIAT. Geometric foundations of rough paths. In *Geometry, analysis and dynamics on sub-Riemannian manifolds. Vol. II*, EMS Ser. Lect. Math., 171–210. Eur. Math. Soc., Zürich, 2016. doi:10.4171/163-1/3.

[FG16b]   P. K. FRIZ and B. GESS. Stochastic scalar conservation laws driven by rough paths. *Ann. Inst. H. Poincaré Anal. Non Linéaire* **33**, no. 4, (2016), 933–963. doi:10.1016/j.anihpc.2015.01.009.

[FG19]   M. FURLAN and M. GUBINELLI. Weak universality for a class of 3d stochastic reaction-diffusion models. *Probab. Theory Related Fields* **173**, no. 3-4, (2019), 1099–1164. doi:10.1007/s00440-018-0849-6.

[FGGR16]   P. K. FRIZ, B. GESS, A. GULISASHVILI, and S. RIEDEL. The Jain-Monrad criterion for rough paths and applications to random Fourier series and non-Markovian Hörmander theory. *Ann. Probab.* **44**, no. 1, (2016), 684–738. arXiv:1307.3460. doi:10.1214/14-AOP986.

[FGL15]   P. FRIZ, P. GASSIAT, and T. LYONS. Physical Brownian motion in a magnetic field as a rough path. *Trans. Amer. Math. Soc.* **367**, no. 11, (2015), 7939–7955. arXiv:1302.2531. doi:10.1090/S0002-9947-2015-06272-2.

[FGLS17]   P. K. FRIZ, P. GASSIAT, P.-L. LIONS, and P. E. SOUGANIDIS. Eikonal equations and pathwise solutions to fully non-linear SPDEs. *Stoch. Partial Differ. Equ. Anal. Comput.* **5**, no. 2, (2017), 256–277. arXiv:1602.04746. doi:10.1007/s40072-016-0087-9.

[FGP18]   P. K. FRIZ, P. GASSIAT, and P. PIGATO. Precise asymptotics: robust stochastic volatility models. *arXiv e-prints* (2018), 1–34. arXiv:1811.00267.

[FHL16]   G. FLINT, B. HAMBLY, and T. LYONS. Discretely sampled signals and the rough Hoff process. *Stochastic Process. Appl.* **126**, no. 9, (2016), 2593–2614. doi:10.1016/j.spa.2016.02.011.

[FHL20]   P. K. FRIZ, A. HOCQUET, and K. LÊ. Rough Markov diffusions and stochastic differential equations, 2020. In preparation.

[FK20]   P. K. FRIZ and T. KLOSE. Precise Laplace Asymptotics for Singular Stochastic Partial Differential Equations: The case of the 2D generalised Parabolic Anderson Model, 2020. In preparation.

[FLS06]   P. FRIZ, T. LYONS, and D. STROOCK. Lévy's area under conditioning. *Ann. Inst. H. Poincaré Probab. Statist.* **42**, no. 1, (2006), 89–101. doi:10.1016/j.anihpb.2005.02.003.

[FNC82]   M. FLIESS and D. NORMAND-CYROT. Algèbres de Lie nilpotentes, formule de Baker-Campbell-Hausdorff et intégrales itérées de K. T. Chen. In *Seminar on Probability, XVI*, vol. 920 of *Lecture Notes in Math.*, 257–267. Springer, Berlin-New York, 1982.

[FO09]   P. FRIZ and H. OBERHAUSER. Rough path limits of the Wong-Zakai type with a modified drift term. *J. Funct. Anal.* **256**, no. 10, (2009), 3236–3256. doi:10.1016/j.jfa.2009.02.010.

[FO10]   P. FRIZ and H. OBERHAUSER. A generalized Fernique theorem and applications. *Proc. Amer. Math. Soc.* **138**, (2010), 3679–3688. doi:10.1090/S0002-9939-2010-10528-2.

[FO11]   P. FRIZ and H. OBERHAUSER. On the splitting-up method for rough (partial) differential equations. *J. Differential Equations* **251**, no. 2, (2011), 316–338. doi:10.1016/j.jde.2011.02.009.

[FO14]   P. FRIZ and H. OBERHAUSER. Rough path stability of (semi-)linear SPDEs. *Probab. Theory Related Fields* **158**, no. 1-2, (2014), 401–434. doi:10.1007/s00440-013-0483-2.

[Föl81]   H. FÖLLMER. Calcul d'Itô sans probabilités. In *Seminar on Probability, XV (Univ. Strasbourg, Strasbourg, 1979/1980) (French)*, vol. 850 of *Lecture Notes in Math.*, 143–150. Springer, Berlin, 1981. doi:10.1007/bfb0088364.

[FP18]   P. K. FRIZ and D. J. PRÖMEL. Rough path metrics on a Besov-Nikolskii-type scale. *Trans. Amer. Math. Soc.* **370**, no. 12, (2018), 8521–8550. doi:10.1090/tran/7264.

[FR11]   P. FRIZ and S. RIEDEL. Convergence rates for the full Brownian rough paths with applications to limit theorems for stochastic flows. *Bull. Sci. Math.* **135**, no. 6-7, (2011), 613–628. doi:10.1016/j.bulsci.2011.07.006.

[FR13]   P. FRIZ and S. RIEDEL. Integrability of (non-)linear rough differential equations and integrals. *Stoch. Anal. Appl.* **31**, no. 2, (2013), 336–358. doi:10.1080/07362994.2013.759758.

[FR14]   P. FRIZ and S. RIEDEL. Convergence rates for the full Gaussian rough paths. *Ann. Inst. Henri Poincaré Probab. Stat.* **50**, no. 1, (2014), 154–194. doi:10.1214/12-AIHP507.

[Fri05]   P. K. FRIZ. Continuity of the Itô-map for Hölder rough paths with applications to the support theorem in Hölder norm. In *Probability and partial differential equations in modern applied mathematics*, vol. 140 of *IMA Vol. Math. Appl.*, 117–135. Springer, New York, 2005. doi:10.1007/978-0-387-29371-4_8.

[FS06]   W. H. FLEMING and H. M. SONER. *Controlled Markov processes and viscosity solutions*, vol. 25 of *Stochastic Modelling and Applied Probability*. Springer, New York, second ed., 2006, xviii+429. doi:10.1007/0-387-31071-1.

[FS13]   P. FRIZ and A. SHEKHAR. Doob-Meyer for rough paths. *Bull. Inst. Math. Acad. Sin. (N.S.)* **8**, no. 1, (2013), 73–84. arXiv:1205.2505.

[FS17]   P. K. FRIZ and A. SHEKHAR. General rough integration, Lévy rough paths and a Lévy-Kintchine-type formula. *Ann. Probab.* **45**, no. 4, (2017), 2707–2765. arXiv:1212.5888. doi:10.1214/16-AOP1123.

[FT17]   P. K. FRIZ and H. TRAN. On the regularity of SLE trace. *Forum Math. Sigma* **5**, (2017), e19, 17. doi:10.1017/fms.2017.18.

[FV05]   P. FRIZ and N. VICTOIR. Approximations of the Brownian rough path with applications to stochastic analysis. *Ann. Inst. H. Poincaré Probab. Statist.* **41**, no. 4, (2005), 703–724. doi:10.1016/j.anihpb.2004.05.003.

[FV06a]   P. FRIZ and N. VICTOIR. A note on the notion of geometric rough paths. *Probab. Theory Related Fields* **136**, no. 3, (2006), 395–416. doi:10.1007/s00440-005-0487-7.

[FV06b]   P. FRIZ and N. VICTOIR. A variation embedding theorem and applications. *J. Funct. Anal.* **239**, no. 2, (2006), 631–637. doi:10.1016/j.jfa.2005.12.021.

[FV07]   P. FRIZ and N. VICTOIR. Large deviation principle for enhanced Gaussian processes. *Ann. Inst. H. Poincaré Probab. Statist.* **43**, no. 6, (2007), 775 – 785. doi:10.1016/j.anihpb.2006.11.002.

[FV08a]   P. FRIZ and N. VICTOIR. The Burkholder-Davis-Gundy inequality for enhanced martingales. In *Séminaire de probabilités XLI*, vol. 1934 of *Lecture Notes in Math.*, 421–438. Springer, Berlin, 2008. doi:10.1007/978-3-540-77913-1_20.

[FV08b]   P. FRIZ and N. VICTOIR. Euler estimates for rough differential equations. *J. Differential Equations* **244**, no. 2, (2008), 388–412. doi:10.1016/j.jde.2007.10.008.

[FV08c]   P. FRIZ and N. VICTOIR. On uniformly subelliptic operators and stochastic area. *Probab. Theory Related Fields* **142**, no. 3-4, (2008), 475–523. doi:10.1007/s00440-007-0113-y.

[FV10a]   P. FRIZ and N. VICTOIR. Differential equations driven by Gaussian signals. *Ann. Inst. H. Poincaré Probab. Statist.* **46**, no. 2, (2010), 369–413. doi:10.1214/09-AIHP202.

[FV10b]   P. FRIZ and N. VICTOIR. *Multidimensional Stochastic Processes as Rough Paths*, vol. 120 of *Cambridge Studies in Advanced Mathematics*. Cambridge University Press, Cambridge, 2010, xiv+670. doi:10.1017/CBO9780511845079.

[FV11]   P. FRIZ and N. VICTOIR. A note on higher dimensional p-variation. *Electron. J. Probab.* **16**, (2011), 1880–1899. doi:10.1214/EJP.v16-951.

[FZ18]   P. K. FRIZ and H. ZHANG. Differential equations driven by rough paths with jumps. *J. Differential Equations* **264**, no. 10, (2018), 6226–6301. doi:10.1016/j.jde.2018.01.031.

[FZK20]   P. K. FRIZ and P. ZORIN-KRANICH. Rough semimartingales and p-variation estimates for martingale transforms, 2020. In preparation.

[Gas20]   P. GASSIAT. Non-uniqueness for reflected rough differential equations. *arXiv e-prints* (2020), 1–25. arXiv:2001.11914.

[GGLS20]   P. GASSIAT, B. GESS, P.-L. LIONS, and P. E. SOUGANIDIS. Speed of propagation for hamilton–jacobi equations with multiplicative rough time dependence and convex hamiltonians. *Probab. Theory Related Fields* **176**, no. 1, (2020), 421–448. doi:10.1007/s00440-019-00921-5.

[GH19]   A. GERASIMOVICS and M. HAIRER. Hörmander's theorem for semilinear spdes. *Electron. J. Probab.* **24**, (2019), 56 pp. doi:10.1214/19-EJP387.

[GHN19]   A. GERASIMOVICS, A. HOCQUET, and T. NILSSEN. Non-autonomous rough semilinear PDEs and the multiplicative sewing lemma. *arXiv e-prints* (2019), 1–48. arXiv:1907.13398.

[GIP15]   M. GUBINELLI, P. IMKELLER, and N. PERKOWSKI. Paracontrolled distributions and singular PDEs. *Forum Math. Pi* **3**, (2015), e6, 75. doi:10.1017/fmp.2015.2.

[GIP16]   M. GUBINELLI, P. IMKELLER, and N. PERKOWSKI. A Fourier analytic approach to pathwise stochastic integration. *Electron. J. Probab.* **21**, (2016), Paper No. 2, 37. doi:10.1214/16-EJP3868.

[GJ14]   P. GONÇALVES and M. JARA. Nonlinear fluctuations of weakly asymmetric interacting particle systems. *Arch. Ration. Mech. Anal.* **212**, no. 2, (2014), 597–644. doi:10.1007/s00205-013-0693-x.

[GL09]   M. GUBINELLI and J. LÖRINCZI. Gibbs measures on Brownian currents. *Comm. Pure Appl. Math.* **62**, no. 1, (2009), 1–56. doi:10.1002/cpa.20260.

[GL20]   P. GASSIAT and C. LABBÉ. Existence of densities for the dynamic Φ_3^4 model. *Ann. Inst. Henri Poincaré Probab. Stat.* **56**, no. 1, (2020), 326–373. doi:10.1214/19-AIHP963.

[GLP99]   G. GIACOMIN, J. L. LEBOWITZ, and E. PRESUTTI. Deterministic and stochastic hydrodynamic equations arising from simple microscopic model systems. In *Stochastic partial differential equations: six perspectives*, vol. 64 of *Math. Surveys Monogr.*, 107–152. Amer. Math. Soc., Providence, RI, 1999. doi:10.1090/surv/064/03.

[GOT19]   B. GESS, C. OUYANG, and S. TINDEL. Density bounds for solutions to differential equations driven by gaussian rough paths. *J. Theoret. Probab.* (2019). doi:10.1007/s10959-019-00967-0.

[GP15]   M. GUBINELLI and N. PERKOWSKI. *Lectures on singular stochastic PDEs*, vol. 29 of *Ensaios Matemáticos [Mathematical Surveys]*. Sociedade Brasileira de Matemática, Rio de Janeiro, 2015, 89.

[GP17]   M. GUBINELLI and N. PERKOWSKI. KPZ reloaded. *Comm. Math. Phys.* **349**, no. 1, (2017), 165–269. arXiv:1508.03877. doi:10.1007/s00220-016-2788-3.

[GP18]   M. GUBINELLI and N. PERKOWSKI. Energy solutions of KPZ are unique. *J. Amer. Math. Soc.* **31**, no. 2, (2018), 427–471. doi:10.1090/jams/889.

[GPS16]   B. GESS, B. PERTHAME, and P. E. SOUGANIDIS. Semi-discretization for stochastic scalar conservation laws with multiple rough fluxes. *SIAM J. Numer. Anal.* **54**, no. 4, (2016), 2187–2209. doi:10.1137/15M1053670.

[GS15]   B. GESS and P. E. SOUGANIDIS. Scalar conservation laws with multiple rough fluxes. *Commun. Math. Sci.* **13**, no. 6, (2015), 1569–1597. doi:10.4310/CMS.2015.v13. n6.a10.

[GS17]   B. GESS and P. E. SOUGANIDIS. Stochastic non-isotropic degenerate parabolic-hyperbolic equations. *Stochastic Process. Appl.* **127**, no. 9, (2017), 2961–3004. doi:10.1016/j.spa.2017.01.005.

[GT10]   M. GUBINELLI and S. TINDEL. Rough evolution equations. *Ann. Probab.* **38**, no. 1, (2010), 1–75. doi:10.1214/08-AOP437.

[Gub04]   M. GUBINELLI. Controlling rough paths. *J. Funct. Anal.* **216**, no. 1, (2004), 86–140. doi:10.1016/j.jfa.2004.01.002.

[Gub10]   M. GUBINELLI. Ramification of rough paths. *J. Differential Equations* **248**, no. 4, (2010), 693–721. doi:10.1016/j.jde.2009.11.015.

[Gub12]   M. GUBINELLI. Rough solutions for the periodic Korteweg–de Vries equation. *Commun. Pure Appl. Anal.* **11**, no. 2, (2012), 709–733. doi:10.3934/cpaa.2012. 11.709.

[Hai11a]   M. HAIRER. On Malliavin's proof of Hörmander's theorem. *Bull. Sci. Math.* **135**, no. 6-7, (2011), 650–666. doi:10.1016/j.bulsci.2011.07.007.

[Hai11b]   M. HAIRER. Rough stochastic PDEs. *Comm. Pure Appl. Math.* **64**, no. 11, (2011), 1547–1585. doi:10.1002/cpa.20383.

[Hai13]   M. HAIRER. Solving the KPZ equation. *Ann. of Math. (2)* **178**, no. 2, (2013), 559–664. arXiv:1109.6811. doi:10.4007/annals.2013.178.2.4.

[Hai14a]   M. HAIRER. Singular stochastic PDEs. In *Proceedings of the International Congress of Mathematicians—Seoul 2014. Vol. IV*, 49–73. Kyung Moon Sa, Seoul, 2014. arXiv:1403.3353.

[Hai14b]   M. HAIRER. A theory of regularity structures. *Invent. Math.* **198**, no. 2, (2014), 269–504. doi:10.1007/s00222-014-0505-4.

[Hai15]   M. HAIRER. Introduction to regularity structures. *Braz. J. Probab. Stat.* **29**, no. 2, (2015), 175–210. doi:10.1214/14-BJPS241.

[Hai16]   M. HAIRER. The motion of a random string. *arXiv e-prints* (2016), 1–20. arXiv: 1605.02192.

[Hep69]   K. HEPP. On the equivalence of additive and analytic renormalization. *Comm. Math. Phys.* **14**, (1969), 67–69. doi:10.1007/BF01645456.

[HH10]   K. HARA and M. HINO. Fractional order Taylor's series and the neo-classical inequality. *Bull. Lond. Math. Soc.* **42**, no. 3, (2010), 467–477. doi:10.1112/blms/ bdq013.

[HH18]   A. HOCQUET and M. HOFMANOVÁ. An energy method for rough partial differential equations. *J. Differential Equations* **265**, no. 4, (2018), 1407–1466. doi:10.1016/ j.jde.2018.04.006.

[HK15]   M. HAIRER and D. KELLY. Geometric versus non-geometric rough paths. *Ann. Inst. Henri Poincaré Probab. Stat.* **51**, no. 1, (2015), 207–251. doi:10.1214/ 13-AIHP564.

[HL07]   K. HARA and T. LYONS. Smooth rough paths and applications for Fourier analysis. *Rev. Mat. Iberoam.* **23**, no. 3, (2007), 1125–1140. doi:10.4171/RMI/526.

[HL10]   B. HAMBLY and T. LYONS. Uniqueness for the signature of a path of bounded variation and the reduced path group. *Ann. of Math. (2)* **171**, no. 1, (2010), 109–167. doi:10.4007/annals.2010.171.109.

[HL19]   M. HAIRER and X.-M. LI. Averaging dynamics driven by fractional Brownian motion. *arXiv e-prints* (2019), 1–42. Ann. Probab., to appear. arXiv:1902.11251.

[HLN19]   M. HOFMANOVÁ, J.-M. LEAHY, and T. NILSSEN. On the Navier-Stokes equation perturbed by rough transport noise. *J. Evol. Equ.* **19**, no. 1, (2019), 203–247. doi: 10.1007/s00028-018-0473-z.

[HM11]   M. HAIRER and J. C. MATTINGLY. A theory of hypoellipticity and unique ergodicity for semilinear stochastic PDEs. *Electron. J. Probab.* **16**, (2011), no. 23, 658–738. doi:10.1214/EJP.v16-875.

[HM12]   M. HAIRER and J. MAAS. A spatial version of the Itô-Stratonovich correction. *Ann. Probab.* **40**, no. 4, (2012), 1675–1714. doi:10.1214/11-AOP662.

[HMW14] M. HAIRER, J. MAAS, and H. WEBER. Approximating rough stochastic PDEs. *Comm. Pure Appl. Math.* **67**, no. 5, (2014), 776–870. doi:10.1002/cpa.21495.

[HN07]   Y. HU and D. NUALART. Differential equations driven by Hölder continuous functions of order greater than 1/2. In *Stochastic analysis and applications*, vol. 2 of *Abel Symp.*, 399–413. Springer, Berlin, 2007. doi:10.1007/978-3-540-70847-6_17.

[HN09]   Y. HU and D. NUALART. Rough path analysis via fractional calculus. *Trans. Amer. Math. Soc.* **361**, no. 5, (2009), 2689–2718. doi:10.1090/S0002-9947-08-04631-X.

[HNS20] A. HOCQUET, T. NILSSEN, and W. STANNAT. Generalized Burgers equation with rough transport noise. *Stochastic Process. Appl.* **130**, no. 4, (2020), 2159–2184. doi:10.1016/j.spa.2019.06.014.

[Hof06] B. HOFF. *The Brownian Frame Process as a Rough Path*. Ph.D. thesis, University of Oxford, 2006. arXiv:math/0602008.

[Hof16] M. HOFMANOVÁ. Scalar conservation laws with rough flux and stochastic forcing. *Stoch. Partial Differ. Equ. Anal. Comput.* **4**, no. 3, (2016), 635–690. doi:10.1007/s40072-016-0072-3.

[Hof18] M. HOFMANOVÁ. On the rough Gronwall lemma and its applications. In *Stochastic partial differential equations and related fields, in Honor of Michael Röckner, Bielefeld 2016*, vol. 229 of *Springer Proc. Math. Stat.*, 333–344. Springer, Cham, 2018. doi:10.1007/978-3-319-74929-7_2.

[Hop50] E. HOPF. The partial differential equation $u_t + uu_x = \mu u_{xx}$. *Comm. Pure Appl. Math.* **3**, (1950), 201–230. doi:10.1002/cpa.3160030302.

[Hör67] L. HÖRMANDER. Hypoelliptic second order differential equations. *Acta Math.* **119**, (1967), 147–171. doi:10.1007/bf02392081.

[HP13]   M. HAIRER and N. S. PILLAI. Regularity of laws and ergodicity of hypoelliptic SDEs driven by rough paths. *Ann. Probab.* **41**, no. 4, (2013), 2544–2598. doi:10.1214/12-AOP777.

[HP15]   M. HAIRER and E. PARDOUX. A Wong-Zakai theorem for stochastic PDEs. *J. Math. Soc. Japan* **67**, no. 4, (2015), 1551–1604. doi:10.2969/jmsj/06741551.

[HQ18]   M. HAIRER and J. QUASTEL. A class of growth models rescaling to KPZ. *Forum Math. Pi* **6**, (2018), e3, 112. arXiv:1512.07845. doi:10.1017/fmp.2018.2.

[HS90]   W. HEBISCH and A. SIKORA. A smooth subadditive homogeneous norm on a homogeneous group. *Studia Mathematica* **96**, no. 3, (1990), 231–236. doi:10.4064/sm-96-3-231-236.

[HS17]   M. HAIRER and H. SHEN. A central limit theorem for the KPZ equation. *Ann. Probab.* **45**, no. 6B, (2017), 4167–4221. doi:10.1214/16-AOP1162.

[HS19]   M. HAIRER and P. SCHÖNBAUER. The support of singular stochastic PDEs. *arXiv e-prints* (2019), 1–147. arXiv:1909.05526.

[HSV07] M. HAIRER, A. M. STUART, and J. VOSS. Analysis of SPDEs arising in path sampling. II. The nonlinear case. *Ann. Appl. Probab.* **17**, no. 5-6, (2007), 1657–1706. doi:10.1214/07-AAP441.

[HT13]   Y. HU and S. TINDEL. Smooth density for some nilpotent rough differential equations. *J. Theoret. Probab.* **26**, no. 3, (2013), 722–749. doi:10.1007/s10959-011-0388-x.

[HT19]   F. A. HARANG and S. TINDEL. Volterra equations driven by rough signals. *arXiv e-prints* (2019), 1–51. arXiv:1912.02064.

[HW13]   M. HAIRER and H. WEBER. Rough Burgers-like equations with multiplicative noise. *Probab. Theory Related Fields* **155**, no. 1-2, (2013), 71–126. doi:10.1007/s00440-011-0392-1.

[HW15] M. HAIRER and H. WEBER. Large deviations for white-noise driven, nonlinear stochastic PDEs in two and three dimensions. *Ann. Fac. Sci. Toulouse Math. (6)* **24**, no. 1, (2015), 55–92. doi:10.5802/afst.1442.

[HX19] M. HAIRER and W. XU. Large scale limit of interface fluctuation models. *Ann. Probab.* **47**, no. 6, (2019), 3478–3550. doi:10.1214/18-aop1317.

[IK06] Y. INAHAMA and H. KAWABI. Large deviations for heat kernel measures on loop spaces via rough paths. *J. London Math. Soc. (2)* **73**, no. 3, (2006), 797–816. doi:10.1112/S0024610706022654.

[IK07] Y. INAHAMA and H. KAWABI. Asymptotic expansions for the Laplace approximations for Itô functionals of Brownian rough paths. *J. Funct. Anal.* **243**, no. 1, (2007), 270–322. doi:10.1016/j.jfa.2006.09.016.

[IKN18] S. ISHIWATA, H. KAWABI, and R. NAMBA. Central limit theorems for non-symmetric random walks on nilpotent covering graphs: Part ii. *arXiv e-prints* (2018), 1–41. arXiv:1808.08856.

[IM85] A. INOUE and Y. MAEDA. On integral transformations associated with a certain Lagrangian—as a prototype of quantization. *J. Math. Soc. Japan* **37**, no. 2, (1985), 219–244. doi:10.2969/jmsj/03720219.

[IN19] Y. INAHAMA and N. NAGANUMA. Asymptotic expansion of the density for hypoelliptic rough differential equation. *arXiv e-prints* (2019), 1–33. arXiv:1902.05219.

[Ina06] Y. INAHAMA. Laplace's method for the laws of heat processes on loop spaces. *J. Funct. Anal.* **232**, no. 1, (2006), 148–194. doi:10.1016/j.jfa.2005.06.006.

[Ina10] Y. INAHAMA. A stochastic Taylor-like expansion in the rough path theory. *J. Theor. Probab.* **23**, (2010), 671–714. doi:10.1007/s10959-010-0287-6.

[Ina13] Y. INAHAMA. Laplace approximation for rough differential equation driven by fractional Brownian motion. *Ann. Probab.* **41**, no. 1, (2013), 170–205. doi:10.1214/11-AOP733.

[Ina14] Y. INAHAMA. Malliavin differentiability of solutions of rough differential equations. *J. Funct. Anal.* **267**, no. 5, (2014), 1566–1584. doi:10.1016/j.jfa.2014.06.011.

[Ina15] Y. INAHAMA. Large deviation principle of Freidlin-Wentzell type for pinned diffusion processes. *Trans. Amer. Math. Soc.* **367**, no. 11, (2015), 8107–8137. doi:10.1090/S0002-9947-2015-06290-4.

[Ina16a] Y. INAHAMA. Large deviations for rough path lifts of Watanabe's pullbacks of delta functions. *Int. Math. Res. Not. IMRN* **2016**, no. 20, (2016), 6378–6414. doi:10.1093/imrn/rnv349.

[Ina16b] Y. INAHAMA. Short time kernel asymptotics for rough differential equation driven by fractional Brownian motion. *Electron. J. Probab.* **21**, (2016), Paper No. 34, 29. doi:10.1214/16-EJP4144.

[INY78] N. IKEDA, S. NAKAO, and Y. YAMATO. A class of approximations of Brownian motion. *Publ. Res. Inst. Math. Sci.* **13**, no. 1, (1977/78), 285–300. doi:10.2977/prims/1195190109.

[IT17] Y. INAHAMA and S. TANIGUCHI. Short time full asymptotic expansion of hypoelliptic heat kernel at the cut locus. *Forum Math. Sigma* **5**, (2017), e16, 74. doi:10.1017/fms.2017.14.

[IW89] N. IKEDA and S. WATANABE. *Stochastic differential equations and diffusion processes*. North-Holland Publishing Co., Amsterdam, second ed., 1989, xvi+555.

[JLM85] G. JONA-LASINIO and P. K. MITTER. On the stochastic quantization of field theory. *Comm. Math. Phys.* **101**, no. 3, (1985), 409–436. doi:10.1007/bf01216097.

[JM83] N. C. JAIN and D. MONRAD. Gaussian measures in B_p. *Ann. Probab.* **11**, no. 1, (1983), 46–57. doi:10.1214/aop/1176993659.

[Kal02] O. KALLENBERG. *Foundations of modern probability*. Probability and its Applications (New York). Springer-Verlag, New York, second ed., 2002, xx+638. doi:10.1007/978-1-4757-4015-8.

[Kel16] D. KELLY. Rough path recursions and diffusion approximations. *Ann. Appl. Probab.* **26**, no. 1, (2016), 425–461. doi:10.1214/15-aap1096.

[KM16]   D. KELLY and I. MELBOURNE. Smooth approximation of stochastic differential
            equations. *Ann. Probab.* **44**, no. 1, (2016), 479–520. doi:10.1214/14-AOP979.

[KM17]   D. KELLY and I. MELBOURNE. Deterministic homogenization for fast–slow systems
            with chaotic noise. *J. Funct. Anal.* **272**, no. 10, (2017), 4063–4102. doi:10.1016/
            j.jfa.2017.01.015.

[Koh78]   J. J. KOHN. Lectures on degenerate elliptic problems. In *Pseudodifferential operator
            with applications (Bressanone, 1977)*, 89–151. Liguori, Naples, 1978.

[KPP95]   T. G. KURTZ, E. PARDOUX, and P. PROTTER. Stratonovich stochastic differential
            equations driven by general semimartingales. *Ann. Inst. H. Poincaré Probab. Statist.*
            31, no. 2, (1995), 351–377.

[KPZ86]   M. KARDAR, G. PARISI, and Y.-C. ZHANG. Dynamic scaling of growing interfaces.
            Phys. Rev. Lett. **56**, no. 9, (1986), 889–892. doi:10.1103/PhysRevLett.56.889.

[KR77]   N. V. KRYLOV and B. L. ROZOVSKII. The Cauchy problem for linear stochastic
            partial differential equations. *Izv. Akad. Nauk SSSR Ser. Mat.* **41**, no. 6, (1977),
            1329–1347, 1448. doi:10.1070/im1977v011n06abeh001768.

[KRT07]   I. KRUK, F. RUSSO, and C. A. TUDOR. Wiener integrals, Malliavin calculus and
            covariance measure structure. *J. Funct. Anal.* **249**, no. 1, (2007), 92–142. doi:
            10.1016/j.jfa.2007.03.031.

[KS84]   S. KUSUOKA and D. STROOCK. Applications of the Malliavin calculus. I. In *Stochas-
            tic analysis (Katata/Kyoto, 1982)*, vol. 32 of *North-Holland Math. Library*, 271–306.
            North-Holland, Amsterdam, 1984. doi:10.1016/S0924-6509(08)70397-0.

[KS85]   S. KUSUOKA and D. STROOCK. Applications of the Malliavin calculus. II. *J. Fac.
            Sci. Univ. Tokyo Sect. IA Math.* **32**, no. 1, (1985), 1–76.

[KS87]   S. KUSUOKA and D. STROOCK. Applications of the Malliavin calculus. III. *J. Fac.
            Sci. Univ. Tokyo Sect. IA Math.* **34**, no. 2, (1987), 391–442.

[Kun82]   H. KUNITA. Stochastic partial differential equations connected with nonlinear filtering.
            In *Nonlinear filtering and stochastic control (Cortona, 1981)*, vol. 972 of *Lecture
            Notes in Math.*, 100–169. Springer, Berlin, 1982. doi:10.1007/BFb0064861.

[Kun84]   H. KUNITA. First order stochastic partial differential equations. In K. ITÔ, ed.,
            Stochastic Analysis, vol. 32 of *North-Holland Mathematical Library*, 249 – 269.
            Elsevier, 1984. doi:10.1016/S0924-6509(08)70396-9.

[Kup16]   A. KUPIAINEN. Renormalization group and stochastic PDEs. *Ann. Henri Poincaré*
            17, no. 3, (2016), 497–535. doi:10.1007/s00023-015-0408-y.

[Kus01]   S. KUSUOKA. Approximation of expectation of diffusion process and mathematical
            finance. In *Taniguchi Conference on Mathematics Nara '98*, vol. 31 of *Adv. Stud. Pure
            Math.*, 147–165. Math. Soc. Japan, Tokyo, 2001. doi:10.2969/aspm/03110147.

[KZK19]   V. KOVAČ and P. ZORIN-KRANICH. Variational estimates for martingale paraprod-
            ucts. *Electron. Commun. Probab.* **24**, (2019), Paper No. 48, 14. doi:10.1214/
            19-ecp257.

[LCL07]   T. J. LYONS, M. CARUANA, and T. LÉVY. *Differential equations driven by rough
            paths*, vol. 1908 of *Lecture Notes in Mathematics*. Springer, Berlin, 2007, xviii+109.
            Lectures from the 34th Summer School on Probability Theory held in Saint-Flour,
            July 6–24, 2004, With an introduction concerning the Summer School by Jean Picard.
            doi:10.1007/978-3-540-71285-5.

[Lê18]   K. LÊ. A stochastic sewing lemma and applications. *arXiv e-prints* (2018). Electron.
            J. Probab., to appear. arXiv:1810.10500.

[Led96]   M. LEDOUX. Isoperimetry and Gaussian analysis. In *Lectures on probability theory
            and statistics (Saint-Flour, 1994)*, vol. 1648 of *Lecture Notes in Math.*, 165–294.
            Springer, Berlin, 1996. doi:10.1007/bfb0095676.

[Lej06]   A. LEJAY. Stochastic differential equations driven by processes generated by diver-
            gence form operators. I. A Wong-Zakai theorem. *ESAIM Probab. Stat.* **10**, (2006),
            356–379. doi:10.1051/ps:2006015.

[Lej12]   A. LEJAY. Global solutions to rough differential equations with unbounded vector
            fields. In *Séminaire de Probabilités XLIV*, vol. 2046 of *Lecture Notes in Math.*,
            215–246. Springer, Heidelberg, 2012. doi:10.1007/978-3-642-27461-9_11.

[Lep76]   D. LEPINGLE. La variation d'ordre p des semi-martingales. *Z. Wahrscheinlichkeits-theorie und Verw. Gebiete* **36**, no. 4, (1976), 295–316. doi:10.1007/BF00532696.

[LL03]   A. LEJAY and T. J. LYONS. On the Importance of the Levy Area for Studying the Limits of Functions of Converging Stochastic Processes. Application to Homogenization. In D. BAKRY, L. BEZNEA, G. BUCUR, and M. RÖCKNER, eds., *Current Trends in Potential Theory*, vol. 7 of *Current Trends in Potential Theory Conference Proceedings, Bucharest, September 2002 and 2003*. The Theta foundation / American Mathematical Society, Bucarest, 2003.

[LL06]   X.-D. LI and T. J. LYONS. Smoothness of Itô maps and diffusion processes on path spaces. I. *Ann. Sci. École Norm. Sup. (4)* **39**, no. 4, (2006), 649–677. doi: 10.1016/j.ansens.2006.07.001.

[LLQ02]   M. LEDOUX, T. LYONS, and Z. QIAN. Lévy area of Wiener processes in Banach spaces. *Ann. Probab.* **30**, no. 2, (2002), 546–578. doi:10.1214/aop/1023481002.

[LN15]   T. LYONS and H. NI. Expected signature of Brownian motion up to the first exit time from a bounded domain. *Ann. Probab.* **43**, no. 5, (2015), 2729–2762. doi: 10.1214/14-AOP949.

[LO18]   O. LOPUSANSCHI and T. ORENSHTEIN. Ballistic random walks in random environment as rough paths: convergence and area anomaly. *arXiv e-prints* (2018), 1–15. arXiv:1812.01403.

[LP18]   C. LIU and D. J. PRÖMEL. Examples of Itô càdlàg rough paths. *Proc. Amer. Math. Soc.* **146**, no. 11, (2018), 4937–4950. doi:10.1090/proc/14142.

[LPS13]   P.-L. LIONS, B. PERTHAME, and P. E. SOUGANIDIS. Scalar conservation laws with rough (stochastic) fluxes. *Stoch. Partial Differ. Equ. Anal. Comput.* **1**, no. 4, (2013), 664–686. doi:10.1007/s40072-013-0021-3.

[LPS14]   P.-L. LIONS, B. PERTHAME, and P. E. SOUGANIDIS. Scalar conservation laws with rough (stochastic) fluxes: the spatially dependent case. *Stoch. Partial Differ. Equ. Anal. Comput.* **2**, no. 4, (2014), 517–538. doi:10.1007/s40072-014-0038-2.

[LQ98]   T. LYONS and Z. QIAN. Flow of diffeomorphisms induced by a geometric multiplicative functional. *Probab. Theory Related Fields* **112**, no. 1, (1998), 91–119. doi:10.1007/s004400050184.

[LQ02]   T. LYONS and Z. QIAN. *System control and rough paths*. Oxford Mathematical Monographs. Oxford University Press, Oxford, 2002, x+216. Oxford Science Publications. doi:10.1093/acprof:oso/9780198506485.001.0001.

[LQZ02]   M. LEDOUX, Z. QIAN, and T. ZHANG. Large deviations and support theorem for diffusion processes via rough paths. *Stochastic Process. Appl.* **102**, no. 2, (2002), 265–283. doi:10.1016/S0304-4149(02)00176-X.

[LS84]   P.-L. LIONS and A.-S. SZNITMAN. Stochastic differential equations with reflecting boundary conditions. *Comm. Pure Appl. Math.* **37**, no. 4, (1984), 511–537. doi: 10.1002/cpa.3160370408.

[LS98a]   P.-L. LIONS and P. E. SOUGANIDIS. Fully nonlinear stochastic partial differential equations. *C. R. Acad. Sci. Paris Sér. I Math.* **326**, no. 9, (1998), 1085–1092. doi: 10.1016/S0764-4442(98)80067-0.

[LS98b]   P.-L. LIONS and P. E. SOUGANIDIS. Fully nonlinear stochastic partial differential equations: non-smooth equations and applications. *C. R. Acad. Sci. Paris Sér. I Math.* **327**, no. 8, (1998), 735–741. doi:10.1016/S0764-4442(98)80161-4.

[LS00a]   P.-L. LIONS and P. E. SOUGANIDIS. Fully nonlinear stochastic PDE with semilinear stochastic dependence. *C. R. Acad. Sci. Paris Sér. I Math.* **331**, no. 8, (2000), 617–624. doi:10.1016/S0764-4442(00)00583-8.

[LS00b]   P.-L. LIONS and P. E. SOUGANIDIS. Uniqueness of weak solutions of fully nonlinear stochastic partial differential equations. *C. R. Acad. Sci. Paris Sér. I Math.* **331**, no. 10, (2000), 783–790. doi:10.1016/S0764-4442(00)01597-4.

[LS01]   W. V. LI and Q.-M. SHAO. Gaussian processes: inequalities, small ball probabilities and applications. In *Stochastic processes: theory and methods*, vol. 19 of *Handbook of Statist.*, 533–597. North-Holland, Amsterdam, 2001. doi:10.1016/s0169-7161(01)19019-x.

[LS17]   O. LOPUSANSCHI and D. SIMON. Area anomaly in the rough path Brownian scaling limit of hidden Markov walks. *arXiv e-prints* (2017), 1–27. arXiv:1709.04288.

[LS18]   O. LOPUSANSCHI and D. SIMON. Lévy area with a drift as a renormalization limit of Markov chains on periodic graphs. *Stochastic Process. Appl.* **128**, no. 7, (2018), 2404–2426. doi:10.1016/j.spa.2017.09.004.

[LV04]   T. LYONS and N. VICTOIR. Cubature on Wiener space. *Proc. R. Soc. Lond. Ser. A Math. Phys. Eng. Sci.* **460**, no. 2041, (2004), 169–198. Stochastic analysis with applications to mathematical finance. doi:10.1098/rspa.2003.1239.

[LV06]   A. LEJAY and N. VICTOIR. On (p, q)-rough paths. *J. Differential Equations* **225**, no. 1, (2006), 103–133. doi:10.1016/j.jde.2006.01.018.

[LV07]   T. LYONS and N. VICTOIR. An extension theorem to rough paths. *Ann. Inst. H. Poincaré Anal. Non Linéaire* **24**, no. 5, (2007), 835–847. doi:10.1016/j.anihpc.2006.07.004.

[LX13]   T. J. LYONS and W. XU. A uniform estimate for rough paths. *Bull. Sci. Math.* **137**, no. 7, (2013), 867–879. doi:10.1016/j.bulsci.2013.04.004.

[LX17]   T. J. LYONS and W. XU. Hyperbolic development and inversion of signature. *J. Funct. Anal.* **272**, no. 7, (2017), 2933–2955. doi:10.1016/j.jfa.2016.12.024.

[LX18]   T. J. LYONS and W. XU. Inverting the signature of a path. *J. Eur. Math. Soc. (JEMS)* **20**, no. 7, (2018), 1655–1687. doi:10.4171/JEMS/796.

[LY02]   F. LIN and X. YANG. *Geometric measure theory—an introduction*, vol. 1 of *Advanced Mathematics (Beijing/Boston)*. Science Press Beijing, Beijing, 2002, x+237.

[LY13]   T. J. LYONS and D. YANG. The partial sum process of orthogonal expansions as geometric rough process with Fourier series as an example—an improvement of Menshov-Rademacher theorem. *J. Funct. Anal.* **265**, no. 12, (2013), 3067–3103. doi:10.1016/j.jfa.2013.08.032.

[LY15]   T. J. LYONS and D. YANG. The theory of rough paths via one-forms and the extension of an argument of Schwartz to rough differential equations. *J. Math. Soc. Japan* **67**, no. 4, (2015), 1681–1703. doi:10.2969/jmsj/06741681.

[LY16]   T. LYONS and D. YANG. Recovering the pathwise Itô solution from averaged Stratonovich solutions. *Electron. Commun. Probab.* **21**, (2016), Paper No. 7, 18. doi:10.1214/16-ECP3795.

[Lyo91] T. LYONS. On the nonexistence of path integrals. *Proc. Roy. Soc. London Ser. A* **432**, no. 1885, (1991), 281–290. doi:10.1098/rspa.1991.0017.

[Lyo94] T. LYONS. Differential equations driven by rough signals. I. An extension of an inequality of L. C. Young. *Math. Res. Lett.* **1**, no. 4, (1994), 451–464. doi:10.4310/MRL.1994.v1.n4.a5.

[Lyo95] T. J. LYONS. The interpretation and solution of ordinary differential equations driven by rough signals. In *Stochastic analysis (Ithaca, NY, 1993)*, vol. 57 of *Proc. Sympos. Pure Math.*, 115–128. Amer. Math. Soc., Providence, RI, 1995. doi:10.1090/pspum/057/1335466.

[Lyo98] T. J. LYONS. Differential equations driven by rough signals. *Rev. Mat. Iberoamericana* **14**, no. 2, (1998), 215–310. doi:10.4171/RMI/240.

[Lyo14] T. LYONS. Rough paths, signatures and the modelling of functions on streams. In *Proceedings of the International Congress of Mathematicians—Seoul 2014. Vol. IV*, 163–184. Kyung Moon Sa, Seoul, 2014. arXiv:1405.4537.

[LZ99]   T. LYONS and O. ZEITOUNI. Conditional exponential moments for iterated Wiener integrals. *Ann. Probab.* **27**, no. 4, (1999), 1738–1749. doi:10.1214/aop/1022677546.

[Mal78] P. MALLIAVIN. Stochastic calculus of variations and hypoelliptic operators. *Proc. Intern. Symp. SDE* (1978), 195–263.

[Mal97] P. MALLIAVIN. *Stochastic analysis*, vol. 313 of *Grundlehren der Mathematischen Wissenschaften [Fundamental Principles of Mathematical Sciences]*. Springer-Verlag, Berlin, 1997, xii+343. doi:10.1007/978-3-642-15074-6.

[McK69] H. P. MCKEAN, JR. *Stochastic integrals*. Probability and Mathematical Statistics, No. 5. Academic Press, New York-London, 1969, xiii+140. doi:10.1090/chel/353.

[McS72]   E. J. MCSHANE. Stochastic differential equations and models of random processes.
           In *Proceedings of the Sixth Berkeley Symposium on Mathematical Statistics and
           Probability (Univ. California, Berkeley, Calif., 1970/1971), Vol. III: Probability theory*,
           263–294. Univ. California Press, Berkeley, Calif., 1972.

[Mey92]   Y. MEYER. *Wavelets and operators*, vol. 37 of *Cambridge Studies in Advanced Math-
           ematics*. Cambridge University Press, Cambridge, 1992, xvi+224. Translated from
           the 1990 French original by D. H. Salinger. doi:10.1017/cbo9780511623820.

[Mon02]   R. MONTGOMERY. *A tour of subriemannian geometries, their geodesics and appli-
           cations*, vol. 91 of *Mathematical Surveys and Monographs*. American Mathematical
           Society, Providence, RI, 2002, xx+259. doi:10.1090/surv/091.

[MP18]   J. MARTIN and N. PERKOWSKI. A Littlewood-Paley description of modelled distri-
           butions. *arXiv e-prints* (2018), 1–25. arXiv:1808.00500.

[MR06]   M. B. MARCUS and J. ROSEN. *Markov processes, Gaussian processes, and local
           times*, vol. 100 of *Cambridge Studies in Advanced Mathematics*. Cambridge University
           Press, Cambridge, 2006, x+620. doi:10.1017/CBO9780511617997.

[MSS06]   A. MILLET and M. SANZ-SOLÉ. Large deviations for rough paths of the fractional
           Brownian motion. *Ann. Inst. H. Poincaré Probab. Statist.* **42**, no. 2, (2006), 245–271.
           doi:10.1016/j.anihpb.2005.04.003.

[MST18]   V. MAGNANI, E. STEPANOV, and D. TREVISAN. A rough calculus approach to level
           sets in the Heisenberg group. *J. Lond. Math. Soc., II. Ser.* **97**, no. 3, (2018), 495–522.
           doi:10.1112/jlms.12115.

[MW18]   A. MOINAT and H. WEBER. Space-time localisation for the dynamic Φ_3^4 model.
           arXiv e-prints (2018), 1–27. arXiv:1811.05764.

[Nor86]   J. NORRIS. Simplified Malliavin calculus. In *Séminaire de Probabilités, XX, 1984/85*,
           vol. 1204 of *Lecture Notes in Math.*, 101–130. Springer, Berlin, 1986. doi:10.1007/
           BFb0075716.

[NP88]   D. NUALART and É. PARDOUX.   Stochastic calculus with anticipating inte-
           grands. *Probab. Theory Related Fields* **78**, no. 4, (1988), 535–581. doi:10.1007/
           BF00353876.

[NT11]   D. NUALART and S. TINDEL. A construction of the rough path above fractional
           Brownian motion using Volterra's representation. *Ann. Probab.* **39**, no. 3, (2011),
           1061–1096. doi:10.1214/10-AOP578.

[Nua06]   D. NUALART. *The Malliavin calculus and related topics*. Probability and its Ap-
           plications (New York). Springer-Verlag, Berlin, second ed., 2006, xiv+382. doi:
           10.1007/3-540-28329-3.

[OSSW18]   F. OTTO, J. SAUER, S. SMITH, and H. WEBER. Parabolic equations with rough
           coefficients and singular forcing. *arXiv e-prints* (2018), 1–93. arXiv:1803.07884.

[Par79]   E. PARDOUX. Stochastic partial differential equations and filtering of diffusion pro-
           cesses. *Stochastics* **3**, no. 2, (1979), 127–167. doi:10.1080/17442507908833142.

[Pic08]   J. PICARD. A tree approach to p -variation and to integration. *Ann. Probab.* **36**, no. 6,
           (2008), 2235–2279. doi:10.1214/07-AOP388.

[PL11]   A. PAPAVASILIOU and C. LADROUE. Parameter estimation for rough differential
           equations. *Ann. Statist.* **39**, no. 4, (2011), 2047–2073. doi:10.1214/11-AOS893.

[Pro05]   P. E. PROTTER. *Stochastic integration and differential equations*, vol. 21 of *Stochastic
           Modelling and Applied Probability*. Springer-Verlag, Berlin, 2005, xiv+419. Second
           edition. Version 2.1, Corrected third printing. doi:10.1007/978-3-662-10061-5.

[PS08]   G. A. PAVLIOTIS and A. M. STUART. *Multiscale methods*, vol. 53 of *Texts in Applied
           Mathematics*. Springer, New York, 2008, xviii+307. Averaging and homogenization.
           doi:10.1007/978-0-387-73829-1.

[PT16]   D. J. PRÖMEL and M. TRABS. Rough differential equations driven by signals
           in Besov spaces. *J. Differential Equations* **260**, no. 6, (2016), 5202–5249. doi:
           10.1016/j.jde.2015.12.012.

[PT18]   D. J. PRÖMEL and M. TRABS. Paracontrolled distribution approach to stochastic
           Volterra equations. *arXiv e-prints* (2018), 1–39. arXiv:1812.05456.

[PW81]   G. PARISI and Y. S. WU. Perturbation theory without gauge fixing. *Sci. Sinica* **24**, no. 4, (1981), 483–496. doi:10.1360/ya1981-24-4-483.

[Qua11]   J. QUASTEL. Introduction to KPZ. *Current Developments in Mathematics* **2011**, (2011), 125–194. doi:10.4310/cdm.2011.v2011.n1.a3.

[Ras38]   P. RASHEVSKII. About connecting two points of complete non-holonomic space by admissible curve (in Russian). *Uch. Zapiski ped. inst. Libknexta* **2**, (1938), 83–94.

[Ree58]   R. REE. Lie elements and an algebra associated with shuffles. *Ann. of Math. (2)* **68**, (1958), 210–220. doi:10.2307/1970243.

[Rie17]   S. RIEDEL. Transportation–cost inequalities for diffusions driven by gaussian processes. *Electron. J. Probab.* **22**, (2017), 26 pp. doi:10.1214/17-EJP40.

[RS17]   S. RIEDEL and M. SCHEUTZOW. Rough differential equations with unbounded drift term. *J. Differential Equations* **262**, no. 1, (2017), 283–312. doi:10.1016/j.jde.2016.09.021.

[RX13]   S. RIEDEL and W. XU. A simple proof of distance bounds for Gaussian rough paths. *Electron. J. Probab.* **18**, (2013), no. 108, 1–18. doi:10.1214/EJP.v18-2387.

[RY99]   D. REVUZ and M. YOR. *Continuous martingales and Brownian motion*, vol. 293 of *Grundlehren der Mathematischen Wissenschaften [Fundamental Principles of Mathematical Sciences]*. Springer-Verlag, Berlin, third ed., 1999, xiv+602. doi:10.1007/978-3-662-06400-9.

[Rya02]   R. A. RYAN. *Introduction to tensor products of Banach spaces*. Springer Monographs in Mathematics. Springer-Verlag London Ltd., London, 2002, xiv+225. doi:10.1007/978-1-4471-3903-4_1.

[Sch18]   P. SCHÖNBAUER. Malliavin calculus and density for singular stochastic partial differential equations. *arXiv e-prints* (2018), 1–63. arXiv:1809.03570.

[See18a]   B. SEEGER. Approximation schemes for viscosity solutions of fully nonlinear stochastic partial differential equations. *arXiv e-prints* (2018), 1–40. arXiv:1802.04740.

[See18b]   B. SEEGER. Perron's method for pathwise viscosity solutions. *Comm. Partial Differential Equations* **43**, no. 6, (2018), 998–1018. doi:10.1080/03605302.2018.1488262.

[Sim97]   L. SIMON. Schauder estimates by scaling. *Calc. Var. Partial Differential Equations* **5**, no. 5, (1997), 391–407. doi:10.1007/s005260050072.

[Sip93]   E.-M. SIPILÄINEN. *A pathwise view of solutions of stochastic differential equations*. Ph.D. thesis, University of Edinburgh, 1993.

[Sou19]   P. E. SOUGANIDIS. Pathwise solutions for fully nonlinear first- and second-order partial differential equations with multiplicative rough time dependence. In F. FLANDOLI, M. GUBINELLI, and M. HAIRER, eds., *Singular Random Dynamics : Cetraro, Italy 2016*, 75–220. Springer International Publishing, Cham, 2019. doi:10.1007/978-3-030-29545-5_3.

[ST78]   V. N. SUDAKOV and B. S. TSIREL'SON. Extremal properties of half-spaces for spherically invariant measures. *J. Sov. Math.* **9**, no. 1, (1978), 9–18. doi:10.1007/BF01086099.

[ST18]   H. SINGH and J. TEICHMANN. An elementary proof of the reconstruction theorem. *arXiv e-prints* (2018), 1–25. arXiv:1812.03082.

[Str11]   D. W. STROOCK. *Probability theory*. Cambridge University Press, Cambridge, second ed., 2011, xxii+527. An analytic view. doi:10.1017/cbo9780511974243.

[Sus78]   H. J. SUSSMANN. On the gap between deterministic and stochastic ordinary differential equations. *Ann. Probability* **6**, no. 1, (1978), 19–41. doi:10.1214/aop/1176995608.

[Sus91]   H. J. SUSSMANN. Limits of the Wong-Zakai type with a modified drift term. In *Stochastic analysis, Proc. Conf. Honor Moshe Zakai 65th Birthday, Haifa/Isr.*, 475–493. Academic Press, Boston, MA, 1991.

[SV72]   D. W. STROOCK and S. R. S. VARADHAN. On the support of diffusion processes with applications to the strong maximum principle. In *Proceedings of the Sixth Berkeley*

Symposium on Mathematical Statistics and Probability, Volume 3: Probability Theory,
333–359. University of California Press, Berkeley, Calif., 1972.

[SV73]   D. W. STROOCK and S. R. S. VARADHAN. Limit theorems for random walks on Lie
          groups. *Sankhyā Ser. A* **35**, no. 3, (1973), 277–294.

[Tan84]   H. TANAKA. Limit theorems for certain diffusion processes with interaction. In
          Stochastic analysis (Katata/Kyoto, 1982), vol. 32 of *North-Holland Math. Library*,
          469–488. North-Holland, Amsterdam, 1984. doi:10.1016/S0924-6509(08)
          70405-7.

[TC15]   S. TINDEL and K. CHOUK. Skorohod and Stratonovich integration in the plane.
          Electron. J. Probab. **20**, (2015), 39. Id/No 39. doi:10.1214/EJP.v20-3041.

[Tei11]   J. TEICHMANN. Another approach to some rough and stochastic partial dif-
          ferential equations. *Stoch. Dyn.* **11**, no. 2-3, (2011), 535–550. doi:10.1142/
          S0219493711003437.

[Tow02]   N. TOWGHI. Multidimensional extension of L. C. Young's inequality. *JIPAM. J.
          Inequal. Pure Appl. Math.* **3**, no. 2, (2002), Article 22, 13 pp. (electronic).

[TZ18]   N. TAPIA and L. ZAMBOTTI. The geometry of the space of branched Rough Paths.
          arXiv e-prints (2018), 1–35. Proc. LMS, to appear. arXiv:1810.12179.

[Um74]   G. S. UM. On normalization problems of the path integral method. *J. Math. Phys.* **15**,
          no. 2, (1974), 220–224. doi:10.1063/1.1666626.

[Unt10]   J. UNTERBERGER. A rough path over multidimensional fractional Brownian motion
          with arbitrary Hurst index by Fourier normal ordering. *Stochastic Process. Appl.* **120**,
          no. 8, (2010), 1444–1472. doi:10.1016/j.spa.2010.04.001.

[Vic04]   N. VICTOIR. Levy area for the free Brownian motion: existence and non-existence.
          J. Funct. Anal. **208**, no. 1, (2004), 107–121. doi:10.1016/S0022-1236(03)
          00063-6.

[Wer12]   B. M. WERNESS. Regularity of Schramm-Loewner evolutions, annular crossings,
          and rough path theory. *Electron. J. Probab.* **17**, (2012), no. 81, 21. doi:10.1214/
          EJP.v17-2331.

[Wil01]   D. R. E. WILLIAMS. Path-wise solutions of stochastic differential equations driven
          by Lévy processes. *Rev. Mat. Iberoamericana* **17**, no. 2, (2001), 295–329. doi:
          10.4171/RMI/296.

[WZ65]   E. WONG and M. ZAKAI. On the convergence of ordinary integrals to stochastic
          integrals. *Ann. Math. Statist.* **36**, no. 5, (1965), 1560–1564. doi:10.1214/aoms/
          1177699916.

[Yas18]   P. YASKOV. Extensions of the sewing lemma with applications. *Stochastic Processes
          Appl.* **128**, no. 11, (2018), 3940–3965. doi:10.1016/j.spa.2017.09.023.

[You36]   L. C. YOUNG. An inequality of the Hölder type, connected with Stieltjes integration.
          Acta Math. **67**, no. 1, (1936), 251–282. doi:10.1007/BF02401743.

[Zim69]   W. ZIMMERMANN. Convergence of Bogoliubov's method of renormalization in
          momentum space. *Comm. Math. Phys.* **15**, (1969), 208–234. doi:10.1007/
          BF01645676.

Index

$\|\cdot\|_\alpha$, 68
$\mathcal{BC}$, 214
$\mathcal{BUC}$, 228
$\mathscr{C}^\alpha$, 16
$\mathscr{C}^\alpha_g$, 20
$\mathscr{C}^{0,\alpha}_g$, 30
$\mathscr{C}^{0,\alpha}_{g,0}$, 160
C^γ, 246
$\mathscr{C}^{p\text{-var}}_g$, 187
$\mathscr{C}^{p\text{-var}}$, 187
$\mathscr{D}^{2\alpha}_X$, 70
$\mathscr{D}^\gamma_\alpha$, 265
$\mathscr{D}^\gamma(V)$, 264
$G(\!(\mathbf{R}^d)\!)$, 29
$\mathscr{M}$, 250
ϱ_α, 18
$T(\!(V)\!)$, 28
$T_{\geq\alpha}$, 265
T_h, 36
$\mathcal{W}^1$, 186

admissible models, 272

Baker–Campbell–Hausdorff formula, 23
Borell's inequality, 190
Bouleau–Hirsch criterion, 196
bracket of a rough path, 93
Brownian motion, 186
    Banach-valued, 54
    fractional, 175, 177
    Hölder roughness, 114
    Hilbert-valued, 54
    in magnetic field, as rough path, 46
    Itô, as rough path, 43
    physical, 46, 56
    Stratonovich, as rough path, 44
Brownian rough path, 40, 45

Burkholder–Davis–Gundy inequality, 58

Cameron–Martin
    embedding theorem, 186
    paths, 186
    space, 186
    theorem for Brownian rough path, 160
    variation embedding, 186
    variation embedding, improved, 202
Carnot–Carathéodory
    norm, 24
Cass–Litterer–Lyons estimates, 195
Chen's relation, 15, 21, 25, 28
Chen–Strichartz formula, 149
complementary Young regularity, 185
concentration of measure, 190
continuity equation, 211
controlled rough paths, 70
    composition with regular functions, 121
    integration, 71
    of low regularity, 127
    operations on, 119
    relation to rough paths, 119
covariance function, 165
cubature formula, 52, 57
cubature on Wiener space, 50

Davie's lemma, 64
differential equations
    Young, 132
dilation, 17
division property, 122
Doob–Meyer
    decomposition, 107
    for rough paths, 110

enhanced Brownian motion, 40

© Springer Nature Switzerland AG 2020
P. K. Friz, M. Hairer, *A Course on Rough Paths*, Universitext,
https://doi.org/10.1007/978-3-030-41556-3

fast-slow system, 163
Fawcett's formula, 51
Fernique theorem, 190
  for Gaussian rough paths, 191
  generalised, 190
Feynman–Kac formula, 215
filtering, 236
flow, 148
fractional Brownian motion, 116, 165, 177
  as rough path, 175
Freidlin–Wentzell large deviations, 156, 218

Gaussian rough paths, 165
group-like, 29
Gubinelli derivative, 63, 70
  uniqueness, 109

Hölder roughness, 112
  of Brownian motion, 114
  of fractional Brownian motion, 116
Hölder space, 246, 260
Hörmander's theorem, 196
  rough path proof, 200
Hölder–Zygmund spaces, 271
Hamilton–Jacobi equation, 241
harmonic analysis, 38, 58, 88, 253
Heisenberg group, 24
homogenisation, 163
hybrid Itô-rough differential equation, 236

integrability of rough integrals, 192
integral
  of controlled rough paths, 71
  rough, 63, 67
  Skorokhod, 103
  stochastic rough, 86
  Stratonovich anticipating, 103
integration
  backward Itô, 97
  Itô, 89
  of controlled rough paths, 71
  rough, 63, 67, 86
  stochastic rough, 86
  Stratonovich, 91
interpolation, 30
Itô's formula, 92, 94
  controlled rough path point of view, 125
Itô–Föllmer formula, 92
Itô–Lyons map, 141

Kallianpur–Striebel formula, 237
Kolmogorov type criteria, 40, 42, 52, 165
KPZ equation, 289
  Hopf–Cole solution, 291

solution via regularity structures, 293
solution via rough paths, 315

Lépingle's BDG inequality, 58
Lévy's stochastic area, 39
Laplace method, 156, 162
large deviations, 156
large deviations of Schilder type, 160
law of the iterated logarithm, 111
Lie algebra, 23
Lie group, 21, 26
  free nilpotent, 26
lift
  BPHZ lift, 308
  canonical lift, 296

Malliavin calculus, 196
Malliavin covariance matrix, 196
model, 249
modelled distribution, 250
  composition with regular function, 264
  differentiation, 263
  singular, 85
multiplicative functional, 64
  almost, 64

neo-classical inequality, 82
Norris' lemma, 112

one-form, 62
Ornstein–Uhlenbeck process, 178, 179, 232

p-variation, 185
polymer measure, 163

quadratic variation
  in the sense of Föllmer, 95

random dynamical system, 183
random walk, 53
reconstruction theorem, 251, 256
regularity structure, 243, 245
  model for, 249
  polynomial structure, 246
  rough path structure, 247
renormalisation
  BPHZ renormalisation, 308
  renormalisation group, 307
Riemann–Stieltjes sum, 61
  compensated, 62
robustness
  of filtering, 236
  of maximum likelihood estimation, 102
  of rough integration, 74

rough
 continuity equation, 211
 convolution, 87
 Hamilton–Jacobi equation, 241
 scalar conservation law, 240
 transport equation, 207
 truly, 109
rough differential equation, 131, 134, 137
 calculus of variations, 197
 Davie's definition, 143
 driven by Gaussian signal
  Hörmander theory, 196
  Malliavin calculus, 196
 Euler approximation, 143
 explicit solution, 149
 explosion, 149
 flows, 148
 Hörmander's theorem, 200
 in the sense of Davie, 144
 linear, 145
 Lyons' definition, 144
 Milstein approximation, 143
 partial, 207
 partial, Feynman–Kac formula, 214
 Peano existence, 151
 Picard iteration, 151
 stochastic, 216
 with drift, 149
rough Gronwall lemma, 145
rough integral, 63
 improper, 85
 integrability, 192
rough integration, 63
rough partial differential equations, 207
rough path, 16
 bracket, 93
 branched, 27, 38
 Brownian, large deviations, 160
 Brownian, support, 160
 càdlàg, 38
 Cameron–Martin theorem for, 160
 controlled, 63, 70
 controlled, of lower regularity, 76
 convergence, via interpolation, 30
 convergence, via Kolmogorov, 42
 discrete, 38
 Donsker theorem, 53
 Doob–Meyer for, 110
 extension theorem, 81, 150
 Fernique theorem, 190
 for Gaussian process, 175
 for Ornstein-Uhlenbeck process, 178
 for physical Brownian motion, 46
 for stochastic heat equation, 230

 from random walk, 53
 Gaussian, 165
  Fernique theorem for, 191
  Malliavin calculus for, 196
 geometric, 20
 integral, 67
 integral of convolution type, 86
 Kolmogorov criterion, 39
 Kolmogorov tightness criterion, 52
 Lévy–Kintchine formula, 56
 Lyons lift, 81, 150
 Lyons–Victoir extension, Lyons–Victoir
  extension, 34, 261
 metric, 18
 mildly controlled, 129
 norm
  homogeneous, 17
  homogeneous p-variation, 187
 Norris' lemma for, 112
 pure area, 31, 193
 reduced, 93
 relation to controlled rough paths, 119
 space-time, 149
 spaces, separability, 30
 spatial, 230
 time reversal, 29
 translation, 33, 36
 translation operator, 36, 188
 weakly geometric, 20, 38
 with jumps, 38
rough path norm, 17
rough transport equation, 207
rough viscosity solutions, 228

scalar conservation law, 240
Schauder estimates, 272
sewing lemma, 64, 65
 stochastic, 78, 85, 86
 with semigroups, 86
shuffle algebra, 128
shuffle product, 28
stability
 flows of rough differential equations, 148
 functions of controlled rough paths, 122
 rough differential equations, 141
 rough integration, 74
 viscosity solutions, 229
statistics
 applications to, 102, 183
stochastic contuity equation, 211
stochastic differential equation, 153
 Freidlin–Wentzell large deviations, 156
 in Itô sense, 153
 in Stratonovich sense, 153

Stroock–Varadhan support theorem, 155
  with jumps, 163
  with singular drift, 163
  Wong–Zakai approximations, 154
stochastic heat equation as rough path, 230
stochastic integration
  anticipating, 103
  backward Itô, 97
  Itô, 89
  Stratonovich, 91
stochastic partial differential equation, 207
  Burger-like, 230
  Feynman–Kac formula, 214
  KPZ, 289
  linear stochastic heat equation, 232
  singular semilinear, 289
  spatial Itô–Stratonovich correction, 237
  stochastic HJB equation, 230
  Zakai equation, 230
stochastic transport equation, 207
Stroock–Varadhan support theorem, 155, 218

tensor algebra
  truncated, 21
tensor norm
  injective, 55
 ⋅ projective, 10, 55
tensor series, 28
translation of a rough path, 36, 188
translation operator, 36

higher order, 33
  second order, 33
transport equation, 207
true roughness
  as condition for Hörmander's theorem, 201
  of Brownian motion, 111
truly rough, 109

universal limit theorem, 151

variation
  $2D$ ϱ-variation, 168
  controlled ϱ-variation, 169
  regularity, 185

wavelets, 256
Wiener–Itô chaos, 186
Wong–Zakai theorem
  for Brownian rough path, 45
  for SDEs, 154, 161
  for singular SPDEs, 161
  for SPDEs, 218
word, 28

Young
  2D maximal inequality, 168
  differential equations, 133
  inequality, 62
  integral, 61

Printed in the United States
By Bookmasters